Spatial Variation of Seismic Ground Motions

Modeling and Engineering Applications

ADVANCES IN ENGINEERING

A SERIES OF REFERENCE BOOKS, MONOGRAPHS, AND TEXTBOOKS

Series Editor

Haym Benaroya

Department of Mechanical and Aerospace Engineering
Rutgers University

Published Titles:

Spatial Variations of Seismic Ground Motions: Modeling and Engineering Applications, *Aspasia Zerva*

Fundamentals of Rail Vehicle Dynamics: Guidance and Stability, *A. H. Wickens*

Advances in Nonlinear Dynamics in China: Theory and Applications, *Wenhu Huang*

Virtual Testing of Mechanical Systems: Theories and Techniques, *Ole Ivar Sivertsen*

Nonlinear Random Vibration: Analytical Techniques and Applications, *Cho W. S. To*

Handbook of Vehicle-Road Interaction, *David Cebon*

Nonlinear Dynamics of Compliant Offshore Structures, *Patrick Bar-Avi and Haym Benaroya*

Upcoming Titles:

Handbook of Space Engineering, Archaeology and Heritage, *Ann Darrin and Beth O'Leary*

Lunar Settlements, *Haym Benaroya*

Spatial Variation of Seismic Ground Motions

Modeling and Engineering Applications

Aspasia Zerva

CRC Press
Taylor & Francis Group
Boca Raton London New York

CRC Press is an imprint of the
Taylor & Francis Group, an **informa** business

CRC Press
Taylor & Francis Group
6000 Broken Sound Parkway NW, Suite 300
Boca Raton, FL 33487-2742

© 2009 by Taylor & Francis Group, LLC
CRC Press is an imprint of Taylor & Francis Group, an Informa business

International Standard Book Number-13: 978-0-8493-9929-9 (Hardcover)

Library of Congress Cataloging-in-Publication Data

Zerva, Aspasia.
   Spatial variation of seismic ground motions : modeling and engineering applications / Aspasia Zerva.
      p. cm. -- (Advances in engineering series ; 1)
   Includes bibliographical references and index.
   ISBN 978-0-8493-9929-9 (hardback : alk. paper)
   1. Lifeline earthquake engineering--Mathematical models. 2. Public utilities--Earthquake effects--Mathematical models. I. Title. II. Series.

TA654.6.Z47 2008
624.1'762--dc22
                                             2008014225

Visit the Taylor & Francis Web site at
http://www.taylorandfrancis.com

and the CRC Press Web site at
http://www.crcpress.com

Dedication

To all the earthquake victims throughout the world and through the ages . . .

may their suffering not have been in vain . . .

may their suffering have contributed

to the advancement of our knowledge

for the safety of humankind in the years to come . . .

Contents

Foreword

It gives me great pleasure for a number of reasons to write these introductory remarks for Professor Zerva's monograph. It is a wonderful book, unique in its focus on spatial variations of seismic ground motions. We all know the critical importance of engineering for seismic environments, one of the most challenging engineering disciplines. Lives and fortunes are at stake, with earthquake engineers deeply aware that their assumptions and choices have the most serious implications. Thus, this work is a welcome addition to the literature, and will be often sought by novice and expert trying to better understand earthquake mechanics and seismic design.

Another source of pleasure for me is that this monograph marks the first new book in this series under my editorship. The reader should understand that the effort required to write such a book, even after many years of experience and great knowledge, is a major undertaking. There is the gathering and distilling the literature, in this case over six hundred papers and books. Then there is the effort at organizing the community's understanding, along with one's own work and insight, into a coherent volume that leads the reader in a logical way through the subject. This is truly a massive undertaking. The reader needs to realize that writing such a book is a major service to the earthquake engineering community and to the general population it serves. The application of this knowledge will save lives. That is the reason such a book is written. The financial rewards are minimal when compared to the number of hours, days, months and years spent in its preparation. But we are grateful.

The last source of pleasure for me personally is that I have known Aspasia Zerva for many years. I consider her a good friend, and an exceptionally talented colleague. I value her opinion and consider myself fortunate to be in close geographic proximity so that we can meet and discuss many aspects of the profession and of the academic world in which we live.

I am confident that this volume will be much sought after, and it will herald other exceptional works by talented people. My congratulations to Aspa on this fantastic achievement.

Haym Benaroya

Preface

The spatial variation of seismic ground motions denotes the differences in the seismic time histories at various locations on the ground surface. This book focuses on the spatial variability of the motions that is caused by the propagation of the waveforms from the earthquake source through the earth strata to the ground surface. In recent years, the modeling of the spatial variation of the seismic ground motions and its influence on the response of lifeline systems, such as pipelines, tunnels, bridges and dams, attracted significant interest, and, currently, design codes incorporate its effects in their provisions. The topic is being addressed by the engineering seismology community for the modeling of the spatial variability, the probabilistic engineering community for the simulation of spatially variable seismic time series and the incorporation of spatial variability in random vibration approaches, and the earthquake engineering community for examining its effects on lifeline systems and utilizing it in the design of the structures.

This effort brings together the various aspects of the spatial variation of seismic ground motions. Topics covered in this book include: the evaluation of the spatial variability from seismic data recorded at dense instrument arrays by means of signal processing techniques; the presentation of the most widely used parametric coherency models along with brief descriptions of their derivation; the illustration of the causes underlying the spatial variation of the motions and its physical interpretation; the estimation of seismic ground-surface strains from single station data, spatial array records as well as analytical methods; the introduction of the concept of random vibrations as applied to discrete-parameter and continuous structural systems on multiple supports; the generation of simulations and conditional simulations of spatially variable seismic ground motions; an overview of the effects of the spatial variability of the seismic motions on the response of long structures; and the brief description of selective seismic codes that incorporate spatial variability issues in their design recommendations.

This book was written for graduate students, researchers and practicing engineers interested in advancing the current state-of-knowledge in the analysis and modeling of the spatial variation of seismic ground motions, or utilizing spatially variable excitations in the seismic response evaluation of long structures either by means of random vibrations or Monte Carlo simulations. Even though the book was conceived as an entity, its chapters are mostly self-contained and serve as tutorial and/or reference on the relevant topics. Where appropriate, example applications illustrate concepts and derivations; these examples were either taken from the literature or developed by the writer for the purposes of the book.

Acknowledgments

The writing of this book was made feasible by the efforts of organizations in installing, operating and making public the data of spatial instrument arrays, and the efforts of researchers in analyzing the data, modeling the spatial variation of the seismic ground motions and examining its effects on the response of a wide variety of lifeline systems. These efforts are gratefully acknowledged. Many thanks go to Prof. Haym Benaroya, the editor of the monograph series, for encouraging the writer to venture into this task.

The writer is indebted to Dr. Norman A. Abrahamson, Dr. Serafim Arzoumanidis, Prof. Haym Benaroya, Dr. David M. Boore, Prof. Carl J. Costantino, Prof. George Deodatis, Prof. Amr Elnashai, Dr. Richard J. Fragaszy, Prof. Douglas A. Foutch, Dr. Joan Gomberg, Dr. Stephen Hartzell, Dr. Sissy Nikolaou, Dr. Joy Pauschke, Prof. Athina P. Petropulu, Prof. Gerhart I. Schuëller, Dr. W.R. (Bill) Stephenson, Prof. Mihailo D. Trifunac and Prof. Y.K. Wen for their most valuable comments, suggestions and recommendations; their critical and insightful reviews of different parts of this book contributed to the improvement of its quality. The writer is also most appreciative of her former and current Ph.D. students, Dr. Songtao Liao, Dr. Lei Lou and Mr. Junjie Huang, who went diligently over the many versions of this book and provided her with their thoughtful comments and suggestions. A special word of thanks goes to David Molten, M.D., for his comments on the dedication and to Dr. Grace Hsuan for her artistic ideas on the cover. Ms. Sharon Stokes and Ms. Amanda Gonzalez made editorial suggestions and pointed out typos; their efforts are greatly appreciated. Drexel University also granted the writer partial leave of absence during the 2006 fall quarter to assist her with the preparation of the manuscript.

Many tables and figures herein were reproduced/reprinted from the literature. The writer thanks Dr. Norman A. Abrahamson, Dr. David M. Boore, Prof. José Domínguez, Dr. Joan Gomberg, Dr. Stephen Hartzell, Prof. Orlando Maeso, Prof. Fabio Sabetta, Dr. W.R. (Bill) Stephenson, Prof. Mihailo D. Trifunac and Prof. Y.K. Wen, who shared with her figures from their work for inclusion in the book. The wording of the acknowledgments in the captions (and their footnotes) of the reproduced/reprinted figures and tables is in accordance with the requests of the copyright owners.

On a personal basis – Foremost, my thanks go to my family, my mother, Emilia, my sister, Loukia, my brother, Konstantinos, and, especially, my father, the late Vassilios Zervas, one of the most outstanding Civil Engineers I have ever known. I am thankful to my friends in the U.S. and Greece for their encouragement. My most sincere appreciation goes to S.A. for his help and support during this endeavor, and for his insightful advice on what it takes to write a book. I am indebted to P.I. for her forever shining light. And last, but not at all least, my gratitude goes to F.M. for, without his companionship and endurance, this effort would never have been accomplished.

Biographical Sketch

Aspasia Zerva earned her Diploma with Honors from the Department of Civil Engineering at the Aristoteleion University in Thessaloniki, Greece, her M.S. from the Department of Theoretical and Applied Mechanics at the University of Illinois at Urbana-Champaign, and her Ph.D. from the Department of Civil Engineering also at the University of Illinois at Urbana-Champaign. Her principal expertise is in the analysis of seismic array data, modeling of spatially variable seismic ground motions, linear and nonlinear dynamic response of lifelines, system identification, simulation techniques, and signal processing. She has published widely on these topics and is an active consultant to the government on issues of the spatial variation of seismic ground motions.

Dr. Zerva served on national and international panels on Earthquake Engineering and Engineering Seismology, and as Program Director of the Earthquake Engineering Research Centers at the National Science Foundation. She is Professor in the Department of Civil, Architectural and Environmental Engineering, and Affiliated Professor in the Department of Electrical and Computer Engineering at Drexel University in Philadelphia.

1 Introduction

The term "spatial variation of seismic ground motions" denotes the differences in the amplitude and phase of seismic motions recorded over extended areas. The spatial variation of the seismic ground motions can result from the relative surface-fault motion for sites located on either side of a causative fault, soil liquefaction, landslides, and from the general transmission of the waves from the source through the different earth strata to the ground surface. This book concentrates on the latter cause for the spatial variation of surface ground motions.

The spatial variation of seismic ground motions has an important effect on the response of long structures (lifelines), such as pipelines, tunnels, dams and bridges. Because these structures extend over long distances parallel to the ground, their supports undergo different motions during an earthquake. Since the 1960's, pioneering studies analyzed the influence of the spatial variation of the motions on above-ground and buried structures. At the time, the differential motions at the structures' supports were attributed to the wave passage effect, i.e., it was considered that the ground motions propagate with a constant velocity on the ground surface without any change in their shape. The spatial variation of the motions was then described by the deterministic time delay required for the waveforms to reach the further-away supports of the structures. In these early studies, it was recognized that the consideration of inclined, rather than vertical, propagation of plane waves is beneficial for the translational response of the large, mat, rigid foundations of nuclear power plants, induces seismic strains in buried pipelines and torsion in building structures, and affects the response of bridges.

After the installation of the first dense seismograph arrays in the late 1970's - early 1980's, the modeling of the spatial variation of the seismic ground motions and its effect on the response of various structural systems attracted extensive research interest. One of the first arrays installed was the El Centro differential array [87] that recorded the 1979 Imperial Valley earthquake. The linear accelerometer array, illustrated in Fig. 1.1, consisted of six stations DA1-DA6, with separation distances from the reference (southernmost) station of 18, 55, 128, 213 and 305 m (60, 180, 420, 700 and 1000 ft), respectively. Each array element consisted of a three component set of force-balanced accelerometers; an analog-recording SMA-1 accelerometer, denoted by EDA in the figure, was installed south of DA1 [499]. Station DA6 was not triggered by the main event, but recorded its aftershocks. The array was located 24 km from the epicenter of the earthquake with the closest point of rupture being only 5 km away from its location (Fig. 1.1). The array, however, which provided an abundance of data for small and large magnitude events that have been extensively studied by engineers and seismologists, is the Strong Motion Array in Taiwan-Phase 1 (SMART-1), located in Lotung, in the north-east corner of Taiwan [68]. This two-dimensional array, which started being operational in 1980, consisted of 37 force-balanced triaxial accelerometers arranged on three concentric circles, the inner denoted by I, the middle by M and the outer by O (Fig. 1.2), with radii of 0.2, 1 and 2 km, respectively.

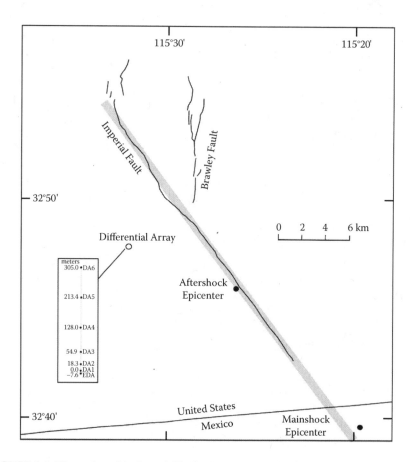

FIGURE 1.1 Illustration of the Imperial Valley indicating the Imperial and Brawley faults and the location of the El Centro differential array. The configuration of the differential array is shown in the insert. The figure also indicates the location of the epicenters of the earthquake of October 15, 1979, and its 23:19 aftershock. Station DA6 was not triggered by the main event, but recorded its aftershocks. (Reproduced from P. Spudich and E. Cranswick, "Direct observation of rupture propagation during the 1979 Imperial Valley earthquake using a short baseline accelerometer array," *Bulletin of the Seismological Society of America*, Vol. 74, pp. 2083–2114, Copyright ©1984 Seismological Society of America.)

Twelve equispaced stations, numbered 1-12, were located on each ring, and station C00 was located at the center of the array. Two additional stations, E01 and E02, were added to the array in 1983, at distances of 2.8 and 4.8 km, respectively, south of C00. The array was located in a recent alluvial valley, primarily rice fields with the water level being at or close to the ground surface, except for station E02 that was located on a slate outcrop [11]. A smaller scale, three-dimensional array, the Lotung Large Scale Seismic Test (LSST) array was installed in 1985 within the southwest quandrant of the SMART-1 array (Fig. 1.2). The LSST array consisted of 15 free-surface and 8 downhole triaxial accelerometers [13], [14]. On the ground surface (Fig. 1.3(a)), the array extended radially along three arms at 120° intervals, with each arm extending

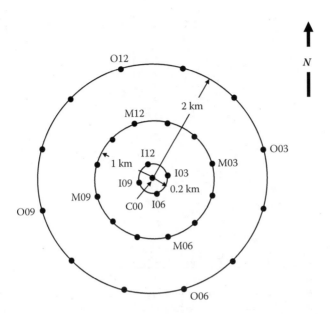

FIGURE 1.2 Configuration of the SMART-1 array. The figure illustrates the location of the center station (C00), and the inner (I01-I12), middle (M01-M12) and outer (O01-O12) ring stations of the array with radii of 0.2, 1 and 2 km, respectively. Stations E01 and E02, located at distances of 2.8 and 4.8 km, respectively, south of C00 are not presented in the figure (after Bolt *et al.* [68]).

approximately 50 m. Two models of a reactor containment vessel, at 1/4 and 1/12 scale, were also constructed at the site and instrumented with 14 accelerometers and 20 strain gauges [11]. Figure 1.3(b) presents the configuration of the downhole arrays, designated as DHA and DHB, which were located underneath the northern arm (Fig. 1.3(a)) and reached a depth of 47 m; the figure also shows the location of the 1/4 scale containment vessel. The array provided data that complement the data recorded at the SMART-1 array in that they permit spatial variability evaluations at short (as short as a few meters) separation distances.

Spatial variability studies based on array data started appearing in the literature almost as soon as the array records became available. These data provided valuable information on the physical causes underlying the variation of the motions over extended areas and the means for its modeling. Figure 1.4 illustrates schematically the effects leading to the spatial variation of the motions after Abrahamson [7]. Part (a) of the figure presents the early, and most commonly recognized, cause for the spatial variation of the motions, namely the wave passage effect: A plane wave impinges the site, and, due to its inclined incidence, causes time delays in the arrival of the waveforms at stations 1 and 2 on the ground surface. Part (b) presents the extended source effect: As rupture propagates along an extended fault, it transmits energy that arrives delayed on the ground surface, resulting in variability in the shape of the waveforms at the various locations. Part (c) of the figure illustrates the scattering effect: Waves propagating away from the source encounter scatterers along their path

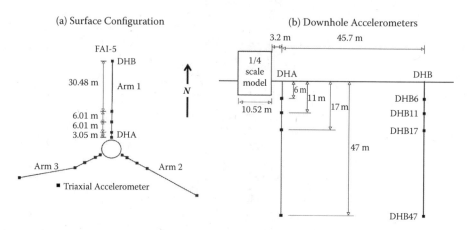

FIGURE 1.3 Configuration of the LSST array located in the southwest quadrant of the SMART-1 array (Fig. 1.2). Part (a) of the figure presents its surface configuration; the circle in the middle indicates the location of the reactor containment vessel model. The array extended radially along three arms, each with 5 triaxial accelerometers, at 120° intervals. Part (b) shows the location of the downhole accelerometers underneath Arm 1, indicated by DHA and DHB in the two subplots. (Reprinted from *Structural Safety*, Vol. 10, N.A. Abrahamson, J.F. Schneider and J.C. Stepp, "Spatial coherency of shear waves from the Lotung, Taiwan large-scale seismic test," pp. 145–162, Copyright ©1991, with permission from Elsevier; courtesy of N.A. Abrahamson.)

that modify their waveform and direction of propagation, and cause differences in the appearance of the seismograms at the various locations on the ground surface. Part (d) of the figure presents the attenuation effect of the waves as they travel away from the source, which, however, for most man-made structure, is not significant.

The early studies investigating the trend of seismic data recorded over extended areas also revealed that the correlation of the motions decreases as the frequency and the separation distance between the stations increase, and pointed to the probabilistic nature of this phenomenon. Signal processing techniques were then utilized to describe the spatial variability of the seismic data, generally, during the strong motion shear-wave window, by means of the coherency. A large number of parametric coherency models, that were fitted to the decay of the recorded data with frequency and separation distance, appeared in the literature. The modeling of the spatial variability of the seismic motions permitted, in turn, the evaluation of the response of a wide range of above-ground and buried structural systems subjected to these excitations. An extensive number of publications analyzed the effects of the spatial variation of the seismic ground motions on the response of pipelines, tunnels, dams, suspension, cable-stayed and highway bridges, nuclear power plants, as well as on rotation and pounding of conventional building structures. Currently, the topic is actively investigated and considerations of its effects have appeared, in various forms, in design recommendations. This book addresses the issue of the spatial variation of the seismic ground motions from its modeling to its applications in earthquake engineering; its organization is as follows.

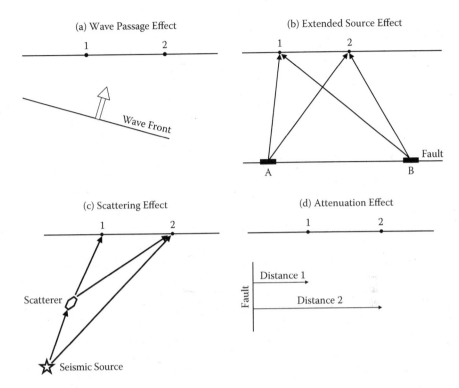

FIGURE 1.4 Illustration of the physical causes underlying the spatial variation of the seismic ground motions. Part (a) of the figure presents, schematically, the wave passage effect, part (b) the extended source effect, part (c) the scattering effect, and part (d) the attenuation effect (after Abrahamson [7]).

Chapter 2 highlights the conventional approach for the estimation of the space-time random field of the seismic ground motions from data recorded at dense instrument arrays. The time and frequency domain descriptions of stochastic processes, stochastic vector processes and random fields are presented; the assumption of Gaussianity is invoked in this chapter, and used, essentially, throughout the book. The spatial coherency is derived from the characterization of bivariate stochastic processes, and its absolute value (lagged coherency) and its complex part (phase spectrum) are discussed. The derivations are based on signal processing methodologies, following mostly the work of Jenkins and Watts [243]. The material is presented without proofs, unless a proof is necessary for the clarification of a concept. Data recorded at the SMART-1 array (Fig. 1.2) during Event 5, the earthquake of January 21, 1981, in the north-south direction, are used as an example application of the derivations. For uniformity and continuity, the same set of data is also utilized for the illustration of methodologies in subsequent chapters.

Chapter 3 presents the parameterization of the space-time random field. Functional forms for the power spectral density of the seismic ground motions that are commonly utilized in engineering applications are described first. Early studies on spatial variability, which recognized that the correlation of the seismic data decreases as frequency and station separation distance increase, are illustrated next. Physical insights on the causes for the spatial variation of the motions (Fig. 1.4), its directional dependence and its behavior at shorter and longer separation distances are highlighted. Selective models from the extensive number of empirical, semi-empirical and analytical coherency models that have appeared in the literature are also presented and discussed. The description of an alternative methodology that parameterizes simultaneously the point and spatial estimates of the random field concludes this chapter.

Chapter 4 presents a physical, rather than statistical, analysis of the spatial variability of the seismic ground motions. It begins with an introduction of frequency-wavenumber spectral estimation techniques that identify the propagation characteristics of the seismic motions at every frequency during each window analyzed. It then proceeds with the examination of the spatial variability from the differences between the spatially recorded data and a representative time history determined from their average characteristics. The analysis indicates that there exist correlation patterns in the amplitude and phase variation of the spatially recorded data around the amplitude and phase of their representative time series. The implications of this trend for the qualitative physical characterization of the spatial coherency are briefly described.

Chapter 5 addresses the topic of seismic ground-surface strains. The seismic ground deformations constitute, essentially, the seismic loads for buried pipelines and tunnels. The chapter presents an overview of semi-empirical and analytical approaches for the estimation of the transient seismic strain field (strains and rotations), and highlights the various causes underlying the ground deformation. It introduces the "traveling-wave" assumption, that was proposed by Newmark [366] for the estimation of peak seismic ground strains, and is still used, in various forms, in the design of buried structures. The chapter also examines the adequacy of the traveling-wave assumption in reproducing the actual ground deformations by comparing its strain estimates with seismic strains evaluated from array data, strainmeter data and strain simulations.

Chapter 6 presents an introduction to linear random vibrations for multiply-supported structures. It begins with the description of random vibrations of a single-degree-of-freedom oscillator subjected to an external force or a base excitation, and progresses to the illustration of random vibrations of multi-degree-of-freedom systems subjected to multi-support excitations. The effect of the exponential decay of commonly used coherency models on the linear response of simple structural models is presented in some detail, so that insight into the complex behavior of realistic structures subjected to spatially variable excitations is obtained. This chapter also describes the pioneering work of Hindy and Novak [221] on the effect of the loss of coherency in the seismic excitations on the random-vibration response of continuous, buried pipelines. The chapter concludes with the presentation of a random-vibration, response-spectrum-compatible approach for the examination of spatial variability effects on lifeline systems, and a brief illustration of nonlinear random vibrations.

Chapter 7 describes the concept of simulations of seismic ground motions concentrating on the spectral representation method. It begins with the simulation of time series at a single location on the ground surface as stationary, and then modifies the generated time series with intensity and frequency modulating functions, that give the simulated motions the appearance of seismic records. The chapter proceeds with the generation of spatially variable ground motions, compares the properties of the simulations with the characteristics of the target random field and with the trend of recorded data, and illustrates how the exponential decay of the coherency models affects the shape of simulated displacement time series. The chapter concludes with a commonly used approach for the generation of non-stationary, non-homogeneous and response-spectrum-compatible simulations.

Extending the concept of simulations from Chapter 7, Chapter 8 presents conditional simulations of spatially variable seismic ground motions, i.e., simulated time series that obey a coherency model but are also compatible with a predefined time history at the site, that can be, e.g., a recorded accelerogram. A commonly used conditional simulation scheme is first described in the chapter. It is noted that, generally, simulation or conditional simulation schemes provide spatially variable acceleration time series. The evaluation of the seismic response of lifeline systems by means of commercial finite element codes, however, generally requires displacement time series as input excitations at the structures' supports, if the spatial variability of the seismic ground motions is taken into consideration. Simulations, whether conditional or unconditional, require processing, as do recorded data, before they are integrated to yield velocities and displacements. This chapter proceeds with the description of a recently proposed scheme for the processing of simulated spatially variable ground motions. Two example applications of conditionally simulated acceleration, velocity and displacement time series, one resulting in zero and the other in non-zero residual displacements, illustrate the approach.

The last chapter, Chapter 9, presents example applications of the effects of spatially variable ground motions on the response of long structures. It begins with the analysis of large, mat, rigid foundations, such as those of nuclear power plants, subjected to spatially variable excitations, and concentrates, mainly, on the classic work of Luco and Wong [324]. The influence of the conventional modeling of the spatial variation of the motions on the response of an embankment and a concrete gravity dam, as well as the significance of the complex 3-D canyon topography on the incident seismic ground motion field and, subsequently, the response of arch dams are illustrated next. The chapter continues with the presentation of the effects of spatially variable ground motions on the response of suspension and cable-stayed bridges, concentrating, basically, on the pioneering and insightful work of Abdel-Ghaffar et al. [1], [2], [3], [362]. The seismic response of highway bridges subjected to spatially variable excitations has attracted by far the most wide-spread interest in the community, and an abundance of research efforts on this topic appeared in the literature; an overview of representative publications, that address, mainly, the effect of the spatial variation of the motions on the nonlinear structural response, is also presented in Chapter 9. Where applicable, selective design recommendations, that incorporate spatial variability effects in their provisions, are briefly described. Finally, some general remarks conclude the chapter.

2 Stochastic Estimation of Spatial Variability

Seismic data recorded at dense instrument arrays permit the probabilistic (stochastic[1]) estimation and modeling of the spatial variability of the ground motions. Signal processing techniques are first applied to the data to evaluate their stochastic estimators in the time or, more commonly, the frequency domain. These techniques are described in this chapter. Once the estimators are obtained from the data, parametric models are fitted to the estimates. This process, as well as the description of a number of available spatial variability models, are presented in Chapter 3. The parametric models are then used either directly in random vibration analyses of the structures (Chapters 6 and 9) or in Monte Carlo simulations for the generation of spatially variable motions (Chapters 7 and 8) to be applied as input excitations at the supports of lifelines (Chapter 9).

The procedure for the estimation of the stochastic spatial variation of seismic motions from recorded data considers that the motions are realizations of space-time random fields, i.e., multi-dimensional and multivariate random functions of location, expressed as the position vector with respect to a selected origin, $\vec{r} = \{x, y, z\}^T$, superscript T indicating transpose, and time, t. At each location, the acceleration time history in each direction (north-south, east-west or vertical), $a(\vec{r}, t)$, is a stochastic process of time t, i.e., at each specific time t, $a(\vec{r}, t)$ is the realization of a random variable. This chapter begins with some basic definitions for random variables in Section 2.1. The concept of stochastic processes is then presented in Section 2.2. This section describes the time and frequency domain characterization of the process through its mean, autocovariance and power spectral density functions, and introduces the assumptions of stationarity and ergodicity. The derivations are presented for ground motion components recorded in a single direction, i.e., it is indirectly assumed that the motions in the three orthogonal directions can be analyzed separately; this assumption will be revisited in Section 3.3.2. The phase properties of recorded accelerograms are also highlighted in this section. Section 2.3 describes the joint characteristics of the time histories at two discrete locations on the ground surface. The time histories are now considered to be realizations of a bivariate stochastic process, also, often, termed bivariate vector process. This section presents the joint descriptors of the bivariate process, namely the cross covariance function in the time domain and the cross spectral density in the frequency domain. Section 2.4 introduces the concept of the complex-valued coherency in the frequency domain, and defines the lagged coherency as its absolute value, and the phase spectrum as its

---

[1] Merriam-Webster Dictionary: Stochastic, from the Greek στοχαστικός (stochastikos) skillful in aiming, from στοχάζεσθαι (stochazesthai) to aim at, guess at, from στόχος (stochos) target, aim, guess. Wikipedia: Stochastic, from the Greek στόχος (stochos) or "goal," means of, relating to, or characterized by conjecture; conjectural; random.

phase. The interpretation of the lagged coherency and the phase spectrum of seismic data is presented in this section. The correspondence between the cross correlation function and the coherency and the definition of the plane-wave coherency are also illustrated in Section 2.4. The derivations for the bivariate stochastic vector processes set the basis for the description of the characteristics of multivariate processes, which are highlighted in Section 2.5. The multivariate stochastic process describes the joint characteristics of the recorded data at all considered discrete locations (stations) on the ground surface. This section also presents the additional assumptions of homogeneity and isotropy, which will be utilized in Chapter 3 for the parameterization of the spatially variable ground motions. The section concludes with a brief description of the concept of the random field viewed as the extension of the multivariate vector process to the continuous case, for which the stochastic characteristics of the motions are provided at all locations on the ground surface. The bias and variance characteristics of the stochastic estimators, as well as the necessity for their smoothing either in the time or the frequency domain are also presented in Sections 2.2–2.4. To illustrate the concepts and derivations of the stochastic estimators described in this chapter, the data recorded at the SMART-1 array (Fig. 1.2) during Event 5 are utilized in example applications.

The literature on the evaluation of the stochastic estimators of deterministic and random signals is quite extensive, including a large number of books, as, e.g., among many others, those by Bartlett [43], Benaroya and Han [51], Bendat [52], Bendat and Piersol [53], Blackman and Tukey [60], Bloomfield [61], Bracewell [80], Brillinger [81], Davenport [126], Jenkins and Watts [243], Oppenheim and Schafer [383], Papoulis [393], Parzen [398], Porat [408], [409], Rosenblatt [427], Yaglom [569], [570], and Vanmarcke [551]. The signal processing derivations in this chapter follow mostly the work of Jenkins and Watts [243]. The material is presented without proofs, unless a proof is necessary for clarification, but provides the necessary information for the evaluation of the stochastic space-time descriptors of spatially recorded array data. The interested reader is referred to the aforementioned literature for further insight in the derivations and for the proofs of the concepts.

2.1   BASIC DEFINITIONS

This section highlights basic definitions for random variables, as an introduction to the description of random processes, vectors and fields.

Following, e.g., Ang and Tang [28], let X denote a (continuous) random variable with cumulative distribution function (CDF):

$$F_X(x) = P(X \leq x) \quad \text{for all} \quad x \tag{2.1}$$

where x indicates the value of the random variable X and P stands for probability. The probability density function (PDF) of X is defined as:

$$f_X(x) = \frac{d F_X(x)}{dx} \tag{2.2}$$

provided that the derivative on the right-hand side of the equation exists. The mean, μ_X, and variance, var(X), are, respectively:

$$\mu_X = E(X) = \int_{-\infty}^{+\infty} x f_X(x)\, dx \tag{2.3}$$

$$\text{var}(X) = \sigma_X^2 = \int_{-\infty}^{+\infty} (x - \mu_X)^2 f_X(x)\, dx = E(X^2) - \mu_X^2 \tag{2.4}$$

where E denotes expectation and σ_X indicates the standard deviation.

The joint probability distribution function of two random variables X and Y is defined as:

$$F_{X,Y}(x, y) = P(X \le x, Y \le y) \tag{2.5}$$

and the joint probability density function as:

$$f_{X,Y}(x, y) = \frac{\partial^2 F_{X,Y}(x, y)}{\partial x \partial y} \tag{2.6}$$

provided, again, that the partial derivatives exist. The covariance of X and Y is given by:

$$\begin{aligned} \text{cov}(X, Y) &= E[(X - \mu_X)(Y - \mu_Y)] \\ &= E(XY) - \mu_X \mu_Y \end{aligned} \tag{2.7}$$

with μ_Y indicating the mean value of Y, and the joint second moment of X and Y being:

$$E(XY) = \int_{-\infty}^{+\infty} \int_{-\infty}^{+\infty} xy f_{X,Y}(x, y)\, dx\, dy \tag{2.8}$$

The physical interpretation of Eq. 2.7 is as follows [28]: When cov(X, Y) is large and positive, both the values of X and Y tend to be either large or small relative to their corresponding mean values. When cov(X, Y) is large and negative, the values of X tend to be large relative to μ_X, whereas the values of Y tend to be small relative to μ_Y, and vice versa. When cov(X, Y) is small or zero, then there is little or no linear relationship between X and Y.

The normalized covariance, or correlation coefficient, ρ_{XY} is defined as:

$$\rho_{XY} = \frac{\text{cov}(X, Y)}{\sigma_X \sigma_Y} \tag{2.9}$$

which assumes values $-1 \le \rho_{XY} \le +1$. Similar to the characteristics of the covariance (Eq. 2.7), when $\rho_{XY} = 1$, X and Y are linearly related, and the slope of the line is positive. When $\rho_{XY} = -1$, X and Y are, again, linearly related, but the slope is now negative. As $|\rho_{XY}|$ decreases, the values of the X and Y pairs start "scattering" around the straight line, with the scatter increasing as $|\rho_{XY}|$ decreases. $\rho_{XY} = 0$ indicates that either X and Y are uncorrelated, i.e., the values of the X and Y pairs will spread out over the entire $X - Y$ plane, or, alternatively, that there exists a nonlinear functional relationship between the two.

In many applications, including the evaluation of the stochastic descriptors of the seismic motions from array data as well as the simulation of seismic ground motions, the assumption of Gaussianity is invoked. For completeness, the probability density function of a Gaussian random variable and the joint probability density function of two Gaussian random variables are provided in the following equations. The Gaussian (or normal) probability density function is given by:

$$f_X(x) = \frac{1}{\sigma_X \sqrt{2\pi}} \exp\left[-\frac{1}{2} \left(\frac{x - \mu_X}{\sigma_X} \right)^2 \right] \qquad -\infty < x < +\infty \qquad (2.10)$$

and the joint probability density function of two random variables by:

$$f_{X,Y}(x, y) = \frac{1}{2\pi \sigma_X \sigma_Y \sqrt{1 - \rho_{XY}^2}} \exp\left[-\frac{1}{2(1 - \rho_{XY}^2)} \left\{ \left(\frac{x - \mu_X}{\sigma_X} \right)^2 \right. \right.$$
$$\left. \left. - 2\rho_{XY} \left(\frac{x - \mu_X}{\sigma_X} \right) \left(\frac{y - \mu_Y}{\sigma_Y} \right) + \left(\frac{y - \mu_Y}{\sigma_Y} \right)^2 \right\} \right] \qquad (2.11)$$
$$-\infty < x < +\infty; \quad -\infty < y < +\infty$$

for correlated Gaussian random variables, or

$$f_{X,Y}(x, y) = \frac{1}{2\pi \sigma_X \sigma_Y} \exp\left[-\frac{1}{2} \left\{ \left(\frac{x - \mu_X}{\sigma_X} \right)^2 + \left(\frac{y - \mu_Y}{\sigma_Y} \right)^2 \right\} \right] \qquad (2.12)$$
$$-\infty < x < +\infty; \quad -\infty < y < +\infty$$

for uncorrelated (statistically independent) Gaussian random variables.

The joint probability distribution and density functions of more than two variables can be defined by extending Eqs. 2.5 and 2.6 for random variables following any probability distribution, or Eqs. 2.11 and 2.12 for correlated and uncorrelated Gaussian random variables.

2.2 STOCHASTIC PROCESSES

A stochastic process is a sequence of an infinite number of random variables, $X_1, \ldots, X_n$, one for each time $t_1, \ldots, t_n$. The statistical properties of a real stochastic process $x(t)$ are completely determined from its n-th order joint probability distribution function [393]:

$$F_{X_1, \ldots, X_n}(x_1, \ldots, x_n; t_1, \ldots, t_n) = P(x(t_1) \leq x_1, \ldots, x(t_n) \leq x_n) \qquad (2.13)$$

which, as indicated in the previous section, is an extension of Eq. 2.5. This information, however, is rarely provided, and the random process is, generally, characterized by its first and second moments. These are the mean value of the process, defined, as in Eq. 2.3, by [393]:

$$\mu_x(t) = E[x(t)] = \int_{-\infty}^{+\infty} x f_X(x, t) \, dx \qquad (2.14)$$

and its autocovariance function, which from Eq. 2.7 becomes [393]:

$$R_{xx}(t_1, t_2) = E\{[x(t_1) - \mu_x(t_1)][x(t_2) - \mu_x(t_2)]\}$$
$$= E[x(t_1)x(t_2)] - \mu_x(t_1)\mu_x(t_2) \qquad (2.15)$$

with $E[x(t_1)x(t_2)]$ given, through Eq. 2.8, by:

$$E[x(t_1)x(t_2)] = \int_{-\infty}^{+\infty} \int_{-\infty}^{+\infty} x_1 x_2 f_{X_1,X_2}(x_1, x_2; t_1, t_2) \, dx_1 \, dx_2 \qquad (2.16)$$

It is noted that lower case subscripts in the expressions for the moments, as, e.g., in Eqs. 2.14 and 2.15, indicate that the estimates refer to random processes, as suggested by Porat [408]. Generally, upper case subscripts in the moment expressions (e.g., in Eqs. 2.3 and 2.9) indicate estimates for random variables.

The following subsections describe the process for obtaining the stochastic descriptors of the time series at a single location on the ground surface along with the necessary assumptions for their evaluation, and illustrate the concepts with example applications in the time and frequency domains.

2.2.1 MEAN AND AUTOCOVARIANCE FUNCTIONS

For the evaluation of the stochastic descriptors of the seismic ground motions at each specific location $\vec{r}_j = \{x_j, y_j, z_j\}^T$, $j = 1, \ldots, N$, on the ground surface, it is considered that the acceleration time history, $a(\vec{r}_j, t)$, is a stochastic process of time t, i.e., at each specific time t, $a(\vec{r}_j, t)$ is the realization of a random variable. As indicated earlier, the process is generally described by its first and second moments, i.e., its mean value and autocovariance functions (Eqs. 2.14 and 2.15), which take the form:

$$\mu_a(\vec{r}_j, t) = E[a(\vec{r}_j, t)] \qquad (2.17)$$

$$\tilde{R}_{aa}(\vec{r}_j, t_1, t_2) = E[a(\vec{r}_j, t_1)a(\vec{r}_j, t_2)] - \mu_a(\vec{r}_j, t_1)\mu_a(\vec{r}_j, t_2) \qquad (2.18)$$

In the following, the subscript a will be dropped from the notation of the descriptors in Eqs. 2.17 and 2.18, which, instead, will be denoted by:

$$\mu_j(t) = \mu_a(\vec{r}_j, t) \qquad (2.19)$$

$$\tilde{R}_{jj}(t_1, t_2) = \tilde{R}_{aa}(\vec{r}_j, t_1, t_2) \qquad (2.20)$$

The subscript j now indicates dependence of the estimates on the location of the recording station. In Eq. 2.20, a single subscript j would suffice, i.e., $\tilde{R}_{jj}(t_1, t_2) \equiv \tilde{R}_j(t_1, t_2)$. The double subscript is, however, maintained for clarity in notation for the subsequent derivations of the cross covariance and cross spectral density functions in Sections 2.3.1 and 2.3.2, respectively.

The above two moments (Eqs. 2.19 and 2.20) fully characterize a Gaussian random process, and Gaussianity, as indicated earlier, is an assumption commonly made in the evaluation of the characteristics of the random field from recorded data, as well as in its simulation (Chapters 7 and 8). Additional assumptions to that of Gaussianity, however, need to be made in order to extract valuable information from the limited amount of data available, such as the recorded time histories at the array stations during an earthquake. These are the assumptions of stationarity and ergodicity described in the following.

Stationarity and Ergodicity

The assumption of stationarity implies that the stochastic descriptors of the motions do not depend on absolute time, but are functions of time differences (or time lag) only. In the strict sense, stationarity requires that all joint probability density functions of the process are the same if time is shifted by a (positive or negative) time lag τ, i.e., from Eq. 2.13:

$$F_{X_1,\ldots,X_n}(x_1, \ldots, x_n; t_1, \ldots, t_n) = F_{X_1,\ldots,X_n}(x_1, \ldots, x_n; t_1+\tau, \ldots, t_n+\tau) \quad (2.21)$$

In the wide sense, the assumption of stationarity applies to the mean and the autocovariance functions of the process. It is noted that, under the assumption of Gaussianity, the first two moments fully characterize the process, and, hence, stationarity in the wide sense for a Gaussian process implies stationarity in the strict sense [243]. The mean value, Eq. 2.19, of a stationary process becomes then a constant. For processed acceleration time histories, this mean value is, generally, zero, and, if it is not, the time histories are "demeaned", i.e., the mean value is subtracted from the process. (Additional information on processing of seismic data is provided in Section 8.2.) Hence, for all practical purposes, it can be considered that $\mu_j(t) = 0$, $j = 1, \ldots, N$, and this will be assumed in the subsequent derivations. For a stationary process, the autocovariance function of the acceleration time history at a station j is no longer a function of the two time instants, t_1 and t_2, but independent of absolute time, t, and function of the time lag, $\tau = t_2 - t_1$, only. Equation 2.20 then becomes:

$$\tilde{R}_{jj}(t_1, t_2) = \tilde{R}_{jj}(t, t + \tau) = \tilde{R}_{jj}(\tau) \quad (2.22)$$

The stationarity assumption carries a peculiar characteristic: Since there is no dependence of the moments on absolute time but only the time lag, the time histories have neither a beginning nor an end, and maintain the same stochastic characteristics throughout their (infinite) duration. This characteristic is unrealistic, as, obviously, seismic ground motions have an absolute starting and ending time. Still, for ease of derivation, one may argue its validity for the following reason: Generally, the stochastic characteristics of seismic ground motions for engineering applications are evaluated from the strong motion shear (S-) wave window, i.e., a segment of the actual seismic time history, which, however, maintains the same properties throughout its duration. This strong motion window can be viewed as a segment of an infinite series with uniform characteristics through time, i.e., a stationary process. It is noted that the assumption of stationarity is relaxed in the simulation of random processes and fields (Chapters 7 and 8) for their subsequent use in engineering applications.

The consideration of a finite segment of the time history in the evaluations suggests that the original time series is actually multiplied by a rectangular window of duration T. The selection of T affects the resolution of the estimates: For example, to correctly identify peaks in the signal at two frequencies that are apart by Δf, the duration of the rectangular window should be greater than $1/\Delta f$ [243]. Generally, instead of the rectangular window, a cosine-tapered window is utilized, with cosine functions at both its ends. The duration of the cosine functions is a small percentage of the overall duration T of the segment [10], and results in a smooth transition region in the analyzed segment from zero acceleration values to finite ones at the beginning of the segment, and from finite ones to zero at its end.

It is also assumed that the stationary time histories at the recording stations are ergodic. A stationary process is ergodic, if averages taken along any realization of the process over its infinite duration are identical to the ensemble averages, i.e., the information contained in each realization is sufficient for the full description of the process. Obviously, the information from the time history is available only for a finite duration window, which, presumably, represents a segment of the stationary process. The consideration of ergodicity, however, is important, especially in the subsequent parametric modeling of the stochastic descriptors of the motions: The evaluation of these parametric models for the random processes and fields would require, ideally, records at the same site from many earthquakes with similar characteristics, so that an ensemble of data can be analyzed, and averages of the ensemble evaluated (Eqs. 2.17–2.22). However, in reality, there is only one realization of the random process or the random field, i.e., one time history at each recording station or one set of recorded data at an array for an earthquake with specific characteristics. The assumption of ergodicity permits the use of the characteristics of a single realization over its duration as representative of the ensemble characteristics, to which parametric models can be fitted, as will be illustrated in Chapter 3. These models are necessary for random vibration analyses (Chapter 6) or for the generation of artificial time histories in Monte Carlo simulations (Chapters 7 and 8).

Considering then the available information at each recording station over the duration T of the strong motion S-wave window, the autocovariance function of Eq. 2.22 becomes [243]:

$$\hat{R}_{jj}(\tau) = \begin{cases} \dfrac{1}{T} \displaystyle\int_0^{T-|\tau|} a_j(t)\,a_j(t+\tau)\,dt & |\tau| \le T \\[6pt] 0 & |\tau| > T \end{cases} \tag{2.23}$$

It can be recognized from the equation that the autocovariance is an even function of the time lag τ, i.e.,

$$\hat{R}_{jj}(\tau) = \hat{R}_{jj}(-\tau) \tag{2.24}$$

It is noted that the accent $\hat{\ }$ in Eqs. 2.23 and Eq. 2.24, as well as in subsequent equations herein, indicates time averages of the estimators based on the single available time history at each station, whereas the accent $\tilde{\ }$ in Eqs. 2.18, 2.20 and 2.22 indicates ensemble averages.

Equation 2.23, as well as the following derivations in this chapter, evaluate the stochastic descriptors of the time histories as continuous functions of time. In reality, the ground motions are sampled at discrete times with a time step Δt, depending on the recording instrument. The sampled time histories can then be regarded as the product of the continuous ones, $a(\vec{r}_j, t), j = 1, \ldots, N$, used in the derivations herein, and a train of Dirac delta functions at times $t_k = k\Delta t; k = 0, \ldots, T/\Delta t$ [243].

Example Applications

Data recorded at the SMART-1 array (Fig. 1.2) in Lotung, Taiwan, during Event 5 are utilized to illustrate the various concepts described herein. Event 5 occurred on January 29, 1981, and had a magnitude of $M_L = 6.3$. Its distance from Lotung was 30 km and its focal depth 25 km. The epicentral direction of the earthquake almost coincided with the diameter from station O06 to station O12, i.e., close to the north-south direction (Fig. 1.2).

Figure 2.1 presents the time histories recorded in the north-south direction at the center station C00, four inner ring stations (I03, I06, I09 and I12) and four middle ring stations (M03, M06, M09 and M12) during the strong motion, S-wave window of the data. The duration of the window is 5.12 sec; the time step for the array instruments was $\Delta t = 0.01$ sec, which implies that 512 data points are considered in the analysis. In the figure, the time scale corresponds to actual time in the records, i.e., the window is the $7.0 - 12.12$ sec segment of the series. The signals are tapered with a double cosine function of duration, at each end, 15% of the window's length [10], [613]. It can be seen that, for the time histories at the close-by (inner ring) stations (Fig. 2.1(a)), the peaks and valleys of the signals occur at approximately the same times except for station I06, at which the waves appear to be arriving earlier than for the other stations. This is consistent with the propagation of the waveforms in the epicentral direction, which was close to the direction from O06 to O12 (Fig. 1.2). Figure 2.1(b) presents the time histories at C00 and the selected middle ring stations. For these further-away stations, the differences in the arrival times of peaks and valleys appear to be significant, and are caused by the propagation of the waves in the epicentral direction as well as arrival time perturbations that are particular for each recording station. These effects are further discussed in Sections 2.4.1 and 4.2. The time history at the center station C00 is presented in both Figs. 2.1(a) and 2.1(b) as a measure of comparison of the variability in the data between the close-by (inner ring) and further-away (middle ring) stations, and, also, because it will serve as the reference station in subsequent illustrations of the concepts described herein.

The autocovariance estimators of the motions at the nine stations are evaluated from Eq. 2.23 and presented in Fig. 2.2. The estimators are sharply peaked at $\tau = 0$, decay quickly with the time lag τ, and exhibit symmetry around the zero axis (Eq. 2.24). It is noted that, for the center and inner ring stations (Fig. 2.2(a)), the autocovariance functions appear more similar than those of the center and middle ring stations (Fig. 2.2(b)). The most significant differences occur between station M12 and the rest of the stations (Fig. 2.2(b)), as will be further discussed in the following sections.

Bias and Variance of Autocovariance Functions

Equation 2.23 is a biased estimator of the true autocovariance function, and becomes asymptotically unbiased as the duration T of the analyzed segment tends to infinity. (Parenthetically, it is noted that bias indicates the difference between the mean value of the estimator and the true value of the quantity to be estimated.) An unbiased estimate for the autocovariance function could have been obtained if T in the denominator of Eq. 2.23 were substituted by $T - |\tau|$ [243]. This substitution, however, would increase

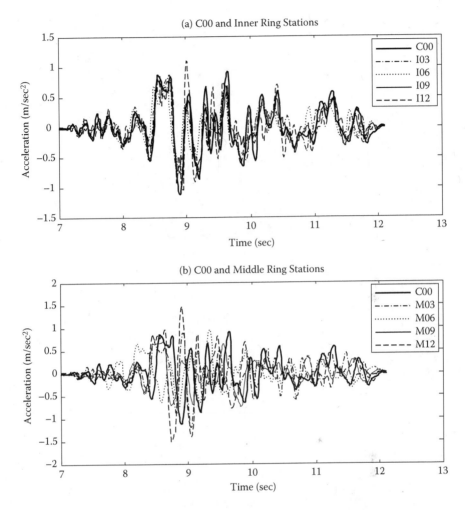

FIGURE 2.1 Tapered time histories recorded at selected stations of the SMART-1 array (Fig. 1.2) during the strong motion, S-wave window of Event 5 in the north-south direction; the duration of the window is 5.12 sec. Part (a) presents the time series at station C00 and stations I03, I06, I09 and I12 of the inner ring (radius of 200 m), and part (b) those at station C00 and stations M03, M06, M09 and M12 of the middle ring (radius of 1 km). The data at station C00 are presented in both subfigures, because they serve as reference for comparisons and derivations.

the mean-square error of the estimator:

$$\text{Mean-Square Error} = \text{Bias}^2 + \text{Variance} \qquad (2.25)$$

which is a compromise between bias and variance [243]. Jenkins and Watts [243] also note that, in general, the correlation between adjacent autocovariance function values tends to be quite large. This implies, e.g., that a large value of $\hat{R}_{jj}(\tau)$ will be

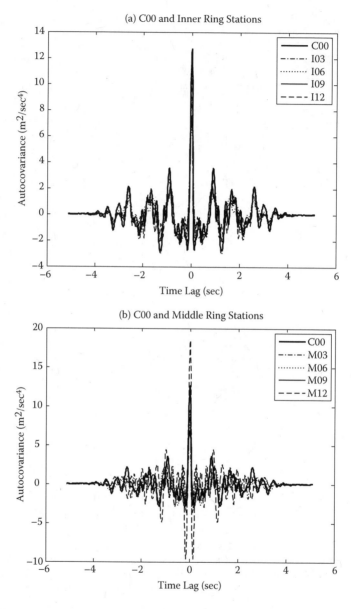

FIGURE 2.2 Autocovariance functions (Eq. 2.23) of the time histories of Fig. 2.1 at the selected stations of the SMART-1 array: Part (a) illustrates the results at C00 and the inner ring stations I03, I06, I09 and I12, and part (b) those at C00 and the middle ring stations M03, M06, M09 and M12.

followed by a large value at the next time step, i.e., $\hat{R}_{jj}(\tau + \Delta t)$. If ground motions are simulated based on this autocovariance estimator, the resulting autocovariance function of the simulations will exhibit large values as τ increases, when the true value of the estimator will already have died out (Fig. 2.2).

It is customary, for stationary random processes, to work in the frequency rather than the time domain. The next section defines the Fourier transform as will be utilized herein for the analysis of spatially variable ground motions, and is followed by the evaluation of the power spectral density of random processes and the description of its characteristics.

2.2.2 DEFINITION OF FOURIER TRANSFORM

The Fourier transform of deterministic and random signals is defined in three different ways in the literature including the references provided herein. For clarification purposes, the definitions are summarized in the following.

The derivations by, e.g., Jenkins and Watts [243], which are, essentially, followed herein, utilize the following Fourier transformation:

$$x(t) = \int_{-\infty}^{+\infty} X(f)e^{i2\pi ft}df \qquad (2.26)$$

$$\Longleftrightarrow X(f) = \int_{-\infty}^{+\infty} x(t)e^{-i2\pi ft}\,dt \qquad (2.27)$$

with $i = \sqrt{-1}$ and f being the frequency (in Hz).

Alternatively, e.g., Papoulis [393] and Porat [408] use:

$$x(t) = \frac{1}{2\pi} \int_{-\infty}^{+\infty} X(\omega)e^{i\omega t}\,d\omega \qquad (2.28)$$

$$\Longleftrightarrow X(\omega) = \int_{-\infty}^{+\infty} x(t)e^{-i\omega t}\,dt \qquad (2.29)$$

where ω is the frequency (in rad/sec). Equation 2.28 results from the transformation of variables $\omega = 2\pi f$ in Eq. 2.26.

On the other hand, e.g., Parzen [398] and Vanmarcke [551] consider that:

$$x(t) = \int_{-\infty}^{+\infty} X(\omega)e^{i\omega t}\,d\omega \qquad (2.30)$$

$$\Longleftrightarrow X(\omega) = \frac{1}{2\pi} \int_{-\infty}^{+\infty} x(t)e^{-i\omega t}\,dt \qquad (2.31)$$

where the factor $1/(2\pi)$ in Eq. 2.31 accommodates the fact that Eq. 2.30 is a multiple (by 2π) of Eq. 2.26. Because Eqs. 2.30 and 2.31 are more commonly used in random vibration analyses and in simulation of random processes and fields, they will also be used herein.

2.2.3 POWER SPECTRAL DENSITY

With the aforementioned definition of the Fourier transform (Eqs. 2.30 and 2.31), the autocovariance function, $\hat{R}_{jj}(\tau)$, and the power spectral density (or power spectrum),

$\hat{S}_{jj}(\omega)$, of the process are Wiener-Khinchine transformation pairs [551], namely:

$$\hat{R}_{jj}(\tau) = \int_{-\infty}^{+\infty} \hat{S}_{jj}(\omega)e^{i\omega\tau}\,d\omega \tag{2.32}$$

$$\Longleftrightarrow \hat{S}_{jj}(\omega) = \frac{1}{2\pi}\int_{-\infty}^{+\infty} \hat{R}_{jj}(\tau)e^{-i\omega\tau}\,d\tau \tag{2.33}$$

Equation 2.33 then suggests that the power spectrum reflects how the variance of the time series is distributed over frequency [243].

Alternatively, the power spectral estimate of Eq. 2.33 can be evaluated directly in the frequency domain as follows: Let $\mathcal{A}_j(\omega) = \Lambda_j(\omega)\exp[i\,\Phi_j(\omega)]$ denote the Fourier transform, defined as in Eq. 2.31, of the time history $a_j(t)$. The power spectrum of the series becomes then:

$$\hat{S}_{jj}(\omega) = \frac{2\pi}{T}\mathcal{A}_j^*(\omega)\,\mathcal{A}_j(\omega) = \frac{2\pi}{T}\Lambda_j^2(\omega) \tag{2.34}$$

where * denotes complex conjugate. Equation 2.34 indicates that the power spectral density is real-valued. It further provides the physical interpretation of the power spectrum, namely that it is a scaled square of the Fourier amplitudes of the time history at the recording station during the analyzed window.

The estimate of Eq. 2.34 is commonly referred to as periodogram. The periodogram is an asymptotically (i.e., as $T \to +\infty$) unbiased estimator of the power spectrum, but, since T is finite in all applications, $\hat{S}_{jj}(\omega)$ is biased. The expected value of the periodogram is, actually, the convolution of the true spectrum with the Fourier transform of the time window used to truncate the series into finite duration [243]. Obviously, the longer the analyzed segment, the smaller the bias of the estimate. Additionally, the periodogram is an inconsistent estimate of the power spectral density: Its variance at every frequency is approximately equal to the value of the true spectrum at that frequency [243], [408], and cannot be decreased by increasing the duration of the analyzed segment. The variance of the periodogram can only be reduced by smoothing in the time or frequency domain [43], [243], [408].

Smoothed Power Spectral Estimator

The smoothed spectral estimator of Eq. 2.33 is obtained from [243]:

$$\bar{S}_{jj}(\omega) = \frac{1}{2\pi}\int_{-\infty}^{+\infty} w(\tau)\hat{R}_{jj}(\tau)e^{-i\omega\tau}\,d\tau$$

$$= \frac{1}{2\pi}\int_{-\infty}^{+\infty} \bar{R}_{jj}(\tau)e^{-i\omega\tau}\,d\tau \tag{2.35}$$

where

$$\bar{R}_{jj}(\tau) = w(\tau)\hat{R}_{jj}(\tau) \tag{2.36}$$

is the smoothed autocovariance function of Eq. 2.23, and $w(\tau)$ is a lag window with properties:

$$w(0) = 1$$
$$w(\tau) = w(-\tau) \qquad\qquad\qquad (2.37)$$
$$w(\tau) = 0 \qquad |\tau| \geq L\Delta t, \quad L\Delta t < T$$

The last equality in the above equation indicates that the autocovariance needs only be evaluated up to time lag $L\Delta t$ instead of T. It is noted that, in Eqs. 2.35 and 2.36, as well as subsequent equations herein, the accent $\bar{}$ indicates smoothed estimators, whereas the accent $\hat{}$, as, e.g., in Eqs. 2.32–2.34, denotes unsmoothed ("raw") ones.

Equivalently, the smoothed spectral estimate can be evaluated directly in the frequency domain through the following convolution expression [243]:

$$\bar{S}_{jj}(\omega) = \int_{-\infty}^{+\infty} W(u)\hat{S}_{jj}(\omega - u)\,du \qquad\qquad (2.38)$$

where the spectral window, $W(\omega)$, and the lag window, $w(\tau)$, are Fourier transforms of each other, i.e.,

$$w(\tau) = \int_{-\infty}^{+\infty} W(\omega)e^{i\omega\tau}\,d\omega$$
$$\Longleftrightarrow W(\omega) = \frac{1}{2\pi}\int_{-\infty}^{+\infty} w(\tau)e^{-i\omega\tau}\,d\tau \qquad\qquad (2.39)$$

from which it follows that the spectral window, $W(\omega)$, has the properties:

$$\int_{-\infty}^{+\infty} W(\omega)\,d\omega = w(0) = 1$$
$$W(\omega) = W(-\omega) \qquad\qquad (2.40)$$

For discrete frequencies, Eq. 2.38 takes the form:

$$\bar{S}_{jj}^{M}(\omega_n) = \sum_{m=-M}^{+M} W(m\Delta\omega)\,\hat{S}_{jj}(\omega_n + m\Delta\omega)$$
$$= \frac{2\pi}{T}\sum_{m=-M}^{+M} W(m\Delta\omega)\,\Lambda_j^2(\omega_n + m\Delta\omega) \qquad\qquad (2.41)$$

where $\Delta\omega = 2\pi/T$ is the frequency step, $\omega_n = n\Delta\omega$ is the discrete frequency, $W(m\Delta\omega)$ is the spectral window, $2M + 1$ the number of frequencies over which the averaging is performed, the superscript M indicates the dependence of the estimate on the length of the smoothing window, and the last equality in Eq. 2.41 results from Eq. 2.34. Examples of smoothing windows include the rectangular, Bartlett, Tukey, Parzen, Hanning and Hamming windows (e.g., [43], [60], [243], [408]). It is noted that different smoothing windows yield similar results, as long as the bandwidth of the windows is the same [243].

Equations 2.38 and 2.41 suggest that the smoothed spectral estimate is an average-over-frequency of the "raw" spectral estimate [60]. With the smoothing operation, the variance of the estimate is reduced to a fraction of that of the periodogram depending on the spectral bandwidth of the window, which is defined as:

$$\text{Bandwidth} = \mathcal{B} = \left[\int_{-\infty}^{+\infty} w^2(\tau)\,d\tau \right]^{-1} = \left[2\pi \int_{-\infty}^{+\infty} W^2(\omega)\,d\omega \right]^{-1} \tag{2.42}$$

Indeed [243]:

$$\text{Variance} \times \text{Bandwidth} = \text{Constant} \tag{2.43}$$

which suggests that the larger the spectral bandwidth, the smaller the variance, and vice versa. The smoothed estimate (Eq. 2.38) inherits the smoothness properties of the window [81], but this is achieved at the cost of resolution.

A compromise needs then to be made regarding bias, variance and resolution of the estimate [243]. As indicated earlier, its bias is reduced as the duration of the lag window increases, which implies that, in the frequency domain, the bandwidth decreases. Hence, to achieve small bias in the estimate, the bandwidth of the spectral window should be small, which is also associated with resolution: The bandwidth of the smoothing window should be such that the narrowest important detail in the spectrum is captured. On the other hand, a narrow bandwidth implies, according to Eq. 2.43, that the variance of the estimate will be large, and, hence, the estimator unstable. Abrahamson $et\ al.$ [14], in evaluating an optimal window for the estimation of the coherency (Section 2.4.2), suggested an 11-point ($M = 5$) Hamming (spectral) window, if the coherency estimate is to be used in structural analysis, for structural damping coefficient 5% of critical, and for time windows less than approximately 2000 samples. Their suggestion was based on the observation that the choice of the smoothing window should be directed not only from the statistical properties of the coherency estimate, but also from the purpose for which it is derived, so that the required resolution in structural response evaluations is maintained.

Example Applications

With the aforementioned considerations, the power spectral density estimate of a recorded time series, also commonly referred to as the "point" estimate of the motions, is evaluated as follows: The time series at location $\vec{r}_j$ on the ground surface is expressed as $a_j(t_l)$, where $t_l = l\Delta t$, $l = 0, \ldots, N_T$, indicates discrete time. Let $\mathcal{L}_j(\omega_n)$ be a scaled Fourier transform of $a_j(t_l)$ as [606]:

$$\mathcal{L}_j(\omega_n) = \sqrt{\frac{\Delta t}{2\pi N_T}} \sum_{l=0}^{N_T-1} a_j(l\Delta t)\exp[-i\omega_n l\Delta t] \tag{2.44}$$

where $N_T = T/\Delta t$. The expression:

$$\hat{S}_{jj}(\omega_n) = [\mathcal{L}_j(\omega_n)]^2 \tag{2.45}$$

is equivalent to the unsmoothed power spectral density estimates of Eqs. 2.33 and 2.34. Figure 2.3 presents the periodograms of the analyzed segments (Fig. 2.1) at the

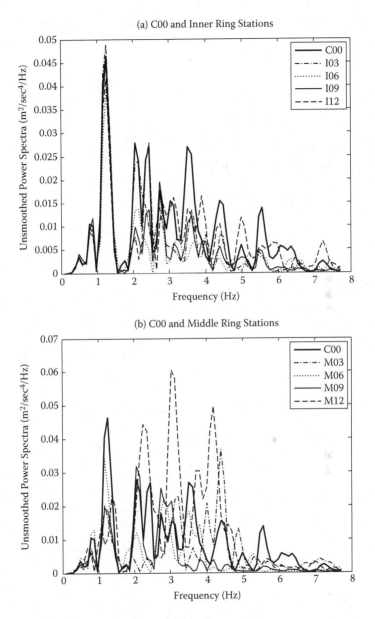

FIGURE 2.3 Unsmoothed power spectral density functions (Eq. 2.34) of the time histories of Fig. 2.1 at the selected stations of the SMART-1 array: Part (a) presents the periodograms at C00 and the inner ring stations I03, I06, I09 and I12, and part (b) those at C00 and the middle ring stations M03, M06, M09 and M12.

center and inner ring stations (Fig. 2.3(a)), and the center and middle ring stations (Fig. 2.3(b)). It can be clearly recognized from Figs. 2.2 and 2.3 that the power spectral density (as a scaled square of the Fourier amplitudes of the data) is a clearer descriptor of the characteristics of the seismic motions than the autocovariance function.

Figure 2.3(a) suggests that the power spectral densities at C00, I03, I06, I09 and I12 have fairly similar frequency content, which is consistent with the shape of their autocovariance functions (Fig. 2.2(a)). The peaks in the spectra occur at the same frequencies with the highest peak at approximately 1.3 Hz and essentially the same amplitude at all stations. The amplitudes of the spectra decay thereafter with the data at C00 producing the highest peaks. On the other hand, more significant differences are observed in the spectra of the center and middle ring stations in Fig. 2.3(b). The peak at approximately 1.3 Hz is not so pronounced for the middle ring stations (Fig. 2.3(b)) as it is for C00 and the inner ring ones (Fig. 2.3(a)), and the middle ring station spectra peak at higher frequencies, with the highest peak occurring at 3 Hz for the motions at station M12. The differences between the spectra at the inner and middle ring stations can be attributed to the distance between the stations: It is recalled that the radius of the inner ring of the SMART-1 array (Fig. 1.2) is 200 m, which implies that the maximum distance between the stations in Fig. 2.3(a) is 400 m, whereas the radius of the middle ring is 1 km, i.e., the maximum separation distance between the stations of Fig. 2.3(b) is 2 km. As the waves propagate over longer distances, they encounter larger variability in the soil strata along their path, resulting in more pronounced differences between the motions at the further-away stations. Additionally, there may be characteristics in the soil strata and the propagation pattern of the waves that are particular for each recording station. For example, the data at station M12 result in an autocovariance function (Fig. 2.2(b)) and a power spectral density (Fig. 2.3(b)) that differ significantly from the data at the other stations of the array.

With the scaled Fourier transform of the time series (Eq. 2.44), the smoothed spectral estimate of the motions at each recording station is evaluated from the following expression:

$$\bar{S}_{jj}^M(\omega_n) = \sum_{m=-M}^{+M} W(m\Delta\omega) \, [\mathcal{L}_j(\omega_n + m\Delta\omega)]^2 \tag{2.46}$$

Equation 2.46 is equivalent to Eqs. 2.35, 2.38 and 2.41. From the available smoothing windows, the Hamming window is most commonly used for smoothing the seismic spectral estimates [10]. Its expression (in samples) is given by:

$$W(m) = 0.54 - 0.46\cos\left(\frac{\pi(m+M)}{M}\right) \qquad m = -M, \dots, M \tag{2.47}$$

and its graphical representation for $M = 1, 3, 5, 7$ and 9 is illustrated in Fig. 2.4. It can be seen from the figure that all windows have unit value at the 0-th sample, and, as M increases, the sharpness of the window decreases. The width of the window (or, equivalently, the number of "points" of the window) is given by $2M + 1$, which, for $M = 1, 3, 5, 7$ and 9, assume the values $2M + 1 = 3, 7, 11, 15$ and 19, respectively. The area underneath the Hamming window is $1.08M$. If the window is used in the frequency domain, as is most commonly the case in the analysis of seismic data, it ought to satisfy the characteristics of spectral windows (Eq. 2.40), and, hence, the area underneath the window (Fig. 2.4) needs to be equal to unity, i.e., the right-hand side of Eq. 2.47 needs to be divided by $1.08M$.

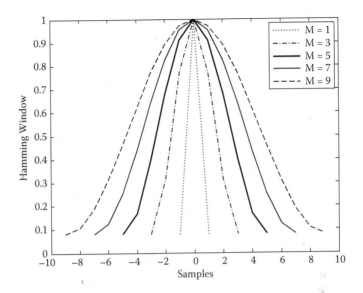

FIGURE 2.4 Illustration of the Hamming window (Eq. 2.47) as function of samples for window lengths of $2M + 1 = 3, 7, 11, 15$ and 19, that correspond, respectively, to $M = 1, 3, 5, 7$ and 9 in the figure caption.

Figure 2.5 presents the power spectral density at C00 smoothed with the Hamming (spectral) window for various values of M ($M = 1, 3, 5, 7$ and 9). The smoothed spectral density with $M = 1$ is almost indistinguishable from the unsmoothed estimate of the spectrum (Fig. 2.3). It can be seen from Fig. 2.5 that, as M increases, the resolution of the smoothed spectrum decreases very sharply for $M \leq 5$, but less so for the higher values of M. As indicated earlier, an 11-point Hamming spectral window has been suggested by Abrahamson *et al.* [14] as optimal for maintaining resolution and reducing the variance of coherency estimates (Section 2.4.2). Figure 2.6 then presents the power spectral densities of the data at C00 and the inner and middle ring stations of Fig. 2.3 smoothed with an 11-point Hamming window. It can be clearly recognized from the comparison of Figs. 2.3 and 2.6 that the "sharpness" of the original spectra (Fig. 2.3) is considerably reduced by the smoothing process (Fig. 2.6). It is, however, recalled that the variance of the smoothed estimates is reduced to a fraction of the variance of the unsmoothed ones (Eqs. 2.42 and 2.43).

2.2.4 PHASE CHARACTERISTICS OF (SINGLE) RECORDED ACCELEROGRAMS

The power spectral density of recorded data (Figs. 2.3 and 2.6) provides information about the scaled square Fourier amplitude of the motions, but does not retain information about their Fourier phase. Ohsaki [380], in analyzing phases and phase differences of recorded accelerograms, noted that the probability distribution of the Fourier phases of the series appears to be uniform. This property of the phases, i.e., uniform distribution between $[0, 2\pi)$, is most commonly used in simulations of

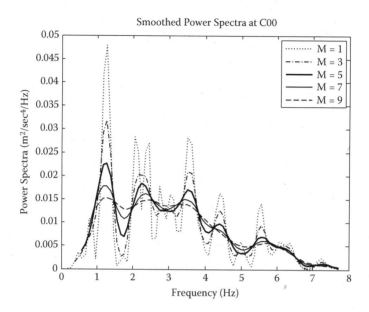

FIGURE 2.5 Effect of smoothing by means of the frequency domain Hamming window (Fig. 2.4) with various window lengths ($M = 1, 3, 5, 7$ and 9) on the power spectral density (Eq. 2.46) of the seismic motions at station C00 of the SMART-1 array.

seismic ground motions, as will be further elaborated upon in Chapter 7. Because this property does not reveal any additional information or pattern, the phases of a seismic record are sometimes referred to as "characterless" [372].

Additionally, however, Ohsaki [380] pointed out that information is contained in the Fourier phase differences: The histogram of the Fourier phase differences resembles the shape of the envelope function of the acceleration time history. The phase differences are basically the finite difference approximation of the derivative of the phase with respect to the frequency, i.e., $d\Phi_j(\omega)/d\omega$, which has units of time. The phase derivative with respect to frequency is termed group delay, or group delay time [441], or envelope delay [71]. Katsukura *et al.* [262] investigated analytically and numerically the relation between the group delay time spectrum and the envelope of the time history and reached results similar to Ohsaki's [380] observations. Nigam [372] derived analytically the phase derivative properties of a class of random processes, and concluded that the group delay time spectrum reflects the non-stationary characteristics of a modulated white-noise random process. Hence, the mean and standard deviation of the group delay time spectrum correspond to those of the envelope function of the time series, with the standard deviation being directly related to its duration [441]. The properties of the phase of the accelerograms have been used in studies of dispersive waves, e.g., [16], the analysis of source, path and site characteristics, e.g., [350], [351], [441], and in simulation of seismic ground motions, e.g., [71], [346], [440], [471], [472], [521], [527], [567]. A treatise on phase derivatives has been recently presented by Boore [71].

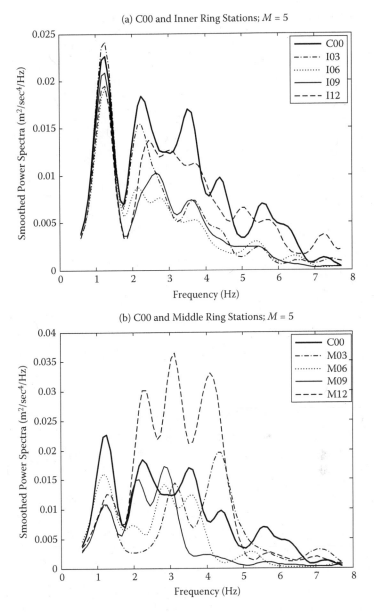

FIGURE 2.6 Smoothed power spectral density functions, derived from Eq. 2.46 and the $M = 5$ Hamming window, of the time histories of Fig. 2.1 at the selected SMART-1 array stations: Part (a) presents the power spectra at C00 and the inner ring stations I03, I06, I09 and I12, and part (b) those at C00 and the middle ring stations M03, M06, M09 and M12.

The following should be emphasized at this point: The phase differences described in this subsection and the corresponding group delay time spectrum represent the variation with frequency of the phase properties of a single accelerogram. As such, they are to be distinguished from the phase spectra introduced in the following section, which reflect, at each frequency, the difference between the Fourier phases of time histories recorded at two stations.

2.3 BIVARIATE STOCHASTIC PROCESSES

Section 2.2 dealt with the evaluation of the "point" estimates of the motions, for which it was considered that the time series at each recording station is a realization of a random process, and the data at the recording stations were analyzed independently of one another. However, seismic data recorded at an array of sensors are "interrelated." This section evaluates the joint stochastic characteristics of the motions between two stations, considering that the records at the two stations are realizations of a bivariate (vector) process. As was the case for the evaluation of the estimators of random processes, bivariate processes are also, generally, described by their second moments, i.e., the cross covariance function in the time domain or the cross spectral density in the frequency domain. The approach presented in this section can be readily extended to multivariate stochastic processes and fields, the characteristics of which are highlighted in Section 2.5.

2.3.1 CROSS COVARIANCE FUNCTION

Similar to the definition of the autocovariance function (Eq. 2.18), the cross covariance function between the time histories $a(\vec{r}_j, t_1)$ and $a(\vec{r}_k, t_2)$ at two locations, $\vec{r}_j$ and $\vec{r}_k$, on the ground surface, and times t_1 and t_2 is defined as:

$$\tilde{R}_{aa}(\vec{r}_j, t_1; \vec{r}_k, t_2) = \mathrm{E}\{[a(\vec{r}_j, t_1) - \mu_a(\vec{r}_j, t_1)][a(\vec{r}_k, t_2) - \mu_a(\vec{r}_k, t_2)]\}$$
$$= \mathrm{E}[a(\vec{r}_j, t_1)\, a(\vec{r}_k, t_2)] - \mu_a(\vec{r}_j, t_1)\mu_a(\vec{r}_k, t_2) \qquad (2.48)$$

and, again, as in Eq. 2.20, the subscripts j and k will be used to indicate location dependence, i.e.,

$$\tilde{R}_{jk}(t_1, t_2) = \tilde{R}_{aa}(\vec{r}_j, t_1; \vec{r}_k, t_2) \qquad (2.49)$$

Considering next that the time histories have been demeaned, and, in addition, that they are jointly stationary, the cross covariance function of Eq. 2.49 can be re-written as:

$$\tilde{R}_{jk}(t_1, t_2) = \tilde{R}_{jk}(t, t + \tau) = \tilde{R}_{jk}(\tau) \qquad (2.50)$$

i.e., the cross covariance between the motions at the two stations is independent of the actual time and varies only with the time lag τ, as was also the case for the stationary autocovariance function (Eq. 2.22).

Example Applications

With information available only from a single record at each station during the earthquake and with the assumption of ergodicity, the ensemble cross covariance estimate of Eq. 2.50 assumes a form similar to that of the autocovariance function (Eq. 2.23), namely:

$$\hat{R}_{jk}(\tau) = \begin{cases} \dfrac{1}{T} \displaystyle\int_{0}^{T-|\tau|} a_j(t)\, a_k(t+\tau)\, dt & |\tau| \leq T \\[4mm] 0 & |\tau| > T \end{cases} \tag{2.51}$$

where T indicates again the duration of the analyzed segment of the series. It is noted that a more rigorous definition of the cross covariance function is provided by [243]:

$$\hat{R}_{jk}(\tau) = \begin{cases} \dfrac{1}{T} \displaystyle\int_{-T/2}^{T/2-\tau} a_j(t)\, a_k(t+\tau)\, dt & 0 \leq \tau \leq T \\[4mm] \dfrac{1}{T} \displaystyle\int_{-T/2-\tau}^{T/2} a_j(t)\, a_k(t+\tau)\, dt & -T \leq \tau \leq 0 \end{cases} \tag{2.52}$$

where time t is allowed to vary between $-T/2$ and $T/2$, and it is understood that the estimate of Eq. 2.52 is equal to zero for $|\tau| > T$.

Figure 2.7 presents the cross covariance function of the data at C00 and the inner ring stations (Fig. 2.7(a)) and C00 and the middle ring stations (Fig. 2.7(b)). Contrary to the autocovariance functions of the data (Fig. 2.2), the cross covariance functions are not peaked at $\tau = 0$. The peaks of the cross covariance functions between C00 and the inner ring stations (I03, I06, I09 and I12) in Fig. 2.7(a) occur at time lags τ_0 close to $\tau = 0$ with similar amplitudes, whereas, for the cross covariance functions between C00 and the middle ring stations (M03, M06, M09 and M12) in Fig. 2.7(b), the locations of the peaks as well as their amplitudes vary, with the most distinguishable differences occurring for the station pairs C00-M12 and C00-M06. The shift of the peak values of the cross covariance functions from $\tau = 0$ is associated with the propagation of the waves, and has a significant effect on the bias of the spectral estimates, as will be further elaborated upon in Section 2.3.2. Additionally, it can be recognized from Fig. 2.7 that the cross covariance functions are not symmetric about the $\tau = 0$ axis. An equivalent symmetry property of the cross covariance function, which can be derived from Eq. 2.51 or Eq. 2.52, is:

$$\hat{R}_{jk}(\tau) = \hat{R}_{kj}(-\tau) \tag{2.53}$$

Bias and Variance

The bias and variance characteristics of the cross covariance estimators are similar to those of the autocovariance estimates [243]: If T in the denominator of Eq. 2.51 were substituted by $T - |\tau|$, the estimate would be unbiased, but this substitution would increase the mean-square error (Eq. 2.25). Additionally, the neighboring values of the cross covariance function also tend to be highly correlated. Hence, it is again preferable to work in the frequency rather than the time domain. The frequency domain estimators, in this case, the cross spectral densities of the motions between the stations, are described in Section 2.3.2.

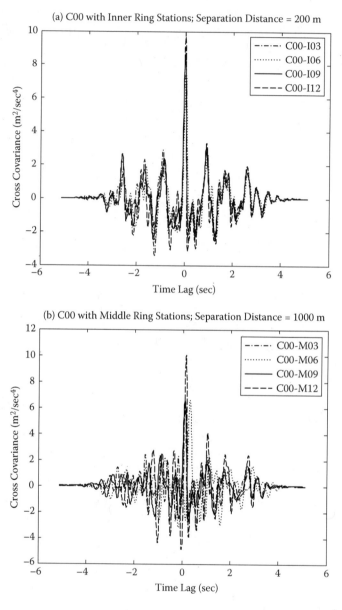

FIGURE 2.7 Cross covariance functions (Eq. 2.51) of the time histories of Fig. 2.1: Part (a) illustrates the cross covariance between C00 and the inner ring stations I03, I06, I09 and I12, at a separation distance of 200 m, and part (b) between C00 and the middle ring stations M03, M06, M09 and M12, at a separation distance of 1000 m.

Cross Correlation Function

The spatial variation of the seismic ground motions is expressed, in the time domain, by the cross correlation function, which is defined from Eqs. 2.23 and 2.51 as:

$$\hat{\rho}_{jk}(\tau) = \frac{\hat{R}_{jk}(\tau)}{\sqrt{\hat{R}_{jj}(0)\hat{R}_{kk}(0)}} \qquad (2.54)$$

i.e., it is the cross covariance function normalized with respect to the square root of the product of the peak values of the autocovariance functions of the motions at the two stations. In the early studies of spatial variability, the cross correlation function has been extensively evaluated from recorded data, as will be illustrated in Chapter 3. Its correspondence to the more commonly used spatial coherency is presented in Section 2.4.3.

2.3.2 CROSS SPECTRAL DENSITY

The cross spectral density (or cross spectrum) between the time histories at two stations is defined as the Fourier transform (Eq. 2.31) of the cross covariance function (Eq. 2.51), i.e.,

$$\hat{S}_{jk}(\omega) = \frac{1}{2\pi} \int_{-\infty}^{+\infty} \hat{R}_{jk}(\tau)e^{-i\omega\tau}\, d\tau \qquad (2.55)$$

Equation 2.55 indicates that the cross spectrum is complex-valued with the property $\hat{S}_{jk}(\omega) = \hat{S}_{jk}^*(-\omega)$. As was the case for the pair of the autocovariance function and the power spectral density (Eqs. 2.32 and 2.33), the inverse Fourier transform of Eq. 2.55 also holds, i.e.,

$$\hat{R}_{jk}(\tau) = \int_{-\infty}^{+\infty} \hat{S}_{jk}(\omega)e^{i\omega\tau}\, d\omega \qquad (2.56)$$

Alternatively, the cross spectral estimate of Eq. 2.55 can be evaluated directly in the frequency domain as follows: Let $\mathcal{A}_j(\omega) = \Lambda_j(\omega)\exp[i\,\Phi_j(\omega)]$ and $\mathcal{A}_k(\omega) = \Lambda_k(\omega)\exp[i\,\Phi_k(\omega)]$ be the Fourier transforms of the time histories $a_j(t)$ and $a_k(t)$, respectively, according to Eq. 2.31. The cross spectrum estimator of Eq. 2.55 becomes:

$$\hat{S}_{jk}(\omega) = \frac{2\pi}{T}\,\mathcal{A}_j^*(\omega)\,\mathcal{A}_k(\omega)$$

$$= \frac{2\pi}{T}\,\Lambda_j(\omega)\,\Lambda_k(\omega)\exp[i\{\Phi_k(\omega) - \Phi_j(\omega)\}] \qquad (2.57)$$

The absolute value of the cross spectral density, i.e.,

$$|\hat{S}_{jk}(\omega)| = \frac{2\pi}{T}\,\Lambda_j(\omega)\,\Lambda_k(\omega) \qquad (2.58)$$

is usually termed the cross amplitude spectrum, and the phase difference in Eq. 2.57, i.e.,

$$\hat{\varphi}_{jk}(\omega) = \tan^{-1}\frac{\Im[\hat{S}_{jk}(\omega)]}{\Re[\hat{S}_{jk}(\omega)]} = \Phi_k(\omega) - \Phi_j(\omega) \qquad (2.59)$$

the phase spectrum [243]. $\Re[\hat{S}_{jk}(\omega)]$ and $\Im[\hat{S}_{jk}(\omega)]$ in Eq. 2.59 represent, respectively, the real and imaginary part of the cross spectrum. The cross amplitude spectrum is controlled by the Fourier amplitudes of the motions at the two stations, and the phase spectrum indicates whether the frequency component of the time history at one station precedes or follows the other time series at that frequency.

The statistical properties of the cross spectral density are similar to those of the power spectrum (Section 2.2.3), especially in that its variance cannot be reduced by increasing the duration of the analyzed segment. This suggests that the cross spectral estimates need also be smoothed.

Smoothed Cross Spectral Density

The smoothed cross spectral density of Eq. 2.55 is obtained, as was the smoothed power spectral density of Eq. 2.35, from [243]:

$$\bar{S}_{jk}(\omega) = \frac{1}{2\pi} \int_{-\infty}^{+\infty} w(\tau)\hat{R}_{jk}(\tau)e^{-i\omega\tau} d\tau$$

$$= \frac{1}{2\pi} \int_{-\infty}^{+\infty} \bar{R}_{jk}(\tau)e^{-i\omega\tau} d\tau \tag{2.60}$$

where

$$\bar{R}_{jk}(\tau) = w(\tau)\hat{R}_{jk}(\tau) \tag{2.61}$$

and $w(\tau)$ is the lag window defined in Eq. 2.37. Equivalently, the smoothed cross spectrum can be evaluated directly in the frequency domain through the convolution:

$$\bar{S}_{jk}(\omega) = \int_{-\infty}^{+\infty} W(u)\hat{S}_{jk}(\omega - u) du \tag{2.62}$$

and the properties of the spectral window are as defined in Eqs. 2.39 and 2.40. For discrete frequencies, Eqs. 2.60 and 2.62 take the form:

$$\bar{S}_{jk}^M(\omega_n) = \frac{2\pi}{T} \sum_{m=-M}^{+M} W(m\Delta\omega) \, \Lambda_j(\omega_n + m\Delta\omega) \, \Lambda_k(\omega_n + m\Delta\omega)$$

$$\times \exp\{i \left[\Phi_k(\omega_n + m\Delta\omega) - \Phi_j(\omega_n + m\Delta\omega)\right]\} \tag{2.63}$$

Equation 2.63 is derived in a manner similar to the smoothed, discrete power spectral density of Eq. 2.41. An important observation that can be recognized from Eq. 2.63 (and also applies to Eqs. 2.60 and 2.62) is that the Fourier phases of the individual time series contribute not only to the phase spectrum of the smoothed estimator, but also to its cross spectral amplitude. For the unsmoothed estimates, the cross amplitude spectrum (Eq. 2.58) is not affected by the phases, which contribute only to the phase spectrum (Eq. 2.59). For clarity and reference in further derivations herein, the smoothed cross amplitude spectrum is given by:

$$\left|\bar{S}_{jk}^M(\omega_n)\right| = \frac{2\pi}{T} \left| \sum_{m=-M}^{+M} W(m\Delta\omega) \, \Lambda_j(\omega_n + m\Delta\omega) \, \Lambda_k(\omega_n + m\Delta\omega) \right.$$

$$\left. \times \exp\{i \left[\Phi_k(\omega_n + m\Delta\omega) - \Phi_j(\omega_n + m\Delta\omega)\right]\} \right| \tag{2.64}$$

and the smoothed phase spectrum by:

$$\bar{\varphi}_{jk}^M(\omega_n) = \tan^{-1} \frac{\Im[\bar{S}_{jk}^M(\omega_n)]}{\Re[\bar{S}_{jk}^M(\omega_n)]} \tag{2.65}$$

It is noted that the last equality of the unsmoothed phase spectrum in Eq. 2.59, i.e., $\hat{\varphi}_{jk}(\omega_n) = \Phi_k(\omega_n) - \Phi_j(\omega_n)$, no longer applies to the smoothed estimate (Eq. 2.65).

With the smoothing process of Eq. 2.60, Eq. 2.62 or Eq. 2.63, the bias and variance of the cross spectral estimator inherit the same characteristics as those of the smoothed power spectral estimate described in Section 2.2.3: Bias is reduced as the duration of the lag window increases and, consequently, the bandwidth of the spectral window decreases, whereas the variance of the estimate decreases as the bandwidth of the spectral window increases (Eq. 2.43). There is, however, a significant difference between the bias of the power and cross spectral densities [243] that results from the fact that the autocovariance function is symmetric with respect to $\tau = 0$ (Eq. 2.24 and Fig. 2.2) and is peaked at $\tau = 0$, whereas the cross covariance function is not symmetric around $\tau = 0$ and is peaked, generally, at a time lag $\tau_0 \neq 0$ (Eq. 2.53 and Fig. 2.7). If the duration of the lag window is selected such that the peak of the cross covariance function at τ_0 is excluded from the evaluation or if the value of the lag window at τ_0 is small, i.e., $w(\tau_0) << 1$, the bias of the estimate will be significantly affected. An approach to remedy this problem is through the alignment of the time histories.

Alignment of Time Series

If the time histories at two stations j and k are identical except for a time shift $|\tau_0|$ in the arrival of the signal between the two stations due to wave propagation, their cross covariance function will be identical to their autocovariance function, except that the cross covariance function will be centered at $\tau_0 = +/-|\tau_0|$, with the "+" or "−" sign depending on the direction of propagation and the order of the stations j and k in the estimation of the cross convariance function (Eqs. 2.51 and 2.53). The phase spectrum of the time series will then be a linear function of frequency, i.e., $\omega\tau_0$. If the time histories are strongly correlated and propagating, then their cross covariance function will still have a significant peak at τ_0, and the phase spectra will again show the linear trend over a significant frequency range. To remove this trend, the time histories can be aligned: τ_0 is evaluated from the peak of the cross covariance function or from the linear trend of the phase spectrum, and the time history at the further-away station is "shifted in time" by an amount τ_0. If, on the other hand the time histories are weakly correlated, their cross covariance function will include many spurious peaks and the identification of τ_0 will become difficult. In this case, the estimation of τ_0 from the linear trend of the phase spectrum may be more appropriate [243]. Alternatively, more refined approaches for the estimation of the arrival time of the waveforms between stations have been proposed by the geophysical community for the analysis of data collected at closely spaced sensors and used for exploration purposes [498]. For example, Rothman [430], [431] proposed an approach that views the problem as nonlinear inversion, estimates the cross correlation functions of the data (Eq. 2.54), but, instead of picking their peak, the correlation functions are transformed into probability distributions. Random numbers are generated from the distributions in an iterative scheme that minimizes the objective function, until convergence to the correct time shifts is achieved. The issue of wave propagation and arrival time perturbations of seismic data recorded at array stations will be further elaborated upon in Sections 2.4.1 and 4.2, as it plays an important role in the analysis, modeling and simulation of spatially variable ground motions.

Figure 2.8 presents the time histories at the 9 stations of the SMART-1 array shown in Fig. 2.1 after each time history at the inner and middle ring stations has been aligned with respect to the center station of the array, C00, by shifting the time histories an amount equal to the time lag corresponding to the peak of the cross covariance functions shown in Fig. 2.7. It can be recognized from Fig. 2.8 that the

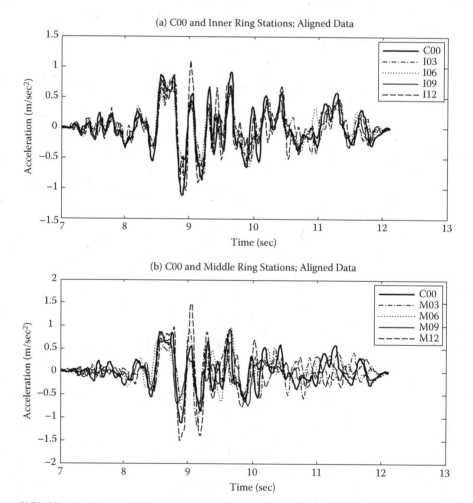

FIGURE 2.8 Aligned, tapered time histories corresponding to the non-aligned data presented in Fig. 2.1. Part (a) presents the aligned time series at the selected inner ring stations I03, I06, I09 and I12, and part (b) those at the middle ring stations M03, M06, M09 and M12. The data at the stations are aligned with respect to the strong motion S-wave window of the ground motion at the center station, C00, which, for comparison purposes, is presented in both subfigures.

peaks and valleys of the aligned time histories occur at approximately the same time, which was not the case for the actual (non-aligned) time histories of Fig. 2.1. Figure 2.9 shows the autocovariance of the aligned time histories. Figure 2.2, which presents the autocovariance functions of the non-aligned data, and Fig. 2.9 are very similar.

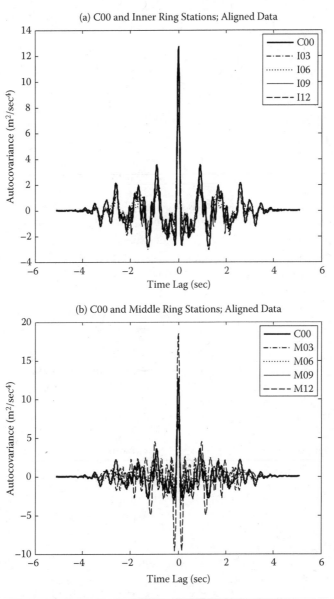

FIGURE 2.9 Autocovariance functions (Eq. 2.23) of the aligned time histories of Fig. 2.8 at the selected stations of the SMART-1 array: Part (a) illustrates the results at C00 and the inner ring stations I03, I06, I09 and I12, and part (b) those at C00 and the middle ring stations M03, M06, M09 and M12. The corresponding results for the non-aligned data are presented in Fig. 2.2.

The small differences between the figures occur because, due to the shifting of the time series by an amount τ_0 (when $\tau_0 \neq 0$), the autocovariance functions of the aligned data in Fig. 2.9 contain the information provided by the signals over the shifted window of the data. Figure 2.10 presents the cross covariance functions of the aligned time histories between C00 and the inner ring stations (Fig. 2.10(a)), and between C00 and the middle ring stations (Fig. 2.10(b)). All cross covariance functions are sharply peaked at $\tau = 0$. It is noted, however, that the cross covariance functions in Fig. 2.10 are not symmetric about $\tau = 0$.

Example Applications of Cross Spectral Densities

With the scaled Fourier transform of Eq. 2.44, the cross spectral densities between seismic motions recorded at array stations can be evaluated as:

$$\hat{S}_{jk}(\omega_n) = \mathcal{L}_j^*(\omega_n)\mathcal{L}_k(\omega_n) \tag{2.66}$$

The above equation is equivalent to Eqs. 2.55 and 2.57. Figure 2.11 presents the unsmoothed cross amplitude spectra (i.e., the absolute value of Eq. 2.66) of the aligned time series (Fig. 2.8) between C00 and the inner ring stations (Fig. 2.11(a)) and C00 and the middle ring stations (Fig. 2.11(b)). As was the case for the unsmoothed power spectral densities (Fig. 2.3), the spectra are sharply peaked, but their variance is large.

The smoothed estimates of the cross spectra by means of the scaled Fourier transform of the time series (Eq. 2.44) are given by:

$$\bar{S}_{jk}^M(\omega_n) = \sum_{m=-M}^{+M} W(m\Delta\omega)\mathcal{L}_j^*(\omega_n + m\Delta\omega)\mathcal{L}_k(\omega_n + m\Delta\omega) \tag{2.67}$$

which is equivalent to Eqs. 2.60, 2.62 and 2.63. Figure 2.12 presents the effect of smoothing with 3-, 7-, 9-, 11-, 15- and 19-point Hamming windows (Fig. 2.4) on the cross amplitude spectra between C00 and I06 (Fig. 2.12(a)) and C00 and M06 (Fig. 2.12(b)). As was the case for the effect of smoothing on the power spectral density at C00 in Fig. 2.5, smoothing decreases the resolution of the spectral estimates, but, also, decreases their variance (Eq. 2.43).

For completeness, Fig. 2.13 presents the smoothed (with an 11-point Hamming window) power spectral densities of the aligned data at C00 and the inner ring stations (Fig. 2.13(a)), and at C00 and the middle ring stations (Fig. 2.13(b)). There are minor differences between the estimators of the non-aligned (Fig. 2.6) and the aligned data (Fig. 2.13) that are caused by the consideration of the shifted time series in the evaluation of the latter, as has already been discussed in the comparison of their respective autocovariance functions in Figs. 2.2 and 2.9. Figure 2.14 presents the smoothed (again with the 11-point Hamming window) cross amplitude spectra of the motions between C00 and I03, I06, I09 and I12 at a separation distance of 200 m (Fig. 2.14(a)), and between C00 and M03, M06, M09 and M12 at a separation distance of 1 km (Fig. 2.14(b)). It can be clearly seen from the figures that, for the shorter separation distance (Fig. 2.14(a)), the similarity in the inner ring station data (Figs. 2.8(a) and 2.13(a)) leads to similar cross amplitude estimates, whereas the more different middle ring data at the longer separation distance (Figs. 2.8(b) and 2.13(b)) yield cross

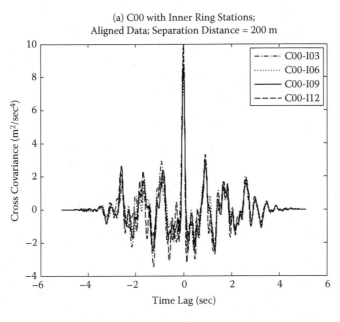

(a) C00 with Inner Ring Stations;
Aligned Data; Separation Distance = 200 m

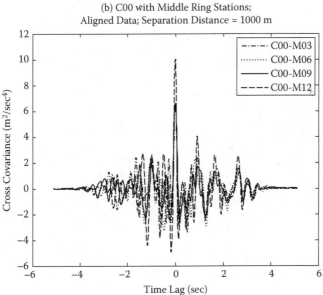

(b) C00 with Middle Ring Stations;
Aligned Data; Separation Distance = 1000 m

FIGURE 2.10 Cross covariance functions (Eq. 2.51) of the aligned time histories of Fig. 2.8: Part (a) presents the cross covariance between C00 and the inner ring stations I03, I06, I09 and I12, at a separation distance of 200 m, and part (b) between C00 and the middle ring stations M03, M06, M09 and M12, at a separation distance of 1000 m. The corresponding results for the non-aligned data are presented in Fig. 2.7.

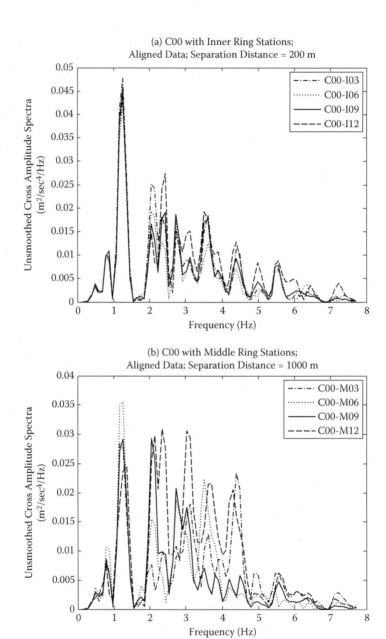

FIGURE 2.11 Unsmoothed cross amplitude spectra (Eq. 2.58) of the aligned time histories of Fig. 2.8: Part (a) presents the results between C00 and the inner ring stations I03, I06, I09 and I12, at a separation distance of 200 m, and part (b) between C00 and the middle ring stations M03, M06, M09 and M12, at a separation distance of 1000 m.

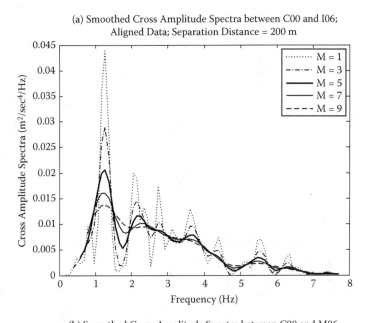

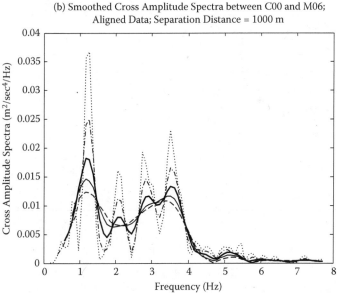

FIGURE 2.12 Effect of smoothing by means of the frequency domain Hamming window (Fig. 2.4) with various window lengths ($M = 1, 3, 5, 7$ and 9) on selected cross amplitude spectra (Eq. 2.64) of the aligned motions of Fig. 2.8: Part (a) indicates this effect on the cross amplitude spectrum between C00 and the inner ring station I06 at a separation distance of 200 m, and part (b) on the cross amplitude spectrum between C00 and the middle ring station M06 at a separation distance of 1000 m.

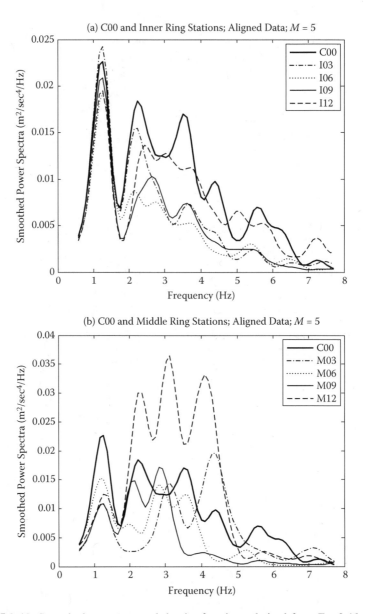

FIGURE 2.13 Smoothed power spectral density functions, derived from Eq. 2.46 and the $M = 5$ Hamming window, of the aligned time histories of Fig. 2.8 at the selected SMART-1 array stations: Part (a) presents the power spectra at C00 and the inner ring stations I03, I06, I09 and I12, and part (b) those at C00 and the middle ring stations M03, M06, M09 and M12. The corresponding results for the non-aligned data are presented in Fig. 2.6.

amplitude spectra with significant differences in their frequency content. Figures 2.13 and 2.14 provide the reference figures for the evaluation of the lagged coherency in example applications of the following section. The phase spectra of the cross spectral estimates (Eqs. 2.59 and 2.65) will be further elaborated upon in Section 2.4.1.

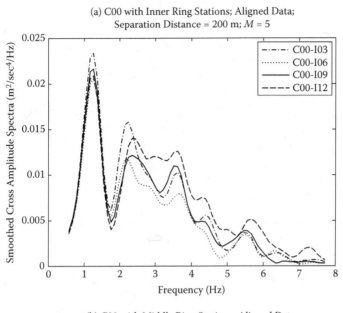

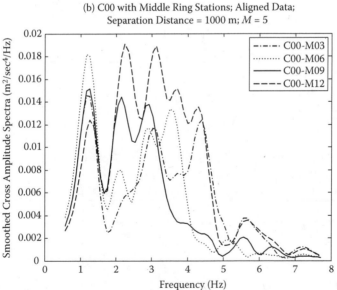

FIGURE 2.14 Smoothed cross amplitude spectra, derived from Eq. 2.64 and the $M = 5$ Hamming window, of the aligned time histories of Fig. 2.8 at the selected SMART-1 array stations: Part (a) presents the results between C00 and the inner ring stations I03, I06, I09 and I12, at a separation distance of 200 m, and part (b) those between C00 and the middle ring stations M03, M06, M09 and M12, at a separation distance of 1000 m.

2.4 COHERENCY

The cross spectral density estimators of the last section fully describe the joint characteristics of the processes at two locations on the ground surface. It is, however, customary to work with the coherency of spatially variable ground motions, which is described in this section, rather than directly with the cross spectral density.

The coherency of the seismic motions is obtained from the smoothed cross spectrum of the time series between the two stations j and k (Eq. 2.63), normalized with respect to the corresponding power spectra (Eq. 2.41) as, e.g., [14], [208], [597], [606]:

$$\bar{\gamma}_{jk}^{M}(\omega) = \frac{\bar{S}_{jk}^{M}(\omega)}{\sqrt{\bar{S}_{jj}^{M}(\omega)\,\bar{S}_{kk}^{M}(\omega)}} \qquad (2.68)$$

and is a complex number; the square of the absolute value of the coherency, the coherence:

$$\left|\bar{\gamma}_{jk}^{M}(\omega)\right|^{2} = \frac{|\bar{S}_{jk}^{M}(\omega)|^{2}}{\bar{S}_{jj}^{M}(\omega)\,\bar{S}_{kk}^{M}(\omega)} \qquad (2.69)$$

is a real number assuming values $0 \le \left|\bar{\gamma}_{jk}^{M}(\omega)\right|^{2} \le 1$. The absolute value of the coherency:

$$\left|\bar{\gamma}_{jk}^{M}(\omega)\right| = \frac{|\bar{S}_{jk}^{M}(\omega)|}{\sqrt{\bar{S}_{jj}^{M}(\omega)\,\bar{S}_{kk}^{M}(\omega)}} \qquad (2.70)$$

is termed the lagged coherency. Its real part, $\Re[\bar{\gamma}_{jk}^{M}(\omega)]$, is commonly referred to as the unlagged coherency, and its phase:

$$\bar{\varphi}_{jk}^{M}(\omega) = \tan^{-1}\left(\frac{\Im[\bar{\gamma}_{jk}^{M}(\omega)]}{\Re[\bar{\gamma}_{jk}^{M}(\omega)]}\right) = \tan^{-1}\left(\frac{\Im[\bar{S}_{jk}^{M}(\omega)]}{\Re[\bar{S}_{jk}^{M}(\omega)]}\right) \qquad (2.71)$$

is the smoothed phase spectrum. It is noted that the above equation is the same as the phase spectrum of the smoothed cross spectral estimator of Eq. 2.65. The complex coherency of Eq. 2.68 can then be alternatively expressed as:

$$\bar{\gamma}_{jk}^{M}(\omega) = \left|\bar{\gamma}_{jk}^{M}(\omega)\right| \exp[i\bar{\varphi}_{jk}^{M}(\omega)] \qquad (2.72)$$

2.4.1 PHASE SPECTRUM

The phase spectrum of the complex coherency (Eq. 2.71), or, equivalently, of the cross spectral density (Eq. 2.65) incorporates two effects, that of wave propagation across the array (or between two stations) and that of random phase variability at each station. The effect of wave passage on the phase spectrum is presented first, and followed by a discussion on the deviations in the arrival times of the waves at the stations from the overall pattern of uniform propagation of a plane wave across the array. The evaluation of the phase spectra of non-aligned and aligned seismic motions is presented next, and the contributions of wave passage and random phase variability to the spectra are illustrated. A brief characterization of the statistical properties of the phase spectrum concludes this section.

Wave Passage

Consider, e.g., that the ground motions consist of a waveform that propagates unchanged with velocity $\vec{c}$ on the ground surface along a line connecting two stations. The vector of the separation distance between the stations is denoted by $\vec{\xi}_{jk} = \vec{r}_k - \vec{r}_j$. The analytical coherency expression of Eq. 2.68 for this type of motion would be [551]:

$$\gamma_{jk}(\xi_{jk}, \omega) = \exp\left[-i \, \frac{\omega \, (\vec{c} \cdot \vec{\xi}_{jk})}{|\vec{c}|^2}\right] = \exp\left[-i \, \frac{\omega \, \xi_{jk}}{c}\right] \qquad (2.73)$$

where $\xi_{jk} = |\vec{\xi}_{jk}|$, $c = |\vec{c}|$, and it has been implicitly assumed that the waveform propagates from station j to station k. Equation 2.73 implies that $|\gamma_{jk}(\xi_{jk}, \omega)| = 1$ and the complex term in the equation describes the wave passage effect, i.e., the delay in the arrival of the waveforms at the further-away station caused solely by their propagation. The phase spectrum of such motions is then a linear function of frequency described by:

$$\vartheta_{jk}(\xi_{jk}, \omega) = -\frac{\omega \, \xi_{jk}}{c} \qquad (2.74)$$

where ξ_{jk}/c reflects, actually, the time lag τ_0 defined earlier (Section 2.3.2) in the discussion of the alignment of the time histories. Equation 2.74 is most commonly used to describe the wave passage part of the phase spectrum in the simulation of spatially variable ground motions (Chapters 7 and 8).

The apparent propagation velocity of the seismic motions across an array can be estimated by means of signal processing techniques, such as the conventional method [10], [93], the high resolution method [93], or the multiple signal characterization method [171], [172]. These techniques, which are described in some detail in Section 4.1, evaluate, in different forms, the frequency-wavenumber (F-K) spectrum of the motions, and identify the propagation characteristics of the waveforms from the locations of the peaks of the F-K spectrum. For a single type of wave dominating the motions during the analyzed time segment, as is, generally, the case for the strong motion S-wave window used in spatial variability evaluations, a single peak is identified, the location of which determines the average apparent velocity and direction of propagation of the impinging wave at that particular frequency. Since body waves are non-dispersive, except in highly attenuated media, the location of the peak varies only slightly over the frequency range where the S-wave controls the motions [501]. In this case, the phase spectrum caused by the apparent propagation of waveforms can be approximated by Eq. 2.74 with constant propagation velocity, c, over frequency. However, as frequency increases, additional wave components, and, essentially, scattered energy, start dominating the motions. In this case, the propagation characteristics of the waveforms fluctuate significantly with frequency, and Eq. 2.74 is no longer valid.

Clearly, when more than one wave contribute to the motions and their direction and velocity of propagation vary, Eq. 2.74 is, again, not applicable.

Arrival Time Perturbations

In addition to the above considerations, F-K spectra analyses identify the average pattern of wave propagation across an array. However, the seismic data at the array stations incorporate random time delay fluctuations around the wave passage delay, that are particular for each recording station. These arrival time perturbations are caused by the upward traveling of the waves through the horizontal variations of the geologic structure underneath the array [498] and, also, due to deviations of the prop-agation pattern of the waves from that of plane wave propagation [66]. They can be so significant, that the apparent propagation pattern of the wavefronts between sets of stations can be very different from the estimated overall pattern of constant apparent propagation across the array [205]. For example, it will be shown in Section 4.2 that the seismic ground motions of Event 5 during the strong motion S-wave window prop-agate along the epicentral direction of the earthquake, which, essentially, coincides with the direction from O06 to O12 (Fig. 1.2) with an apparent propagation velocity on the ground surface of 4.5 km/sec. For the stations considered in the example illus-trations of this chapter, this suggests that the waveforms arrive first at M06, then I06, almost simultaneously at I03, C00 and I09, then at I12 and, finally, at M12. Consid-ering again C00 as the reference station, as was the case for the alignment of the time histories, the analyzed time segment of the motions at M06 should arrive earlier than at C00 by (1 km/4.5 km/sec =) 0.22 sec, by (0.2 km/4.5 km/sec =) 0.04 sec at I06, arrive simultaneously at stations I03, C00 and I09, and arrive later, by 0.04 sec, at I12 and, also, later, by 0.22 sec at M12. The alignment process, however, revealed that the motions at M06 arrive earlier than at C00 by 0.32 sec, at I06 by 0.04 sec, at I03 by 0.02 sec, they arrive simultaneously at I09 and C00, but, they also arrive earlier than at C00 both at I12, by 0.01 sec, and at M12, by 0.14 sec. Similar results, though for a longer window during the same event, were reported by Harichandran [202]. It is noted that station M12, for which a significant reversed propagation pattern from the average one has been identified through the alignment process, also exhibited a frequency content different than the other stations (Fig. 2.13(b)). This may be caused by a peculiarity either at the location or the surrounding medium of this station.

Boissières and Vanmarcke [66] modeled these arrival time fluctuations from SMART-1 data. They considered an extension of the closure property [208], in which "closure" is checked by relating the lags of all triplets of stations taken two by two. Their model considers that the time lag between two stations is given by:

$$\Delta t_{jk} = \Delta t_{jk}^{wp} + \Delta t_{jk}^{r} \qquad (2.75)$$

in which $\Delta t_{jk}^{wp} = \xi_{jk}/c$ is the time lag between the two stations j and k due to the average pattern of wave propagation across the array, and Δt_{jk}^{r} represents the random fluctuations between individual stations. From their analysis of the SMART-1 data they concluded that Δt_{jk}^{r} is a normally distributed random variable with zero mean and standard deviation equal to $2.7 \times 10^{-2} + 5.41 \times 10^{-5}\xi_{jk}$, in which ξ_{jk} is measured in m.

Evaluation of Phase Spectra

Figure 2.15 presents the unwrapped[2] phase spectra between C00 and the inner ring stations, in part (a), and between C00 and the middle ring stations, in part (b), of the non-aligned data (Fig. 2.1), for which the cross spectral estimates were smoothed with an 11-point Hamming window. It can be clearly seen from the figure that, over the dominant frequency range of the motions (up to 4-5 Hz in Fig. 2.6), the phase spectra have a linear trend with frequency. This, as indicated earlier, is directly related to the propagation pattern of the single broadband S-wave that controls the motions in their dominant frequency range: For the close-by stations (Fig. 2.15(a)), the slope of the linear trend is small, which is associated with the short distance between the stations, and, consequently, a small value for τ_0, i.e., the deviation of the peak of the cross covariance functions from the origin (Fig. 2.7(a)). The maximum delay for the arrival of the waves at these inner ring stations relative to C00 was, as evaluated previously, 0.04 sec for station I06, and, indeed, the phase spectrum for the station pair C00-I06 exhibits the larger slope of the phase spectra of the non-aligned data in Fig. 2.15(a). For the further-away stations, the slope of the phase spectra increases, which, again, is consistent with the location of the peaks of the cross covariance function (Fig. 2.7(b)), with the largest slope occurring for station pair C00-M06, for which the time delay due to the propagation of the waves was 0.32 sec. At higher frequencies, the phase spectra deviate from their linear trend, indicating that the S-wave no longer controls the motions. It is noted that the scale of the phase spectra in Figs. 2.15(a) and 2.15(b) is different.

Figure 2.16 presents the unwrapped phase spectra of the aligned data (Fig. 2.8) between C00 and the inner ring stations (Fig. 2.16(a)), and C00 and the middle ring stations (Fig. 2.16(a)), for which the cross spectral estimates were, again, smoothed with an 11-point Hamming window. It can be seen from Fig. 2.16 that, where there was a linear trend for the non-aligned data (Fig. 2.15), the phase spectra of the aligned data fluctuate around the zero axis (except for station pair C00-M06 that exhibits an additional linear trend between 4 Hz and 6 Hz). This fluctuation constitutes the contribution of the random variability in the phases at each station. In Fig. 2.15, the fluctuation due to the random phase variability occurs around the linear trend of the phase spectra. It is noted that, again, the scale of the phase spectra in Figs. 2.16(a) and 2.16(b) is different.

The phase spectra of the aligned data fluctuate fairly "smoothly" around the zero axis, which is somewhat counterintuitive with the notion of random variability of the phases at the individual stations. This "smoothness" of the phase spectra results exactly from the fact that the cross spectral estimates have been smoothed over 11 near-neighbor frequencies. The actual phase differences of the aligned motions between station pairs are presented in Fig. 2.17: Part (a) of the figure presents the phase differences between the data at C00 and the inner ring stations, and part (b) those between C00 and the middle ring stations. Wrapped, rather than unwrapped, phase differences

---

[2] The unwrapped phase is not restricted in the range $[-\pi, +\pi)$ or $[0, 2\pi)$, as is the case for the wrapped phase. If absolute jumps between consecutive wrapped phases in the spectrum are greater than π, then multiples of $\pm 2\pi$ are added to the wrapped phase to yield the unwrapped phase.

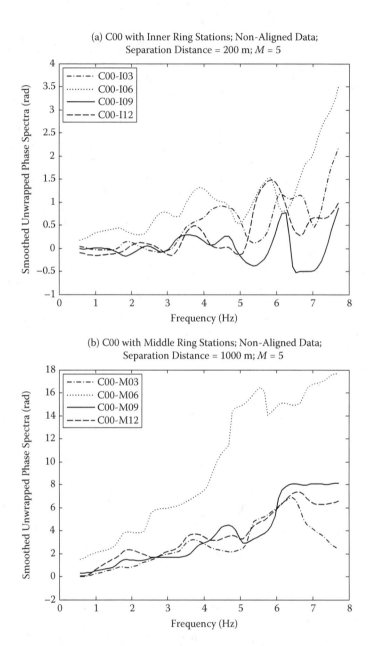

FIGURE 2.15 Smoothed, unwrapped phase spectra, derived from Eq. 2.71 and the $M = 5$ Hamming window, of the non-aligned time histories of Fig. 2.1: Part (a) presents the phase spectra between C00 and the inner ring stations I03, I06, I09 and I12, at a separation distance of 200 m, and part (b) those between C00 and the middle ring stations M03, M06, M09 and M12, at a separation distance of 1000 m. The scale of the phase spectra in the two subplots is not the same.

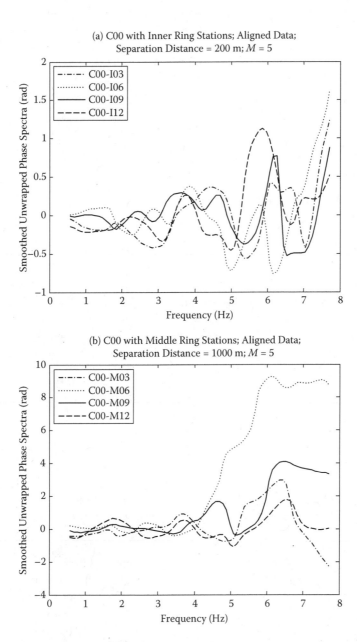

FIGURE 2.16 Smoothed, unwrapped phase spectra, derived from Eq. 2.71 and the $M = 5$ Hamming window, of the aligned time histories of Fig. 2.8: Part (a) presents the phase spectra between C00 and the inner ring stations I03, I06, I09 and I12, at a separation distance of 200 m, and part (b) those between C00 and the middle ring stations M03, M06, M09 and M12, at a separation distance of 1000 m. The scale of the phase spectra in the two subplots is not the same.

are presented in the two subplots of the figure, as wrapped phase differences will be needed in the next subsection for the interpretation of the coherency. The random variability of the phase differences can be clearly observed in Figs. 2.17(a) and (b). In the following, the notation with capital letters for the phases, $\Phi_j(\omega)$, will be utilized to describe the phases of the non-aligned data at each station $j = 1, \ldots, N$, and the lower case notation, $\phi_j(\omega)$, introduced at this point, will denote the phases of the aligned data, after the propagation effects have been removed, i.e.,

$$\phi_j(\omega) = \Phi_j(\omega) \pm \omega\tau_0 \qquad (2.76)$$

Bias and Variance

Jenkins and Watts [243] provide the variance of the phase spectrum as:

$$\mathrm{var}\left[\bar{\varphi}_{jk}^M(\omega)\right] \approx \frac{1}{2\,\mathcal{B}\,T}\left(\frac{1}{\left|\bar{\gamma}_{jk}^M(\omega)\right|^2} - 1\right) \qquad (2.77)$$

where $\mathcal{B}$ is the bandwidth of the smoothing window (Eq. 2.42) and T is the duration of the analyzed time series segment. Equation 2.77 suggests that the variance of the phase spectrum estimator increases as the lagged coherency (Eq. 2.70) decreases and tends to infinity as the lagged coherency tends to zero. Jenkins and Watts [243] further report that the bias of the phase spectrum estimator is proportional to the second derivative of the phase spectrum and, also, proportional to its first derivative multiplied by the first derivative of the logarithm of the cross amplitude spectrum. They suggest that, generally, the dominant term would be the one involving the first derivative of the phase spectrum, but, because this term is multiplied by the first derivative of the logarithm of the cross amplitude spectrum, the product would be small, resulting in an overall small bias.

2.4.2 LAGGED COHERENCY

The lagged coherency, $\left|\bar{\gamma}_{jk}^M(\omega)\right|$ of Eq. 2.70, is a measure of "similarity" in the seismic motions, and indicates the degree to which the data recorded at the two stations are related by means of a linear transfer function [81]. If, e.g., one process can be obtained by means of linear transformation of another process and in the absence of noise, coherency is equal to one [365]; for uncorrelated processes, coherency becomes zero. In this sense, the lagged coherency reflects, at each frequency, the correlation of the motions, as was illustrated in Section 2.1 for random variables. It is expected that at low frequencies and short separation distances, the motions will be similar and, therefore, theoretically, the lagged coherency will tend to unity as frequency and station separation distance tend to zero. On the other hand, at large frequencies and long station separation distances, the motions will become uncorrelated, and, theoretically, the lagged coherency will tend to zero. The value of the lagged coherency in-between these extreme cases will decay with frequency and station separation distance. Indeed, the parametric functional forms describing the lagged coherency are, generally, exponential functions decaying with separation distance and frequency (Chapter 3).

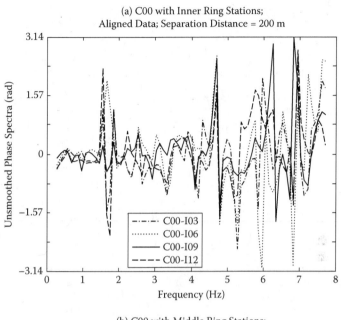

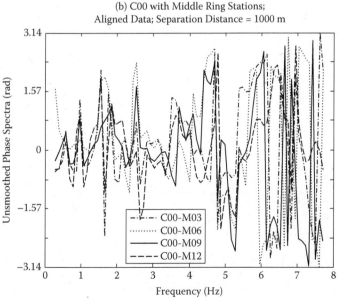

FIGURE 2.17 Unsmoothed (wrapped) phase spectra (Eq. 2.59) of the aligned time histories of Fig. 2.8: Part (a) presents the phase spectra between C00 and the inner ring stations I03, I06, I09 and I12, at a separation distance of 200 m, and part (b) those between C00 and the middle ring stations M03, M06, M09 and M12, at a separation distance of 1000 m.

Unsmoothed Lagged Coherency

The following peculiarity of the lagged coherency estimate needs to be mentioned: Consider that the lagged coherency is not defined by means of the smoothed spectral estimates, as in Eq. 2.70, but, instead, through the unsmoothed estimators of Eqs. 2.34 and 2.57 for the power and cross spectral densities, respectively. The coherency estimate corresponding to Eq. 2.70 would then be:

$$|\hat{\gamma}_{jk}(\omega_n)| = \frac{|\hat{S}_{jk}(\omega_n)|}{\sqrt{\hat{S}_{jj}(\omega_n)\,\hat{S}_{kk}(\omega_n)}}$$

$$= \frac{|\Lambda_j(\omega)\,\Lambda_k(\omega)\exp[i\{\Phi_k(\omega) - \Phi_j(\omega)\}]|}{\sqrt{\Lambda_j^2(\omega)\Lambda_k^2(\omega)}} \equiv 1 \qquad (2.78)$$

i.e., the expression becomes identically equal to unity, and no additional information can be extracted from the data. Hence, the issue of smoothing of the spectral estimators is not only important for the reduction of their variance, but, also, for extracting meaningful information regarding the coherency of the signals. As indicated earlier, Abrahamson *et al.* [14] suggested the 11-point Hamming (spectral) window, so that both the variance of the spectral estimators is reduced and the estimates maintain sufficient resolution for structural response applications.

Statistics of Lagged Coherency

The statistics of the (smoothed) coherency estimate $|\bar{\gamma}_{jk}^M(\omega)|$ (Eq. 2.70) are not simple. Jenkins and Watts [243] provide an approximate expression for the variance of the coherency as:

$$\text{var}\left[|\bar{\gamma}_{jk}^M(\omega)|\right] \approx \frac{1}{2\,B\,T}\left(1 - |\bar{\gamma}_{jk}^M(\omega)|^2\right)^2 \qquad (2.79)$$

where, again, B is the bandwidth of the smoothing window (Eq. 2.42) and T is the duration of the analyzed time segment. The above expression suggests that the variance of the coherency depends on its value, i.e., as $|\bar{\gamma}_{jk}^M(\omega)|$ decreases, its variance increases. ·

On the other hand, the transformation:

$$\text{arctanh}\left[|\bar{\gamma}_{jk}^M(\omega)|\right] = \frac{1}{2}\ln\left[\frac{1 + |\bar{\gamma}_{jk}^M(\omega)|}{1 - |\bar{\gamma}_{jk}^M(\omega)|}\right] \qquad (2.80)$$

has an approximately constant variance ($\approx 1/(2\,B\,T)$) [243]. Considering next that the inverse hyperbolic tangent of the lagged coherency estimate, $\text{arctanh}[|\bar{\gamma}_{jk}^M(\omega)|]$, is approximately normally distributed, its 95% confidence interval can be estimated from the expression [243]:

$$\text{arctanh}\left[|\bar{\gamma}_{jk}^M(\omega)|\right] \pm \frac{1.96}{\sqrt{2\,B\,T}} \qquad (2.81)$$

It is then preferable to use the $\text{arctanh}\left[|\bar{\gamma}_{jk}^M(\omega)|\right]$ transformation for the parametric fitting of coherency models to recorded data (e.g., [13], [206], [303]). However, the statistical properties of the transformation arctanh$\left[\left|\tilde{\gamma}_{jk}^{M}(\omega)\right|\right]$ assume that the spectrum of the process is approximately constant over the frequency bandwidth of the window, so that the bias of the power spectra do not affect that of the coherence (Eq. 2.69) [13], [243]. Jenkins and Watts [243] also point out that the bias of the coherence is proportional to its value (Eq. 2.69) and the square of the first derivative of the phase spectrum (Eq. 2.65 or Eq. 2.71). The latter effect can be quite significant if the time delay between the time series is large, which suggests, as discussed earlier in Section 2.3.2 for the cross spectral estimators, that the time series need to be aligned before their coherency is evaluated.

Further sources of bias in the coherency are introduced when it is estimated from recorded accelerograms: Bias is expected at low frequencies due to the sensitivity characteristics of seismometers and the small intensity of the low frequency components of the ground motions [41], [224], [418], and additional uncertainty can be caused by inaccuracies in recorder synchronization (e.g., [41], [386], [418]), and by imperfect elimination of time lags caused by the wave passage effects [418].

Example Applications

Figure 2.18 presents the effect of smoothing on the lagged coherency (Eq. 2.70) of the aligned data recorded at stations C00 and I06 at a separation distance of 200 m (Fig. 2.18(a)), and stations C00 and M06 at a separation distance of 1000 m (Fig. 2.18(b)). For the evaluation of the lagged coherencies in the figure, the power and cross spectra are smoothed with 3-, 7-, 11-, 15-, and 19-point Hamming windows (i.e., $M = 1, 3, 5, 7$ and 9, respectively, in Eq. 2.70). It is recalled that, if no smoothing is performed on the data, the lagged coherency is identically equal to unity. For $M = 1$ in Fig. 2.18, the estimates are close to unity with sharp drops at certain frequencies. As M increases, the estimators become smoother at the cost of resolution. For both separation distances, the estimated lagged coherency is close to unity in the low frequency range, and decays as frequency increases. At higher frequencies, the estimates have spurious peaks. This can be attributed to the low values of the power spectral densities in this frequency range (Fig. 2.13), which appear in the denominator of the lagged coherency estimate (Eq. 2.70). It should also be noted that the lagged coherency obtained from recorded data has the following two inherent characteristics: (i) At low frequencies (especially as $\omega \to 0$), the theoretical value of the coherency should tend to unity. However, due to smoothing, the coherency estimates will always be less than one. (ii) At high frequencies, especially for the longer separation distances, the theoretical value of the lagged coherency of the S-wave should tend to zero. However, the lagged coherency evaluated from the data will not tend to zero but, rather, to the value of the coherency of noise smoothed with the particular window used in the evaluation, which, for a $M = 5$ Hamming spectral window, is equal to 0.33 [14]. These issues are addressed later herein (Section 2.4.4) in the description of the plane-wave coherency [12], [14], and in the presentation of empirical coherency models developed from seismic array data (Section 3.4.1).

Figure 2.19(a) presents the arctanh lagged coherency (Eq. 2.80) between the aligned time histories at stations C00 and I06 along with its 95% confidence interval (Eq. 2.81). The effect of the constant variance of the estimator in the arctanh lagged

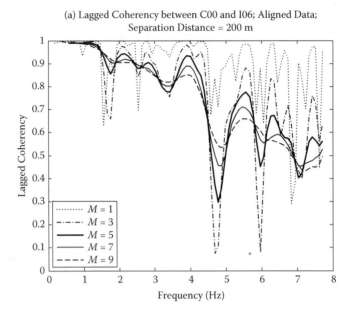

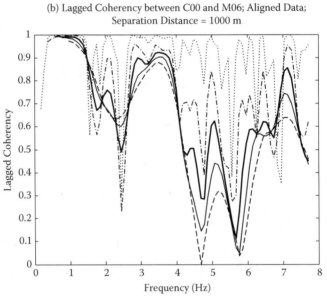

FIGURE 2.18 Effect of smoothing by means of the frequency domain Hamming window (Fig. 2.4) with various window lengths ($M = 1, 3, 5, 7$ and 9) on selected lagged coherency estimates (Eq. 2.70) of the aligned motions of Fig. 2.8: Part (a) presents this effect on the lagged coherency between C00 and the inner ring station I06 at a separation distance of 200 m, and part (b) on the lagged coherency between C00 and the middle ring station M06 at a separation distance of 1000 m.

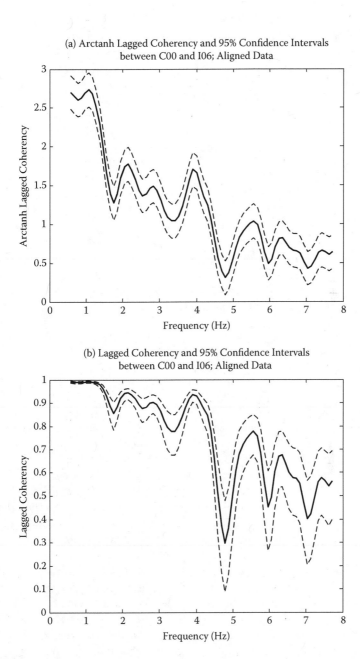

FIGURE 2.19 Confidence intervals of the lagged coherency of the aligned time histories (Fig. 2.8) between stations C00 and I06 at a separation distance of 200 m smoothed with a $M = 5$ Hamming window: Part (a) presents the arctanh transformation of the lagged coherency (Eq. 2.80) and its 95% confidence intervals, and part (b) the lagged coherency (Eq. 2.70) and its 95% confidence intervals.

coherency domain is obvious from the figure. Figure 2.19(b) presents the same information as Fig. 2.19(a), but in the lagged coherency domain. The confidence intervals now become narrow when the lagged coherency assumes high values in the low frequency range, but their width increases significantly as the value of the coherency decreases (Eq. 2.79).

Figure 2.20 presents the lagged coherency of the aligned data between C00 and the inner ring stations (Fig. 2.20(a)), and C00 and the middle ring stations (Fig. 2.20(b)). The spectral estimates were smoothed with an 11-point ($M = 5$) Hamming window. Both subfigures indicate that lagged coherency decays with frequency, with a slower decay for the close-by stations (separation distance of 200 m in Fig. 2.20(a)) than for the further-away stations (separation distance of 1000 m in Fig. 2.20(b)). For the close-by stations, the value of the coherency remains above 0.8 up to a frequency of approximately 4.5 Hz, which is also the dominant frequency range of the motions (Fig. 2.13(a)). Past this frequency, it decays further, and, past 6 Hz, it starts exhibiting spurious peaks at certain frequencies, as illustrated earlier in Fig. 2.18. For the further-away stations of Fig. 2.20(b), the values of the lagged coherency decrease more quickly with frequency, and suggest that the coherency estimates can be lower than unity even at the very low frequencies. The station pair C00-M12 (Fig. 2.20(b)) appears to produce the lowest value for the coherency in the lower frequency range, which can be anticipated, given the differences in the motions at this station compared to the other array stations. Figure 2.20 then suggests that lagged coherency decays with both frequency and separation distance, and that it can be appropriately modeled as an exponentially decaying function of frequency and station separation distance (Chapter 3).

Interpretation of Coherency

A significant characteristic of the lagged coherency, which does not become apparent directly from Eq. 2.70, is that it is only minimally affected by the amplitude variability between the motions at the two stations. The numerator of the coherency (Eqs. 2.64 and 2.70) contains information about the phases of the cross spectrum in addition to information about the amplitudes. The information about the differences in the phases of the motions at the two stations is introduced in the estimate indirectly through the smoothing process. The denominator of the lagged coherency expression (Eq. 2.70) contains information about the amplitudes of the motions at the two stations after the power spectral densities have been smoothed over the selected frequency band around ω_n (Eq. 2.41). Hence, the effect of the amplitudes of the motions at the two stations on coherency is reduced. Figure 2.21 illustrates this point: The solid lines in the two subfigures present the lagged coherency of the smoothed estimators of the aligned data for station pairs C00-I06 (Fig. 2.21(a)) and C00-M06 (Fig. 2.21(b)). These estimates were obtained from Eqs. 2.41, 2.64 and 2.70, though for aligned data, and are identical to the lagged coherencies for these two station pairs presented in Fig. 2.20. The dashed lines in Fig. 2.21 were obtained with the consideration of only the phases of the cross spectral densities of the aligned data, i.e.,

$$\left| \bar{\gamma}_{jk}^{M}(\omega_n) \right| = \left| \sum_{m=-M}^{+M} W(m\Delta\omega) \, \exp\{i \, [\phi_k(\omega_n + m\Delta\omega) - \phi_j(\omega_n + m\Delta\omega)]\} \right| \quad (2.82)$$

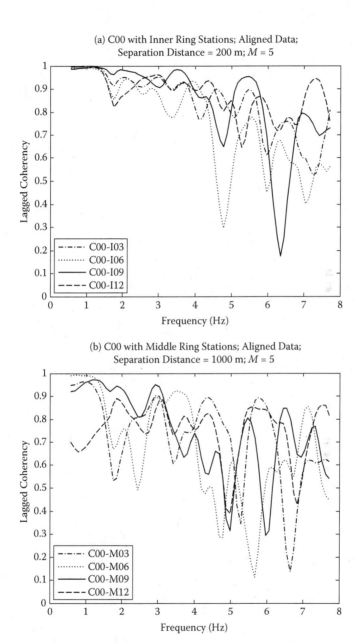

FIGURE 2.20 Smoothed lagged coherency estimates, derived from Eq. 2.70 with the $M = 5$ Hamming window, of the aligned time histories of Fig. 2.8: Part (a) illustrates the lagged coherency between C00 and the inner ring stations I03, I06, I09 and I12, at a separation distance of 200 m, and part (b) the lagged coherency between C00 and the middle ring stations M03, M06, M09 and M12, at a separation distance of 1000 m.

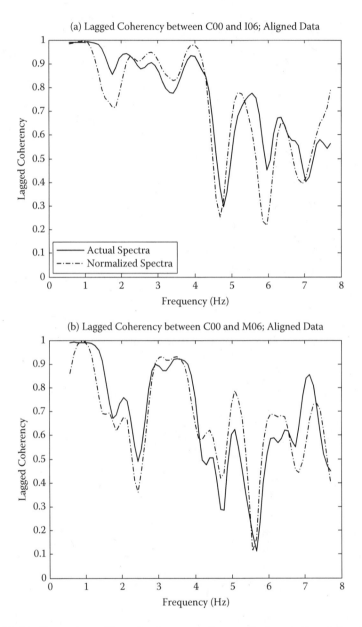

FIGURE 2.21 Comparison of lagged coherency estimates of selected aligned time histories of Fig. 2.8 utilizing the definition of the lagged coherency with the actual spectral densities (Eq. 2.70) and its approximation with normalized spectra (Eq. 2.82): Part (a) illustrates the results between stations C00 and I06 at a separation distance of 200 m, and part (b) those between stations C00 and M06 at a separation distance of 1000 m.

where the lower case description for the phases, $\phi_j(\omega)$ and $\phi_k(\omega)$, indicates that they incorporate only the random phase variability (Eq. 2.76). Both solid and dashed lines in the subfigures of Fig. 2.21 have the same trend and look similar, which suggests that, indeed, the lagged coherency is minimally affected by the amplitude variability in the data. This leads to the apparent paradox that lagged coherency, although the modulus of a complex number, is attributed more to the phase variability in the motions rather than their amplitude. However, this observation also permits the interpretation of coherency.

Abrahamson [6] presented the relation between the lagged coherency and the random phase variability, which is summarized in the following: Let $\phi_j(\omega)$ and $\phi_k(\omega)$ reflect, as in Eq. 2.82, the phases at two stations j and k on the ground surface, after the wave passage effects have been removed (Eq. 2.76). The relation between $\phi_j(\omega)$ and $\phi_k(\omega)$ can be expressed as:

$$\phi_k(\omega) - \phi_j(\omega) = \beta_{jk}(\omega)\pi \; \epsilon_{jk}(\omega) \tag{2.83}$$

in which $\epsilon_{jk}(\omega)$ are random numbers uniformly distributed between $[-1, +1)$, and $\beta_{jk}(\omega)$ is a deterministic function of the frequency ω and assumes values between 0 and 1. $\beta_{jk}(\omega)$ indicates the fraction of random phase variability (between $[-\pi, +\pi)$ from the product $\pi\epsilon_{jk}(\omega)$) that is present in the Fourier phase differences of the time histories. For example, if $\beta_{jk}(\omega) = 0$, there is no phase difference between the stations, and the phases at the two stations are identical and fully deterministic. In the other extreme case, i.e., when $\beta_{jk}(\omega) = 1$, the phase difference between stations is completely random. Based on Eq. 2.83, and neglecting the amplitude variability of the data (e.g., Eq. 2.82), Abrahamson [6] noted that the mean value of the lagged coherency can be expressed as:

$$E[|\gamma_{jk}(\omega)|] = \frac{\sin(\beta_{jk}(\omega)\pi)}{\beta_{jk}(\omega)\pi} \tag{2.84}$$

It is easy to verify from Eq. 2.84 that when the lagged coherency tends to one, $\beta_{jk}(\omega)$ is a small number, i.e., only a small fraction of randomness appears in the phase differences between the motions at the two stations; as coherency decreases, $\beta_{jk}(\omega)$ increases, and, for zero coherency, $\beta_{jk}(\omega) = 1$. Figure 2.22 illustrates the relation presented in Eq. 2.84. The dotted lines in Fig. 2.22 are phases generated as random numbers uniformly distributed between $[-1, +1)$ and multiplied by $\beta_{jk}(\omega)\pi$ defined in Eq. 2.83. The dashed lines in Fig. 2.22 present the "envelope" function $\beta_{jk}(\omega)\pi$, and the thicker solid lines the mean value of the lagged coherency of Eq. 2.84. Figure 2.22(a) illustrates the case where the lagged coherency is equal to one at low frequencies ($\omega \to 0$), and the envelope function tends to zero in this frequency range. This is a theoretically anticipated but not readily achievable case in the analysis of recorded data, because, in reality, the phases of the data at the individual stations will be different due to, e.g., segmentation and tapering. Case II in Fig. 2.22(b) presents an envelope function with finite values at low frequencies, which reflects into coherency values less than one. At higher frequencies in both cases, the envelope function is equal to π ($\beta_{jk}(\omega) = 1$), which renders random distribution of phases between $[-\pi, +\pi)$, and a lagged coherency that is equal to zero. It is noted, again, that the lagged coherency estimated from recorded data cannot assume zero values

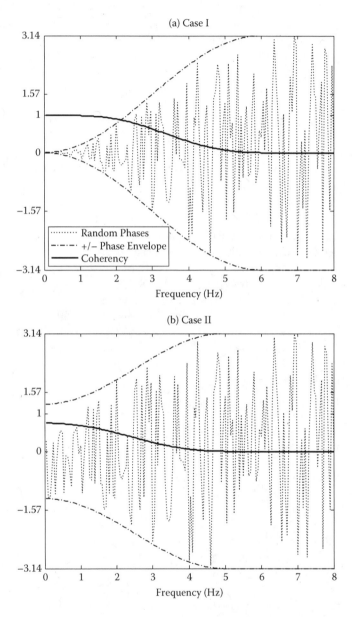

FIGURE 2.22 Relation between random phase differences, the envelope function $\beta(\omega)\pi$ and the mean value of the lagged coherency according to Eqs. 2.83 and 2.84: Part (a) presents the variation of random phase differences leading to perfect correlation at low frequencies, whereas part (b) the variation of random phase differences that yields partial correlation in the low frequency range.

in the higher frequency range; coherency at higher frequencies assumes the value of the coherency of noise smoothed with the window used in the evaluation [14]. However, the validity of Eq. 2.84 and Fig. 2.22(b) can be confirmed with the variation of the phase differences of the actual data in Fig. 2.17: In the low frequency range, especially for the close-by stations (Fig. 2.17(a)), the phase differences of the data vary randomly, but at a close distance from the zero axis. This distance increases with frequency until it reaches the value of $\pm\pi$, i.e., the phases vary randomly within the bounds of an envelope function defined as in Eq. 2.83. The same occurs for the further-away stations (Fig. 2.17(b)), only in this case, the distance of the envelope function from the zero axis is wider even at low frequencies. The spurious peaks at a frequency less than approximately 2 Hz are associated with the low values of the power spectral densities of the data at this frequency (Figs. 2.11 and 2.13). The phase variability in the data and its association with amplitude variability will be further elaborated upon in Section 4.2.

2.4.3 CROSS CORRELATION FUNCTION AND COHERENCY

The necessity for the smoothing of the coherency, so that its value is not identically equal to unity (Eq. 2.78), gives the estimate the appearance of "artificiality." This is not, however, the case, as is illustrated next in some detail. The derivations in the following are based on the approach presented by Harichandran [202], [204].

Consider that the time series at two stations j and k have been bandpassed over a frequency range $\delta\Omega$ centered at $\pm\Omega$. The harmonic components of the series at, e.g., station j over these frequency bands are then expressed as:

$$a_j^\Omega(t) = \int_{\Omega-\delta\Omega/2}^{\Omega+\delta\Omega/2} \mathcal{A}_j(\omega)e^{i\omega t}\, d\omega + \int_{-\Omega-\delta\Omega/2}^{-\Omega+\delta\Omega/2} \mathcal{A}_j(\omega)e^{i\omega t}\, d\omega \qquad (2.85)$$

Following Eq. 2.51, the frequency-dependent cross covariance function, $\hat{R}_{jk}^\Omega(\tau)$, is given by:

$$\hat{R}_{jk}^\Omega(\tau) = \begin{cases} \dfrac{1}{T}\displaystyle\int_0^{T-|\tau|} a_j^\Omega(t)\,a_k^\Omega(t+\tau)\, dt & |\tau| \le T \\[2ex] 0 & |\tau| > T \end{cases} \qquad (2.86)$$

and the frequency-dependent cross correlation function, $\hat{\rho}_{jk}^\Omega(\tau)$, becomes, according to Eq. 2.54:

$$\hat{\rho}_{jk}^\Omega(\tau) = \frac{\hat{R}_{jk}^\Omega(\tau)}{\sqrt{\hat{R}_{jj}^\Omega(0)\hat{R}_{kk}^\Omega(0)}} \qquad (2.87)$$

Since the harmonic components of the time series, $a_j^\Omega(t)$ in Eq. 2.85, contain contributions only from the frequency ranges $[|\Omega|\pm\delta\Omega/2]$, their cross spectrum (Eq. 2.57) is given by:

$$\hat{S}_{jk}^\Omega(\omega) = \begin{cases} \hat{S}_{jk}(\omega) & \Omega-\delta\Omega/2 \le |\omega| \le \Omega+\delta\Omega/2 \\[1ex] 0 & \text{otherwise} \end{cases} \qquad (2.88)$$

or, alternatively,

$$\hat{S}_{jk}^{\Omega}(\omega) = \delta\Omega \; W_R(\Omega - |\omega|) \; \hat{S}_{jk}(\omega) \tag{2.89}$$

where $W_R(\omega)$ is the rectangular window, defined as:

$$W_R(\omega) = \begin{cases} 1/\delta\Omega & |\omega| \le \delta\Omega/2 \\ 0 & |\omega| > \delta\Omega/2 \end{cases} \tag{2.90}$$

Based on the definition of Eq. 2.56, the frequency-dependent cross covariance function (Eq. 2.86) becomes:

$$\hat{R}_{jk}^{\Omega}(\tau) = \int_{-\infty}^{+\infty} \hat{S}_{jk}^{\Omega}(\omega) e^{i\omega\tau} \, d\omega$$

$$= 2\delta\Omega \int_{0}^{+\infty} W_R(\Omega - \omega) \left\{ \Re\left[\hat{S}_{jk}^{\Omega}(\omega) \right] \cos(\omega\tau) - \Im\left[\hat{S}_{jk}^{\Omega}(\omega) \right] \sin(\omega\tau) \right\} \, d\omega \tag{2.91}$$

where the last equality in the above expression follows from the fact that $\hat{S}_{jk}(\omega) = \hat{S}_{jk}^{*}(-\omega)$.

If $\delta\Omega$ is small, Eq. 2.91 can be approximated by:

$$\hat{R}_{jk}^{\Omega}(\tau) \approx 2\delta\Omega \left\{ \cos(\Omega\tau) \int_{\Omega-\delta\Omega/2}^{\Omega+\delta\Omega/2} W_R(\Omega - \omega) \, \Re\left[\hat{S}_{jk}^{\Omega}(\omega) \right] d\omega \right.$$

$$\left. - \sin(\Omega\tau) \int_{\Omega-\delta\Omega/2}^{\Omega+\delta\Omega/2} W_R(\Omega - \omega) \, \Im\left[\hat{S}_{jk}^{\Omega}(\omega) \right] d\omega \right\} \tag{2.92}$$

which, with a change in the variables of integration, becomes:

$$\hat{R}_{jk}^{\Omega}(\tau) \approx 2\delta\Omega \left\{ \cos(\Omega\tau) \int_{-\delta\Omega/2}^{\delta\Omega/2} W_R(u) \, \Re\left[\hat{S}_{jk}^{\Omega}(\Omega - u) \right] du \right.$$

$$\left. - \sin(\Omega\tau) \int_{-\delta\Omega/2}^{\delta\Omega/2} W_R(u) \, \Im\left[\hat{S}_{jk}^{\Omega}(\Omega - u) \right] du \right\} \tag{2.93}$$

According to Eq. 2.62 and the properties of the rectangular window (Eq. 2.90), the first integral on the right-hand side of Eq. 2.93 is the real part of the cross spectrum, $\Re[\bar{S}_{jk}(\Omega)]$, smoothed over the frequency range $[|\Omega| \pm \delta\Omega/2]$, and the second integral is, correspondingly, the imaginary part of the cross spectrum, $\Im[\bar{S}_{jk}(\Omega)]$, smoothed over the same frequency range. Hence, Eq. 2.93 can be rewritten as:

$$\hat{R}_{jk}^{\Omega}(\tau) \approx 2\delta\Omega \left\{ \cos(\Omega\tau)\Re[\bar{S}_{jk}(\Omega)] - \sin(\Omega\tau)\Im[\bar{S}_{jk}(\Omega)] \right\}$$

$$= 2\delta\Omega |\bar{S}_{jk}(\Omega)| \cos[\Omega\tau + \bar{\varphi}_{jk}(\Omega)] \tag{2.94}$$

with $\bar{\varphi}_{jk}(\Omega) = \tan^{-1}\{\Im[\bar{S}_{jk}(\Omega)]/\Re[\bar{S}_{jk}(\Omega)]\}$ being, as in Eq. 2.65, the smoothed phase spectrum.

For discrete frequencies, as, e.g., $\Omega = \omega_n$ and $\delta\Omega = 2M\Delta\omega$, Eq. 2.94 becomes:

$$\hat{R}_{jk}^{\omega_n}(\tau) \approx 2M\Delta\omega \left|\bar{S}_{jk}^{M}(\omega_n)\right| \cos\left[\omega_n\tau + \bar{\varphi}_{jk}^{M}(\omega_n)\right] \qquad (2.95)$$

In a similar manner, it can be shown that:

$$\hat{R}_{jj}^{\omega_n}(0) \approx 2M\Delta\omega \left|\bar{S}_{jj}^{M}(\omega_n)\right| \qquad (2.96)$$

and the frequency-dependent cross correlation function, Eq. 2.87, becomes:

$$\hat{\rho}_{jk}^{\omega_n}(\tau) = \frac{\hat{R}_{jk}^{\omega_n}(\tau)}{\sqrt{\hat{R}_{jj}^{\omega_n}(0)\hat{R}_{kk}^{\omega_n}(0)}}$$

$$\approx \frac{\left|\bar{S}_{jk}^{M}(\omega_n)\right|}{\sqrt{\bar{S}_{jj}^{M}(\omega_n)\bar{S}_{kk}^{M}(\omega_n)}} \cos[\omega_n\tau + \bar{\varphi}_{jk}^{M}(\omega_n)]$$

$$= \left|\bar{\gamma}_{jk}^{M}(\omega_n)\right| \cos\left[\omega_n\tau + \bar{\varphi}_{jk}^{M}(\omega_n)\right] \qquad (2.97)$$

where $\left|\bar{\gamma}_{jk}^{M}(\omega_n)\right|$ is the smoothed lagged coherency (Eq. 2.70). Equation 2.97 then indicates that the unsmoothed cross correlation function derived from bandpassed time series is equivalent to the Fourier transform of the coherency smoothed over the same frequency range.

If, on the other hand, the spectral estimates are smoothed over the entire frequency range, then it can be easily shown that, because of the following equalities resulting from Eqs. 2.32 and 2.56, namely:

$$\hat{R}_{jj}(0) = \int_{-\infty}^{+\infty} \hat{S}_{jj}(\omega)\,d\omega = \bar{S}_{jj}^{\infty}; \quad \hat{R}_{jk}(0) = \int_{-\infty}^{+\infty} \hat{S}_{jk}(\omega)\,d\omega = \bar{S}_{jk}^{\infty} \qquad (2.98)$$

the coherency estimate becomes:

$$\bar{\gamma}_{jk}^{\infty} = \frac{\bar{S}_{jk}^{\infty}}{\sqrt{\bar{S}_{jj}^{\infty}\bar{S}_{kk}^{\infty}}} = \frac{\hat{R}_{jk}(0)}{\sqrt{\hat{R}_{jj}(0)\hat{R}_{kk}(0)}} = \hat{\rho}_{jk}(0) \qquad (2.99)$$

i.e., it is equal to the value of the cross correlation function (Eq. 2.54) at zero time lag.

2.4.4 PLANE-WAVE COHERENCY

Abrahamson *et al.* [12], [14] noted that the lagged coherency (Eq. 2.70) describes only the deviations of the ground motions from plane wave propagation at each frequency, but does not consider the deviations of the motions from a single plane wave at all frequencies. In other words, if the analyzed segment contains wave components in addition to the plane wave, as is most commonly the case at the higher frequencies where scattered energy or noise contribute significantly to the records, the correlation of these additional wave components is reflected in the lagged coherency as if they were part of the plane wave. To express the departure of the data from that of plane wave propagation at all frequencies, Abrahamson *et al.* [12], [14] introduced the concept of the plane-wave coherency, which is briefly illustrated in the following.

To account for the different wave components contributing to the motions, the phase spectrum of the coherency was partitioned as [12], [14]:

$$\exp[i\varphi_{jk}(\omega)] = h(\xi_{jk}^r, \omega)\exp\left[-i\frac{\omega\xi_{jk}^r}{c}\right] + [1 - h(\xi_{jk}^r, \omega)]\exp[i\chi(\omega)] \quad (2.100)$$

where $h(\xi_{jk}^r, \omega)$ is the relative power of the coherent wavefield that can be described by a plane wave at all frequencies, and $\chi(\omega)$ is a random term. ξ_{jk}^r in Eq. 2.100 is the separation distance between the two stations projected along the direction of propagation of the plane wave, and c is, as in Eq. 2.73, the apparent propagation velocity of the plane wave. However, Eq. 2.73 considers that the entire segment of the analyzed time series propagates with the constant velocity of the plane wave, whereas Eq. 2.100 differentiates between the pattern of propagation of the plane-wave and the scatterd energy components in the motions. It is also noted that the use of ξ_{jk}^r in Eq. 2.100 is more general than the use of ξ_{jk} in Eq. 2.73, because it relaxes the condition that the wave propagates along the line connecting the stations.

The complex plane-wave coherency incorporates then the first summand of Eq. 2.100 in its phase spectrum and is defined as [14]:

$$\gamma_{jk}^{pw}(\xi_{jk}, \xi_{jk}^r, \omega) = \left|\bar{\gamma}_{jk}^M(\xi_{jk}, \omega)\right| h(\xi_{jk}^r, \omega)\exp\left[-i\frac{\omega\xi_{jk}^r}{c}\right] \quad (2.101)$$

where $|\bar{\gamma}_{jk}^M(\xi_{jk}, \omega)|$ is the lagged coherency. From Eq. 2.101, the lagged plane-wave coherency becomes:

$$\left|\gamma_{jk}^{pw}(\xi_{jk}, \xi_{jk}^r, \omega)\right| = \left|\bar{\gamma}_{jk}^M(\xi_{jk}, \omega)\right| h(\xi_{jk}^r, \omega) \quad (2.102)$$

Because $h(\xi_{jk}^r, \omega) \leq 1$, the plane-wave coherency, $|\gamma_{jk}^{pw}(\xi_{jk}, \xi_{jk}^r, \omega)|$, is smaller than the lagged coherency, $|\bar{\gamma}_{jk}^M(\xi_{jk}, \omega)|$. Illustrations of the plane-wave coherency after Abrahamson *et al.* [7], [12], [14] are presented in Section 3.4.1.

2.5 MULTIVARIATE STOCHASTIC PROCESSES AND STOCHASTIC FIELDS

The statistical characterization of bivariate stochastic (vector) processes can be readily extended to multivariate ones. The cross spectral density of a zero-mean N-variate stochastic process is described in matrix form as:

$$\bar{\mathbf{S}}(\omega) = \begin{bmatrix} \bar{S}_{11}(\omega) & \bar{S}_{12}(\omega) & \cdots & \bar{S}_{1N}(\omega) \\ \bar{S}_{21}(\omega) & \bar{S}_{22}(\omega) & \cdots & \bar{S}_{2N}(\omega) \\ \vdots & \vdots & \vdots & \vdots \\ \bar{S}_{N1}(\omega) & \bar{S}_{N2}(\omega) & \cdots & \bar{S}_{NN}(\omega) \end{bmatrix} \quad (2.103)$$

where the superscript M has been dropped for convenience, the diagonal elements of the matrix, $\bar{S}_{jj}(\omega)$, $j = 1, \ldots N$, are the power spectral density estimates at the various locations given by Eqs. 2.35, 2.38 or 2.46, and the off-diagonal elements, $\bar{S}_{jk}(\omega)$, $j, k = 1, \ldots N$, are the cross spectral density estimates provided by Eqs. 2.60, 2.62 or 2.63.

An important characteristic of the cross spectral density matrix is that it is Hermitian, i.e.,

$$\bar{S}(\omega) = [\bar{S}^*(\omega)]^T$$
$$\bar{S}_{ij}(\omega) = \bar{S}_{ji}^*(\omega) \tag{2.104}$$

Additionally, the cross spectral density matrix is positive semidefinite, i.e., every principal minor of Eq. 2.103 is non-negative [243]. If $N = 2$ in Eq. 2.103, this implies that

$$\begin{vmatrix} \bar{S}_{11}(\omega) & \bar{S}_{12}(\omega) \\ \bar{S}_{21}(\omega) & \bar{S}_{22}(\omega) \end{vmatrix} \geq 0 \tag{2.105}$$

which yields the bounds of the coherence estimate as:

$$0 \leq |\bar{\gamma}_{12}(\omega)|^2 = \frac{|\bar{S}_{12}(\omega)|^2}{\bar{S}_{11}(\omega)\,\bar{S}_{22}(\omega)} \leq 1 \tag{2.106}$$

It can be clearly recognized that, for the example applications presented herein, Eq. 2.103 fully describes the second moment characteristics of the motions at the considered array stations. In this case, $N = 9$, the diagonal terms of the matrix are given by Eq. 2.46 and Fig. 2.13, and the off-diagonal terms by Eq. 2.63 with their cross amplitude and phase spectra illustrated in Figs. 2.14 and 2.16, respectively. Additional assumptions, however, are required when parametric models are fitted to the estimators from recorded data. These are the assumptions of homogeneity and isotropy described below.

2.5.1 HOMOGENEITY AND ISOTROPY

The concept of homogeneity is similar to that of stationarity, discussed earlier in Section 2.2.1. Homogeneity, generally, refers to the space variables and implies that the stochastic descriptors of the motions are functions of the separation distance vector, $\vec{\xi}_{jk} = \vec{r}_k - \vec{r}_j$, but not of the absolute location of the stations, the same way that stationarity, which refers to the time variable, implies dependence on the time lag only but not the absolute time. Physically, homogeneity suggests that the power spectral densities of the seismic motions at different recording stations do not vary significantly, i.e., they are station independent. Clearly, the power spectral densities of Fig. 2.13 differ at the various stations of the SMART-1 array, but, on the average, their frequency content follows the same trend with fluctuations (more so for station M12 than the other stations in Fig. 2.13). It is, generally, considered that seismic data recorded at dense instrument arrays, which are located on uniform site conditions, are homogeneous. The assumption of homogeneity, however, is not valid for seismic ground motions recorded at stations on different local site conditions, as will be illustrated in Section 3.3.5 for the data recorded at the Parkway array in the Wainuiomata Valley, New Zealand [303], [506]. Variable site conditions can have a significant effect on, e.g., the seismic response of highway bridges crossing narrow sediment valleys, when their lateral supports rest on rock outcrop and their middle supports on the sediments. An example of simulated time series at variable site conditions is presented in Section 7.3.2, and their effect on the response of highway bridges in Section 9.4.

In addition to the assumption of homogeneity of data recorded at uniform site conditions, the assumption of isotropy is commonly invoked. Isotropy further assumes that the stochastic descriptors of the seismic motions are rotationally invariant, i.e., independent of the direction of the vector of the separation distance between the stations, but functions of its absolute value only, i.e., $\xi_{jk} = |\vec{\xi}_{jk}|$ [551]. The assumption of isotropy was indirectly utilized in the plots of Fig. 2.20, where coherency estimates between stations with the same separation distance were grouped together irrespective of the orientation of the station pairs. For example, for the coherency estimates of Fig. 2.20(a), the line connecting stations I03, C00 and I09 is perpendicular to that connecting stations I06, C00 and I12 (Fig. 1.2). The assumption of isotropy will be further elaborated upon in Section 3.3.3.

2.5.2 RANDOM FIELDS

The multivariate stochastic process (Eq. 2.103) describes the joint characteristics of the recorded data at all considered (discrete) stations on the ground surface. Herein, the concept of the random field of the seismic ground motions will denote a multivariate process where information is available for all locations on the ground surface, i.e., the separation distance between stations, ξ, now becomes a continuous variable. The stochastic descriptors of the random field in the time and frequency domains, i.e., covariance functions, spectral densities and coherency, become then:

$$\bar{R}(\tau) = \bar{R}_{jj}(\tau) \Longleftrightarrow \bar{S}(\omega) = \bar{S}_{jj}(\omega)$$
$$\bar{R}(\xi, \tau) = \bar{R}_{jk}(\tau) \Longleftrightarrow \bar{S}(\xi, \omega) = \bar{S}_{jk}(\omega) \tag{2.107}$$
$$\bar{\gamma}(\xi, \omega) = \frac{\bar{S}(\xi, \omega)}{\bar{S}(\omega)} = \bar{\gamma}_{jk}(\omega)$$

and encompass all discrete estimates at stations $j, k = 1, \ldots, N$. The time history at any location on the ground surface becomes then a realization of the random field.

Once the spectral estimators are obtained from seismic data (Eq. 2.103), parametric models are fitted to the estimates. These parametric models are continuous functions of ξ, and, hence, represent a random field (Eq. 2.107). The parametric modeling of spatially variable ground motions is described in the next chapter.

3 Parametric Modeling of Spatial Variability

Lifeline engineering applications, i.e., random vibrations (Chapter 6) and Monte Carlo simulations (Chapters 7 and 8), require the parametric modeling of the spatially variable seismic ground motion random field (Eq. 2.107). Its parametric description is, generally, given by the cross spectral density of the motions, $S(\xi, \omega)$:

$$S(\xi, \omega) = S(\omega) \, \gamma(\xi, \omega) \tag{3.1}$$

$$|S(\xi, \omega)| = S(\omega) \, |\gamma(\xi, \omega)| \tag{3.2}$$

$$\gamma(\xi, \omega) = |\gamma(\xi, \omega)| \, \exp[i\vartheta(\xi, \omega)] \tag{3.3}$$

where, as previously defined, ξ is the separation distance between stations, ω is frequency in [rad/sec], $S(\omega)$ is the power spectral density of the motions and $|S(\xi, \omega)|$ is their cross amplitude spectrum, $\gamma(\xi, \omega)$ is the complex coherency with $|\gamma(\xi, \omega)|$ indicating the lagged coherency, and the exponential term in the last expression reflects the apparent propagation of the motions on the ground surface.

The simplest approximation for the modeling of the wave passage effect (Eq. 3.3), which is also commonly used in simulations (Chapters 7 and 8), is through Eqs. 2.73 and 2.74, repeated here for convenience:

$$\vartheta(\xi, \omega) = -\frac{\omega(\vec{c} \cdot \vec{\xi})}{|\vec{c}|^2} = -\frac{\omega\xi}{c} \tag{3.4}$$

where $\vec{c}$ reflects the apparent propagation of the motions, and the last equality in the above expression is valid if the direction of propagation of the waveforms and the direction of the station separation vector coincide. When a non-dispersive body wave dominates the analyzed time window of the motions, c can be considered as constant over the frequency range of the wave (Section 2.4.1). Past this frequency range, however, where scattered energy dominates, c is no longer a constant, as has been elaborated upon in Sections 2.4.1 and 2.4.4 and illustrated in Figs. 2.15 and 2.16. Additionally, when more waves than one arrive at the array from different directions, the apparent propagation of the motions cannot be approximated by Eq. 3.4. Early studies on spatial variability, e.g., [208], evaluated the "gross" apparent propagation velocity of a single wave dominating the motions by estimating the relative arrival time delays, τ_0, of the motions at the array stations with respect to a reference station through alignment (Sections 2.3.2 and 2.4.1), and fitting a straight line to the ξ/τ_0 data. More rigorously, the direction of propagation of the incoming waves and an estimate of their apparent propagation velocity at each frequency can be obtained through frequency-wavenumber analyses that are described in Section 4.1. Phase estimates will not be addressed further in this chapter, which concentrates on the parameterization of the power spectral density and the lagged coherency of the motions.

In the process of parameterization, nonparametric power spectral densities, cross spectral densities and coherencies are first estimated from the data by means of the approaches described in Chapter 2. Functional forms are then fitted to the nonparametric estimates and the parameters of the analytical functions are identified, most commonly, through deterministic linear or nonlinear regression analyses. The process is, generally, performed in two steps. In the first step, the point estimates of the motions are identified: Nonparametric power spectral densities (Eq. 2.46) are evaluated from the strong motion segment of the ground motions at all stations of interest. Invoking the assumption of homogeneity (Section 2.5.1), i.e., that the random field is not a function of absolute location, a single parametric form is fitted to the nonparametric spectra of the motions at the stations considered. The second step parameterizes the spatial estimates of the motions: In this step, nonparametric smoothed lagged coherency estimates (Eq. 2.70) are obtained from the data for each station pair. Studies that utilize the assumption of isotropy (Section 2.5.1), i.e., the directional invariance of the random field for each component of the motions, average the nonparametric estimates over all station pairs having the same separation distance. Studies that assume that the random field is anisotropic consider, instead, projected distances of the station separation vector in the directions along and normal to the direction of propagation of the waves. In either case, an analytical expression is then fitted to the nonparametric estimates.

This chapter begins, in Section 3.1, with the description of parametric models most commonly used to describe the power spectral density of the random field, which are the Kanai-Tajimi spectrum [253], [514] and its modified version presented by Clough and Penzien [107]. These spectra have received wide acceptance by the engineering community for the representation of the point characteristics of the motions. Their parameters for soft, medium and firm soil conditions, often utilized in simulations of seismic ground motions, are also presented in the section. The section concludes with a brief description of alternative seismological spectra, which are further described in Chapter 8. Section 3.2 presents some early studies on spatial variability from data recorded at the first few installed spatial arrays, and, especially, the SMART-1 array in Lotung, Taiwan [68]. These studies concentrated mostly on the evaluation of the behavior of seismic ground motions recorded over extended areas rather than the establishment of parametric coherency models. As the number of spatial arrays and the number of events recorded at each array increased over the years, some physical dependencies of coherency started becoming apparent; these dependencies are presented in Section 3.3. Section 3.4 presents parametric models for the spatial coherency, beginning in Section 3.4.1 with empirical coherency models: Because, essentially, every investigating group proposed their own functional form for the lagged coherency and used their best rationale for the optimal smoothing process of the nonparametric spectra, there exists a wide range of functional forms for the lagged coherency and an even wider range of model parameters; the section presents some of the most commonly used ones. It is followed, in Section 3.4.2, with the description of some semi-empirical coherency models, the most widely used one being the coherency model of Luco and Wong [324], and, in Section 3.4.3, with a brief description of efforts in analytically modeling the spatial coherency. It is emphasized that the list of publications on the topic is quite extensive, and the material presented herein is by no means exhaustive. Finally, Section 3.5 presents a methodology that parameterizes the random field in a single step, bypasses the requirement of spectral smoothing of the nonparametric estimates in the parametric model evaluation, and embeds the approach in a statistical system identification framework.

3.1   PARAMETRIC POWER SPECTRAL DENSITIES

The most commonly used parametric forms for the power spectral density are the Kanai-Tajimi spectrum [253], [514], or its extension presented by Clough and Penzien [107]. The physical basis of the Kanai-Tajimi spectrum is that it passes a white process through a soil filter with frequency ω_g and damping ζ_g. The resulting expression for the power spectral density of ground accelerations, as will be derived later in Eqs. 6.23 and 6.24, becomes:

$$S_{KT}(\omega) = S_\circ \frac{1 + 4\zeta_g^2(\frac{\omega}{\omega_g})^2}{[1 - (\frac{\omega}{\omega_g})^2]^2 + 4\zeta_g^2(\frac{\omega}{\omega_g})^2} \tag{3.5}$$

in which $S_\circ$ is the amplitude of the bedrock excitation spectrum, considered to be a white process. A deficiency of Eq. 3.5 is that the spectrum produces infinite variances for the ground velocity and displacement. For any stationary process, the power spectral densities of velocity, $S_v(\omega)$, and displacement, $S_d(\omega)$ are related to the power spectral density of acceleration, $S(\omega)$, through the expressions:

$$S_v(\omega) = \frac{1}{\omega^2} S(\omega); \quad S_d(\omega) = \frac{1}{\omega^4} S(\omega) \tag{3.6}$$

It is apparent from Eqs. 3.5 and 3.6 that the velocity and displacement spectra of the Kanai-Tajimi acceleration spectrum are not defined as $\omega \to 0$. Clough and Penzien [107] passed the Kanai-Tajimi spectrum (Eq. 3.5) through an additional filter with parameters ω_f and ζ_f, and described the power spectrum of ground accelerations as:

$$S_{CP}(\omega) = S_\circ \frac{1 + 4\zeta_g^2(\frac{\omega}{\omega_g})^2}{[1 - (\frac{\omega}{\omega_g})^2]^2 + 4\zeta_g^2(\frac{\omega}{\omega_g})^2} \frac{(\frac{\omega}{\omega_f})^4}{[1 - (\frac{\omega}{\omega_f})^2]^2 + 4\zeta_f^2(\frac{\omega}{\omega_f})^2} \tag{3.7}$$

which yields finite variances for velocities and displacements.

The Kanai-Tajimi and Clough-Penzien spectra (Eqs. 3.5 and 3.7) have been extensively used in the parameterization of the point estimates of the motions. Examples of their use in the two-step process for the parameterization of the spatially variable random field include the work of Hao *et al.* [200] and Harichandran [205], [206]. Hao *et al.* [200] modeled the power spectral density of the motions with the Kanai-Tajimi spectrum, and identified its parameters for two events at the SMART-1 array: Event 24 (magnitude 7.2, epicentral distance 92 km and depth 25 km) and Event 45 (magnitude 7.0, epicentral distance 79 km and depth 7 km). They analyzed the three components (east-west, north-south and vertical) of the motions during three different time windows. Harichandran [205] utilized both the Kanai-Tajimi and the Clough-Penzien spectra for the identification of the parameters of the point estimates of the motions during Event 20 (magnitude 6.9, epicentral distance 117 km and depth 31 km) and

Event 24 at the SMART-1 array in the radial and tangential directions. For the strong motion window of the motions recorded during Event 24, the parameters of the Kanai-Tajimi spectrum reported by Hao *et al.* [200] were $\omega_g = 6.91$ rad/sec and $\zeta_g = 0.26$ for the north-south component of the motions, and $\omega_g = 7.54$ rad/sec and $\zeta_g = 0.3$ for the east-west component. The parameters of the Kanai-Tajimi spectrum reported by Harichandran [205] during the same event were $\omega_g = 7.21$ rad/sec and $\zeta_g = 0.22$ for the radial component of the motions, and $\omega_g = 8.53$ rad/sec and $\zeta_g = 0.36$ for the tangential component; the epicentral direction of this event coincided with the direction from station O05 to station O11 (Fig. 1.2). The values identified by both Hao *et al.* [200] and Harichandran [205] are in fairly good agreement, which suggests stability in the estimation of the power spectral density of the motions, in spite of the different processing of the seismic data performed by the two investigating teams. On the other hand, their proposed coherency models differ significantly, as will be discussed in Section 3.4.1.

The Kanai-Tajimi and Clough-Penzien spectra (Eqs. 3.5 and 3.7) have also been extensively used in the simulation of seismic ground motions. Proposed values for their parameters that result in spectra for various soil conditions can be found, e.g., in Refs. [138], [148] and [221]. Table 3.1(a) presents the parameters of the spectra suggested by Hindy and Novak [221] in their pioneering study of the effect of spatially variable seismic ground motions on the response of buried pipelines, which will be described in Section 6.3.1. The parameters of "Spectrum 2" were obtained by fitting the parametric form of Eq. 3.7 to the spectra evaluated by Ruiz and Penzien [432] from

---

TABLE 3.1
Parameters of the Kanai-Tajimi (Eq. 3.5) and the Clough-Penzien (Eq. 3.7) spectra for various soil conditions. (Part (a) after A. Hindy and M. Novak, "Pipeline response to random ground motion," *Journal of the Engineering Mechanics Division, ASCE*, Vol. 106, pp. 339–360, 1980; reproduced with permission from ASCE. Part (b) after A. Der Kiureghian and A. Neuenhofer, "Response spectrum method for multisupport seismic excitation," *Earthquake Engineering and Structural Dynamics*, Vol. 21, pp. 713–740, Copyright © 1992 John Wiley & Sons Limited; reproduced with permission.)

(a) After Hindy and Novak [221]			
	ω_g[rad/sec]	ζ_g	ω_f[rad/sec]	ζ_f
"Spectrum 1"	2π	0.4	0.2π	0.4
"Spectrum 2"	5π	0.6	0.5π	0.6
"Spectrum 3"	10π	0.8	π	0.8

(b) After Der Kiureghian and Neuenhofer [138]			
Soil Type	ω_g[rad/sec]	ζ_g	ω_f[rad/sec]	ζ_f
soft	5.0	0.2	0.5	0.6
medium	10.0	0.4	1.0	0.6
firm	15.0	0.6	1.5	0.6

a number of horizontal acceleration components, for which the earthquake magnitude, source-site distance and site conditions were similar. Hindy and Novak [221] noted that the parameter values ω_g and ζ_g of their "Spectrum 2" were also in agreement with the values proposed by Tajimi [514]. Acceleration, velocity and displacement spectra based on the Clough-Penzien formulation (Eq. 3.7) with parameters as reported by Hindy and Novak [221] for the Spectrum 2 soil conditions (Table 3.1(a)) are illustrated in Fig. 7.1, and used in Chapter 7 for the simulation of random processes and fields. Hindy and Novak [221] also introduced "Spectrum 1" and "Spectrum 3" (Table 3.1(a)) such that the frequency content of the motions varies from sharply peaked at low frequencies (Spectrum 1) to a fairly uniform distribution over a significant frequency range (Spectrum 3). Later, Der Kiureghian and Neuenhofer [138] reported parameters for the spectra of Eqs. 3.5 and 3.7 for soil conditions classified as "soft", "medium" and "firm", which are presented in Table 3.1(b). It can be seen from the comparison of Tables 3.1(a) and 3.1(b) that the parameters of "Spectrum 1" and "Spectrum 2" introduced by Hindy and Novak [221] are compatible with the parameters of the spectra for soft and firm soil conditions, respectively, proposed by Der Kiureghian and Neuenhofer [138].

Figure 3.1(a) presents the Kanai-Tajimi (Eq. 3.5) and Fig. 3.1(b) the Clough-Penzien (Eq. 3.7) acceleration spectra for the three soil conditions of Table 3.1(b); the spectra were normalized with respect to the amplitude of the white bedrock excitation spectrum, S_o. It is clear from the figures that, essentially, the only difference between the two formulations occurs at the very low frequencies, where the Kanai-Tajimi spectra (Fig. 3.1(a)) tend to a finite value, whereas the Clough-Penzien spectra (Fig. 3.1(b)) tend to zero as $\omega^4 \rightarrow 0$ (Eq. 3.7). The effect of the different soil conditions can also be clearly recognized from the figures: The shape of the power spectral densities for firm soil conditions is similar to that of a band-limited white-noise process with a small peak at 1.95 Hz. The spectra for medium soil conditions exhibit a clearer peak at 1.42 Hz for both formulations, that is shifted to the left of the peak of the spectra for firm soil conditions. The spectra for soft soil conditions are sharply peaked at the low frequency of 0.77 Hz. Velocity and displacement power spectral densities derived from the Clough-Penzien spectra by means of Eq. 3.6 will have frequency content lower than that of accelerations (Fig. 3.1(b)) because of the respective division by ω^2 and ω^4 (Eq. 3.6); this effect is illustrated later herein in Fig. 7.1.

It should be noted at this point, that the Kanai-Tajimi and Clough-Penzien spectra model only the effect of S-waves propagating vertically from the bedrock, through a horizontal layer, to the ground surface; the bedrock excitation (S_o in Eqs. 3.5 and 3.7) is assumed to be a white process. More refined descriptions for the point estimates of the motions are provided by "seismological" spectra, which take into account the effects of the rupture at the fault and the transmission of waves through the media from the fault to the ground surface. For example, Joyner and Boore [246] presented the following description for the stochastic seismic ground motion spectrum:

$$S(\omega) = CF \cdot SF(\omega) \cdot AF(\omega) \cdot DF(\omega) \cdot IF(\omega) \qquad (3.8)$$

in which, CF represents a scaling factor, which is a function of the radiation pattern, the free surface effect, and the material density and S-wave velocity in the near source region; $SF(\omega)$ is a source factor that depends on the moment magnitude and

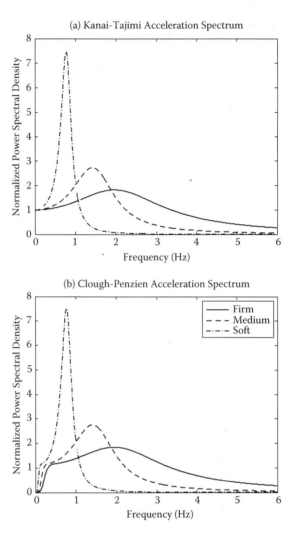

FIGURE 3.1 Illustration of the Kanai-Tajimi spectra [253], [514] of Eq. 3.5 in part (a) and the Clough-Penzien spectra [107] of Eq. 3.7 in part (b) for firm, medium and soft soil conditions. The parameters of the spectra are the ones reported by Der Kiureghian and Neuenhofer [138] and presented in Table 3.1(b). The spectra are normalized with respect to the amplitude of the white bedrock excitation spectrum, $S_\circ$.

the rupture characteristics; $AF(\omega)$ is the amplification factor described either by a frequency-dependent transfer function as, e.g., the filters in the Kanai-Tajimi or Clough-Penzien spectra (Eqs. 3.5 and 3.7), or, in terms of the site impedance $\sqrt{(\rho_0 V_0)/(\rho_r V_r)}$, where ρ_0 and V_0 are the density and S-wave velocity in the source region, and ρ_r and V_r the corresponding quantities near the recording station; $DF(\omega)$ is a diminution factor, that accounts for the attenuation of the waveforms; and $IF(\omega)$ is a filter used to shape the resulting spectrum so that it represents the seismic ground motion quantity of interest. Seismological spectra that incorporate these effects can be used instead of Eqs. 3.5 and 3.7 in engineering applications; one of the models (Ref. [436]) is used in Section 8.3.1 for the conditional simulation of spatially variable ground motions.

3.2   EARLY STUDIES ON SPATIAL VARIABILITY

As indicated in the Introduction, the spatial variation of seismic ground motions started being analyzed extensively after the installation of dense instrument arrays. Early studies on spatial variability of seismic ground motions were based, e.g., on data recorded at the El Centro differential array (Fig. 1.1) during the 1979 Imperial Valley earthquake [371], [480], [499], the Chusal differential array in the Garm region of the former USSR [269], the Colwick array at the Nevada Test Site [340], and, then extensively, the SMART-1 array (Fig. 1.2) in Lotung, Taiwan [68], [201], [317]. Some of these initial investigations of the spatial variability of seismic ground motions were geared towards a better understanding of the phenomenon of "spatial averaging" of the incident excitations that is caused by large, rigid, mat foundations of structures, such as power plants. Spatial averaging results in the reduction of the high frequency translational components of the motions, as will be further illustrated in Section 9.1. Examples of early studies on spatial variability are presented in the following.

3.2.1   EL CENTRO DIFFERENTIAL ARRAY

Smith *et al.* [480] analyzed the data recorded at the El Centro differential array (Fig. 1.1) during the 1979 Imperial Valley earthquake. The epicenter of this $M_S = 6.9$ earthquake was approximately 24 km from the array, but the closest point of rupture was only about 5 km [480]. The analysis consisted of the evaluation of spectral ratios and normalized autocovariance and cross covariance functions during the P-wave window of the vertical component and the S-wave window of the two horizontal components of the motions over three frequency ranges: the entire frequency range of the data, and the frequency bands between $5 - 15$ Hz and $15 - 25$ Hz. Their results for the peak normalized cross covariance (or cross correlation) of the three ground motion components are presented in Fig. 3.2. It is noted from the figures that the vertical motions (Fig. 3.2(a)) are more highly correlated than the horizontal ones (Figs. 3.2(b) and 3.2(c)). Additionally, the station separation distance, at which the cross correlation of the data becomes low, is significantly shorter for the horizontal motions (Figs. 3.2(b) and 3.2(c)) than the vertical ones (Fig. 3.2(a)), especially for the higher frequency ranges. Smith *et al.* [480] attributed these differences to the shorter time window of the P-wave component in the vertical motions, the higher frequency content of the P-wave, and the fact that the P-wave window consisted, mostly, of a coherent wave propagating with an apparent velocity of 6.8 km/sec. On the other hand, the horizontal motion window was longer, the frequency content of the S-waves was lower than the frequency content of the P-waves, and, also, the horizontal motions consisted of a group of waves arriving at the array from different source regions due to the propagation of rupture across the extended fault close to the array. Figure 3.2, however, clearly indicates that the correlation of all three components of the data decreases with frequency and station separation distance.

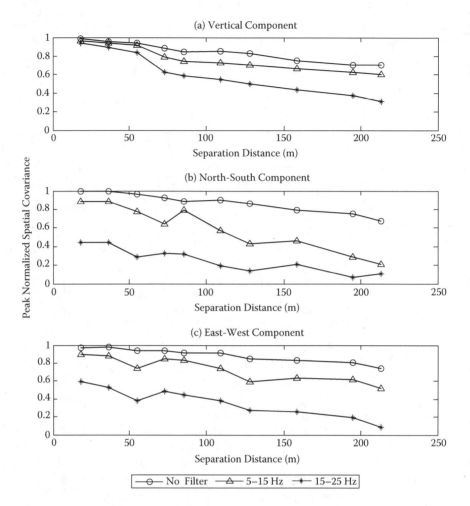

FIGURE 3.2 Variation of the peak values of the normalized spatial covariance functions with station separation distance evaluated by Smith *et al.* [480] from the data recorded at the El Centro differential array. Part (a) presents the covariance functions of the vertical component of the data, part (b) those of the north-south component, and part (c) the results for the east-west component of the motions. The covariance functions were evaluated for three frequency windows. (Reproduced from S.W. Smith, J.E. Ehrenberg and E.N. Hernandez, "Analysis of the El Centro differential array for the 1979 Imperial Valley earthquake," *Bulletin of the Seismological Society of America*, Vol. 72, pp. 237–258, Copyright ©1982 Seismological Society of America.)

3.2.2 CHUSAL ARRAY

King [269] and King and Tucker [270] conducted an analysis similar to the one by Smith *et al.* [480] utilizing data recorded at the El Centro and the Chusal differential arrays. The Chusal array was located in Garm, Tajikistan, and consisted of a strong and a weak motion differential array: The strong motion array consisted of three triaxial accelerometers at a maximum separation distance of 128 m, and the weak motion array of 12 intermediate-period seismometers with an average separation distance of 15 m, which were deployed for recording north-south motions only.

King [269] and King and Tucker [270] evaluated the peak normalized spatial covariance of the data recorded at both the El Centro and Chusal differential arrays, which they bandpassed through three frequency ranges: 1 – 3 Hz, 3 – 10 Hz and 10 – 30 Hz. The peak normalized spatial covariance of the ground motions recorded during the Imperial Valley earthquake in both horizontal directions, and the data of two events recorded in the north-south direction at the weak motion Chusal array (Events 4 and 5, both with $M_L = 2.2$) during the S-wave window are presented in Fig. 3.3. Clearly, Fig. 3.3 indicates that the values of the normalized covariance evaluated from the data at both arrays decay as frequency and separation distance between the stations increase. King and Tucker [270], however, noted that the peak

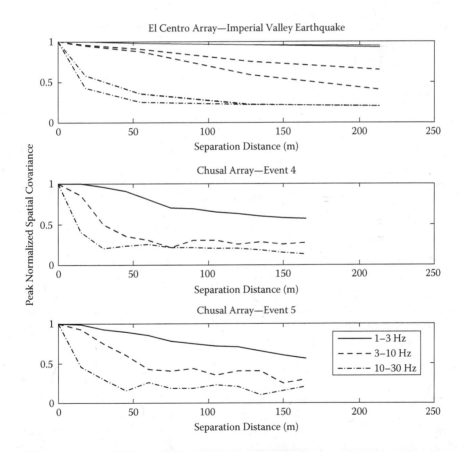

FIGURE 3.3 Comparison of the variation of the peak normalized spatial covariance functions with station separation distance evaluated by King and Tucker [270] for three frequency windows between data recorded at the El Centro differential array in both horizontal directions during the Imperial Valley earthquake and data recorded at the Chusal weak-motion differential array in the north-south direction during two events (after King and Tucker [270]).

normalized covariances obtained from the Chusal array data decay faster with station separation distance than those at the El Centro differential array, and attributed this difference to the different site conditions of the two arrays: Whereas the El Centro differential array was located on a flat, homogeneous site [87], the Chusal differential arrays were located on soil containing rocks and boulders and with the depth to the crystalline bedrock changing rapidly from 5 m to 55 m over a distance of 170 m [270].

3.2.3 SMART-1 ARRAY

The data recorded at the SMART-1 array during Event 5, the earthquake of January 29, 1981, have already been utilized in Chapter 2 for the illustration of the derivations of the nonparametric stochastic estimates of seismic ground motions. The data from this celebrated array have provided unique means for the investigation of the spatial variability of seismic ground motions by a large number of researchers, especially because they were quickly and broadly disseminated by the Seismographic Station of the University of California at Berkeley and the Institute of Earth Sciences of the Academia Sinica in Taipei. Selected early studies utilizing SMART-1 data are presented in the following.

Frequency-Dependent Cross Correlation

Loh *et al.* [317] analyzed data from the SMART-1 array soon after the first few events had been recorded. Figure 3.4 presents their results for the frequency-dependent cross correlation function of the east-west component of the data at stations C00 and I06 (Fig. 1.2) during the earthquake of November 14, 1980, in part (a), and the earthquake of January 29, 1981, in part (b). The results were obtained using a moving window approach in the frequency domain. The technique first evaluates the harmonic components of the time histories at the two stations for a frequency window centered at each frequency (Eq. 2.85), and then utilizes these harmonic components to evaluate the frequency-dependent cross correlation functions (Eq. 2.87), as described in Section 2.4.3. The frequency band used in the evaluation of the estimates for both events was 0.488 Hz. Figure 3.4(a) suggests that the motions of the November 14, 1980, event are highly correlated for periods greater than 1 sec, but correlation falls off rapidly at shorter periods. On the other hand, the cross correlation of the January 29, 1981, data in Fig. 3.4(b) appears to be low throughout the period range considered, even for these close-by stations (at a separation distance of 200 m).

Relative Coherency

Abrahamson [4] also analyzed the data recorded at the SMART-1 array during the earthquake of January 29, 1981. He suggested that a quantitative measure of the relative coherency is the peak power of the conventional frequency-wavenumber spectrum (defined in Section 4.1.1), after the cross spectral densities of the data have been normalized. The evaluation of the relative coherency was conducted for a 4-sec P-wave window of the vertical motions, and a 5-sec S-wave window of the radial component of the horizontal motions. For each direction, two sets of analyses were performed: The first analysis utilized the data at the center, inner and middle ring stations of the

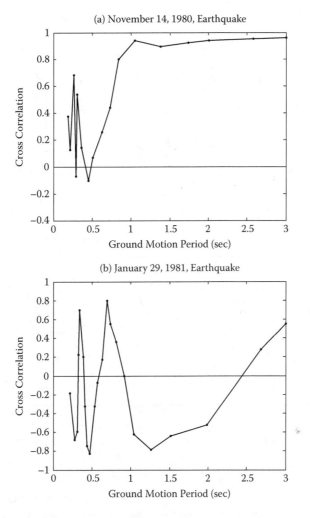

(a) November 14, 1980, Earthquake

(b) January 29, 1981, Earthquake

FIGURE 3.4 Frequency-dependent cross correlation functions of the east-west component of the motions recorded during the earthquakes of November 14, 1980, in part (a), and of January 29, 1981, in part (b), at stations C00 and I06 of the SMART-1 array, as reported by Loh *et al.* [317]; the width of the moving frequency window used in the evaluation was 0.488 Hz. (After C.H. Loh, J. Penzien and Y.B. Tsai, "Engineering analysis of SMART-1 array accelerograms," *Earthquake Engineering and Structural Dynamics*, Vol. 10, pp. 575–591, Copyright © 1982 John Wiley & Sons Limited; reproduced with permission.)

array that recorded the event (total of 17 stations) at a maximum separation distance of 2 km, and the second the data at the center, inner, middle and outer ring stations (total of 27 stations) at a maximum separation distance of 4 km (Fig. 1.2). Figure 3.5(a) presents the relative coherency of the motions at the 17 stations, and Fig. 3.5(b) the relative coherency of the motions at the 27 stations. The horizontal lines, at 0.4 and 0.26, indicate the 95% confidence level of white noise for the 17- and the 27-station

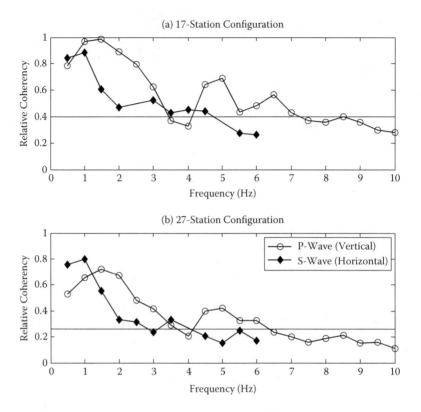

FIGURE 3.5 Relative coherency of the vertical component of the motions during the P-wave window and the horizontal component of the motions during the S-wave window of data recorded during Event 5 at the SMART-1 array, as reported by Abrahamson [4]. Part (a) illustrates the relative coherency from data recorded at 17 stations of the array (center, inner and middle ring stations) at a maximum separation distance of 2 km. Part (b) presents the relative coherency from data recorded at 27 stations of the array (center, inner, middle and outer ring stations) at a maximum separation distance of 4 km. The horizontal lines in the figures indicate the 95% confidence level of white noise for the two station configurations (after Abrahamson [4]).

configurations, respectively, and for the amount of smoothing performed on the data. Abrahamson [4] noted that when the relative coherency assumes values close to unity, a plane wave dominates the motions, and when the relative coherency is above 0.4 for the 17-station configuration and above 0.26 for the 27-station configuration, its value is significant at the 95% confidence level. Figure 3.5 indicates that the relative coherency of the motions recorded at shorter separation distances (17-station configuration) is higher than the relative coherency of the motions recorded at longer separation distances (27-station configuration). The relative coherency of the vertical motions decays sharply between 3 Hz and 4 Hz, increases between 4 Hz and 5 Hz, and becomes lower than the 95% confidence level at, approximately, 7 Hz and 6.5 Hz for the 17- and 27-station configurations, respectively. The horizontal motions during the

S-wave window are less correlated than the vertical ones over, essentially, the entire frequency range. The relative coherency in this case decays at 2 Hz for both station configurations, and remains at or below the 95% confidence level for the 27-station configuration, whereas, for the 17-station configuration, it drops below that level at, approximately, 5 Hz.

Spatial and Directional Cross Correlation Functions

Zerva *et al.* [589], [599] developed an analytical model for the estimation of the spatial coherency. The model was based on the assumption that the excitation at the source can be approximated by a stationary random process, which is transmitted to the ground surface by means of frequency transfer functions, different for each station. The frequency transfer functions are the Fourier transform of impulse response functions determined from an analytical wave propagation scheme, the method of self-similar potentials [457], and a system identification technique in the time domain [47]. The resulting surface ground motions are then stationary random processes specified by their power and cross spectral densities, from which the spatial variation of the motions can be obtained.

The model was utilized to reproduce the stochastic characteristics of the data recorded at the SMART-1 array during Event 5 [589], [598]. Figure 3.6 presents the analytically derived spatial cross correlation of the radial and tangential components of the motions as functions of station separation distance. Also shown in the figure are the empirical results reported in the early studies of Harada [201] and Loh [313] based on their analyses of the recorded data. The results reported by Harada [201] were for the north-south component of the motions; this direction almost coincided with the epicentral direction of this event. The results presented by Loh [313] were for the radial and tangential components. Again, these early empirical and analytical studies indicate that the spatial correlation of the motions decreases with distance.

Figure 3.7 presents the mean value of the directional cross correlation function between the radial and tangential components of the motions evaluated with the analytical model and compared with the empirical results of Loh *et al.* [314]. The mean value was obtained from the directional cross correlation of the motions at stations O06, C00 and O12 (Fig. 1.2), and then averaged over four frequency ranges. It is noted that the cross correlation between different ground motion components at the same station or between different ground motion components at different stations is, generally, not analyzed. Instead, it is considered that the vertical, radial and tangential components of the seismic motions are uncorrelated, as will be further elaborated upon in Section 3.3.2. Figure 3.7 also suggests that, for the ground motions of Event 5 at the SMART-1 array, the radial and tangential components of the motions are only slightly correlated.

It should be mentioned at this point that the work of Harichandran and Vanmarcke [208] based on SMART-1 seismic data also falls within the time frame of these early studies of spatial variability. However, because their derivations belong more to the conventional estimation of coherency, as described in Chapter 2, their model is presented together with other empirical coherency models in Section 3.4.1.

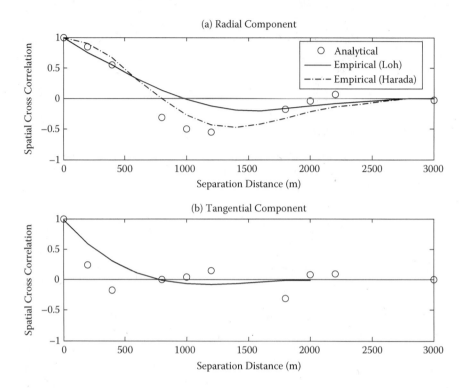

FIGURE 3.6 Spatial cross correlation functions of the radial, in part (a), and tangential, in part (b), components of the motions recorded at the SMART-1 array during Event 5. The empirical results of Loh [313] in parts (a) and (b) are in the corresponding directions. The empirical results of Harada [201] in part (a) are for the component of the motions in the north-south direction, which is close to the epicentral direction for this event. The results labeled "analytical" in the subplots were determined by Zerva *et al.* [589], [599] based on an analytical source-site model used to reproduce the data of this event. (Reproduced from *Probabilistic Engineering Mechanics*, Vol. 1, A. Zerva, A. H-S. Ang and Y.K. Wen, "Development of differential response spectra for lifeline seismic analysis," pp. 208–218, Copyright ©1987, with permission from Elsevier.)

3.3   DEPENDENCE OF COHERENCY ON PHYSICAL PARAMETERS

The evaluation of the spatial coherency relies on the analysis of data recorded at dense instrument arrays by means of signal processing techniques, as described in Chapter 2. Furthermore, coherency reflects, essentially, the phase variability in the seismic data (Section 2.4.2), which cannot be readily attributed to physical causes. However, extensive research conducted with spatial array data, as well as physical insights, revealed the physical causes underlying the spatial coherency, which were schematically illustrated in Fig. 1.4. This section discusses the effect of earthquake magnitude and source-site distance on coherency, the validity of the assumption of its rotational invariance, the behavior of coherency at uniform (soil and rock) and variable site conditions, and the difference in its exponential decay at shorter and longer station separation distances.

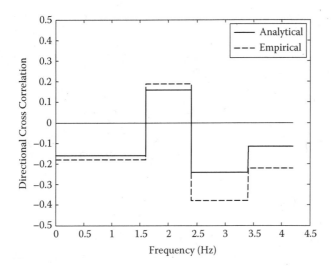

FIGURE 3.7 Mean value of the directional cross correlation function between the radial and tangential component of the motions at stations O06, C00 and O12 of the SMART-1 array during Event 5 averaged over four frequency ranges. The empirical results were reported by Loh *et al.* [314] from the analysis of the recorded data and the analytical results were developed by Zerva [589] based on a source-site model used to reproduce the data of this event (after Zerva [589]).

3.3.1 EARTHQUAKE MAGNITUDE AND SOURCE-SITE DISTANCE

Somerville *et al.* [486], [487] attributed the physical causes of the spatial variability of seismic ground motions to the wave propagation effect, the finite source effect, the effect of scattering of the seismic waves as they propagate from the source to the site, and the local site effects; the first two contributions were grouped together into the "source effect" and the latter two into the "scattering effect". A schematic diagram of these contributions is presented in Fig. 3.8: The top left figure shows ray paths extending from multiple locations on a finite fault (with dimensions that are larger than the source-to-site distance) through a homogeneous medium to two adjacent locations 1 and 2 of a site. The ray paths are shown in more detail in the top right figure along with an illustration of potential seismograms at the two locations. In the two seismograms, the waveforms transmitted from each fault subregion are similar, but arrive delayed at the further-away station due to the wave passage effect. The bottom left figure shows ray paths extending from a point source through a scattering medium to the two stations of the site. The bottom right figure presents the ray paths in more detail along with an illustration of the seismograms at the two locations; Δt in these plots indicates the time delay in the arrival time of the waves at the further-away station. The seismograms in the bottom right figure differ in shape due to the effect of scattering and arrive delayed (by Δt) at station 2 due to the wave passage effect. According to this model, the spatial variability of the ground motions caused by a small earthquake, that can be approximated by a point source, is attributed to the scattering of the waves from the source to the site and the local site effects.

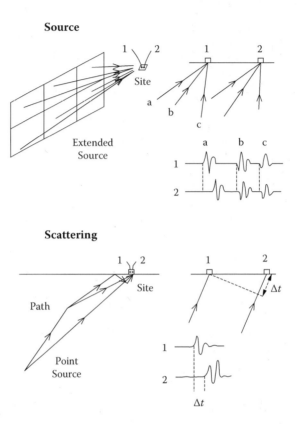

FIGURE 3.8 Schematic diagram of the source and scattering contributions to the spatial variability of the seismic motions after Somerville *et al.* [487]: The top part of the figure indicates the source contribution that incorporates the wave passage and the extended source effects. The bottom part of the figure illustrates the scattering contribution that incorporates the effects of scattering of the seismic waves as they propagate from a point source to the site and the local site effects. (Reprinted from *Structural Safety*, Vol. 10, P.G. Somerville, J.P. McLaren, M.K. Sen and D.V. Helmberger, "The influence of site conditions on the spatial incoherence of ground motions," pp. 1–13, Copyright © 1991, with permission from Elsevier.)

On the other hand, for a large event, each fault subregion will transmit waveforms as if it were a point source (scattering effects), but the total seismic ground motions at each location will be the superposition of the transmitted waveforms from all fault subregions (source effect). Hence, the spatial coherency of the ground motions of a large event should be expected to be smaller than the coherency of the seismic motions of a small event.

The dependence of coherency on earthquake magnitude and distance was investigated by Abrahamson *et al.* [12], [13], [14] using data recorded at the LSST array in Lotung, Taiwan (Fig. 1.3). Initially, Abrahamson *et al.* [13] analyzed four events recorded at the LSST array: the mainshock and aftershock of a near-field event and the mainshock and aftershock of a far-field event. This first investigation suggested that coherency estimates from the mainshock and the aftershock of the far-field event were comparable, whereas, for the near-field event, the aftershock coherency was lower than the mainshock coherency at moderate frequencies, which was attributed to possible nonlinear soil effects and near-field source-dependent effects. It was also observed that the near-field mainshock coherency was lower than the far-field mainshock coherency at low frequencies, which was interpreted as a source effect, whereas the reverse was true for higher frequencies, which was interpreted as a path effect [13].

Extending the investigation further, Abrahamson et al. [14] examined 15 events recorded at the LSST array (Fig. 1.3) with magnitudes ranging between 3.0 and 7.8, and source-site distances between 5 km and 113 km. The earthquakes were grouped as small magnitude ($M \leq 5$) and large magnitude ($M \geq 6$) events, and the source-site distances as long (> 40 km) and short (< 15 km). The magnitude and distance dependence was investigated by means of the 90% confidence intervals of the residuals between the data for each magnitude and distance group with the coherency model developed by Abrahamson et al. [14] for all data sets (presented later in Eq. 3.16). This more extensive LSST data analysis suggested that there was no consistent trend indicating dependence of coherency on extended faults and source-site distances [14], [449]. Abrahamson [7] later noted that, even though, theoretically, the effect of source finiteness on the ground surface coherency is expected, it does not appear to be significant. Spudich [498] gave a possible explanation of why the source finiteness may not considerably affect coherency estimates: For large earthquakes of unilateral rupture propagation, the waves radiated from the source originate from a spatially compact region that travels with the rupture front, and, thus, at any time instant, a relatively small fraction of the total rupture area radiates. Unilateral rupture at the source constitutes the majority of earthquakes [498]; Spudich [498] cautioned, however, that bilateral rupture effects on the spatial coherency are yet unmeasured.

The statistical analysis of the LSST data by Abrahamson et al. [14], the evaluation of cross correlation functions for the 1979 Imperial Valley earthquake at the El Centro differential array by Somerville et al. [486], presented later in Fig. 3.22, and Spudich's [498] explanation suggest that the significance of the source effect is small in comparison to the scattering effect. It is recalled that source effects, scattering effects, as well as linear/nonlinear soil behavior may constructively or destructively interfere in the coherency decay, as observed by Abrahamson et al. [13] from their detailed analysis of the pairs of mainshock-aftershock coherency estimates at the LSST array. Further quantification of the relative contribution of these effects on coherency can be obtained through the investigation of such pairs of mainshock-aftershock coherency estimates from near- and far-field events.

3.3.2 PRINCIPAL DIRECTIONS

Arias [35], Penzien and Watabe [400] and Hadjian [188] noted the existence of the "principal axes" of seismic ground motions recorded in three orthogonal directions at a single station. Along the orthogonal set of principal axes, the component variances have maximum, minimum and intermediate values and their cross covariances are equal to zero.

Considering that each component of the seismic ground motions in the three orthogonal directions can be approximated by the product of a stationary process with deterministic intensity modulating functions (described in Section 7.1.2), Penzien and Watabe [400] showed that the procedure for the transformation of the axes of the ground motions in the evaluation of their principal directions is identical to the approach used for stresses. This approach was then applied to data recorded during three earthquakes in California (at Long Beach in 1933, El Centro in 1940, and Taft in 1952) and three events in Japan (at Tokachi-Oki in 1968, Hiddaka-Sankei in 1970, and Izu-Hanto-Oki in 1974). Penzien and Watabe [400] determined the direction of the principal axes of the motions by selecting successive time intervals over the ground motion duration. They noted that, when the time interval was very short, the major principal axis coincided with the instantaneous resultant acceleration vector, which rapidly changed direction. This instantaneous principal direction estimate will be utilized in Section 5.1.1 for the evaluation of the apparent propagation velocity of body waves on the ground surface [384]. On the other hand, Penzien and Watabe [400] noted that, when the time interval used in the identification of the principal axes was sufficiently long so that the estimate stabilized, the major principal axis pointed in the general direction of the epicenter of the earthquakes and the minor principal axis was nearly vertical.

The majority of the parametric spatial coherency models consider ground motion components in the epicentral (radial), normal to epicentral (tangential) and vertical directions, and assume, explicitly or implicitly, that these components are uncorrelated. Even though principal directions were identified from the three components of the motions at a single station [400], one may argue that, for the strong motion S-wave window and for seismic ground motions propagating in the general epicentral direction, the principal axes at each array station would coincide. Hence, under these conditions, the assumption that the components of the motions are uncorrelated in the "physical" (epicentral, normal to epicentral and vertical) directions, rather than the three "geometric" ones (north-south, east-west and vertical), can be considered valid for all array stations. The low values of the directional correlations between the radial and tangential component of the motions from the early studies of the data recorded during Event 5 at the SMART-1 array (Fig. 3.7) also support this observation.

3.3.3 ISOTROPY

Most lagged coherency estimates assume that the random field for each seismic ground motion component, in addition to being homogeneous, is, also, isotropic. This assumption, as indicated in Section 2.5.1, implies that the rotation of the random field on the ground surface will not affect its joint probability density functions, and, hence, the lagged coherency becomes a function of separation distance only, $\xi = |\vec{\xi}|$, and not direction, $\vec{\xi}$.

The validity of this assumption was investigated by a number of researchers. Some concluded that the ground motion random field is anisotropic, and their proposed coherency models account for the directional dependence of the spatial variation of the motions: For example, the coherency model of Hao $et\ al.$ [200] (presented in Section 3.4.1, Eq. 3.14) utilizes the projected distances, ξ_l and ξ_t, of the station separation vector in the directions along and normal to the direction of the propagation of the waves, respectively. Loh and Lin [316] evaluated isocoherence maps for two events recorded at the SMART-1 array, and observed that they were not axisymmetric. They concluded that the random field is not isotropic, and their proposed coherency model (Section 3.4.1, Eq. 3.11) reflects this anisotropy. Ramadan and Novak [418], in order to preserve the simpler representation of the random field as isotropic, modeled its weak anisotropy characteristics by defining the separation distance as $\xi = \xi_l + \mu_t \xi_t$, in which μ_t is a separation reduction factor to account for the directional variability in the data. They considered the models of Hao *et al.* [200] and Loh and Lin [316] and the set of data from which these models were developed, and concluded that this simple separation distance transformation with variable μ_t preserves the isotropy of the field [418]. Abrahamson *et al.* [14] also observed directional dependence of the coherency from SMART-1 data, and suggested that a possible explanation for this effect is that scattering in the forward direction tends to be in phase with the incident wave, whereas scattering to the side tends to loose phase [106].

On the other hand, for the shorter separation distances of the LSST data, Abrahamson *et al.* [14] noted that there was no directional dependence in the coherency estimates when the angle between the station separation vector and the epicentral direction ranged between $0°$ and $75°$. As this angle increased (in the range $75° - 90°$), there was an indication of the coherency decreasing, suggesting, as did the previously discussed studies, directional dependence of the estimate. Abrahamson *et al.* [14], however, concluded that this effect was not significant enough to be included in their parametric modeling of the coherency based on the LSST data, which is presented in Eq. 3.16 (Section 3.4.1). Generally, the assumption of isotropy is considered valid in the simulation of spatially variable seismic ground motions (Chapters 7 and 8).

3.3.4 UNIFORM (SOIL AND ROCK) SITE CONDITIONS

Somerville *et al.* [487] observed that there were large differences in coherencies of seismic motions recorded on flat sedimentary sites, such as the sites of the El Centro differential array and the SMART-1 array, and those recorded on folded sedimentary rocks, such as the Coalinga anticline in California. Using data from the $M_L = 5.1$ aftershock of the 1979 Imperial Valley earthquake recorded at the array, Somerville *et al.* [487] noted that coherency decreased smoothly with frequency and separation distance, which they attributed to wave scattering in a laterally homogeneous, horizontally layered sediment site. On the other hand, the data recorded at a temporary array in Coalinga during the $M_L = 5.1$ aftershock of the 1983 Coalinga earthquake yielded coherencies that did not show strong dependence on separation distance and frequency, which Somerville *et al.* [487] related to the pattern of wave propagation in a medium having strong lateral heterogeneities.

Schneider *et al.* [449] conducted an extensive study of spatial coherency estimates (Eq. 2.70) evaluated from data at a number of dense arrays at various sites, classified broadly as "soil" and "rock" sites. The characteristics of the arrays (array name, location, site classification, number of surface stations and range of surface station spacing) and the characteristics of the data (number of earthquakes at each array, range of earthquake magnitudes, range of distances of the arrays from the faults, and peak ground acceleration) utilized in these studies are presented in Tables 3.2 and 3.3,

TABLE 3.2

Characteristics of dense arrays used in the studies of Abrahamson *et al.* [14] and Schneider *et al.* [449]. (After J.F. Schneider, J.C. Stepp and N.A. Abrahamson, "The spatial variation of earthquake ground motion and effects of local site conditions," *Proceedings of the Tenth World Conference on Earthquake Engineering*, Madrid, Spain, 1992 Copyright © Balkema; reproduced with permission.)

Array	Location	Site Classification	Number of Surface Stations	Station Spacing Range (m)
EPRI LSST	Taiwan	Soil	15	3 – 85
EPRI Parkfield	CA, USA	Rock	13	10 – 191
Chiba	Japan	Soil	15	5 – 319
USGS Parkfield	CA, USA	Rock	14	25 – 952
El Centro Diff.[1]	CA, USA	Soil	5	18 – 213
Hollister Diff.[1]	CA, USA	Soil	4	61 – 256
Stanford (temp.[2])	CA, USA	Soil	4	32 – 185
Coalinga (temp.[2])	CA, USA	Rock	7	48 – 313
UCSC ZAYA (temp.[2])	CA, USA	Rock	6	25 – 300
Pinyon Flat (temp.[2])	CA, USA	Rock	58	7 – 340

[1]Differential
[2]temporary

TABLE 3.3

Characteristics of dense array data used in the studies of Abrahamson *et al.* [14] and Schneider *et al.* [449]. (After J.F. Schneider, J.C. Stepp and N.A. Abrahamson, "The spatial variation of earthquake ground motion and effects of local site conditions," *Proceedings of the Tenth World Conference on Earthquake Engineering*, Madrid, Spain, 1992 Copyright © Balkema; reproduced with permission.)

Array	Number of Events	Magnitude Range	Distance Range (km)	PGA[1]
EPRI LSST	15	3.0 – 7.8	5 – 113	0.26 g
EPRI Parkfield	12	3.0 – 3.9	13 – 15	0.04 g
Chiba	19	4.8 – 6.7	61 – 105	0.41 g
USGS Parkfield	9	2.2 – 3.5	18 – 45	0.04 g
El Centro Diff.	2	5.1 – 6.5		0.89 g
Hollister Diff.	1	5.3	17	0.20 g
Stanford (temp.)	4	< 4	≈ 40	0.007 g
Coalinga (temp.)	1	5.1	12	0.21 g
UCSC ZAYA (temp.)	3	2.3 – 3.0	9 – 19	
Pinyon Flat (temp.)	6	2.0 – 3.6	14 – 39	

[1]Peak Ground Acceleration

respectively. Figures 3.9 and 3.10 present the results of their analyses for the soil and the rock sites, respectively. In both figures, part (a) plots the lagged coherency at each site for a station separation distance range of 15 – 30 m, and part (b) the lagged coherency at each site for a station separation distance range of 50 – 80 m. When more events than one have been recorded at an array, as, e.g., 19 events for the Chiba array in Fig. 3.9 and 9 events for the USGS Parkfield array in Fig. 3.10, the lagged coherency plotted in the figures is the average coherency for all events recorded at the site. For the Hollister array in Fig. 3.9(b) and the Coalinga array in Fig. 3.10(b), which recorded a single event, the estimated coherency reflects the characteristics of the data of the particular event. The numbers in parentheses in the legends of Figs. 3.9 and 3.10 indicate the number of events (Table 3.3) that were used in the evaluation of the average lagged coherency at each site.

Figure 3.9 suggests that, even though the lagged coherency for each soil array follows its own pattern, the trend of the decay of the data with frequency at the various arrays is fairly similar: Coherency is close to unity in the low frequency range, more so for the close-by stations in Fig. 3.9(a) than the further-away stations in Fig. 3.9(b). As frequency increases, coherency decreases consistently for the data at the majority of the arrays. Exemptions are, for the longer separation distance range in Fig. 3.9(b), the coherency from the four events recorded at the Stanford array, which assumes low values at approximately 4 Hz, and the coherency from the single event recorded at the Hollister array, the motions of which appear to be more highly correlated at higher frequencies than the data at the other arrays.

On the other hand, the coherency estimates evaluated at the rock sites (Fig. 3.10) show significant variability. The most notable differences occur for the average coherency evaluated from the 9 events recorded at the USGS Parkfield array and the average coherency of the three events recorded at the ZAYA array, which have been attributed to topographic effects at the two sites [8], [162]. For the data at the remaining three arrays in Fig. 3.10, i.e., EPRI Parkfield, Pinyon (or Piñon) Flat and Coalinga, coherency decays with frequency in a consistent manner, but, for the longer separation distance range considered, the degree of its exponential decay at the three sites differs (Fig. 3.10(b)).

The comparison of Figs. 3.9 and 3.10 suggests that coherencies estimated at rock sites (Fig. 3.10) are lower at low frequencies than coherencies at soil sites (Fig. 3.9), and, also, that the decay with frequency of the rock-site coherencies is "flatter" and slower than the decay of the coherencies at the soil sites. The differences in the behavior of coherency at rock and soil sites may be explained as follows: If it is considered that coherency is minimally affected by magnitude and source-site distance, then the loss of coherency in the motions can be attributed mostly to path and local site (scattering) effects (Section 3.3.1). However, since scattering at rock and soil sites differs, the exponential decay of the lagged coherency at these different site conditions will also be different. This observation was made in the studies of Toksöz et al. [524] and Menke et al. [341]: Toksöz et al. [524] analyzed data at three arrays in Fennoscandia, which were constructed for research into nuclear test ban treaty verification. The arrays are the FINESA array in southern Finland, the NORESS array in southern Norway and the ARCESS array in northern Norway, and consist of vertical seismometers placed on crystalline bedrock with maximum array spacings

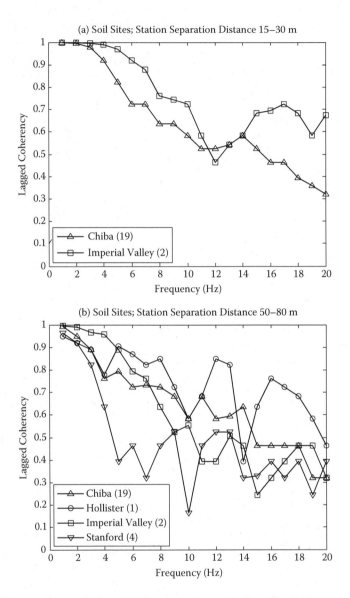

FIGURE 3.9 Nonparametric lagged coherency estimates evaluated by Schneider *et al.* [449] from data sets recorded at various arrays located on soil sites (Tables 3.2 and 3.3). The estimates are grouped in two ranges for the station separation distance: between 15 m and 30 m in part (a), and between 50 m and 80 m in part (b). The numbers in parentheses indicate the number of events recorded at each array, over which the coherency estimates were averaged (adapted from Schneider *et al.* [449]).

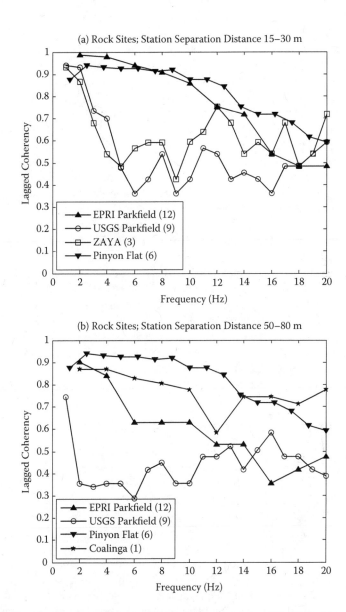

FIGURE 3.10 Nonparametric lagged coherency estimates evaluated by Schneider *et al.* [449] from data sets recorded at various arrays located on rock sites (Tables 3.2 and 3.3). The estimates are grouped in two ranges for the station separation distance: between 15 m and 30 m in part (a), and between 50 m and 80 m in part (b). The numbers in parentheses indicate the number of events recorded at each array, over which the coherency estimates were averaged (adapted from Schneider *et al.* [449]).

of 1500 – 3000 m. The data analyzed in this evaluation were S-waves from quarry blasts at distances of 240 – 350 km. Toksöz *et al.* [524] observed high coherency for separation distances up to 3000 m and frequencies of 2 – 10 Hz, and noted that the coherency at hard rock sites was higher than the coherency at soil sites. They further suggested that the loss of coherency at the rock sites can be attributed to scattering from heterogeneities within the crystalline basement and irregular interfaces within the crust. Menke *et al.* [341] analyzed the coherency of data recorded at two arrays, the ECO and DBM arrays, located on hard rock sites at the Adironack mountains. Coherency estimates at separation distances of 7 – 734 m from four earthquakes and a chemical explosion suggested a similar trend for the exponential decay of the data for all three ground motion components (vertical, radial and tangential), and all epicentral azimuth and distance ranges. Menke *et al.* [341] concluded that their results imply the existence of point scatterers randomly distributed in the shallow crust; their proposed coherency model is presented later herein in Eq. 3.25.

It may also be postulated that, because the soft soil deposits serve as a filter, band-passing the lower frequency components of the incident bedrock excitation (Fig. 3.1), coherency at these sites decreases faster with frequency than the coherency of the motions recorded at rock sites, which exhibit a flatter trend with frequency (Figs. 3.9 and 3.10). This hypothesis may be corroborated with the results of Steidl *et al.* [505], who reported that rock sites have a site response, due to near-surface rock weathering and cracking, that amplifies the incident excitation at frequencies above 2 – 5 Hz. Their observations were based on the analyses of data recorded during local and regional earthquakes at the surface and in depth of two rock sites (Pinyon Flat and Keenwild) in California. An excellent treatise on the seismic response of rock sites can be found in the work of Cranswick [121].

3.3.5 VARIABLE SITE CONDITIONS

The majority of spatial arrays are located at uniform site conditions, and, essentially, all available coherency models have been developed for such uniform sites. In 1995, a temporary dense array of digital seismographs was deployed in the Parkway Valley, Wainuiomata, New Zealand [506], [507]. This flat-floored valley has a width of approximately 400 m and is surrounded by greywacke outcrops. The configuration of the array and the Parkway Valley are shown in Fig. 3.11, with the line indicating the boundary of the soft soil sediments. Four of the array stations (stations 22-25) were deployed on the weathered greywacke, and another 19 stations (stations 2-12 and 14-21) on the soft sediments of the valley; station 13 was never installed. Station 1 was located on hard rock at approximately 2 km to the north-east of the basin. The data recorded at stations 2-25 of the dense array permit the investigation of spatial coherency at sites with irregular subsurface topography at distances pertinent for engineering applications: The minimum and maximum station separation distances between these stations were 22.8 m and 665.7 m, respectively.

Liao *et al.* [298], [303] evaluated coherency estimates from data recorded during two events at the site. The results of their analyses for the earthquake of August 11th, 1995 (magnitude 4.9, depth 28 km and epicentral distance 80 km from the array), recorded at 17 stations are presented herein to illustrate the effect of variable site

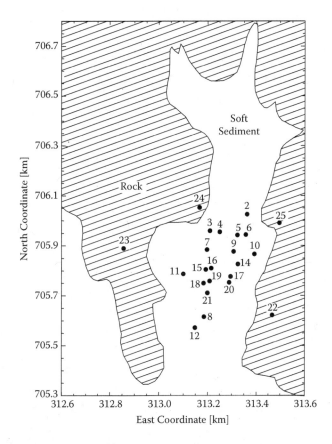

FIGURE 3.11 The Parkway, Wainuiomata, New Zealand, array, after Stephenson [507]. Stations 2-21 are located on the soft sediments of the valley and stations 22-25 on the surrounding weathered greywacke; station 13 was never installed. The line in the figure indicates the boundary between the soft soil sediments and the underlying rock. (Reproduced from *Soil Dynamics and Earthquake Engineering*, Vol. 27, W.R. Stephenson, "Visualisation of resonant basin response at the Parkway array, New Zealand," pp. 487–496, Copyright ©2007, with permission from Elsevier; courtesy of W.R. Stephenson.)

conditions and irregular subsurface topography on coherency estimates. Figures 3.12 and 3.13 present the lagged coherency evaluated from a 2.5 sec S-wave window of the east-west component of the motions. A moving-window frequency-wavenumber analysis of this component of the data recorded at the valley stations suggested that, during the S-wave window, the motions propagate down the valley. The velocity seismograms at the rock stations were not considered in this evaluation, because they would contaminate the identification of the propagation pattern of the waves in the valley. Figure 3.12 presents the lagged coherency of data recorded at close-by (separation distances of less than 60 m) and further-away (separation distances of 200 – 270 m) station pairs, whose orientation (line connecting the two stations) is along the direction of the valley (Fig. 3.11); these are station pairs 17 & 20, 19 & 21 and 16 & 19 for the shorter distance range (Fig. 3.12(a)) and station pairs 3 & 11,

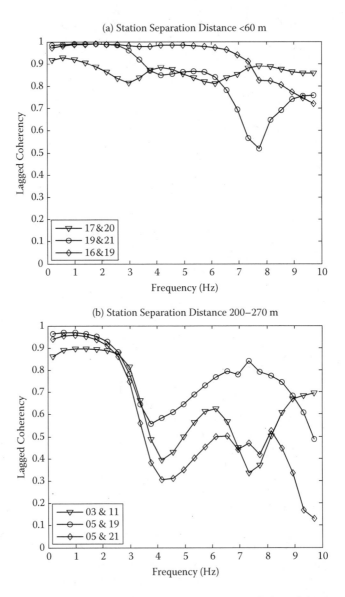

FIGURE 3.12 Lagged coherency of the strong motion S-wave window of the east-west component of the motions recorded at the Parkway, Wainuiomata, New Zealand, array evaluated by Liao *et al.* [298], [303]. All stations are located in the valley and the orientation of the station pairs is along the direction of the valley (Fig. 3.11). Part (a) presents coherency estimates at shorter (< 60 m) and part (b) at longer (200 – 270 m) station separation distances (after Liao [298]).

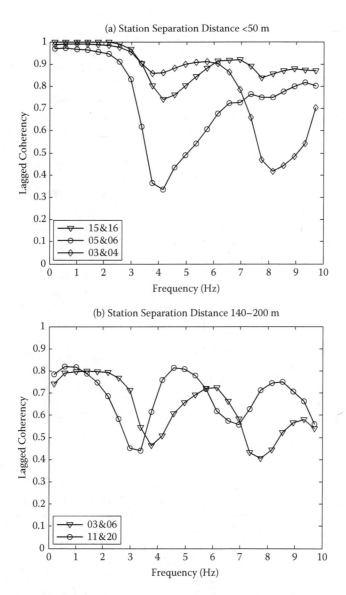

FIGURE 3.13 Lagged coherency of the strong motion S-wave window of the east-west component of the motions recorded at the Parkway, Wainuiomata, New Zealand, array evaluated by Liao *et al.* [298], [303]. All stations are located in the valley and the orientation of the station pairs is across the direction of the valley (Fig. 3.11). Part (a) presents coherency estimates at shorter (< 50 m) and part (b) at longer (140 – 200 m) station separation distances (after Liao [298]).

5 & 19 and 5 & 21 for the longer distance range (Fig. 3.12(b)). Figure 3.13 also presents the lagged coherency of data recorded at close-by (separation distances of less than 50 m) and further-away (separation distances of 140 – 200 m) station pairs, whose orientation, however, is across the direction of the valley (Fig. 3.11); these are station pairs 15 & 16, 5 & 6 and 3 & 4 for the shorter distance range (Fig. 3.13(a)) and station pairs 3 & 6 and 11 & 20 for the longer distance range (Fig. 3.13(b)). For the shorter separation distance range (Figs. 3.12(a) and 3.13(a)), coherencies, overall, assume high values and resemble, in trend, the soil site coherencies developed by Schneider *et al.* [449] in Fig. 3.9. It is noted however, that the frequency range in Fig. 3.9 is up to 20 Hz, whereas, in Figs. 3.12 and 3.13, it is up to 10 Hz. For the longer station separation distance range, the behavior of the lagged coherency for the two station-pair orientations differs: For the station-pair orientation along the valley (Fig. 3.12(b)), coherencies are high in the low frequency range and drop significantly between 3 – 4 Hz. On the other hand, for the station-pair orientation across the valley (Fig. 3.13(b)), coherencies are lower in the lower frequency range, even though the separation distance between these stations (140 – 200 m) is shorter than the separation distance of the stations in Fig. 3.12(b) (200 – 270 m), and tend to have a "flat" trend. Similar differences were observed in the coherency estimates of the north-south component of the motions during the S-wave window, both the north-south and east-west components of the motions during a later-arriving surface-wave window of this earthquake, as well as coherency estimates evaluated from the data of an additional earthquake recorded at the array [298]. It appears then that, as long as the analysis of the two events characterizes the pattern of coherency loss at this array, the assumption of isotropy is not valid for the data recorded at the Parkway Valley. Additionally, it appears that the loss of coherency of the data at the Parkway Valley is greater than the loss of coherency of the data at the SMART-1 array: Figure 2.20(a) presents the loss of coherency of the north-south component of the motions recorded during Event 5 at the SMART-1 array for a station separation distance of 200 m, i.e., comparable to the separation distances of the station pairs of Figs. 3.12(b) and 3.13(b). It can be clearly seen from the comparison of Figs. 2.20(a), 3.12(b) and 3.13(b) that the motions at the alluvial site of the SMART-1 array are more highly correlated than those at the narrow Parkway Valley. It is also noted that, in Fig. 2.20(a), the coherencies for two different station-pair orientations were grouped together, as the line connecting stations I03, C00 and I06 is perpendicular to the line connecting stations I06, C00 and I12 (Fig. 1.2). However, no clear anisotropy trend can be observed from the data at the SMART-1 array in Fig. 2.20(a), as is the case for the data of the Parkway Valley in Figs. 3.12 and 3.13. The differences in the behavior of the coherency at the two sites can be attributed to the complex wave propagation pattern that occurs at sites with irregular subsurface topography, such as the narrow alluvial basin of the Parkway Valley. A visualization of this valley's response as moving pictures of the ground motion in various frequency bands has been recently presented by Stephenson [507].

Figure 3.14 presents the lagged coherency between the data at a station close to the "center" of the valley (station 19) and the data recorded at the rock stations (stations 22-25) of the Parkway Valley array (Fig. 3.11) during the August 11th, 1995, event. Part (a) of the figure shows the lagged coherency for the east-west component of the

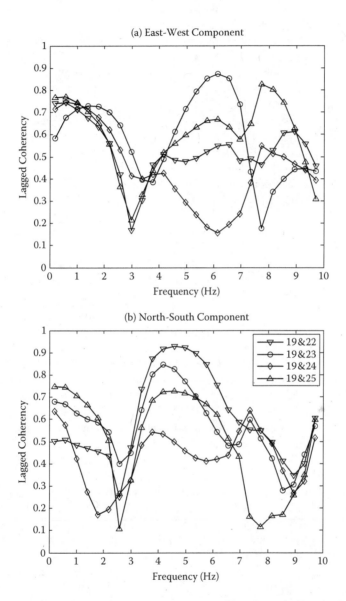

FIGURE 3.14 Lagged coherency between the data recorded at station 19 located in the valley and stations 22 - 25 located on the surrounding rock of the Parkway, Wainuiomata, New Zealand, array (Fig. 3.11) during the strong motion S-wave window evaluated by Liao *et al.* [298], [303]. The separation distance between stations is in the range of 280 – 380 m. Part (a) of the figure presents the coherency estimates of the east-west component of the motions, and part (b) those of the north-south component (after Liao [298]).

motions during the strong motion S-wave window, and part (b) the corresponding coherencies of the north-south component. The range of separation distances is 280 – 380 m. The figure indicates that the motions in both directions are only partially correlated throughout the entire frequency range, which can be, again, attributed to the complex wave propagation pattern at the site. Later in this chapter, Fig. 3.26(b) will present coherency estimates of the north-south component of the motions recorded at the SMART-1 array during Event 5 for comparable station separation distances (400 m). The comparison of Figs. 3.14 and 3.26(b) clearly indicates that the uniform site coherencies are significantly different than those between rock and soil at narrow sediment sites, and that coherency models developed at uniform soil sites cannot be extrapolated to sites with irregular subsurface topography. This observation can have significant consequences for the seismic response of bridges with supports at variable site conditions, as, e.g., rock outcrop for the abutments and sedimentary material for some of the piers, as will be further discussed in Sections 7.3.2, 9.4 and 9.5.

3.3.6 SHORTER AND LONGER SEPARATION DISTANCES

Another characteristic of coherency, which does not become readily apparent, is that its decay is different for shorter and longer separation distances between the recording stations. Data from dense spatial arrays, such as the LSST array (Fig. 1.3), and large arrays, such as the SMART-1 array (Fig. 1.2), recording the same earthquake permitted the evaluation of coherencies at shorter and longer station separation distances [13]. The range of station separation distances for the LSST array is $6 – 85$ m[1], and, for the SMART-1 array, it is $100 – 4000$ m. Figure 3.15 presents the coherency decay of the data recorded during the same earthquakes at the LSST and the SMART-1 arrays as reported by Abrahamson [13]. Part (a) in the figure presents the results from the data of the LSST–Event 7 / SMART-1–Event 40 earthquake ($m_b = 6.1$, depth 16 km and distance from the array 64 km), and part (b) those for the LSST–Event 12 / SMART-1–Event 43 earthquake ($m_b = 5.6$, depth 2 km and distance 4 km). It can be seen from the figure that the decay of the lagged coherency with separation distance at four discrete frequencies of 1.0, 2.0, 3.9 and 7.8 Hz varies between the two arrays. Abrahamson et al. [14] further reported that their extrapolation of the SMART-1 coherencies to shorter separation distances tended to overestimate the true coherency values obtained from the LSST data. Riepl et al. [425] made a similar observation from their analyses of an extensive set of weak motion data recorded at the EUROSEIS site close to Thessaloniki in northern Greece: The loss of coherency with distance for their data was marked by a "cross-over" distance, that distinguished coherency for shorter ($8 – 100$ m) and longer ($100 – 5500$ m) separation distances.

Another point worth noting regarding the behavior of coherency at shorter and longer separation distances is the following: Abrahamson et al. [12], Schneider et al. [449] and Vernon et al. [556] from analyses of data at close separation distances

---

[1] Even though the shortest separation distance between stations on each arm of the array was 3.05 m (Fig. 1.3), Abrahamson et al. [13], [14] indicated that the recorded time histories at the stations closest to the 1/4 scale model were contaminated by soil-structure interaction effects, and, hence, were not considered in the coherency evaluations.

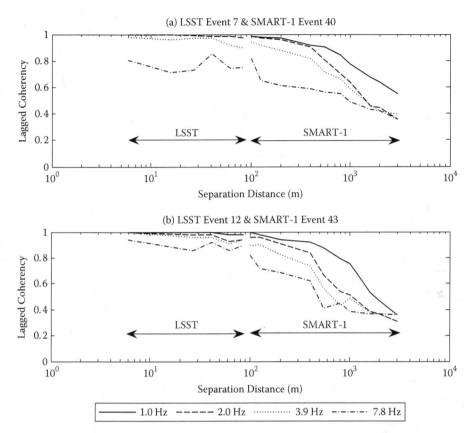

FIGURE 3.15 Illustration of the different behavior of nonparametric lagged coherency at shorter and longer separation distances reported by Abrahamson *et al.* [13]. LSST data were used for the coherency estimates at the shorter separation distance range (6 – 85 m), and SMART-1 data for the longer separation distances (100 – 3000 m). The data are from the same earthquakes recorded at both arrays: Part (a) presents the results for LSST – Event 7/SMART-1 – Event 40, and part (b) those for LSST – Event 12/SMART-1 – Event 43. (Reproduced from *Structural Safety*, Vol. 10, N.A. Abrahamson, J.F. Schneider and J.C. Stepp, "Spatial coherency of shear waves from the Lotung, Taiwan large-scale seismic test," pp. 145–162, Copyright © 1991, with permission from Elsevier.)

(< 100 m) observed that coherency is independent of wavelength, and decays faster with frequency than with separation distance. On the other hand, independent studies of data at longer separation distances (> 100 m) by Novak and his coworkers [375], [418] and Toksöz *et al.* [524] observed that the decay of coherency with separation distance and frequency is the same: Novak [375] and Ramadan and Novak [418], using data from the 1979 Imperial Valley earthquake and Events 20 and 40 at the SMART-1 array, observed that if coherency is plotted as function of a normalized separation distance with wavelength (ξ/λ), with $\lambda = V_s/f$ being the wavelength and V_s an appropriate S-wave velocity, then the coherency plots at various frequencies collapse onto the same curve. Similarly, Toksöz *et al.* [524], from their analysis of data recorded at three arrays in Fennoscandia, described in Section 3.3.4, noted that the curves of the coherency decay with frequency were very similar if the separation distance was scaled with wavelength. The observed coherency dependence on wavelength for the longer separation distances implies that coherency is a function of the product $(\xi\, \omega)$, i.e., that its decay with separation distance and frequency is the same. It appears then that different factors control the loss of coherency of seismic data at shorter and longer separation distances.

3.4 PARAMETRIC COHERENCY MODELING

A mathematical description for the coherency was first introduced in earthquake engineering by Novak and Hindy [376] and Hindy and Novak [221] in 1979–1980, in their pioneering evaluation of the response of a lifeline system (buried pipeline) subjected to seismic motions experiencing loss of coherency by means of random vibrations. The expression, based on wind engineering, was:

$$|\gamma(\xi, \omega)| = \exp\left[-\kappa \left(\frac{\omega\xi}{V_s}\right)^{\nu}\right] \tag{3.9}$$

where κ and ν are constants and V_s is an appropriate S-wave velocity.

The next three subsections present some of the reported empirical, semi-empirical and analytical models for the description of the spatial variation of the seismic ground motions. It is emphasized, however, that the list of models presented is not exhaustive. Throughout this section, the units for the frequency ω will be in [rad/sec], those for the cyclic frequency f in [Hz], and the units for the separation distance, ξ, will be [m] unless otherwise noted. It should also be mentioned that all parametric coherency models are real-valued and positive, and, hence, the absolute value in their notation is redundant; it is kept, however, to clearly distinguish the complex coherency, $\gamma(\xi, \omega) = |\gamma(\xi, \omega)| \exp[i\vartheta(\xi, \omega)]$, from the lagged coherency, $|\gamma(\xi, \omega)|$.

3.4.1 EMPIRICAL COHERENCY MODELS

Because of (i) the variability in seismic data recorded at different sites and during different events; (ii) the differences in the numerical processing of the data used by various investigators; and (iii) the different functional forms used in the regression fitting of a model to data with large scatter, there is a multitude of spatial variability expressions in the literature, as illustrated by the following examples.

Coherency Models from SMART-1 Data

A variety of parametric coherency functional forms have been proposed since data became available at the SMART-1 array, as, e.g., reported by Loh and his co-workers [312], [316], [318]:

$$|\gamma(\xi, \omega)| = \exp(-a(\omega)\xi)$$

$$|\gamma(\xi, \omega)| = \exp\left(-a\frac{\omega\xi}{2\pi c}\right)$$

$$|\gamma(\xi, \omega)| = \exp((-a - b\omega^2)\xi) \tag{3.10}$$

$$|\gamma(\xi, \omega)| = \exp((-a - b\omega)\xi^\nu)$$

with parameters to be determined from the data. Loh and Lin [316], however, suggested that the functional forms of Eqs. 3.10 are valid for the description of one-dimensional (isotropic) coherency estimates. As indicated in Section 3.3.3, their evaluation of isocoherence maps for two SMART-1 events showed anisotropy in the estimates. Based on these observations, they proposed a directionally-dependent coherency model that had the form:

$$|\gamma(\xi, \omega, \theta)| = \exp((-a_1 - b_1\omega^2)|\xi \cos\theta|) \exp((-a_2 - b_2\omega^2)|\xi \sin\theta|) \tag{3.11}$$

with θ indicating the angle between the direction of propagation of the waves and the vector of station separation. Loh and Lin [316] reported the following values for the parameters in the above equation: for Event 40, $a_1 = 0.02, b_1 = 0.005, a_2 = 0.02$ and $b_2 = 0.0011$, and for Event 45, $a_1 = 0.02, b_1 = 0.0025, a_2 = 0.02$ and $b_2 = 0.0012$; in the model of Eq. 3.11, the separation distance, ξ, is measured in [km].

Harichandran and Vanmarcke [208] evaluated nonparametric lagged coherencies from four events at the SMART-1 array, but noted that the assumption of isotropy was valid for the data. Based on this observation, they introduced a coherency model of the form:

$$|\gamma(\xi, f)| = A \exp\left(-\frac{2B\xi}{a\nu(f)}\right) + (1 - A)\exp\left(-\frac{2B\xi}{\nu(f)}\right) \tag{3.12}$$

$$\nu(f) = k\left[1 + \left(\frac{f}{f_0}\right)^b\right]^{-1/2}; \quad B = (1 - A + aA)$$

which is, perhaps, one of the most widely used empirical coherency models in the literature. The model was fitted by a weighted least-squares scheme to the unsmoothed real part of the coherency of the aligned time histories of Event 20 in the radial direction. The values of the identified model parameters were $A = 0.736, a = 0.147$, $k = 5210$ m, $f_0 = 1.09$ Hz and $b = 2.78$. Figure 3.16 presents the variation of the coherency model of Harichandran and Vanmarcke [208] with frequency at separation distances of 100, 300 and 500 m.

Later, Harichandran [205], [206] re-analyzed the data of Event 20, as well as those of Event 24, in both radial and tangential directions, and estimated the parameters of the Kanai-Tajimi (Eq. 3.5) and the Clough-Penzien (Eq. 3.7) spectra, illustrated earlier in Section 3.1, as well as new parameters for the coherency function of Eq. 3.12. In this evaluation, the model of Harichandran and Vanmarcke [208] was fitted to the arctanh transformation of the nonparametric coherency, the properties of which were discussed in Section 2.4.2. Table 3.4 presents the values of the parameters of Eq. 3.12 reported by Harichandran [206] for the radial and tangential components of the motions during the two events. Harichandran [206] pointed out that, in cases, the model identifies $a \to 0$ and $k \to \infty$, but a finite value for their product, in order to fit the estimated nonparametric coherency at long separation distances and high frequencies.

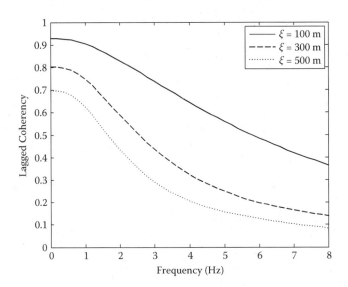

FIGURE 3.16 Variation of the coherency model of Harichandran and Vanmarcke [208] with frequency at separation distances of $\xi = 100, 300$ and 500 m. The functional form of the model is given by Eq. 3.12 and its parameters were determined from the radial component of the data recorded at the SMART-1 array during Event 20.

TABLE 3.4
Parameters of the Harichandran and Vanmacke [208] coherency model (Eq. 3.12) for the radial and tangential components of Events 20 and 24 at the SMART-1 array evaluated by Harichandran [206]. (Reproduced from *Structural Safety*, Vol. 10, R.S. Harichandran, "Estimating the spatial variation of earthquake ground motion from dense array recordings," pp. 219–233, Copyright ©1991, with permission from Elsevier.)

	Direction	A	a	k [m]	f_0 [Hz]	b	ak
Event 20	*Radial*	0.636	0.0186	31200	1.51	2.95	580
	Tangential	0.706	0.00263	257300	0.68	2.15	677
Event 24	*Radial*	0.481	0	∞	0.87	3.41	1992
	Tangential	0.618	0.0173	50100	1.97	5.49	867

Equation 3.12 then degenerates to the form:

$$|\gamma(\xi, f)| = A \exp \left\{ -\frac{2\xi(1-A)}{(a\,k)} \left[1 + \left(\frac{f}{f_0} \right)^b \right]^{1/2} \right\} + (1-A) \qquad (3.13)$$

which involves one less parameter than the original model, as the previously independent parameters a and k are now combined through their product. The radial component of the motions for Event 24 (Table 3.4) is described by this revised model.

The coherency model of Harichandran and Vanmarcke (Eq. 3.12) with the parameters subsequently estimated by Harichandran [206] for the two SMART-1 events (Table 3.4) is presented in Fig. 3.17. Part (a) of the figure shows the lagged coherency of the radial component of the motions and part (b) the lagged coherency of the tangential component as functions of frequency for station separation distances of 100, 300, and 500 m. The comparison of Figs. 3.16 and 3.17(a) for the parametric coherency of the radial component of Event 20 suggests that the two identification schemes produce similar values for the coherency at the low frequencies, but the coherency in Fig. 3.17(a) decays more slowly than the coherency in Fig. 3.16. The differences can be attributed to the different estimates of the nonparametric coherency (real part of coherency vs. arctanh transformation of lagged coherency), that were utilized in the identification process of the parameters in the two approaches. Figure 3.17 also indicates that the parametric coherencies estimated from the data of the two events at the same site exhibit differences in their exponential decay with frequency and separation distance. The model also appears to result in a "corner" frequency for the coherency (Figs. 3.16 and 3.17), which is more pronounced for the tangential component of Event 24 (Fig. 3.17(b)): For frequencies lower than this corner frequency, the coherency values are essentially constant, but, past this frequency, they tend to decrease.

Some general remarks regarding the behavior of parametric lagged coherency models as $\xi \to 0$ and $\omega \to 0$ can be deduced from Eq. 3.12 and Figs. 3.16 and 3.17: It is noted that, correctly, the value of the model is identically equal to one as the separation distance becomes zero, since the motions at a single station should be fully correlated with themselves. On the other hand, even though, theoretically, the value of the coherency should tend to unity as $\omega \to 0$, the model is only partially correlated at zero frequencies. It is recalled, that nonparametric lagged coherency estimates (Eq. 2.70) are evaluated from spectra that are smoothed with a $(2M+1)$-point window. Consequently, the first available nonparametric estimate for the coherency will not be at $\omega = 0$ but at $\omega = M\,\Delta\omega$, with $\Delta\omega$ being the frequency step. Since, due to averaging, the value of the smoothed nonparametric coherency will rarely be equal to one (Section 2.4.2), parametric coherency models that permit values for the coherency other than unity at zero frequencies will result in partial correlation of the motions in the low frequency range. It is noted, however, that Eq. 3.12 assumes a value that is not equal to one (unless its parameters are "forced" to yield a value of one) when frequency is equal to zero. Hence, whether the model is used with unsmoothed nonparametric estimates, as was the case for the development of the model in Fig. 3.16 and will be the case in a subsequent evaluation herein (Section 3.5), or smoothed estimates,

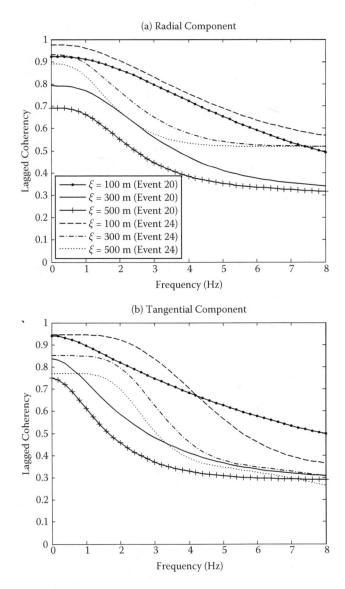

FIGURE 3.17 Variation of the coherency model of Harichandran and Vanmarcke [208] as modified by Harichandran [206] with frequency at separation distances of $\xi = 100, 300$ and 500 m. The functional form of the model is given by Eqs. 3.12 and 3.13, and its parameters were determined from the data recorded at the SMART-1 array during Events 20 and 24 (Table 3.4). Part (a) presents the parametric models for the radial component of the data, and part (b) those for the tangential component.

which were used in the evaluation of the models in Fig. 3.17, coherency will tend to a value that is less than one as $\omega \to 0$.

Hao et al. [200] conducted a thorough investigation of data recorded at the SMART-1 array during Event 24 (one of the events also analyzed by Harichandran [206]) and Event 45. Their extensive study considered the ground motion components in all three directions (east-west, north-south and vertical) during various time windows, and included frequency-wavenumber analyses for the evaluation of the wave propagation pattern of the motions, identification of the parameters of the Kanai-Tajimi spectrum (Eq. 3.5), which were highlighted in Section 3.1, time envelope functions to characterize the non-stationarity of the data (Section 7.1.2), nonparametric spatial coherency estimates and nonparametric directional coherency estimates, i.e., coherency estimates between different ground motion components at the same station. Their investigation of the directional correlations of the data also suggested that their values are low, which is in agreement with observations made earlier in Section 3.3.2. Hao et al. [200] proposed the following functional form for the spatial coherency, which considers that the ground motion random field is anisotropic:

$$|\gamma(\xi_l, \xi_t, f)| = \exp(-\beta_1\xi_l - \beta_2\xi_t)\,\exp\{-[\alpha_1(f)\sqrt{\xi_l} + \alpha_2(f)\sqrt{\xi_t}\,]f^2\} \quad (3.14)$$

with f being frequency, and ξ_l and ξ_t, as indicated in Section 3.3.3, the projected distances of the station separation vector along and normal to the direction of propagation of the motions. Hao et al. [200] estimated the values of β_i, $i = 1, 2$, for the three components of the motions during each event, but provided only graphically the variation of the functions $\alpha_i(f)$, $i = 1, 2$, with frequency. It should be noted that, in this same comprehensive publication, Hao et al. [200] also presented one of the very first approaches for the simulation of spatially variable seismic ground motions incorporating non-stationarity and response-spectrum compatibility in the generated time series, which will be discussed in Chapter 7.

Later, Oliveira et al. [382] proposed a functional form for the variation of $\alpha_i(f)$, $i = 1, 2$, with frequency as:

$$\alpha_i(f) = a_i/f + b_i\,f + c_i; \quad 0.05\,\text{Hz} \le f \le 10\,\text{Hz}; \quad i = 1, 2; \quad (3.15)$$
$$\alpha_i(f < 0.05\,\text{Hz}) = \alpha_i(f = 0.05\,\text{Hz}); \quad \alpha_i(f > 10\,\text{Hz}) = \alpha_i(f = 10\,\text{Hz})$$

and identified the eight parameters of the model (Eqs. 3.14 and 3.15) for 17 events recorded at the SMART-1 array. For the estimation of the nonparametric coherency, Oliveira et al. [382] utilized the entire time series of each component of the motions and each event. In this way, however, the estimated coherency incorporates the effect of the spatial correlation between various wave types that dominate the motions during different time windows, whereas the conventional estimation of the coherency reflects the spatial correlation of a single wave type that dominates the motions over a shorter time window. Oliveira et al. [382] used a 9-point triangular window for the smoothing of the nonparametric spectral estimates, and, before the identification of the model parameters, performed an additional smoothing with a 5-th order polynomial. For illustration purposes, the parameters of Eqs. 3.14 and 3.15 identified by Oliveira et al. [382] for Events 24 and 45, which were earlier analyzed by Hao et al. [200], are presented in Table 3.5 and the resulting coherency functions are plotted in Fig. 3.18.

TABLE 3.5
Parameters of the Hao *et al.* [200] and Oliveira *et al.* [382] coherency model (Eqs. 3.14 and 3.15) for Events 24 and 45 at the SMART-1 array evaluated by Oliveira *et al.* [382]. (Reproduced from *Structural Safety*, Vol. 10, C.S. Oliveira, H. Hao and J. Penzien, "Ground motion modeling for multiple-input structural analysis," pp. 79–93, Copyright © 1991, with permission from Elsevier.)

Event	β_1 ($\times 10^{-4}$)	β_2 ($\times 10^{-5}$)	a_1 ($\times 10^{-3}$)	b_1 ($\times 10^{-6}$)	c_1 ($\times 10^{-5}$)	a_2 ($\times 10^{-3}$)	b_2 ($\times 10^{-6}$)	c_2 ($\times 10^{-4}$)
24	2.622	12.11	3.113	−6.635	2.042	3.286	2.590	−1.050
45	1.109	6.730	3.853	−18.11	11.77	5.163	−7.583	−1.905

The lagged coherency plots in Fig. 3.18(a) were evaluated considering that the stations are located on a line along the direction of propagation of the motions, i.e., $\xi_l = \xi$ and $\xi_t = 0$ in Eq. 3.14, and those in Fig. 3.18(b) considering that the stations are located on a line perpendicular to the direction of propagation of the motions, i.e., $\xi_l = 0$ and $\xi_t = \xi$.

Figure 3.18 suggests that the values of the coherency for the two directions of station orientation and for the two events are very similar, even though the identified parameters for each direction of station orientation and for each event vary considerably (Table 3.5). Furthermore, the figure indicates that coherency is high throughout the entire frequency range plotted in the figure, and, also, remains high at higher frequencies, as can be deduced from Eq. 3.15. The parametric coherencies estimated in this study [382] for the additional 15 events exhibited a pattern similar to that of Fig. 3.18, as observed by Yang and Cheng [574]. There are two possible reasons for the causes underlying the behavior of this model: The first is that, if coherency is independent of earthquake magnitude and source-site distance, then it is affected only by path and local scattering effects (Section 3.3.1). Since all data used in the Oliveira *et al.* [382] investigation have been recorded at the same array, the coherency estimates from all events should be, considering the contribution of local scattering effects only, similar. It is noted, however, that the coherency models of Fig. 3.17 developed by Harichandran [206] for two events at the SMART-1 array exhibited differences in their exponential decay. The second reason for the trend of the model of Oliveira *et al.* [382] in Fig. 3.18 may be due to the use of the entire time histories in the coherency estimation, which conceals the information provided by the data about the coherency of a single wave type dominating the motions over shorter time windows, and, possibly, the utilized smoothing approach and the selected functional form for the model (Eqs. 3.14 and 3.15).

Figures 3.17 and 3.18 also permit the comparison of coherency models evaluated from the same data (Event 24) by different investigators: Different processing of the data, different approaches for the estimation of the nonparametric coherency, and different functional forms for the coherency models can yield a different picture of the exponential decay of the coherency at a site.

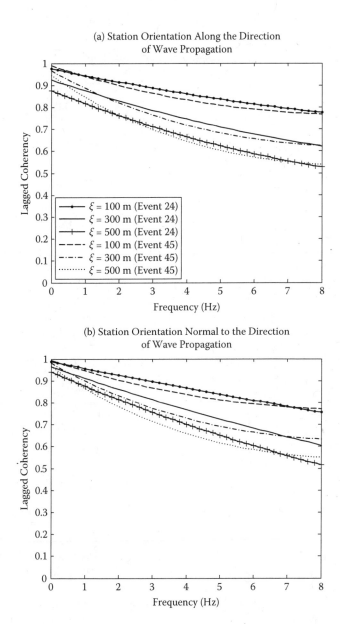

FIGURE 3.18 Variation of the (anisotropic) coherency model of Hao *et al.* [200] as modified by Oliveira *et al.* [382] with frequency at separation distances of $\xi = 100, 300$ and 500 m. The functional form of the model is given by Eqs. 3.14 and 3.15, and its parameters were determined from the data recorded at the SMART-1 array during Events 24 and 45 (Table 3.5). Part (a) presents the variation of the coherency for station-pair orientation along the direction of propagation of the motions, and part (b) for station-pair orientation perpendicular to the direction of propagation of the motions.

Coherency and Amplitude Variability at Short Separation Distances (LSST Data)

Abrahamson et al. [12], [14] followed the rationale that coherency is independent of earthquake magnitude and source-site distance (Section 3.3.1), and suggested that nonparametric coherency estimates evaluated at a site from a number of events can be grouped together for the evaluation of a single parametric coherency model. Based on nonlinear regression analyses of lagged nonparametric coherencies estimated from 15 events recorded at the LSST array (Fig. 1.3), the characteristics of which are presented in the first row of Table 3.3, they proposed the following coherency model for short (≤ 100 m) separation distances:

$$|\gamma(\xi, f)| = \tanh \left\{ (2.54 - 0.012\xi) \left[\exp[(-0.115 - 0.00084\xi)f] \right. \right.$$
$$\left. \left. + \frac{f^{-0.878}}{3} \right] + 0.35 \right\} \qquad (3.16)$$

For the evaluation of the parameters in the above equation, the arctanh transformation of the nonparametric lagged coherency was utilized, which, as indicated in Section 2.4.2, has an approximately constant variance. The constant term (0.35) within the hyperbolic tangent in the equation reflects the effect of noise that is present in the data, and its value was estimated from the median of the nonparametric lagged coherency at frequencies above 50 Hz. Abrahamson et al. [12] reported that the standard errors of Eq. 3.16 between station pairs for the same event (intra-event coherency) and between events for the same station pairs (inter-event coherency) were comparable.

Abrahamson et al. [12], [14] further proposed a functional form for the plane-wave coherency (Eq. 2.102) of the data from the 15 events. To estimate the plane-wave coherency, the ratio of the unlagged plane-wave coherency, i.e., the real part of the complex plane-wave coherency of Eq. 2.101, over the lagged coherency was utilized. The unlagged coherency was evaluated before the time histories were aligned, so that it maintained the characteristics of the wave propagation across the array. The plane-wave coherency correction factor, $h(\xi, f)$ in Eq. 2.102, was then estimated from the expression [12], [14]:

$$\frac{\Re[\gamma(\xi, \xi^r, f)]}{|\gamma(\xi, f)|} = h(\xi, f) \cos \left(\frac{2\pi f \xi^r}{c} \right) \qquad (3.17)$$

where, as in Eqs. 2.101 and 2.102, c is the apparent propagation velocity of the plane wave, and ξ^r the separation distance between the two stations projected along the direction of wave propagation (with the subscripts jk omitted for convenience). From regression analysis of the data, the correction factor, $h(\xi, f)$, for the plane-wave coherency (Eq. 2.102 and 3.17) was found to be a function of frequency, but not of separation distance, as:

$$h(\xi, f) = \left[1 + \left(\frac{f}{19} \right)^4 \right]^{-1} \qquad (3.18)$$

and the apparent propagation velocity c (Eq. 3.17) was estimated as 2.7 km/sec, which was close to the average velocity of 2.17 km/sec evaluated from conventional frequency-wavenumber analyses (Section 4.1.1) of the data from each event.

Figure 3.19(a) presents the lagged coherency and the plane-wave coherency of the LSST data at separation distances of 10, 50 and 100 m. It is noted that this coherency model was evaluated for frequencies greater than 1 Hz and separation distances of 6 m $\leq \xi \leq$ 85 m. Abrahamson *et al.* [12] indicated that the model can be

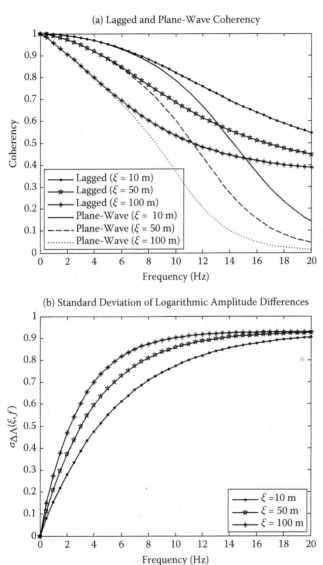

FIGURE 3.19 Part (a) presents the variation with frequency of the lagged and plane-wave coherency models developed by Abrahamson *et al.* [12], [14] from 15 earthquakes recorded at the LSST array. The functional form of the model, which is valid for short station separation distances (< 100 m), is given by Eqs. 2.102, 3.16 and 3.18. Part (b) of the figure presents the model for the differential amplitude variability (Eq. 3.20) reported in the same studies and Ref. [449]. The variation of the models is shown in the subplots for separation distances of $\xi = 10, 50$ and 100 m.

extrapolated to zero frequency, and, as can be seen from Fig. 3.19(a), it tends to unity as $f \to 0$. They noted, however, that the constraint that coherency should tend to unity as $\xi \to 0$ was not enforced into their model, and, hence, it cannot be extrapolated to zero distance. They further noted that the model can be used up to a separation distance of 100 m, but its extrapolation to longer distances would result in the underestimation of the coherency. Regarding the trend of the lagged and the plane-wave coherencies, it can be seen from Fig. 3.19(a) that, at the lower frequencies, where the plane wave dominates the motions, the lagged and plane-wave coherencies are essentially identical ($h(\xi, f) \to 1$). As frequency increases, the lagged coherency, which incorporates the additional, scattered energy contained in the time series, tends to the median value of noise, whereas the plane-wave coherency tends to zero.

As indicated in Section 2.4.2, coherency reflects, essentially, the phase variability in the data. Even though, through the coherency, this phase variability has been widely investigated, the amplitude variability of the seismic motions has not attracted significant attention. In their study of the LSST data and in a subsequent study of the extensive set of data recorded at various arrays (presented herein in Tables 3.2 and 3.3), Abrahamson *et al.* [12] and Schneider et al. [449] examined the variation of the Fourier amplitudes of the motions. They defined $\Delta\Lambda_{jk}(f)$ as the difference of the natural logarithm of the Fourier amplitudes at two stations, $\Lambda_j(f)$ and $\Lambda_k(f)$, i.e.,

$$\Delta\Lambda_{jk}(f) = \ln[\Lambda_j(f)] - \ln[\Lambda_k(f)] \tag{3.19}$$

and evaluated the standard deviation, $\sigma_{\Delta\Lambda}(\xi, \omega)$, of this amplitude difference. The expression for the standard deviation derived from the 15 LSST events had the form [449]:

$$\sigma_{\Delta\Lambda}(\xi, f) = 0.93[1 - \exp(-0.16f - 0.0019f\xi)] \tag{3.20}$$

It is noted that the constraint that $\sigma_{\Delta\Lambda}(\xi, f)$ should tend to zero as $\xi \to 0$, which follows from the definition of $\Delta\Lambda_{jk}(\omega)$ in Eq. 3.19, was not enforced into Eq. 3.20. Figure 3.19(b) presents the variation of $\sigma_{\Delta\Lambda}(\xi, f)$ with frequency at station separation distances of 10, 50 and 100 m. The figure suggests that its value increases with frequency and station separation distance, and tends to a constant value at higher frequencies. The subsequent analyses of the data recorded at the rock and soil sites of the arrays presented in Tables 3.2 and 3.3 by Schneider *et al.* [449] also indicated that, at each site, $\sigma_{\Delta\Lambda}(\xi, f)$ increased with frequency and tended to constant values at higher frequencies.

Coherency from Records at Multiple Sites

Considering that coherency is independent of site conditions and grouping together nonparametric coherency estimates from an extensive set of data recorded at various sites, Abrahamson [7] presented single functional forms for the horizontal and vertical coherency. The data included both the SMART-1 and LSST data, as well as data from ten other arrays. In the model derivation, care was taken to accommodate the different behavior of coherency at shorter and longer station separation distances (Section 3.3.6), and, hence, the model covers both ranges of station separation distances. For the lagged coherency of the horizontal motions, the derived expression was:

$$|\gamma^H(\xi, f)| = \tanh \left\{ \frac{c_1^H(\xi)}{1 + c_2^H(\xi)f + c_4^H(\xi)f^2} \right.$$

$$\left. + \left(4.80 - c_1^H(\xi)\right) \exp\left[c_3^H(\xi)f\right] + 0.35 \right\} \quad (3.21)$$

where the functions $c_1^H(\xi), c_2^H(\xi), c_3^H(\xi)$ and $c_4^H(\xi)$ are given by:

$$c_1^H(\xi) = \frac{3.95}{1 + 0.0077\xi + 0.000023\xi^2} + 0.85 \exp[-0.00013\xi]$$

$$c_2^H(\xi) = \frac{0.4[1 - (1 + (\xi/5)^3)^{-1}]}{[1 + (\xi/190)^8][1 + (\xi/180)^3]}$$

$$c_3^H(\xi) = 3(\exp[-0.05\xi] - 1) - 0.0018\xi$$

$$c_4^H(\xi) = -0.598 + 0.106 \ln(\xi + 325) - 0.0151 \exp[-0.6\xi]$$

The correction factor, $h^H(\xi, f)$, for the evaluation of the plane-wave coherency (Eq. 2.102) was estimated as:

$$h^H(\xi, f) = \left[1 + \left(\frac{f}{c_5^H(\xi)}\right)^6\right]^{-1} \quad (3.22)$$

with

$$c_5^H(\xi) = \exp[8.54 - 1.07 \ln(\xi + 200)] + 100 \exp(-\xi)$$

For the vertical motions, the lagged coherency assumed the form [7]:

$$|\gamma^V(\xi, \omega)| = \tanh \left\{ \frac{c_1^V(\xi)}{1 + c_2^V(\xi)f} + \left(4.65 - c_1^V(\xi)\right) \exp\left[c_3^V(\xi)f\right] + 0.35 \right\} \quad (3.23)$$

where the functions $c_1^V(\xi), c_2^V(\xi)$ and $c_3^V(\xi)$ are given by:

$$c_1^V(\xi) = 3.5 - 0.37 \ln(\xi + 0.04)$$

$$c_2^V(\xi) = 0.65[1 - (1 + \xi/4)^{-1}]$$

$$c_3^V(\xi) = 3(\exp[-0.05\xi] - 1) - 0.0018\xi$$

and the correction factor, $h^V(\xi, f)$, for the evaluation of the plane-wave coherency (Eq. 2.102) was estimated as:

$$h^V(\xi, \omega) = \left[1 + \left(\frac{f}{c_5^V(\xi)}\right)^3\right]^{-1} \quad (3.24)$$

with

$$c_5^V(\xi) = \exp[5.20 - 0.634 \ln(\xi + 0.1)]$$

It is noted that Eq. 3.23 is one of the very few coherency expressions that have been developed for vertical motions.

Figures 3.20 and 3.21 present the variation with frequency at separation distances of $10, 50, 100, 300, 500$ and 1000 m of the horizontal and vertical coherencies, respectively, of the model developed by Abrahamson [7]. Part (a) in the figures presents the lagged coherency and part (b) the plane-wave coherency. As for the LSST coherency model in Fig. 3.19(a), Figs. 3.20(a) and 3.21(a) indicate that the lagged coherency tends to a constant value as frequency and separation distance increase, which is reflected by the constant (0.35) term in the hyperbolic tangent expression of Eqs. 3.21 and 3.23. The plane-wave coherency (Figs. 3.20(b) and 3.21(b)) decays to zero values, since, as discussed earlier in Fig. 3.19(a), it does not contain the contribution of the scattered energy in the motions. In this model, the constraints that the lagged coherency should tend to unity as frequency and/or separation distance tend to zero was enforced. As indicated in Section 3.3.4, however, since scattering at rock and soil sites is different, coherency estimates at different sites cannot be readily grouped together for the evaluation of a single coherency model. This being said, the following two observations should be made regarding the coherency model of Eqs. 3.21–3.24: The first observation is that, since the majority of the data utilized for the development of the model were data at "soil" sites, the exponential decay of the parametric model may not have been overly affected by the slower exponential decay of the data at the "rock" sites, and the model may be viewed as a coherency model for "soil" sites. The second observation is that, from the empirical models presented in this section, the model of Eqs. 3.21–3.24 is the most robust, as it tends to unity when frequency tends to zero (Figs. 3.20 and 3.21), takes into consideration the differences in the exponential decay of the coherency at shorter and longer station separation distances (Section 3.3.6), and, in the high frequency range, the parametric lagged coherency (Figs. 3.20(a) and 3.21(a)) tends to the coherency of noise smoothed with the smoothing window used in the evaluation.

Coherency at Rock Sites

The majority of coherency expressions (e.g., Eqs. 3.10–3.16) were developed for the alluvial site of the SMART-1 array. Coherency models at rock sites are limited. One of the few models reported in the literature was developed by Menke et al. [341] from data recorded at two arrays on hard rock sites at the Adironack mountains, as indicated earlier in Section 3.3.4. The model had the form:

$$|\gamma(\xi, f)| = \exp(-\alpha f \xi) \tag{3.25}$$

with α being in the range of $(0.4 - 0.7) \times 10^{-3}$ sec/m, and is valid for all three ground motion components (vertical, radial and tangential) and all epicentral azimuth and distance ranges.

Recently, Abrahamson [9] proposed a coherency model for rock sites based on 78 earthquakes recorded at the Pinyon Flat array (last entry in Table 3.2). The plane-wave

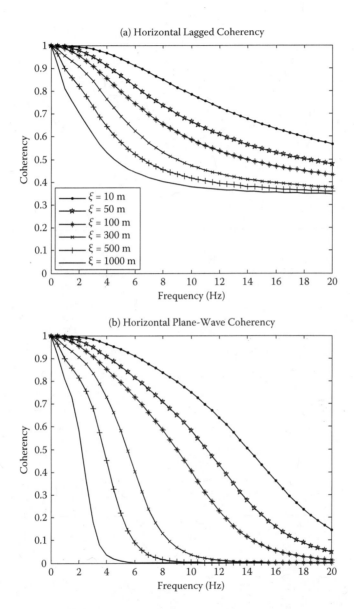

FIGURE 3.20 Horizontal lagged and plane-wave coherency, in parts (a) and (b), respectively, developed by Abrahamson [7] from data sets recorded at 12 arrays. The functional form of the models is given by Eqs. 2.102, 3.21 and 3.22. The models cover both shorter and longer separation distance ranges, and their variation with frequency is shown in the subplots for $\xi = 10, 50, 100, 300, 500$ and 1000 m.

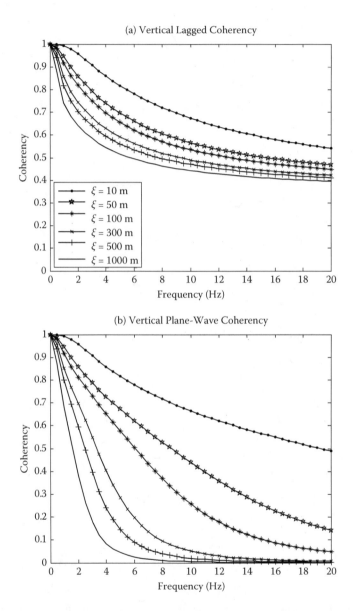

FIGURE 3.21 Vertical lagged and plane-wave coherency, in parts (a) and (b), respectively, developed by Abrahamson [7] from data sets recorded at 12 arrays. The functional form of the models is given by Eqs. 2.102, 3.23 and 3.24. The models cover both shorter and longer separation distance ranges, and their variation with frequency is shown in the subplots for $\xi = 10, 50, 100, 300, 500$ and 1000 m.

coherency was modeled by the following expression:

$$|\gamma^{pw}(\xi, f)| = \left[1 + \left(\frac{f \tanh(a_1\xi)}{f_c(\xi)}\right)^{n_1(\xi)}\right]^{-1/2} \left[1 + \left(\frac{f \tanh(a_1\xi)}{a_2}\right)^{n_2}\right]^{-1/2}$$

(3.26)

with the coefficients of the model for the horizontal motions being: $a_1 = 0.4, a_2 = 40$, $n_1(\xi) = 3.8 - 0.04 \ln(\xi + 1) + 0.0105[\ln(\xi + 1) - 3.6]^2$, $n_2 = 16.4$ and $f_c(\xi) = 27.9 - 4.82 \ln(\xi + 1) + 1.24[\ln(\xi + 1) - 3.6]^2$, and for the vertical motions: $a_1 = 0.4$, $a_2 = 200$, $n_1(\xi) = 2.03 + 0.41 \ln(\xi + 1) - 0.078[\ln(\xi + 1) - 3.6]^2$, $n_2 = 10$ and $f_c(\xi) = 29.2 - 5.20 \ln(\xi + 1) + 1.45[\ln(\xi + 1) - 3.6]^2$. The exponential decay of this model, as illustrated by the coherency of a more limited set of data at the site in Fig. 3.10, is flatter than the coherency models of Figs. 3.20(b) and 3.21(b) for the respective components of the motions.

3.4.2 SEMI-EMPIRICAL MODELS

Semi-empirical models for the spatial variation of the seismic ground motions, i.e., models for which their functional form is based on analytical considerations but their parameter evaluation requires recorded data, have also been introduced.

Somerville *et al.* [486] proposed a comprehensive approach for the site-specific estimation of the spatial variability of seismic ground motions. Their model, which was briefly described in Section 3.3.1 and illustrated in Fig. 3.8, attributes the causes of spatial variability to the wave propagation effect, the finite source effect, the effect of scattering of the seismic waves as they propagate from the source to the site, and the local site effects. The first two effects were grouped into the "source", $C_s(\xi, \omega)$, and the last two into the "scattering", $C_p(\xi, \omega)$, contribution to spatial variability, and are schematically presented in the top and bottom parts, respectively, of Fig. 3.8. In this semi-empirical model, the first contribution (top part of the figure) is evaluated from simulations, that take into account the effect of the rupture from an extended source and the interference of the waves arriving at the site from different subregions of the fault. The second contribution (bottom part of the figure) is evaluated from small earthquakes recorded at the site, for which the rupture area can be considered as a point source. This part incorporates the effects of scattering from the source to the site, local site effects as well as arrival time delays of the waveforms. The site-dependent coherency is determined as the product of the two contributions, with the wave propagation effect being removed from the scattering component so that it is not accounted for twice in the total coherency. Figure 3.22 presents an illustration of the trend of these contributions to spatial coherency, that have been obtained by Somerville *et al.* [486] from the application of their approach to the 1979 Imperial Valley earthquake at the El Centro differential array (Fig. 1.1). Part (a) presents the peak cross correlation of the source contribution bandpass filtered at 3.75, 7.5 and 15.0 Hz as functions of station separation distance; these cross correlation functions were evaluated from simulated motions of the mainshock of the 1979 Imperial Valley earthquake. The peak cross correlation functions in part (b) were evaluated from data recorded during an aftershock of the earthquake, those in part (c) from the

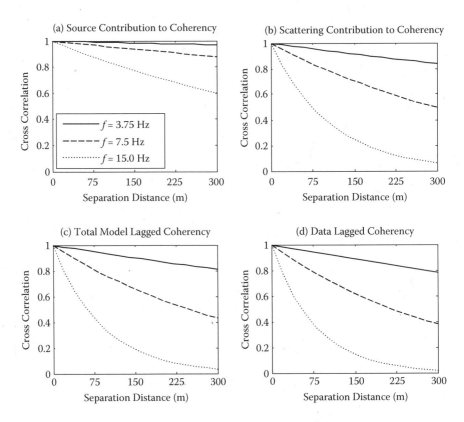

FIGURE 3.22 Illustration of the source and scattering contributions to spatial coherency (Fig. 3.8) of the model of Somerville *et al.* [486] applied to the 1979 Imperial Valley earthquake: Part (a) of the figure indicates the loss of coherency caused by the extended source effect; this contribution was evaluated from simulated ground motions. Part (b) presents the loss of coherency caused by scattering effects, and was determined from the data of an aftershock. Part (c) shows the model coherency resulting from the combined effects of parts (a) and (b), and part (d) the coherency of the data recorded during the mainshock of the earthquake. The coherency shown in the subplots of the figure was obtained from the parametric expressions provided by Somerville *et al.* [486] that fitted data with large scatter, and is presented as function of separation distance at three frequencies (3.75, 7.5 and 15.0 Hz); after Somerville *et al.* [486].

combined effect of the source and scattering contributions, and those in part (d) from the data recorded during the mainshock of this event. It is emphasized that the curves in Fig. 3.22, of functional form $C(\xi, \omega) = \exp[-(a + b\,\omega^2)\xi]$, were obtained by Somerville *et al.* [486] as a best fit to data with large scatter, and merely illustrate the trend of the contributions to spatial coherency. According to this model, the source contribution will always reduce the total coherency, and, as indicated by Somerville *et al.* [487], this contribution (Fig. 3.22(a)) is small. Further discussion on the effect of source finiteness on coherency was presented in Section 3.3.1.

The scattering of the waves along their path from the source to the site can also be evaluated by means of stochastic wave propagation [481], [549]. Luco and

Wong [324], based on the analysis of wave propagation through a random medium by Uscinski [549], presented perhaps the most quoted coherency model, which has the form:

$$|\gamma(\xi, \omega)| = e^{-(\frac{\nu \omega \xi}{V_s})^2} = e^{-\alpha^2 \omega^2 \xi^2} \tag{3.27}$$

$$\nu = \mu \left(\frac{R}{r_0}\right)^{1/2} ; \quad \alpha = \frac{\nu}{V_s}$$

where V_s is an estimate for the elastic S-wave velocity in the random medium, R is the distance in the random medium traveled by the wave, r_0 the scale length of random inhomogeneities along the path, and μ^2 a measure of the relative variation of the elastic properties in the medium. The coherency drop parameter, α, controls the exponential decay of the function: the higher the value of α, the more significant the loss of coherency as separation distance and frequency increase. With appropriate choices for the coherency drop parameter, the model has been shown to fit the spatial variation of recorded data, and has been used extensively by researchers in their evaluation of the seismic response of lifelines (Chapters 6 and 9). Figure 3.23 presents the exponential decay of the model with frequency at separation distances of $100, 300$ and 500 m for a median value $\alpha = 2.5 \times 10^{-4}$ sec/m from the ones suggested by Luco and Wong [324]. It is emphasized at this point that, even though α has the units of inverse velocity, it reflects not only the estimate of the inverse velocity but also the random properties of the medium (Eq. 3.27). It is also noted that the exponential decay with separation distance and frequency of this model is the same, which makes it more valid for the longer separation distance range (Section 3.3.6). Equation 3.27 also indicates that this analytically based model produces coherencies that are equal to unity at zero frequency for any separation distance, and, also, equal to unity for zero separation distance. It is further noted that, since the model contains the contribution of shear waves propagating through a random medium without explicitly considering the scattered energy/noise that is present in recorded data, especially in the higher frequency range, it may be viewed as a plane-wave coherency model. The 3-D version of this model is further described in Section 9.1, along with the work of Luco and Wong [324] on the effect of spatial variability on the response of large, mat, rigid foundations.

Der Kiureghian [135] developed a semi-empirical coherency model based on the theory of random processes. In the approach, the time histories at two stations k and l on the ground surface were considered to be stationary and expressed as:

$$a_k(t) = \sum_{i=1}^{N} A_i \cos(\omega_i t + \phi_i) \tag{3.28}$$

$$a_l(t) = \sum_{i=1}^{N} (p_{kl,i} A_i + q_{kl,i} B_i) \cos[\omega_i (t - \tau_{kl,i}) + \phi_i + \epsilon_{kl,i}] \tag{3.29}$$

Equation 3.28 is the discrete Fourier series of the time history at station k. In Eq. 3.29, B_i are zero-mean, uncorrelated random variables with mean-square values equal to those of A_i (Eq. 3.28), $\tau_{kl,i}$ is the arrival time delay of the wave component from station k to station l, $\epsilon_{kl,i}$ are zero-mean, independent, normally distributed random

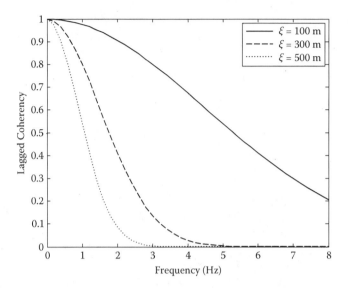

FIGURE 3.23 Variation of the coherency model of Luco and Wong [324] with frequency at separation distances of $\xi = 100, 300$ and 500 m. The functional form of the model is given by Eq. 3.27, and the value of the coherency drop parameter is $\alpha = 2.5 \times 10^{-4}$ sec/m.

phase differences with variance $\alpha^2(\xi_{kl}, \omega_i)$, and $p_{kl,i}$ and $q_{kl,i}$ are deterministic coefficients assuming values between $(0, 1)$ with $p_{kl,i}^2 + q_{kl,i}^2 = 1$. The subscripts kl, i in the parameters $p_{kl,i}$, $q_{kl,i}$, $\tau_{kl,i}$ and $\epsilon_{kl,i}$ indicate their dependence on the separation distance between the stations, ξ_{kl}, and frequency, ω_i. The lagged coherency (Eq. 2.70) of the two processes in Eqs. 3.28 and 3.29 was then evaluated as [135]:

$$|\gamma(\xi_{kl}, \omega_i)| = \cos[\beta(\xi_{kl}, \omega_i)] \exp\left\{ -\frac{1}{2}\alpha^2(\xi_{kl}, \omega_i) \right\} \qquad (3.30)$$

where $\beta(\xi_{kl}, \omega_i) = \tan^{-1}(q_{kl,i}/p_{kl,i})$ is an acute angle. Der Kiureghian [135] termed the above expression the "incoherence" part of the spatial coherency, and considered that it reflects the finite source and scattering effects noted earlier by Somerville *et al.* [486]. The first term on the right-hand side of the equation, $\cos[\beta(\xi_{kl}, \omega_i)]$, results from the assumed form of the amplitude variation at the further-away station (Eq. 3.29), whereas the second term reflects the effect of phase variability. $\cos[\beta(\xi_{kl}, \omega_i)]$ should assume values close to unity, since, as has been observed from the analysis of recorded data (e.g., Fig. 2.21), amplitude variability affects coherency only minimally. Indeed, Der Kiureghian [135] noted that the attenuation of the waveforms, that affects the amplitude of the motions, does not, essentially, contribute to the coherency estimate. The wave passage effect in this model is described by the delay in the arrival of the waves at the further-station ($\tau_{kl,i}$ in Eq. 3.29) caused by their propagation on the ground surface, i.e., $\tau_{kl,i} = \xi_{kl}/c(\omega_i)$, as discussed in the beginning of this chapter regarding Eq. 3.4 and, also, in Section 2.4.1, that presented the conventional estimation of the phase spectrum of the motions. To incorporate the effect of different soil conditions underneath the two stations k and l on spatial variability, Der

Kiureghian [135] considered a 1-D S-wave propagating vertically through two soil columns with different properties. In this way, the variability in the local site conditions does not contribute to the lagged coherency, but introduces only a time delay in the arrival time of waves at the two stations caused by their propagation through two soil columns with different properties. Section 3.3.5, however, illustrated that the lagged coherency of motions recorded at variable site conditions can be very different from the lagged coherency obtained from data recorded at uniform sites due to the complex propagation pattern of the waves at the irregular subsurface topography.

Later, Yang and Cheng [574] provided functional forms for $\alpha(\xi, f)$ and $\beta(\xi, f)$ of Eq. 3.30, with the subscripts k, l and i having been dropped for convenience, which, when substituted back into Eq. 3.30, yielded:

$$|\gamma(\xi, f)| = \left(1 + a_1\xi^{0.25} + a_2(\xi f)^{0.5}\right)^{-0.5} \exp\{-0.5(a_3\xi^{a_4} f^{a_5})^2\} \qquad (3.31)$$

and calibrated the model with the 17 events recorded at the SMART-1 array utilized earlier by Oliveira et al. [382]. The exponential decay of this model is also similar to the decay of the model of Oliveira et al. [382], presented herein in Fig. 3.18. An interesting suggestion was made by Yang and Cheng [574]: They noted that, for structural safety applications, coherency may be expressed as:

$$|\gamma(\xi, f)| = E[|\gamma(\xi, f)|] \pm \mu\sigma_\gamma(\xi, f) \qquad (3.32)$$

where $E[|\gamma(\xi, f)|]$ and $\sigma_\gamma(\xi, f)$ represent, respectively, the mean value and the standard deviation of the lagged coherency estimates for a number of events at a site, and μ is a peak factor related to the reliability of the structure. They recommended that the sign in the equation should be taken as positive if the structural response increases with increasing values of the coherency, and negative otherwise. They then evaluated the mean value of the lagged coherency from the 17 SMART-1 events, which led to the following values for the parameters in Eq. 3.31: $a_1 = 0.115144$, $a_2 = -0.224874 \times 10^{-2}, a_3 = 0.762306 \times 10^{-1}, a_4 = 0.378401$ and $a_5 = 0.220597$. For the variance of the coherency expression in Eq. 3.32, they proposed an expression of the form:

$$\sigma_\gamma(\xi, f) = 0.2 \sin(b_1 f + b_2) + b_3\xi + b_4 f + \frac{b_5}{3f} + b_6 \qquad (3.33)$$

for which the values of the parameters were estimated as: $b_1 = 0.15132, b_2 = -0.87023, b_3 = 0.10736 \times 10^{-3}, b_4 = -0.2596 \times 10^{-1}, b_5 = 0.20221 \times 10^{-3}$ and $b_6 = 0.20716$.

Zerva and Harada [603] presented a semi-empirical, site-specific coherency model that approximated the site topography by a horizontally extended, stochastic layer overlaying a half-space (bedrock). The model included the effect of wave passage with constant velocity on the ground surface (Eqs. 2.73 and 2.74), the loss of coherency of the waves as they travel from the source to the site by means of Luco and Wong's [324] expression (Eq. 3.27), and the local site contribution to coherency that was evaluated from 1-D vertical transmission of S-waves through the stochastic layer. The properties of the layer were determined from a probabilistic analysis of the spatially variable soil characteristics of the site under investigation. The site contribution to the lagged coherency for separation distances of 50, 100 and 300 m is presented in Fig. 3.24(a) and the total lagged coherency, for the same values of the station separation distances, in Fig. 3.24(b). The site contribution to the coherency (Fig. 3.24(a)) is identically equal to one, except for a drop at the stochastic layer predominant frequency (at approximately 0.9 Hz) that is caused by the random variability in the natural frequency of the layer. The total coherency (Fig. 3.24(b)), as the product of the scattering and local site contributions, decays with separation distance and frequency following the model of Luco and Wong (Eq. 3.27), i.e., the loss of coherency due to the scattering of the waves in the bedrock, except, again, for the drop at the layer's predominant frequency. This drop in the coherency of Fig. 3.24(a) can be rationally explained: With everything else being identical, perturbations with small deviations in the layer characteristics will produce the greatest changes in the site response functions. Hence, since coherency is a measure of similarity of the motions, it will be low at the resonant frequency of the layer. The fault of this reasoning, however, as pointed out by Cranswick [121], is that such models do not take into consideration the lateral coupling of the wave propagation at the site. The coherency trough in Fig. 3.24(a) where the spectral amplitude, due to the site effect, is peaked, is not in agreement with observations from recorded data: Abrahamson et al. [13] and Toksöz et al. [524] observed that spectral peaks produce high coherency and spectral troughs low coherency, which were attributed, respectively, to the direct (coherent) energy that dominates spectral peaks and to the scattered energy that dominates spectral troughs. It appears then that the seemingly reasonable and simplifying approximations in the analytical derivations of coherency models do not capture the complex physical causes underlying the spatial variation of seismic ground motions.

From the semi-empirical and empirical models presented in this section and the previous section, respectively, the most commonly used ones are the model of Luco and Wong [324] (Eq. 3.27 and Fig. 3.23), for various values of the coherency drop parameter α, and the model of Harichandran and Vanmarcke [208] (Eq. 3.12 and Fig. 3.16). The differences in the exponential decay of the models with frequency and station separation distance affect the response of lifeline systems: Section 6.4 presents the effect of the selection of a coherency model on the response of generic lifeline systems using random vibrations, Section 7.2.2 illustrates how the exponential decay of the coherency models is reflected in simulated spatially variable seismic ground motions, and Section 9.5 discusses issues regarding the selection of coherency models for the seismic response evaluation of lifeline systems.

3.4.3 ANALYTICAL MODELS

Analytical modeling has been used either to explain physically some of the coherency patterns observed from recorded data or to provide alternative means for the description of the spatial variability of the seismic ground motions. Examples include, but are not limited, to the following studies.

Kausel and Pais [264] and Zendagui et al. [586] analyzed the cross correlation structure of the ground motion random field by considering the effect of body waves arriving at the site from a range of incident angles. Somerville et al. [487] showed that synthetic seismograms, that were generated by means of wave propagation in a complex crustal structure, have low coherency where energy is focused from

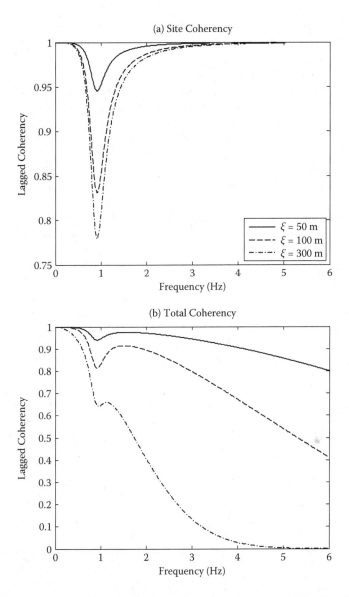

FIGURE 3.24 Illustration of the site-dependent lagged coherency model developed by Zerva and Harada [603]. Part (a) of the figure presents the contribution to the coherency of the variability in the dominant frequency of the random surface layer, and part (b) the total lagged coherency at the site. The variation of the model with frequency is presented at three separation distances of $\xi = 50$, 100 and 300 m.

a variety of ray paths. Toksöz *et al.* [524] used finite difference modeling of layered media with random velocity and/or layer thickness perturbations to explain the significance of near-surface layering and velocity heterogeneities in the resulting ground surface coherency. Zerva and Shinozuka [605] evaluated the effect of random variability in the source parameters on differential ground motions. Kanda [254], using finite element modeling of a layered medium with irregular interfaces and random spatially variable incident motions, analyzed coherency and amplitude variability on the free surface of the site. Horike and Takeuchi [222] used finite difference simulations to explain the differences in the cross correlation and amplitude variation of high-frequency seismic motions observed at different sites. Laouami and Labbe [281] considered that the seismic motions consist of the contributions of a coherent, traveling wave and a zero-mean random factor that introduces the incoherence effect. Zembaty and Rutemberg [585] evaluated spatial response spectra by considering 1-D wave propagation effects. Liao and Li [299] evaluated analytically the spatial coherency at sites with irregular topography based on wave motion finite element simulation techniques and the orthogonal polynomial expansion method. Ding *et al.* [140] evaluated, through numerical simulations, the coherency at a rock site due to a strike-slip fault.

It should be noted, however, that, whereas analytical studies can provide insight into various aspects underlying the spatial variability in the seismic motions, the forward, purely analytical modeling of coherency will be limited by the assumptions that are necessarily made to simplify this complex problem.

3.5 PARAMETRIC CROSS SPECTRUM MODELING

In concluding this chapter, this section illustrates the approach proposed by Zerva and Beck [601], [602] that parameterizes the random field, through its cross amplitude spectrum (Eq. 3.2), in a single step. In this approach, nonparametric power and cross spectral estimates are first evaluated from the data. A parametric cross amplitude spectrum expression is then fitted to these estimates: At zero separation distances, the cross amplitude spectrum collapses onto the power spectral density, and at nonzero separation distances it contains all the information about the coherency. By dealing with the cross amplitude spectra, the approach also bypasses the strict requirement of coherency smoothing and evaluates the parametric coherency indirectly. This ground motion parameter identification was also embedded in a statistical framework, that accounts for both parameter uncertainty and prediction-error uncertainty, and a Bayesian scheme was utilized to update the model with information from recorded data. Because this identification scheme deviates from the usual approaches used in the estimation of the parametric point and spatial characteristics of the random field, it is briefly highlighted in the following. The results of the application of the approach to the north-south component of the motions recorded at the SMART-1 array during the strong motion S-wave window of Event 5 is presented next.

3.5.1 STATISTICAL SYSTEM IDENTIFICATION AND MODEL UPDATING

The statistical system identification scheme used by Zerva and Beck [602] followed the methodology developed by Beck and Katafygiotis [48], [256]. In the approach, a class of models $\mathcal{M}$, describing the system under investigation, is initially specified with the objective to update an initial probabilistic description of the model uncertainties by using data from the system. $\mathcal{M}$ was selected in the study to represent the cross amplitude spectrum of the random field (Eq. 3.2) of aligned seismic data with the Kanai-Tajimi power spectral density (Eq. 3.5) and the coherency model of

Harichandran and Vanmarcke (Eq. 3.12). The parameters of the model become then $\mathbf{a} = \{S_\circ, \omega_g, \zeta_g, A, a, k, f_0, b\}$, the first three for the power spectral density and the last five for the coherency model. The initial parameter uncertainty, that reflects the lack of knowledge concerning which model in $\mathcal{M}$ best represents the system, is described by specifying a probability density function (PDF), $f(\mathbf{a}|\mathcal{M})$.

The stochastic model of the spatially variable ground motions was then embedded in a statistical framework through the introduction of a prediction error, $\mathbf{e}(\xi_j, \omega_n; \mathbf{a})$, describing the difference between the frequency domain model of the ground accelerations, $\mathbf{S}(\xi_j, \omega_n; \mathbf{a})$, and the frequency domain characteristics of the recorded data, $\mathbf{S}_D(\xi_j, \omega_n)$:

$$|\mathbf{S}_D(\xi_j, \omega_n)| = |\mathbf{S}(\xi_j, \omega_n; \mathbf{a})| \ [1 + \mathbf{e}(\xi_j, \omega_n; \mathbf{a})] \tag{3.34}$$

with $\omega_n = n \, \Delta\omega, n = 1, 2, \ldots, N_\omega$, being the discrete frequency, and $\Delta\omega$ the frequency step. ξ_j in Eq. 3.34 is equal to zero for $j = 1$, which corresponds to the power spectral density, and equal to the separation distance between equidistant station pairs for $2 \leq j \leq N_\xi$ for the cross spectral densities.

In Eq. 3.34, the prediction error $\mathbf{e}(\xi_j, \omega_n; \mathbf{a})$ represents the combined effect of errors from sources such as the use of a finite amount of data, measurement noise and model error. The uncertainty in the prediction error is characterized by the selection of a probability model from a set of possible models, $\mathcal{F}$, where each prediction-error probability model in $\mathcal{F}$ is parameterized by σ. The homoscedastic form of the prediction-error model in Eq. 3.34 assumes that the difference between the parametric and nonparametric estimates is proportional to the amplitude of the spectrum at each frequency. Yuen et al. [579] showed that stochastic independence of the spectral estimates at different frequencies is a valid model for a segment of a stationary stochastic process. Hence, the uncertain prediction errors, $\mathbf{e}(\xi_j, \omega_n; \mathbf{a})$, were modeled as independent Gaussian variates with zero mean and constant variance σ^2 over the frequency range of interest, and the parameters in the probability model now become $\Theta = (\mathbf{a}, \sigma)$. The selection of the mathematical description of the ground motion model, defined by $\mathcal{M}$, and the probability distribution of the prediction error, defined by $\mathcal{F}$, specifies the probability distribution of the spectral density estimates through Eq. 3.34 as:

$$f(\mathbf{S}_D|\Theta, \mathcal{M}, \mathcal{F}) = \frac{1}{(\sqrt{2\pi}\sigma)^{N_\xi N_\omega}}$$

$$\times \exp\left[-\frac{1}{2\sigma^2} \sum_{j=1}^{N_\xi} \sum_{n=1}^{N_\omega} \left\{\frac{|\mathbf{S}_D(\xi_j, \omega_n)| - |\mathbf{S}(\xi_j, \omega_n; \mathbf{a})|}{|\mathbf{S}(\xi_j, \omega_n; \mathbf{a})|}\right\}^2\right] \tag{3.35}$$

where $\mathbf{S}_D = \{|\mathbf{S}_D(\xi_j, \omega_n)|, \ j = 1, 2, \ldots, N_\xi \text{ and } n = 1, 2, \ldots, N_\omega\}$. It should be noted that the modeling of the prediction errors as independent random variates requires that the approach be used with unsmoothed spectral estimates only, as smoothing produces correlated spectral values. Hence, the results of the approach in the following application using smoothed, nonparametric spectral estimates are only for comparison purposes with those evaluated using the unsmoothed estimates of the data.

The initial (prior) PDF of the model parameters, $f(\Theta|\mathcal{M}, F)$, for which the assumption of uniform distribution was shown to be adequate [602], was then updated using information from recorded data, $\hat{\mathbf{S}}_D = \{|\hat{\mathbf{S}}_D(\xi_j, \omega_n)|, \ j = 1, 2, \ldots, N_\xi$ and

$n = 1, 2, \ldots, N_\omega\}$, to yield, through Bayes theorem, the updated (posterior) PDF for the model parameters, $f(\Theta|\hat{\mathbf{S}}_{\mathbf{D}}, \mathcal{M}, F)$. For fixed model parameters $\mathbf{a}$, maximizing the logarithm of $f(\Theta|\hat{\mathbf{S}}_{\mathbf{D}}, \mathcal{M}, \mathcal{F})$ with respect to σ yields the most probable estimate for the prediction-error variance for the model specified by $\mathbf{a}$, namely, that it is equal to the mean-square of the observed prediction error [48]:

$$\hat{\sigma}^2(\mathbf{a}) = \frac{1}{N_\xi N_\omega} \sum_{j=1}^{N_\xi} \sum_{n=1}^{N_\omega} \left\{ \frac{|\hat{\mathbf{S}}_D(\xi_j, \omega_n)| - |\mathbf{S}(\xi_j, \omega_n; \mathbf{a})|}{|\mathbf{S}(\xi_j, \omega_n; \mathbf{a})|} \right\}^2 \tag{3.36}$$

The most probable value, $\hat{\mathbf{a}}$, is then given by the minimization of Eq. 3.36 with respect to the model parameters, $\mathbf{a}$, and the corresponding prediction-error variance is $\hat{\sigma}^2 = \hat{\sigma}^2(\hat{\mathbf{a}})$ [602]. It can also be shown [48] that for large amounts of data, $N_\xi N_\omega$, as is the case in this application, $\hat{\mathbf{a}}$ and $\hat{\sigma}$ are, essentially, maximum likelihood estimates.

3.5.2 APPLICATION TO RECORDED SEISMIC DATA

The statistical system identification scheme was applied to data recorded in the north-south direction at the SMART-1 array during the strong motion S-wave window of Event 5. The stations considered were C00, I03, I06, I09, I12, M03, M06, M09 and M12, and the data were aligned with respect to C00 before the application of the approach; the aligned motions at the considered stations were presented in Fig. 2.8.

Figure 3.25 presents the comparison of the nonparametric, smoothed power spectral densities (Eq. 2.41) and cross amplitude spectra (Eq. 2.64) with the parametric ones determined from the statistical system identification; the nonparametric estimates were smoothed with the 11-point, $M = 5$, Hamming window of Eq. 2.47. The power spectra of the data at the center and inner ring stations are shown in part (a) of the figure, and those at the center and middle ring stations in part (b). The cross amplitude spectra of data recorded at two representative station separation distances of 200 m and 1000 m are presented in parts (c) and (d), respectively. The identified model parameters were: $S_o = 79.525 \times 10^{-4}$ m^2/sec^4/Hz, $\omega_g = 16.50$ rad/sec (2.63 Hz), $\zeta_g = 0.50$, $A = 0.860$, $a = 9.355 \times 10^{-6}$, $k = 1.4930 \times 10^8$ m, $f_0 = 3.34$ Hz and $b = 5.72$. The most probable value of the prediction-error standard deviation was $\sigma = 0.424$. The figure indicates that the Kanai-Tajimi spectrum, due to its single peak, matches the overall trend of the data, but not their details. Figure 3.26 compares the nonparametric coherency obtained directly from the data (Eq. 2.70) at four separation distances (200, 400, 800 and 1000 m in parts (a), (b), (c) and (d), respectively) and the coherency model of Harichandran and Vanmarcke (Eq. 3.12) with parameters determined from the cross spectrum identification. It is noted that the coherency estimates (nonparametric and parametric) presented in Fig. 3.26 were not used in the statistical system identification. The coherency model parameters, A, a, k, f_0, b, estimated from the cross spectrum identification, were substituted in Eq. 3.12 and plotted in Fig. 3.26 for the appropriate separation distances to provide the curves labeled "parametric."

Figure 3.27 presents the comparison of the nonparametric, unsmoothed power spectral densities (Eq. 2.34) and cross amplitude spectra (Eq. 2.58) with the parametric

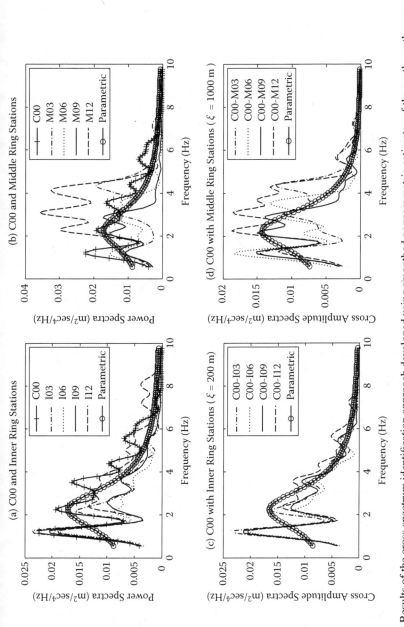

FIGURE 3.25 Results of the cross spectrum identification approach developed using smoothed nonparametric estimates of the north–south component of the data recorded during Event 5 at the SMART-1 array. Parts (a) and (b) present the smoothed nonparametric power spectral densities at C00 and the inner ring stations, and C00 and the middle ring stations, respectively, together with the identified parametric power spectrum. Parts (c) and (d) present selected nonparametric smoothed cross amplitude spectra between C00 and the inner ring stations, and C00 and the middle ring stations, respectively, along with the identified parametric cross amplitude spectrum. (After A. Zerva and J.L. Beck, "Identification of parametric ground motion random fields from spatially recorded data," *Earthquake Engineering and Structural Dynamics*, Vol. 32, pp. 771–791, Copyright ©2003 John Wiley & Sons Limited; reproduced with permission.)

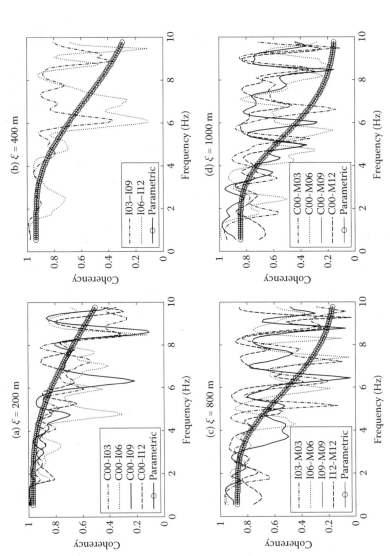

FIGURE 3.26 Nonparametric, smoothed coherency estimates evaluated from the north-south component of the data recorded during Event 5 at the SMART-1 array between station pairs with separation distances of $\xi = 200, 400, 800$ and 1000 m in parts (a), (b), (c) and (d), respectively. The parametric coherency model in the subplots, the functional form of which is given by the Harichandran and Vanmarcke model of Eq. 3.12, was determined indirectly using the cross spectrum identification approach with smoothed spectral estimates (Fig. 3.25). (After A. Zerva and J.L. Beck, "Identification of parametric ground motion random fields from spatially recorded data," *Earthquake Engineering and Structural Dynamics*, Vol. 32, pp. 771–791, Copyright ©2003 John Wiley & Sons Limited; reproduced with permission.)

ones resulting from the system identification. As in Fig. 3.25, the power spectra at the close-by and further-away stations are shown in parts (a) and (b), respectively, and the cross amplitude spectra are presented at the two representative station separation distances of 200 m and 1000 m, in parts (c) and (d), respectively. The identified model parameters in this case were: $S_\circ = 130.92 \times 10^{-4}$ m^2/sec^4/Hz, $\omega_g = 13.57$ rad/sec (2.16 Hz), $\zeta_g = 0.60$; $A = 0.738$, $a = 0.0516$, $k = 5.530 \times 10^4$ m, $f_0 = 3.67$ Hz and $b = 7.09$. The most probable value of the prediction-error standard deviation was $\sigma = 0.573$, which is larger than the corresponding value for the smoothed spectral estimates, as anticipated, due to the larger scatter of the data in the absence of smoothing. Figure 3.28 presents the comparison of the nonparametric coherency estimates obtained from unsmoothed spectra with the parametric ones at separation distances of 200, 400, 800 and 1000 m in parts (a), (b), (c) and (d), respectively. The parametric estimates were evaluated by substituting the identified parameters of the model of Harichandran and Vanmarcke into Eq. 3.12. In all subplots of Fig. 3.28, the nonparametric coherencies are identically equal to one (Eq. 2.78). On the other hand, because the parametric coherency is evaluated from the ratio of the identified cross amplitude spectrum over the power spectral density, it assumes values lower than one. The comparison of Figs. 3.26 and 3.28 indicates that the identified coherency is essentially identical whether smoothed or unsmoothed spectral estimates are used, even though the identified parameters of the models differ. This is more clearly illustrated in Fig. 3.29, which is described next.

    Figures 3.29(a) and 3.29(b) present the results of the cross spectrum identification approach using unsmoothed nonparametric spectra ($M = 0$) and smoothed nonparametric spectra with $M = 1$ and 5. Figure 3.29(a) presents the identified Kanai-Tajimi power spectral density: The spectrum for $M = 1$ is essentially identical to the one for $M = 0$. The figure indicates that the identified parametric spectrum for the lower values of M has a peak amplitude that is higher than that identified for the $M = 5$ parametric estimate, and its dominant frequency is slightly shifted towards lower frequencies. These differences are an expected consequence of the first high peak of the nonparametric spectra that is more pronounced in the unsmoothed estimates (Fig. 3.27) than the smoothed ones (Fig. 3.25). Figure 3.29(b) presents the comparison of the parametric coherency models evaluated for the three smoothing cases ($M = 0, 1$ and 5) at a fixed separation distance of 500 m; the decay of the coherency with frequency in Fig. 3.29(b) is nearly indistinguishable for the three cases. As has been discussed earlier, in the two-step parameterization approach, parametric power spectral densities and parametric coherency models are obtained separately from the corresponding nonparametric estimates. The results of the statistical identification using power spectral densities only are presented in Fig. 3.29(c). The same observations apply here, as in the discussion of Fig. 3.29(a), namely that lower values of M lead to higher amplitudes and a slight shift of the dominant frequency of the parametric spectra towards the lower frequencies. The comparison of Figs. 3.29(a) and 3.29(c) indicates that the Kanai-Tajimi spectrum identified using only the nonparametric power spectral densities leads to higher amplitudes and more narrow-band parametric estimates than those resulting from the cross spectrum identification. This is a consequence of the fact that, in the cross spectrum identification, the parameters of the power spectrum are constrained by the cross amplitude spectra, since the power

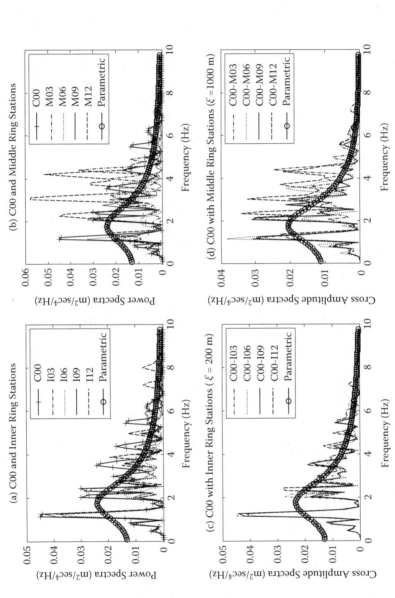

FIGURE 3.27 Results of the cross spectrum identification approach using unsmoothed nonparametric estimates of the data recorded during Event 5 at the SMART-1 array. Parts (a) and (b) present the unsmoothed nonparametric power spectral densities at C00 and the inner ring stations, and C00 and the middle ring stations, respectively, together with the identified parametric power spectrum. Parts (c) and (d) present selected unsmoothed nonparametric cross amplitude spectra between C00 and the inner ring stations, and C00 and the middle ring stations, respectively, along with the identified parametric cross amplitude spectrum. (After A. Zerva and J.L. Beck, "Identification of parametric ground motion random fields from spatially recorded data," *Earthquake Engineering and Structural Dynamics*, Vol. 32, pp. 771–791, Copyright ©2003 John Wiley & Sons Limited; reproduced with permission.)

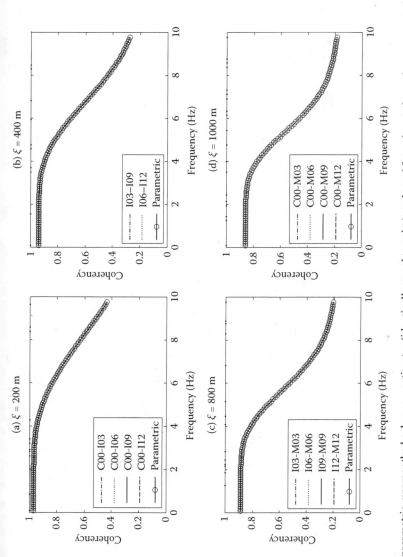

FIGURE 3.28 Nonparametric, unsmoothed coherency estimates (identically equal to unity) evaluated from the north-south component of the data recorded during Event 5 at the SMART-1 array between station pairs with separation distances of ξ = 200, 400, 800 and 1000 m in parts (a), (b), (c) and (d), respectively. The parametric coherency model in the subplots, the functional form of which is given by the Harichandran and Vanmarcke model of Eq. 3.12, was determined indirectly using the cross spectrum identification approach with unsmoothed spectral estimates (Fig. 3.27). (After A. Zerva and J.L. Beck, "Identification of parametric ground motion random fields from spatially recorded data," *Earthquake Engineering and Structural Dynamics*, Vol. 32, pp. 771–791, Copyright ©2003 John Wiley & Sons Limited; reproduced with permission.)

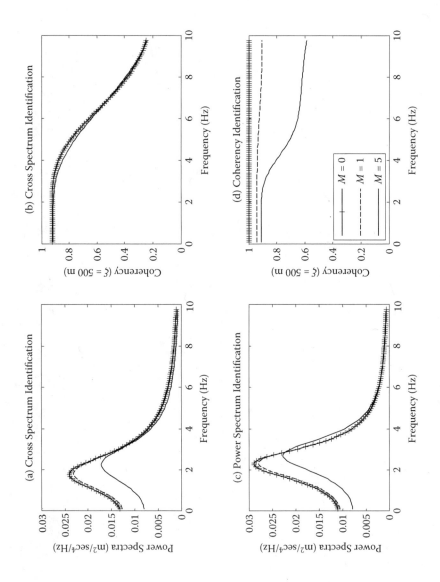

FIGURE 3.29

spectral density is viewed as the special case of the cross spectral density when the separation distance is zero. The differences in the shape of the Kanai-Tajimi spectra identified by the two schemes are not dramatic. This is not the case, however, for the parametric coherency: Figure 3.29(d) presents the results when nonparametric coherency estimates are used directly in the statistical system identification for the evaluation of the parameters of the coherency model of Harichandran and Vanmarcke (Eq. 3.12). As expected, when no smoothing is performed on the nonparametric estimates, coherency is identical to one. The parametric coherency identified for $M = 1$ produces values close to unity throughout the entire frequency range, even for the long separation distance of 500 m presented in Fig. 3.29(d). Figure 3.29(d) also presents the parametric coherency model identified when the nonparametric spectral estimates were smoothed with the 11-point Hamming window. The decay of the parametric coherency models for the three different cases in Fig. 3.29(d) is very different, verifying that coherency models identified from nonparametric coherency estimates depend strongly on the amount of smoothing performed on the data. It is noted that, for $M = 1$ and $M = 5$ in Fig. 3.29(d), coherency tends to high values past the dominant frequency range of the motions, which is a consequence of the behavior of the nonparametric coherency evaluated from the recorded data at higher frequencies (Fig. 3.26), whereas the cross spectrum identification leads to lower values for the parametric coherency in this higher frequency range (Figs. 3.28 and 3.29(b)).

The comparisons presented in Fig. 3.29 indicate that the cross spectrum identification approach provides, at least with the parametric models used in this evaluation, coherency estimates that are independent of the amount of smoothing performed on the spectra. Additionally, this simultaneous evaluation of point and spatial parametric estimates is more consistent with parameter identification in multi-dimensional signal processing, which relies on the random field's frequency-wavenumber spectrum [143]. The estimation of frequency-wavenumber spectra from recorded data is presented in Section 4.1. The cross spectrum approach is also consistent with observations from recorded data that indicate that point and spatial estimates may be interrelated, as will be described in Section 4.2 of the following chapter: Coherency tends to decrease with frequency in the dominant frequency range of the motions, where the broad-band S-wave controls the motions, and approaches zero in the higher frequency range, where scattered energy and/or noise controls the motions [608].

FIGURE 3.29 Comparison between the results of the cross spectrum identification approach and the conventional, separate estimation of point and spatial parametric models of the random field for different amounts of smoothing ($M = 0, 1$ and 5) performed on the nonparametric spectral estimates. Parts (a) and (b) present the parametric power spectral density (described by the Kanai-Tajimi spectrum) and the coherency (described by the model of Harichandran and Vanmarcke and shown for a separation distance of 500 m) identified from the north-south component of the data recorded during Event 5 at the SMART-1 array using the cross spectrum identification scheme. Part (c) illustrates the Kanai-Tajimi spectrum, the parameters of which were identified by fitting the model to the power spectral densities of the data. Part (d) shows the coherency model of Harichandran and Vanmarcke (again, at a separation distance of 500 m), the parameters of which were determined by fitting the model to the lagged coherency estimates of the data. (After A. Zerva and J.L. Beck, "Identification of parametric ground motion random fields from spatially recorded data," *Earthquake Engineering and Structural Dynamics*, Vol. 32, pp. 771–791, Copyright © 2003 John Wiley & Sons Limited; reproduced with permission.)

4 Physical Characterization of Spatial Variability

Spatial arrays are deployed around the world on a permanent (decade or longer) or temporary basis, the latter, mainly, to gain information from aftershock activity after a major event. Examples of such arrays include the ones already mentioned in Chapter 3, as, e.g., the El Centro [480] and the Chusal [269] differential arrays; the Colwick array [340]; the SMART-1 and LSST arrays [10], [13]; the Chiba array [572]; the EPRI Parkfield, USGS Parkfield, Hollister, Stanford, Coalinga, UCSC ZAYA and Pinyon Flat arrays [8], [449]; the FINESA, NORESS, and ARCESS arrays [524]; and the Parkway Valley, Wainuiomata, array [506]. Additional arrays include, but are not limited to: the Garni array in Armenia [349]; the arrays at Tarzana [500], San Fernando Valley [342], Sunnyvale [164] and Garner Valley [33], all in California; at Ashigara Valley, Japan [278]; at Caille [165] and Grenoble [118], both in France; at L'Aquila, Italy, [69], and the EUROSEIS array in Thessaloniki, Greece [330].

Seismic spatial array data provide unique information and serve multiple disciplines in the earthquake engineering, the engineering seismology and the seismology communities. Their deployment led to a better understanding of source, path and site effects on seismic ground motions, and, consequently, to improved capabilities for predicting the seismic hazard at a site. Some illustrations of the usage of spatial array data are highlighted in the following:

Inverse problems in seismology – Spatial array data are used by the seismological community for the estimation of the characteristics of the source, the propagation of waves through the earth strata and the local site effects through inversion processes, as, e.g., Boatwright *et al.* [62], Hartzell [214] and Moya *et al.* [352].

Site effects in ground motions – Complementing borehole data, spatial array data are instrumental in identifying site effects in ground motions and, especially, the contribution of surface waves. Basin induced surface waves can amplify the ground motions (2-D/3-D site effects) beyond the 1-D predictions for site amplification and lengthen their duration, as has been observed among others by, e.g., Chávez-García *et al.* [99], Cornou *et al.* [119], Frankel *et al.* [164], Hartzell *et al.* [215] and Tsuda *et al.* [545].

Spatial variability of seismic ground motions – As has already been described in Chapters 2 and 3, spatial array data permit the estimation and modeling of the spatial variability of seismic ground motions to be used in the seismic response evaluation of long structures, which will be illustrated later herein in Chapters 6 and 9.

Seismic ground strains – The estimation of the seismic strain field is important both from the engineering and the seismological perspective: Seismic ground strains are directly related to the response of buried pipelines and tunnels [217], [538], and seismic stresses developed during an earthquake can be a potential cause of secondary rupture at a fault [502]. Techniques for the estimation of strains borrowed from geodesy require seismic array data, and are highlighted in Section 5.5 of the next chapter.

Wave propagation characteristics – Frequency-wavenumber spectra estimated from spatially recorded data identify the backazimuth and apparent propagation velocity of the impinging waves at the site. Such estimates provide information regarding the path of the waves from the rupture area to the ground surface and allow the evaluation of the contribution of the various wave components to the motions. These techniques have been used extensively in seismology and earthquake engineering, including, among many others, the work of Abrahamson [4], Bokelmann and Baisch [67], Cornou *et al.* [118], [119], Darragh [123], Goldstein and Archuleta [171], [172], Hao *et al.* [200], Hartzell *et al.* [215], Liao [298], McLaughlin [339] and Zerva and Zhang [607], [608].

In spatial variability studies, Frequency-Wavenumber (F-K) spectra estimates are used to identify the direction of propagation and apparent velocity of the incoming waves in order to model the wave passage effect [14], [200], [449], [608], which was discussed in Section 2.4.1. This chapter starts with a brief description of the evaluation of F-K spectra from spatially recorded seismic data. The Conventional (CV) [93], [279], High Resolution (HR) [93], [94] and Multiple Signal Characterization (MUSIC) [170], [447] methods are briefly presented in Sections 4.1.1, 4.1.2 and 4.1.3, respectively. Stacked slowness spectra, that facilitate a clearer identification of broadband waves in seismic motions are presented in Section 4.1.4. The application of these techniques to the low frequency components of the seismic motions recorded during two events at the San Jose seismic array in the Santa Clara Valley, California, [215] is described in Section 4.1.5. The section illustrates the capabilities of F-K spectra in determining the different wave types that contribute to the motions during different time windows and, also, highlights the complex pattern of wave propagation at the site.

The F-K estimates consider that, at each frequency, the waveforms propagate unchanged on the ground surface with the identified average apparent propagation velocity. A first approximation for the representation of spatially variable seismic ground motions (incorporating only the wave passage effect) can then be obtained if a representative time history for the particular site and event simply propagates unchanged on the ground surface. This observation led to an alternative approach for the examination of the spatially variable characteristics of the motions [607], [608], that permits, at least qualitatively, the association of spatial coherency with physical causes. The approach considers that the overall characteristics of seismic motions recorded over extended areas can be described by a "common," coherent component that propagates unchanged on the ground surface. Spatial variability is then viewed as the differences between the characteristics of the common component, i.e., amplitude, wavenumber and phase, and the characteristics of the actual recorded data. Section 4.2 presents the description of this approach and its application to the north-south component of the data recorded during Event 5 at the SMART-1 array. The analysis revealed that there exist correlation patterns in the amplitude and phase variability of the recorded data around the amplitude and phase of the common, coherent component. Hence, through its association with amplitude variability, the phase variability of the data, and, consequently, the spatial coherency, can, also, be qualitatively attributed to physical causes. Some remarks regarding coherency decay and phase and amplitude variability in the data conclude the chapter.

4.1 FREQUENCY-WAVENUMBER (F-K) SPECTRA

The frequency-wavenumber spectrum is defined as [84]:

$$P(\vec{\kappa}, \omega) = |\Psi(\vec{\kappa}, \omega)|^2 \tag{4.1}$$

where $\Psi(\vec{\kappa}, \omega)$ is the triple (time and two space dimensions) Fourier transform of the homogeneous and stationary ground motion random field, ω, again, indicates frequency, and $\vec{\kappa}$ is the horizontal wavenumber. The wavenumber, $\vec{\kappa} = \{\kappa_x, \kappa_y\}^T$, with superscript T indicating transpose, is related to the slowness through the expression:

$$\vec{s} = \vec{\kappa}/\omega \tag{4.2}$$

The absolute value of the slowness is equal to the inverse of the apparent propagation velocity of the wave component and its direction is the opposite of the direction of propagation of the impinging wave. Each point in the $(\vec{\kappa}, \omega)$ domain corresponds to a plane wave in the $(\vec{r}, t)$ domain with a particular orientation (backazimuth), frequency and velocity [93]. If, e.g., the seismic ground motions consisted of a single sinusoid with frequency ω propagating with an apparent velocity c on the ground surface, the F-K spectrum at frequency ω (Eq. 4.1) would be simply a spike in the wavenumber domain at the sinusoid's wavenumber, $\vec{\kappa}$ ($|\vec{\kappa}| = \kappa = \omega/c$). Various methods for the evaluation of F-K spectra from spatially recorded seismic data have appeared in the literature. Among them, the most commonly used ones are the conventional, the high resolution and the multiple signal characterization methods, which are briefly summarized in the following; the details of the derivations and the characteristics of the estimates can be found in the references provided herein.

4.1.1 THE CONVENTIONAL (CV) METHOD

According to the conventional method, the F-K spectral estimate of Eq. 4.1 takes the form [4], [93]:

$$P^{(CV)}(\vec{\kappa}, \omega) = \frac{1}{K^2} \vec{u}^{\dagger}(\vec{\kappa}) \mathbf{S}(\omega) \vec{u}(\vec{\kappa}) \tag{4.3}$$

in which K indicates the number of the recording stations of the array, † denotes conjugate transpose, $\mathbf{S}(\omega)$ is the cross spectral matrix (Eq. 2.103), and $\vec{u}(\vec{\kappa})$ stands for the beamsteering vector with elements

$$u_j(\vec{\kappa}) = \exp[i\vec{\kappa} \cdot \vec{r}_j] \tag{4.4}$$

which have the effect of advancing the phase of the sinusoid observed at station j by an amount corresponding to the time delay with respect to the array origin of a plane wave propagating with wavenumber $\vec{\kappa}$ [4].

4.1.2 THE HIGH RESOLUTION (HR) METHOD

The CV estimate of the F-K spctrum (Eq. 4.3) can be rewritten as [10], [339]:

$$
\begin{aligned}
P^{(CV)}(\vec{\kappa}, \omega) &= \frac{1}{K^2}\, \vec{u}^\dagger(\vec{\kappa})\mathbf{S}(\omega)\vec{u}(\vec{\kappa}) \\
&= \frac{1}{K^2}\, \vec{u}^\dagger(\vec{\kappa})\mathbf{V}(\omega)\mathbf{\Upsilon}(\omega)\mathbf{V}^\dagger(\omega)\vec{u}(\vec{\kappa}) \\
&= \frac{1}{K^2} \sum_{l=1}^{K} \lambda_l(\omega)\vec{u}^\dagger(\vec{\kappa})\vec{V}_l(\omega)\vec{V}_l^\dagger(\omega)\vec{u}(\vec{\kappa})
\end{aligned}
\tag{4.5}
$$

in which $\mathbf{V}(\omega)$ is the matrix of eigenvectors, $\vec{V}_l(\omega)$, of $\mathbf{S}(\omega)$, and $\mathbf{\Upsilon}(\omega)$ is the diagonal matrix of its eigenvalues, $\lambda_l(\omega)$. The description of the CV estimate by means of its singular value decomposition (Eq. 4.5) provides the best means for the comparison between the CV and the HR techniques [4], [339].

For the HR method, the F-K spectral estimate is defined as [93]:

$$
\begin{aligned}
P^{(HR)}(\vec{\kappa}, \omega) &= \left[\sum_{l=1}^{K} \frac{1}{\lambda_l(\omega)}\vec{u}^\dagger(\vec{\kappa})\vec{V}_l(\omega)\vec{V}_l^\dagger(\omega)\vec{u}(\vec{\kappa}) \right]^{-1} \\
&= \frac{1}{\vec{u}^\dagger(\vec{\kappa})[\mathbf{S}(\omega)]^{-1}\vec{u}(\vec{\kappa})}
\end{aligned}
\tag{4.6}
$$

It can be seen from Eqs. 4.5 and 4.6 that the weights in the CV estimate are linearly proportional to the associated eigenvalue, whereas, for the HR estimate, they are inversely proportional to the eigenvalue. The HR estimate passes undistorted any monochromatic plane wave with a specific wavenumber, and suppresses, in an optimum least-squares sense, all other signals with wavenumbers different than the one considered [93]. Consequently, the method provides higher resolution for the identified wave than the conventional method. It is noted, however, that, because $\mathbf{S}(\omega)$ is usually singular, and, thus, noninvertible, prewhitening is, generally, necessary [93].

4.1.3 THE MULTIPLE SIGNAL CHARACTERIZATION (MUSIC) METHOD

The multiple signal characterization (MUSIC) method was developed by Schmidt [447] and was modified by Goldstein [170] and Goldstein and Archuleta [171], [172] for applications to seismic spatial array data. A brief description of the approach is presented in the following.

Let the time histories across an array of K sensors be composed of $q < K$ plane waves at frequency ω:

$$
a(\vec{r}_l, t, \omega) = \sum_{m=1}^{q} \Lambda_m \exp[i(\omega t + \vec{\kappa}_m \cdot \vec{r}_l + \phi_m)] + \eta(\vec{r}_l, t)
\tag{4.7}
$$

in which Λ_m is the amplitude of the m-th plane wave, $\vec{\kappa}_m$ its wavenumber, ϕ_m its phase and $\eta(\vec{r}_l, t)$ indicates random noise. For uncorrelated signals and noise, the cross spectral density matrix (Eq. 2.103) of the ground motions of Eq. 4.7, evaluated, alternatively, from the cross covariance matrix of the time series at each frequency ω, becomes [170]:

$$\mathbf{R} = \mathbf{U}^* \mathbf{Q} \mathbf{U}^T + \sigma^2 \mathbf{I} \tag{4.8}$$

where the dependence of the above equation on ω has been dropped for convenience, * denotes complex conjugate, $\mathbf{I}$ is the identity matrix, σ^2 is the intensity of the noise, and:

$$
\begin{aligned}
\mathbf{U} &= [\vec{u}(\vec{\kappa}_1) \quad\quad \vec{u}(\vec{\kappa}_2) \quad\quad \cdots \quad\quad \vec{u}(\vec{\kappa}_q)] \\
\vec{u}(\vec{\kappa}_m) &= \{\exp[i(\vec{\kappa}_m \cdot \vec{r}_1)] \quad \exp[i(\vec{\kappa}_m \cdot \vec{r}_2)] \quad \cdots \quad \exp[i(\vec{\kappa}_m \cdot \vec{r}_K)]\}^T \\
\mathbf{Q} &= \text{diag}\lceil |\Lambda_1|^2, \quad\quad |\Lambda_2|^2, \quad\quad \cdots \quad\quad |\Lambda_q|^2 \rfloor
\end{aligned}
\tag{4.9}
$$

$\mathbf{U}^* \mathbf{Q} \mathbf{U}^T$ is a positive semi-definite Hermitian matrix with real eigenvalues and complex eigenvectors.

MUSIC is based on the fact that the eigenstructure of $\mathbf{R}$ consists of q larger eigenvalues that correspond to the plane waves and $(K - q)$ smaller ones that correspond to noise, and that the eigenvectors of the noise subspace are orthogonal to the signal direction vectors, $\vec{u}(\vec{\kappa}_m)$, in the signal subspace [170], [447], [448]. The eigenvalues of $\mathbf{R}$ (Eq. 4.8) have then the property $\lambda_R = \{\lambda_1 + \sigma^2 \geq \lambda_2 + \sigma^2 \geq \ldots \geq \lambda_q + \sigma^2 \geq \sigma^2; \lambda_{q+1} = \cdots = \lambda_K = \sigma^2\}$, which suggests that the number of signals in Eq. 4.7 can be determined from the eigenvalues of $\mathbf{R}$ that are greater than σ^2. Let $\mathbf{E}_n$ be the eigenvector matrix of $\mathbf{R}$ associated with the small (noise) eigenvectors $\lambda_i = \sigma^2$, $i > q$, $\mathbf{\Upsilon}_n$ the eigenvalue matrix of the noise components, and $\mathbf{E}_s$ and $\mathbf{\Upsilon}_s$ the matrices of the eigenvectors and eigenvalues associated with the signal subspace. $\mathbf{R}$ (Eq. 4.8) can then be expressed as:

$$\mathbf{R} = \mathbf{E}_s \mathbf{\Upsilon}_s \mathbf{E}_s^\dagger + \mathbf{E}_n \mathbf{\Upsilon}_n \mathbf{E}_n^\dagger \tag{4.10}$$

From the orthogonality between the signal and noise subspaces, it follows that:

$$\mathbf{U}^T \mathbf{E}_n = 0 \tag{4.11}$$

Thus, the signal direction vectors, $\vec{u}(\vec{\kappa}_m)$ (Eq. 4.9), can be determined from the peaks of the directional function:

$$\mathbf{D}(\vec{\kappa}) = \frac{1}{|\vec{\alpha}^T(\vec{\kappa}) \cdot \mathbf{E}_n|^2} \tag{4.12}$$

in which $\vec{\alpha}(\vec{\kappa})$ are the array manifold vectors and $\vec{\kappa}$ can assume any value consistent with the range of velocities appropriate for the site. At the location of the peak $\vec{\alpha}(\vec{\kappa}_m) = \vec{u}(\vec{\kappa}_m)$ and, thus, the propagation characteristics of the plane wave are identified. Furthermore, Eq. 4.12 indicates that when the directional function has a peak at wavenumber $\vec{\kappa}_m$, the propagation characteristics of the plane wave are identified without interference from plane waves with wavenumbers different than $\vec{\kappa}_m$. In this sense, the resolution of the MUSIC approach is better than the resolution of the CV method. Indeed, Zhang [613] showed that, for time series described by Eq. 4.7, the CV estimate of the spectra (Eq. 4.3) contains at each wavenumber $\vec{\kappa}_m$ contributions from waves with wavenumbers different than $\vec{\kappa}_m$.

The directional function (Eq. 4.12) identifies the signal direction vectors, but provides no information about the intensity of the contributing signals. Hence, MUSIC, similar to the HV method (Eq. 4.6), identifies the location of the signals in the F-K domain, but not their amplitude, as both techniques identify the location of the peaks through an inverse evaluation. On the other hand, as can be seen from Eq. 4.3, the CV method provides estimates for both the wavenumber and the amplitude of the signals. However, MUSIC can also be used for the estimation of the signal intensity. Goldstein [170] showed that the diagonal matrix of the square of the signal amplitudes, $\mathbf{Q}$ (Eq. 4.9), can be approximately recovered from the expressions for the cross spectral density matrix (Eqs. 4.8 and 4.10), i.e.,

$$\mathbf{U}^*\hat{\mathbf{Q}}\mathbf{U}^T = \mathbf{E}_s\mathbf{\Upsilon}_s\mathbf{E}_s^\dagger \tag{4.13}$$

With the sign conventions utilized herein, $\hat{\mathbf{Q}}$ takes the form [613]:

$$\hat{\mathbf{Q}} \approx [\mathbf{A}^T\mathbf{A}^*]^{-1}\mathbf{A}^T\mathbf{E}_s\mathbf{\Upsilon}_s\mathbf{E}_s^\dagger\mathbf{A}^*[\mathbf{A}^T\mathbf{A}^*]^{-1} \tag{4.14}$$

in which the columns of $\mathbf{A}$, $\vec{\alpha}(\vec{\kappa}_m)$, $m = 1, \ldots, q$, are the signal direction vectors identified from the peaks of the directional function (Eq. 4.12). The diagonal elements of $\hat{\mathbf{Q}}$ are then estimates of the square of the signal amplitudes.

When the signals q in Eq. 4.7 are correlated, the cross spectral matrix is still given by Eq. 4.8, but $\mathbf{Q}$ in Eq. 4.9 becomes now a full matrix. Goldstein [170] suggested that, for correlated signals, MUSIC can be used with a subarray spectral averaging modification. The approach was based on the work of Evans *et al.* [154] and Shan *et al.* [458] for equispaced arrays, and was revised by Goldstein [170] and Goldstein and Archuleta [171], [172] for two-dimensional linear and equispaced arrays. The modification is performed on the cross spectral density matrix (Eq. 4.8) by averaging each element of $\mathbf{R}$ over its subdiagonal in a submatrix corresponding to linear equispaced sensors. Because the diagonal elements of the matrix are not affected by identical offsets, subarray spectral averaging leaves the diagonal elements unchanged but reduces the values of the off-diagonal elements [171], [172]. To waive the requirement that the array configuration needs to be linear and equispaced, Bokelmann and Baisch [67] suggested an approximate spatial averaging scheme, in which the elements of $\mathbf{R}$ are averaged over all station pairs, for which the approximate equality, $\vec{r}_j - \vec{r}_l \approx$ constant, holds. They also suggested that, for long windows, the possible effects of correlated noise and correlated signal and noise are expected to vanish.

4.1.4 STACKED SLOWNESS (SS) SPECTRA

Spudich and Oppenheimer [501] introduced Stacked Slowness (SS) spectra for better identification of body waves in F-K analyses of seismic ground motions. Because body waves are, essentially, non-dispersive, they have the same slowness vector at all frequencies. Thus, F-K spectra can be stacked according to the expression:

$$P^{(\cdot)}(\vec{s}) = \sum_j P^{(\cdot)}(\vec{s}, \omega_j) \tag{4.15}$$

$$\vec{\kappa}_j = \vec{s}\,\omega_j \tag{4.16}$$

Slowness stacking reduces the effect of spatial aliasing, since peaks due to spatial aliasing have wavenumber shifts that depend inversely on frequency [501]. Hence, broadband body waves can be clearly identified in the slowness spectra plots, since their characteristics remain unchanged over a range of frequencies.

4.1.5 EXAMPLE APPLICATIONS

Hartzell et al. [215] conducted an extensive evaluation of data recorded at the San Jose seismic array in the Santa Clara Valley, California, during 20 events. The 52-station San Jose array is presented in Fig. 4.1. All stations are located on alluvium in the Santa Clara Valley, except for two stations (stations RCK and ROC in Fig. 4.1), which are located on mapped Mesozoic rock in the hills to the east of the valley. The analysis determined the spatial distribution of the site amplification in the valley resulting from all events recorded at the array and a subset of local events only. Hartzell et al. [215] noted that the pattern of site amplification in the two cases was similar, but, for frequencies < 2 Hz, the amplification due to regional events was higher than that for local events due to the generation of surface waves in the valley. They then proceeded with the use of F-K spectra for the identification of the backazimuth and apparent propagation velocity of the impinging waves, and conducted a comparison of the identification capabilities of the CV, HR, MUSIC and SS techniques. Some of their results are presented in the following for illustration of the derivations in the previous subsections.

Figure 4.2 presents the comparison of F-K spectra evaluated by means of the CV, HR, MUSIC and SS techniques during 4-sec windows of the vertical component of the motions recorded at the array during the 7.1 magnitude Hector Mine earthquake. The top part of Fig. 4.3 presents a representative time history of this event; the time history in the figure was recorded at station Q40, which is located close to the center of the array (Fig. 4.1), and was bandpass filtered from 0.125 Hz to 0.5 Hz. The figure also indicates the theoretically estimated arrival times of the P_n-, P_g-, S_n-, S_g-[1] and R-(Rayleigh) waves, as well as the 4-sec windows (letters a-n) that were utilized in the F-K spectra evaluation. The F-K spectra for the CV, HR and MUSIC methods during windows a-l (Fig. 4.2) were evaluated at a frequency of 0.39 Hz and for a wavenumber range of ± 0.7812 /km. This leads to a maximum value for the slowness (Eq. 4.2) of 2.0 sec/km, or, correspondingly, a minimum apparent propagation velocity of 0.5 km/sec. The subplots in Fig. 4.2 are contour plots depicting the variation of the spectra in the two horizontal directions. The horizontal axis represents the east-west coordinate of the wavenumber and the vertical one the north-south coordinate. In the plots, the closer the contours, the higher the elevation of the spectra; the peaks of the spectra identify the wavenumber of the dominant wave(s) controlling the motions at the particular frequency. The stacked slowness (SS) spectra (Eq. 4.15), in the bottom subplots of Fig. 4.2, were based on the CV method and evaluated for a frequency range of

---

[1] Beyond a critical distance, generally in the range of 100 – 200 km, the first waves arriving at a site from seismic sources in the crust have been refracted from the top of the mantle. These waves, called P_n, are followed by the P- (or P_g-) waves that propagate through the crust. Similar definitions apply for the shear waves, S_n and S_g [16].

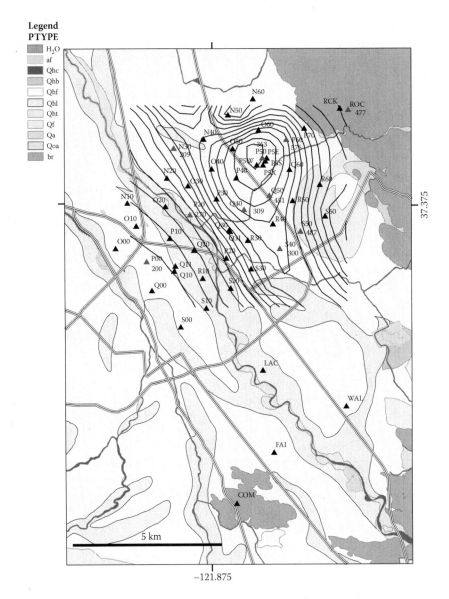

FIGURE 4.1 Detail of the San Jose array with its geological characteristics. The numbers in the figure indicate the average S-wave velocities in the top 30 m obtained from seismic reflection/refraction by Hartzell *et al.* [215]. In the legend: H₂O, water; af, artificial fill; Qhc, modern stream channel deposits; Qhb, Holocene basin deposits; Qhf, Holocene alluvial-fan deposits; Qhl, Holocene alluvial-fan levee deposits; Qht, Holocene stream terrace deposits; Qf, latest Pleistocene to Holocene alluvial-fan deposits; Qa, latest Pleistocene to Holocene alluvium; Qoa, early to late Pleistocene alluvium; br, Mesozoic bedrock. (Reprinted from S. Hartzell, D. Carver, R.A. Williams, S. Harmsen and A. Zerva, "Site response, shallow shear-wave velocity, and wave propagation at the San Jose, California, dense seismic array," *Bulletin of the Seismological Society of America*, Vol. 93, pp. 443–464, Copyright ©2003 Seismological Society of America; courtesy of S. Hartzell.)

0.125 – 0.5 Hz; the SS spectra in the figures are presented over a slowness range of ± 2.0 sec/km. Hartzell *et al.* [215] suggested that all four techniques identified similar values for the backazimuth and the apparent propagation velocity of the waveforms, but selected to use the MUSIC method in their subsequent evaluations because it provided better resolution in windows a, b and c of Fig. 4.2. They also noted that the advantages of the slowness stacking approach (SS in Fig. 4.2) were not realized in their evaluation: Since the study dealt primarily with narrowband surface waves, their dispersion led to broader peaks for the stacked spectra. An earlier study by

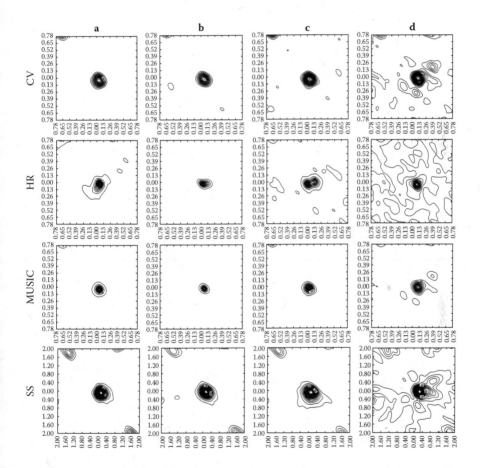

FIGURE 4.2 Comparison of CV, HR and MUSIC F-K spectra and stacked slowness (SS) spectra for selected 4-sec windows (windows a-d shown above and windows e-l presented in the following pages) of the vertical component of the ground motions recorded at the San Jose array during the Hector Mine earthquake. The F-K spectra are shown over a wavenumber range of ± 0.7812 /km, and the SS spectra over a slowness range of ± 2 sec/km. The windows are illustrated in the time history on the top part of Fig. 4.3. (Reprinted from S. Hartzell, D. Carver, R.A. Williams, S. Harmsen and A. Zerva, "Site response, shallow shear-wave velocity, and wave propagation at the San Jose, California, dense seismic array," *Bulletin of the Seismological Society of America*, Vol. 93, pp. 443–464, Copyright © 2003 Seismological Society of America; courtesy of S. Hartzell.)

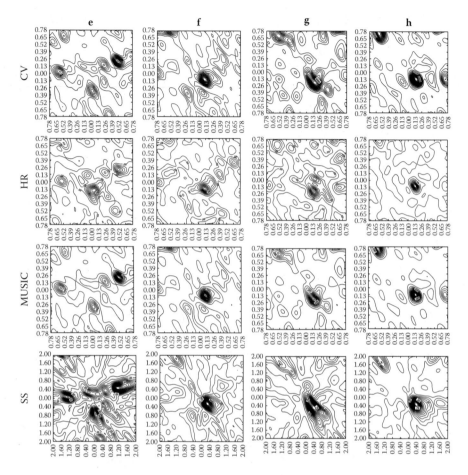

FIGURE 4.2   (Continued)

Zerva and Zhang [607], that used synthetic seismic ground motions resulting from the propagation of broadband body waves in a half-space due to a shear dislocation at the source, also suggested that MUSIC led to the highest resolution, followed by the HR method and then the CV technique, but the CV estimate was always the most robust of the three. Zerva and Zhang [607], however, suggested that slowness stacking significantly increased the identifiability of broadband seismic waves. An example of the application of slowness stacking with the CV and MUSIC techniques applied to the broadband S-wave window of the north-south component of the motions during Event 5 at the SMART-1 array is presented later in Fig. 4.5.

The lower plot of Fig. 4.3 presents the backazimuth and apparent velocity of the motions identified by means of the MUSIC spectra (Fig. 4.2) for the time windows a-n shown on the top part of the figure. This lower plot is a backazimuth and apparent propagation velocity radial grid. In the figure, the outer circle corresponds to an apparent propagation velocity of 0.5 km/sec and the origin to an infinite propagation velocity (i.e., vertical wave incidence). The letters a-n are placed at the appropriate

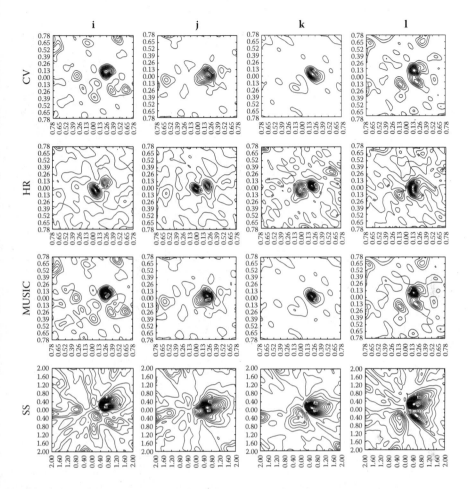

FIGURE 4.2 (Continued)

locations corresponding to the direction of propagation and apparent velocity of the waveforms identified by MUSIC for each time window. The radial grid is superimposed to the generalized geology of the Santa Clara Valley and the backazimuth of the Hector Mine earthquake is labeled "BAZ". Hartzell *et al.* [215] noted that, for the first windows, waves arrive from the general direction of the source with high apparent propagation velocities, which suggests that they are body waves. Later in the record (windows i-l), waves arriving from the north-east direction with surface-wave velocities appear to dominate the records. These waves arrive too early to be direct surface waves from the source region according to the theoretical estimate of the Rayleigh-wave arrival time indicated in the time history on the top part of the figure. Instead, these waves arrive from the closest edge of the valley (Figs. 4.1 and 4.3) and appear to be Rayleigh waves generated from converted S-waves at this boundary [215].

This complex pattern of wave propagation in valleys identified through F-K spectra techniques was further elaborated upon by Hartzell *et al.* [215] using the vertical

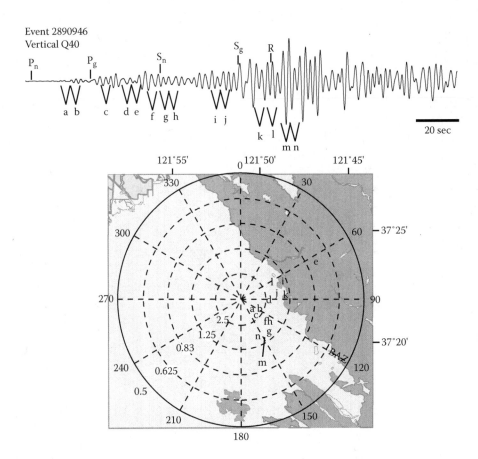

FIGURE 4.3 Top: The vertical acceleration time history at station Q40 recorded during the Hector Mine earthquake and bandpass filtered from 0.125 Hz to 0.5 Hz. The time windows a-n shown on the record form the basis for the F-K analysis (Fig. 4.2). Bottom: The results of the F-K evaluation are given on a radial grid of backazimuth and apparent velocity superimposed on a map view centered on the San Jose array. The corresponding letter for each time window is plotted at a backazimuth and radial distance appropriate for its apparent velocity. The backazimuth of the Hector Mine earthquake is labeled "BAZ." (Reprinted from S. Hartzell, D. Carver, R.A. Williams, S. Harmsen and A. Zerva, "Site response, shallow shear-wave velocity, and wave propagation at the San Jose, California, dense seismic array," *Bulletin of the Seismological Society of America*, Vol. 93, pp. 443–464, Copyright © 2003 Seismological Society of America; courtesy of S. Hartzell.)

component of the motions recorded during an earthquake of magnitude 5.6 near Mammoth Lakes, California, the vertical component of the motions during an earthquake of magnitude 5.2 just north of the San Francisco Bay, and the radial (east-west) component of the motions recorded during an earthquake of magnitude 5.6 near Scotty's Junction, Nevada. The latter results are presented in Fig. 4.4: The top part of the figure presents the time history at station Q40, as well as the theoretical arrival times of the S_n-, S_g- and Rayleigh waves; the time histories were, again, bandpass filtered between 0.125 Hz and 0.5 Hz, and the duration of the analyzed time windows (a-f

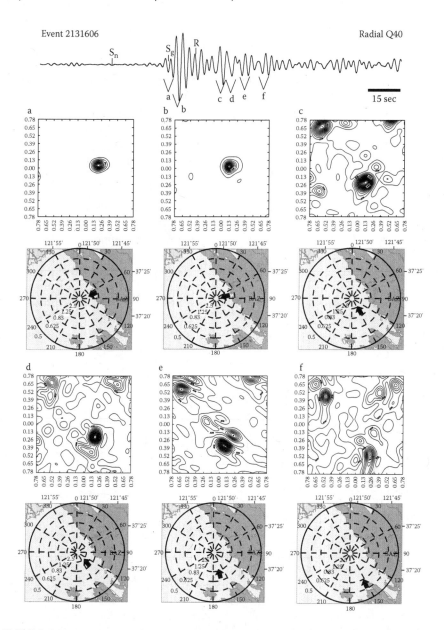

FIGURE 4.4 Top: The radial (east-west) acceleration time history at station Q40 recorded during an earthquake of magnitude 5.6 near Scotty's Junction, Nevada, and bandpass filtered from 0.125 Hz to 0.5 Hz. Bottom: MUSIC spectra and radial grids, superimposed on the valley's geological map, indicating the backazimuth and the apparent velocity of the identified spectral peaks. The arrow on each map indicates the direction of propagation of the waves and is plotted at a radial distance appropriate for its apparent velocity. The backazimuth of the earthquake is labeled "BAZ." (Reprinted from S. Hartzell, D. Carver, R.A. Williams, S. Harmsen and A. Zerva, "Site response, shallow shear-wave velocity, and wave propagation at the San Jose, California, dense seismic array," *Bulletin of the Seismological Society of America*, Vol. 93, pp. 443–464, Copyright © 2003 Seismological Society of America; courtesy of S. Hartzell.)

in the top part of the figure) was 4 sec. The F-K subplots in the bottom part of the figure were generated by means of the MUSIC technique at a frequency of 0.39 Hz and are presented over a wavenumber range of $\pm$ 0.7812 /km. The radial grid of backazimuth and apparent propagation velocity superimposed to the site's geologic map is presented right below each F-K spectrum plot in the figure. The radial grid characteristics of these plots are the same as those of the bottom part of Fig. 4.3, but, in this case, the propagation characteristics of the waveforms are indicated by arrows pointing in the direction of propagation and placed at a distance from the origin that corresponds to the identified apparent velocity; the epicentral direction of the earthquake is also labeled "BAZ" in the figures. The first two windows (a and b) in the figure clearly illustrate waves arriving from the general epicentral direction with high apparent propagation velocities, which suggests that they are body waves. Hartzell *et al.* [215] also reported that between windows b and c, not shown in the figure, waves appeared to be arriving from the general direction of the source with velocities ranging between 1.0 and 2.0 km/sec. However, in the later windows (c-f in the figure), waves arrive from the south-east with velocities consistent with those of surface waves, which suggests that they are surface waves either generated or scattered from the south edge of the valley, an observation that was also made by Frankel *et al.* [163] for this event. Figures 4.2–4.4 then clearly indicate the complex propagation pattern of the waves in the valley and, also, the capabilities of F-K spectra techniques to identify the characteristics of the wave components in the motions.

4.2   AMPLITUDE AND PHASE VARIABILITY

The conventional evaluation of the spatial variability of seismic ground motions, estimated from recorded data as described in Chapter 2 and parameterized as described in Chapter 3, provides functional forms for the description of the coherency, but does not provide insight into its physical causes. Additionally, as illustrated in Section 2.4.2, coherency reflects, essentially, the phase variability in the data, and phase variability is difficult to visualize and attribute to physical causes.

An alternative, more physically insightful approach for the investigation of the spatial coherency of seismic ground motions was proposed by Zerva and Zhang [608]: Frequency-wavenumber spectra estimate, at each frequency, the average propagation characteristics of the waveforms across an array. Generally, for the strong motion S-wave window, which is commonly used in spatial variability studies, a single wave dominates the motions. Furthermore, because body waves are essentially non-dispersive, their apparent propagation velocity characteristics are the same over a range of frequencies [501]. Let then the ground motion time histories be described by the signal component on the right-hand side of Eq. 4.7 with $q = 1$: Since the only parameter in the expression that depends on location is the position vector $\vec{r}_l$, the expression represents, at each frequency ω, a coherent wavetrain that propagates unchanged on the ground surface with the identified velocity. The superposition of the signal components of Eq. 4.7 over the significant frequency range of the motions would then yield a first approximation for spatially variable time histories that incorporates only the wave passage effect (Section 2.4.1). This approximation was termed the "common," coherent component of the motions. Its propagation characteristics were evaluated from F-K techniques applied to the recorded data, and its amplitude and phase at each frequency through a least-squares minimization of the error function between a sinusoidal approximation of the motions (presented later in Eq. 4.17) and the recorded data. The spatial variation of the seismic ground motions was then viewed as the differences between the recorded data and the common, coherent component. These differences include perturbations in the amplitudes, phases and arrival times of the waveforms that are particular for the data at each recording station and constitute the spatial variation of the motions, in addition to the wave passage effect. The methodology is presented in the following and illustrated with the data of the strong motion S-wave window (7.0 – 12.12 sec in the records) of the north-south component of the motions recorded at the SMART-1 array during Event 5. Representative time histories of the north-south component of the motions recorded at C00 and selected inner and middle ring stations of the array during the S-wave window were presented in Fig. 2.1.

4.2.1 APPARENT PROPAGATION CHARACTERISTICS OF THE MOTIONS

Frequency-wavenumber spectra estimation techniques were first applied to the data to evaluate the apparent propagation characteristics of the motions. Figure 4.5 presents the stacked slowness spectra of the data during the strong motion S-wave window utilizing the CV and MUSIC techniques in parts (a) and (b), respectively. In these contour plots, the darkest area corresponds to the highest elevation of the spectra. Both techniques identified the slowness of the dominant wave in the motions as $\vec{s} = \{0.1 \text{ sec/km}, -0.2 \text{ sec/km}\}^T$, i.e., the waves impinge the array at a backazimuth of 153° with an apparent propagation velocity of 4.5 km/sec, which is consistent with the source-site geometry [123], [608] and, also, close to the north-south direction. The advantages of slowness stacking are clearly realized in this broadband wave analysis: The SS spectra in this case were evaluated over a frequency range of 0.195 – 19.5 Hz. For the dominant frequency range of the motions (Fig. 2.13), a peak appeared at essentially every frequency in the vicinity of the broadband S-wave slowness, but was not always the highest. Past the dominant frequency range of the motions, small, spurious peaks appeared in the F-K spectra, which suggests that the motions were controlled by scattered energy. However, because of the low amplitude of the motions in this frequency range (Fig. 2.13), the spurious, scattered energy peaks did not affect the identification of the dominant broadband slowness in the SS spectra (Fig. 4.5). It can also be seen from Fig. 4.5 that, even though both techniques identify the same location for the peak slowness, the resolution of the stacked spectra resulting from MUSIC is higher than the resolution of the CV technique.

4.2.2 SEISMIC GROUND MOTION APPROXIMATION

In the approach [608], [613], the seismic ground motions were approximated by M sinusoids and expressed as:

$$\hat{a}(\vec{r}, t) = \sum_{j=1}^{M} \Lambda(\omega_j) \sin[\omega_j t + \vec{\kappa}(\omega_j) \cdot \vec{r} + \phi(\omega_j)] \qquad (4.17)$$

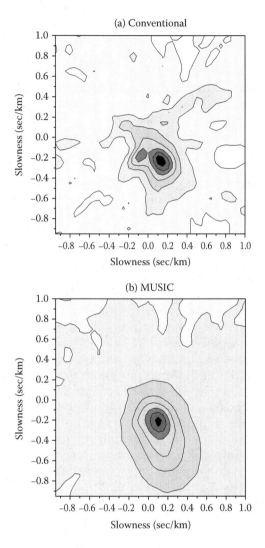

FIGURE 4.5 Stacked slowness spectra evaluated by means of the CV and MUSIC methods in parts (a) and (b), respectively, from the strong motion S-wave window (5.12 sec in the records) of the north-south component of the data recorded at C00 and the inner and middle ring stations of the SMART-1 array (Fig. 1.2). In the contour plots, the darkest area corresponds to the peak of the spectra.

in which $\vec{r}$ indicates location on the ground surface and t is time. Each sinusoidal component is described by its amplitude, $\Lambda(\omega_j)$, phase, $\phi(\omega_j)$, (discrete) frequency, ω_j, and wavenumber $\vec{\kappa}(\omega_j)$, with $\vec{\kappa}(\omega_j)$ being determined from Eq. 4.16 and the slowness identified in Fig. 4.5. It is noted that no noise component is superimposed to the ground motion estimate of Eq. 4.17. The number of sinusoids, M, used in the approach depended on the cut-off frequency, above which the sinusoids did not contribute significantly to the seismic motions. The amplitudes, $\Lambda(\omega_j)$, and phases, $\phi(\omega_j)$, of the sinusoidal components were then determined from the system of equations that result from the least-squares minimization of the error function between the recorded time histories, $a(\vec{r}, t)$, and the approximate ones, $\hat{a}(\vec{r}, t)$ of Eq. 4.17, with respect to the unknowns $\Lambda(\omega_j)$ and $\phi(\omega_j)$, $j = 1, \ldots, M$ [607], [613]. The error function was given by:

$$E = \sum_{l=1}^{L} \sum_{n=1}^{N} [a(\vec{r}_l, t_n) - \hat{a}(\vec{r}_l, t_n)]^2 \qquad (4.18)$$

where $\vec{r}_l$ indicates the location of the recording stations, t_n discrete time and N the number of time steps. Any number of stations, L, ranging from one to the total number of recording stations, can be used for the evaluation of the signal amplitudes and phases. When $L > 1$ in Eq. 4.18, the identified amplitudes and phases represent the common signal characteristics of the motions at the number of stations considered; when $L = 1$, the amplitudes and phases correspond to the motions at the particular station analyzed.

4.2.3 RECONSTRUCTION OF SEISMIC MOTIONS

The data at five stations of the array ($L = 5$) were initially used in Eq. 4.18 for the identification of their common component amplitudes and phases. The stations were C00, I03, I06, I09 and I12 (Fig. 1.2), and the analyzed time segments of the data were presented earlier in Fig. 2.1(a). Once the common component characteristics were identified from the least-squares minimization of Eq. 4.18, they were substituted in Eq. 4.17, and an estimate of the motions, termed "reconstructed" motions, at the stations considered was obtained. The comparison of the recorded motions with the reconstructed ones is presented in Fig. 4.6; no noise (random) component was added to the reconstructed signals. Since the amplitudes and phases of the reconstructed motions at each frequency are identical for all five stations considered, the reconstructed motions represent a coherent waveform that propagates with constant velocity on the ground surface. Figure 4.6 indicates that the reconstructed motions reproduce to a satisfactory degree the recorded data, and, although they consist only of the broadband, coherent body-wave signal (Eq. 4.17), they can describe the major characteristics of the data. The details in the recorded data, that are not matched by the reconstructed motions, constitute the spatially variable nature of the motions, after the (deterministic) wave passage effects have been removed.

4.2.4 VARIATION OF AMPLITUDES AND PHASES

When the data at only one station at a time are used in the evaluation of amplitudes and phases of the motions at different frequencies for that particular station ($L = 1$ in Eq. 4.18), the reconstructed motion is indistinguishable from the recorded one. This does not necessarily mean that the analyzed time history is composed only of the identified broadband wave, but, rather, that the sinusoidal functions of Eq. 4.17 can match the sinusoidally varying recorded time history, i.e., Eq. 4.18 becomes,

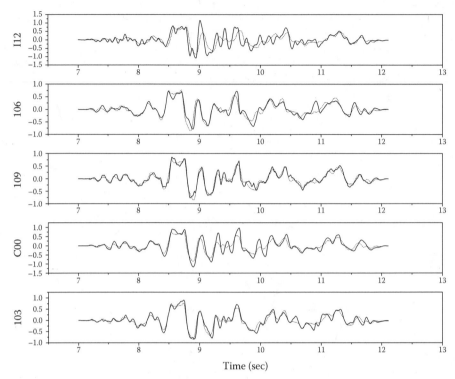

Time (sec)

FIGURE 4.6 Comparison of the recorded (solid lines) and reconstructed (dashed lines) strong ground motions during the S-wave window of the north-south component of Event 5 at the SMART-1 array. The comparison is shown for the center and four inner ring stations of the array. The acceleration units in the subfigures are in m/sec². (After A. Zerva and O. Zhang, "Correlation patterns in characteristics of spatially variable seismic ground motions," *Earthquake Engineering and Structural Dynamics*, Vol. 26, pp. 19–39, Copyright ©1997 John Wiley & Sons Limited; reproduced with permission.)

essentially, compatible to a Fourier transform. The comparison of the identified amplitudes and phases from the data at each individual station with the common component characteristics provides insight into the causes of the spatial variation of the motions.

The top and bottom parts of Fig. 4.7 present the amplitude and phase variation, respectively, of the sinusoidal components of the motions with frequency. The continuous, wider line in these figures, as well as in subsequent figures, indicates the common signal characteristics, namely the contribution of the identified body wave to the motions at all five stations, whereas the thinner, dashed lines represent the corresponding amplitudes and phases when the motions at one station at a time are considered in Eq. 4.18. In the lower frequency range (<1.5 Hz), amplitudes and phases identified from the data at each individual station essentially coincide with those of the common component. In the frequency range of 1.5 – 4.5 Hz, the common component amplitude represents the average of the amplitudes identified at the individual stations and phases start deviating from the common phase. It is noted that

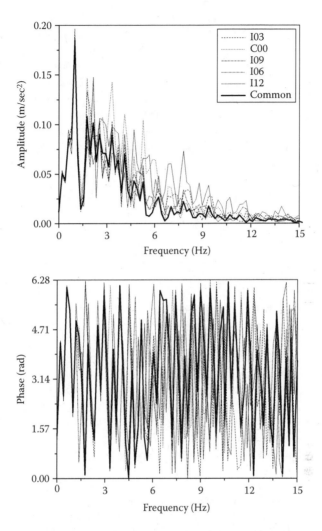

FIGURE 4.7 Comparison of the amplitude and phase of the common component evaluated from the data at C00 and four inner ring stations of the SMART-1 array during the S-wave window of the north-south component of the motions of Event 5 with the amplitude and phase identified from the recorded data at each station. The top subfigure presents the amplitude variability with frequency and the bottom subfigure the phase variability. (After A. Zerva and O. Zhang, "Correlation patterns in characteristics of spatially variable seismic ground motions," *Earthquake Engineering and Structural Dynamics*, Vol. 26, pp. 19–39, Copyright ©1997 John Wiley & Sons Limited; reproduced with permission.)

phases were restricted in the range $[0, 2\pi)$, and, therefore, jumps of approximately 2π do not indicate drastic variations in their values. At higher frequencies, the common component amplitude becomes lower than the amplitudes identified from the data at each station and phases vary randomly.

4.2.5 ALIGNMENT OF TIME HISTORIES

Part of the variabilities in the amplitudes and phases identified from the data at the individual stations around the common component characteristics in Fig. 4.7 are due to the fact that the time history approximation (Eq. 4.17) does not allow for the random arrival time perturbations of the waveforms at the array stations. Their effect, discussed in Sections 2.3.2 and 2.4.1, can be noted in the comparison of the recorded and reconstructed motions in Fig. 4.6, when, e.g., the recorded ground motion at station I12 arrives earlier than the time predicted by the average apparent propagation velocity of the waveforms across the array. To partially eliminate these arrival time perturbations in the subsequent evaluations, the time histories were aligned with respect to C00 with the approach described in Section 2.3.2. The aligned time histories at C00 and the four inner ring stations were presented in Fig. 2.8(a); later in this section, the aligned time histories at C00 and the middle ring stations, shown in Fig. 2.8(b), will also be utilized.

4.2.6 COMMON COMPONENTS IN ALIGNED MOTIONS

For the identification of the amplitudes and phases of the aligned motions, the error function (Eq. 4.18) was used again in the minimization scheme but, in the sinusoidal approximation of the motions (Eq. 4.17), the term $\vec{\kappa}(\omega_j) \cdot \vec{r}$ was set equal to zero, since the aligned motions arrive simultaneously at all array stations. Figure 4.8(a) presents the amplitudes and the phases of the common component, in the top and bottom parts of the figure, respectively, determined from the application of the least-squares minimization scheme to the aligned motions at C00 and the four inner ring stations (I03, I06, I09, I12), together with the amplitudes and phases identified using the aligned motions recorded at one station at a time. Figures 4.7 and 4.8(a) indicate that the variability range of the amplitudes and phases of the data at each station around the common component characteristics is reduced when the motions are aligned, especially at higher frequencies: The amplitude and phase of the common component of the aligned data (Fig. 4.8(a)) appear to represent the average of the amplitudes and phases at the individual stations over a wider frequency range, i.e., up to approximately 6 Hz rather than 4.5 Hz, which was the case in the analysis of the non-aligned data

$\longrightarrow$

FIGURE 4.8 Part (a): Comparison of the amplitude and phase of the common component evaluated from the aligned data at C00 and four inner ring stations during the S-wave window of the north-south component of Event 5 recorded the SMART-1 array with the amplitude and phase identified from the recorded data at each station. The top subfigure presents the amplitude variability with frequency, and the bottom subfigure the phase variability. Part (b): The corresponding variation of amplitudes and phases evaluated from the aligned motions at C00 and four middle ring stations. (After A. Zerva and O. Zhang, "Correlation patterns in characteristics of spatially variable seismic ground motions," *Earthquake Engineering and Structural Dynamics*, Vol. 26, pp. 19–39, Copyright ©1997 John Wiley & Sons Limited; reproduced with permission.)

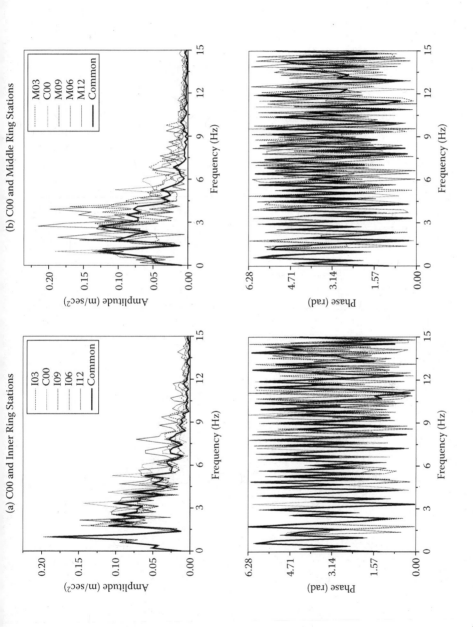

FIGURE 4.8

(Fig. 4.7). Figure 4.8(b) presents the results of the application of the methodology to the aligned seismic motions at the reference station C00 and the four middle ring stations, M03, M06, M09 and M12 (Fig. 2.8(b)). It can be seen from the comparison of parts (a) and (b) in Fig. 4.8 that the range of the variability in the amplitudes and phases identified at the individual stations around the common amplitude and phase for the middle ring data (Fig. 4.8(b)) is wider than the range of the variability for the inner ring data (Fig. 4.8(a)), and that the common component identified from the middle ring data represents the average of the characteristics of the motions over a shorter frequency range (up to approximately 4.5 Hz).

Even though separate analyses were conducted for the data at the shorter and longer separation distances (Figs. 4.8(a) and 4.8(b)), the amplitude and phase of the common components identified from the two sets of data, shown together for comparison purposes in Fig. 4.9, are very similar, especially considering that the longest separation distance for the middle ring stations is 2 km, whereas the maximum separation distance for the inner ring stations is 400 m. Their differences are an expected consequence of the larger scatter in the data at the further-away stations due to the attenuation of the waves, the more significant variations in the site topography and the decreasing wavelength of the wave components at increasing frequencies. The agreement of the amplitudes and phases over an extended area of radius of 1 km in Fig. 4.9 strongly suggests the existence of the common, coherent component in the data. The common component amplitude can be viewed as a "mean value" representing the average amplification of the dominant wave component at the site and is associated with the common phase variability with frequency, that resembles random distribution between $[0, 2\pi)$. The spatial variability of the motions, in addition to their propagation effects already considered, is examined next in terms of normalized differential amplitudes and differential phases.

4.2.7 DIFFERENTIAL AMPLITUDE AND PHASE VARIABILITY IN ALIGNED MOTIONS

The normalized differential amplitudes are obtained by subtracting at each frequency, ω_j, the common component amplitude from the amplitudes identified at the individual stations and dividing by the common component amplitude, i.e.,

$$\Delta\Lambda_l(\omega_j) = \frac{\Lambda_l(\omega_j) - \Lambda_c(\omega_j)}{\Lambda_c(\omega_j)} \tag{4.19}$$

where $\Lambda_l(\omega_j)$ denotes the amplitude identified from the data at station l, and $\Lambda_c(\omega_j)$ is the common component amplitude. The differential phases are obtained by subtracting at each frequency the common component phase from the phases identified at each station, i.e.,

$$\Delta\phi_l(\omega_j) = \phi_l(\omega_j) - \phi_c(\omega_j) \tag{4.20}$$

where $\phi_l(\omega_j)$ is the phase identified from the data at station l, and $\phi_c(\omega_j)$ is the common component phase. Normalized differential amplitudes and differential phases obtained from the data at C00 and the inner ring stations are presented in the top and bottom subplots of Fig. 4.10(a), respectively, and the corresponding results for the

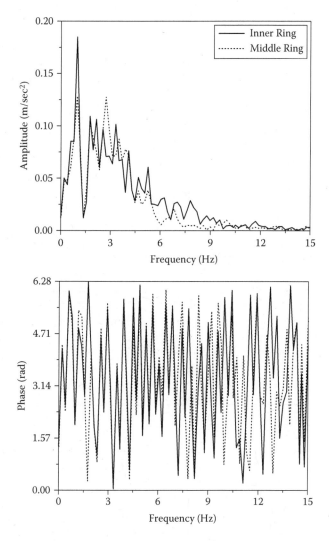

FIGURE 4.9 Comparison of the common component amplitude and phase identified from the analysis of the aligned data during the S-wave window of the north-south component of Event 5 recorded at the inner and the middle ring stations of the SMART-1 array. The top subfigure presents the amplitude variability with frequency, and the bottom subfigure the phase variability. (Reproduced from A. Zerva and V. Zervas, "Spatial variation of seismic ground motions: An overview," *Applied Mechanics Reviews, ASME*, Vol. 55, pp. 271–297, Copyright © 2002, with permission from ASME.)

data at C00 and the middle ring stations in Fig. 4.10(b). The normalized differential amplitudes in the top two subplots of Fig. 4.10 are cut off at a maximum value of 7.0; their actual values, which are not important for the subsequent analysis, can be significantly high, because the common component amplitude can assume low values in the higher frequency range (top subplots of Fig. 4.8). It is also noted that the normalized differential amplitudes, due to their definition (Eq. 4.19), cannot assume values lower than (-1). The differential phases in the bottom two subplots of Fig. 4.10 are allowed to vary between $[-\pi, +\pi)$ rather than between $[0, 2\pi)$, as was the case for the absolute phases in the bottom two subplots of Fig. 4.8 and the bottom part of Fig. 4.9.

Envelope functions, drawn by eye, containing the variability range of the normalized differential amplitudes and the differential phases are also shown in Fig. 4.10 [606]. The phase envelope functions are symmetric with respect to the zero axis. The envelope functions for the amplitudes are symmetric only in the lower frequency range, as, by definition, they cannot become lower than (-1). Isolated peaks within the dominant frequency range of the motions are excluded from both the amplitude and phase envelope functions. The trend of the positive envelopes of the amplitudes and phases is very similar, suggesting that the amplitude and phase variability of the data around their respective common component characteristics are correlated. This observation provides insight into the physical causes underlying the spatial coherency: Amplitude variability is easier visualized and attributed to physical causes than phase variability. If amplitude and phase variability are correlated as shown in Fig. 4.10, it suffices to examine the amplitude variability in order to deduce the physical causes underlying the phase variability, and, consequently, the spatial coherency. For example: In the low frequency range, the envelope functions for both the normalized differential amplitudes and the differential phases are at close distances to the zero axis, with the distances for the middle ring data being longer than the distances for the inner ring data; this can be attributed to the long wavelength of the contributing waves at low frequencies, that do not "see" the site irregularities, particularly for the inner ring stations. As frequency increases within the dominant frequency range of the motions, the distance of the amplitude and phase envelope functions from the zero axis increases gradually with a slower rate for the data of the inner ring stations than the data of the middle ring stations. In this frequency range, the common component amplitudes represent the average of the amplitudes identified from the data at the individual stations (top subplots in Fig. 4.8) implying that the motions are controlled by the broadband signal that is modified in amplitude and phase as it traverses the horizontal variations of the layers underneath the array. The increase in the variabilities of amplitudes and phases around the common component as frequency increases

---

FIGURE 4.10 Part (a): Normalized differential amplitude and differential phase variability with respect to the common component of the aligned motions during the S-wave window of the north-south component of Event 5 recorded at C00 and the inner ring stations of the SMART-1 array. Part (b): The corresponding normalized differential amplitude and differential phase variability with respect to the common component of the aligned motions at C00 and the middle ring stations. The top subfigures present the normalized differential amplitude variability and the bottom subfigures the differential phase variability; the envelope functions in all subfigures are drawn by eye. (Reproduced from A. Zerva and V. Zervas, "Spatial variation of seismic ground motions: An overview," *Applied Mechanics Reviews, ASME*, Vol. 55, pp. 271–297, Copyright © 2002, with permission from ASME.)

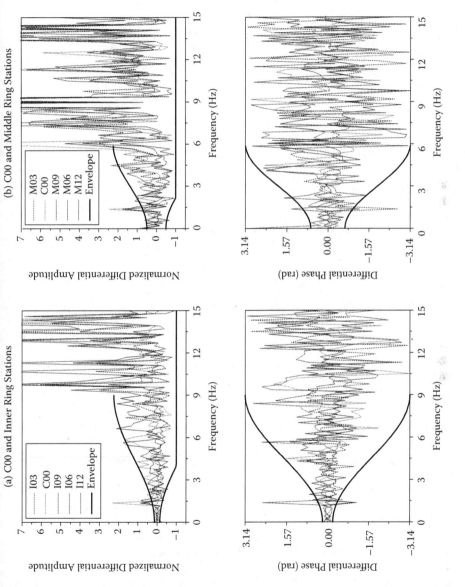

FIGURE 4.10

may also be associated with the decreasing wavelength of the signals at increasing frequencies, which is more pronounced for the middle ring results, and to the more significant contribution of scattered energy. The decreasing wavelength of the signals may also be the reason for the shorter frequency range over which the envelope functions of the normalized differential amplitudes and differential phases of the middle ring data increase with frequency (Fig. 4.10(b)) as compared to the corresponding frequency range of the inner ring data (Fig. 4.10(a)). At higher frequencies, wave components in addition to the broadband signal, and, mainly, scattered energy (noise) dominate the motions. Because these wave components propagate at different velocities, phases at the individual stations (bottom two subplots of Fig. 4.8) deviate significantly from the common phase, and the common signal amplitude no longer represents an average value, but becomes lower than the amplitudes identified at the individual stations (top two subplots of Fig. 4.8). Consequently, in Fig. 4.10, the phase differences in the higher frequency range vary randomly between $[-\pi, +\pi)$, i.e., the differential phase envelopes are parallel to the zero axis at a distance equal to π, and the normalized differential amplitudes assume high values. Noise may also be the cause for the isolated peaks of the normalized differential amplitudes and differential phases in the dominant frequency range of the motions (Fig. 4.10). This variability, however, may not be of significant consequence for the modeling or the simulation of spatially variable seismic ground motions, as it occurs when amplitudes are low within the dominant frequency range of the motions (top subplots of Fig. 4.8).

4.2.8 DIFFERENTIAL PHASE VARIABILITY AND SPATIAL COHERENCY

The envelope functions in Fig. 4.10 contain the variability range of amplitudes and phases around the common component characteristics for motions recorded at separation distances less than the maximum station separation distance for the area considered, i.e., 400 m and 2 km for the inner and middle ring data, respectively. Constrained within the envelope functions, amplitudes and phases vary randomly, suggesting that the variability can be described by the product of the envelope function and a random number uniformly distributed within a specific range. This behavior of the phases is related to the lagged coherency: It is recalled from Eqs. 2.83 and 2.84 that the differences in the phases of the motions between station pairs are directly related to the lagged coherency through envelope functions containing their random variability [6]; this behavior was illustrated in Fig. 2.22. The similarity of Fig. 2.22 and the two bottom subplots in Fig. 4.10 suggests that the differential phase variability identified by means of this methodology is related to conventional coherency estimates.

It should be noted that lagged coherency reflects the differences between the Fourier phases of the motions at two recording stations, rather than the differences between the phases of the data at a station and the common, coherent component phase. It can be easily extrapolated, however, from Figs. 4.8 and 4.10, that the differences between the Fourier phases of the motions recorded at two stations will also vary randomly within bounds of envelope functions with trends similar to the trend of the envelope functions of the differential phase variability with respect to the common component. Figure 4.11(a) presents the unsmoothed phase spectra (Eq. 2.59) of the aligned motions for station pairs C00-I03, C00-I06, C00-I09 and C00-I12, at a

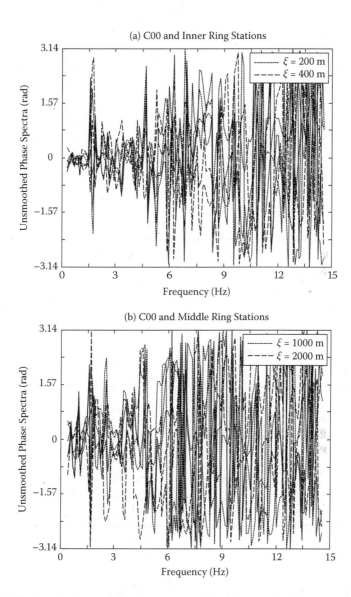

FIGURE 4.11 Part (a): Unsmoothed phase spectra of the aligned motions between station pairs C00-I03, C00-I06, C00-I09 and C00-I12 ($\xi = 200$ m), and I03-I09 and I06-I12 ($\xi = 400$ m) during the S-wave window of the north-south component of Event 5 recorded at the SMART-1 array. Part (b): The corresponding unsmoothed phase spectra of the aligned motions between station pairs C00-M03, C00-M06, C00-M09 and C00-M12 ($\xi = 1$ km), and M03-M09 and M06-M12 ($\xi = 2$ km).

separation distance of 200 m, and I03-I09 and I06-I12, at a separation distance of 400 m, i.e., the figure presents the phase differences of the motions between selected station pairs of the inner ring area of the array. Part (b) of the figure presents the corresponding unsmoothed phase spectra for station pairs C00-M03, C00-M06, C00-M09 and C00-M12, at a separation distance of 1 km, and M03-M09 and M06-M12, at a separation distance of 2 km, i.e., the phase differences of the motions between selected station pairs of the middle ring area of the array. It can be seen from the comparison of Fig. 4.11 and the bottom two subplots of Fig. 4.10 that the trend of the variation of the differential phases with frequency is similar, but the range of the differential phase variability between the data at two stations is wider than the differential phase variability between the data at a station and the common component; this should be anticipated, as the common component represents an "average" estimate of the data at all stations considered.

Figure 4.12 presents the lagged coherency of the motions between the inner and middle ring station pairs that were used in Fig. 4.11 for the evaluation of the unsmoothed phase spectra. The coherency estimates in Fig. 4.12 were obtained from spectral densities smoothed with a $M = 5$ Hamming window. The comparison of the amplitudes and phases of the data and the common component (Fig. 4.8), the envelope functions containing the variability of the normalized differential amplitudes and differential phases (Fig. 4.10), the phase spectra of the motions (Fig. 4.11) and the lagged coherency estimates (Fig. 4.12) relates, qualitatively, coherency with physical causes: In the low frequency range (<1.5 Hz), especially for the inner ring data, the envelope functions for the amplitude and phase variability are close to the zero axis (Fig. 4.10(a)), the phase spectra exhibit small variability around the zero axis (Fig. 4.11(a)), and the lagged coherency assumes values close to one (Fig. 4.12(a)); this behavior is a consequence of the long signal wavelength at low frequencies and the relatively short distance between stations. For the data at the further-away stations, the envelope functions are wider in the low frequency range (Fig. 4.10(b)), the phase spectra vary more significantly about the zero axis (Fig. 4.11(b)) than the phase spectra of the inner ring data, and this is also reflected in the coherency estimates of Fig. 4.12(b), which are lower than the corresponding values in Fig. 4.12(a). The isolated peaks at a frequency of approximately 1.5 Hz in all subplots of Figs. 4.10 and 4.11 appear as smoother troughs in the coherency estimates of Fig. 4.12 due to smoothing. Within the dominant frequency range of the motions (Fig. 4.8), the envelope functions increase with frequency (Fig. 4.10), the variability of the phase spectra about the zero axis also increases (Fig. 4.11), and coherency decreases (Fig. 4.12). This may, again, be attributed to the effect of the decreasing wavelength of the broadband signal over its dominant frequency range, which is more pronounced for the further-away stations, and to the more significant contribution of scattered energy. Past approximately 6 Hz and 4.5 Hz for the inner and the middle ring data, respectively, the common component amplitude no longer represents the average of the amplitudes at the individual stations (top subplots of Fig. 4.8), the envelope functions for the differential phase variability have almost reached the values of $\pm\pi$ (bottom subplots of Fig. 4.10), the phase spectra (Fig. 4.11) start varying randomly between $[-\pi, +\pi)$, and coherency decreases further, more so for the further-away stations than the close-by ones (Fig. 4.12), and starts exhibiting spurious peaks due to the

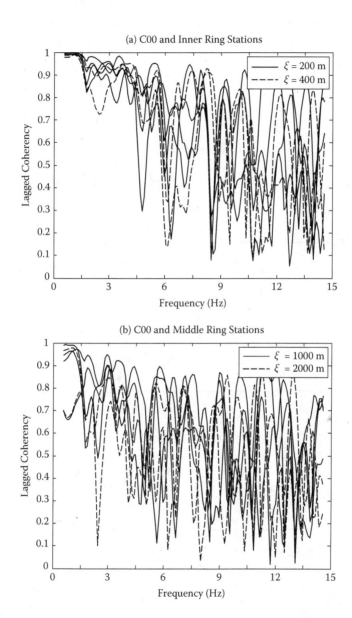

FIGURE 4.12 Part (a): Lagged coherency of the aligned motions between station pairs C00-I03, C00-I06, C00-I09 and C00-I12 ($\xi = 200$ m), and I03-I09 and I06-I12 ($\xi = 400$ m) during the S-wave window of the north-south component of Event 5 recorded at the SMART-1 array. Part (b): The corresponding lagged coherency of the aligned motions between station pairs C00-M03, C00-M06, C00-M09 and C00-M12 ($\xi = 1$ km), and M03-M09 and M06-M12 ($\xi = 2$ km).

low values of the power spectral densities in this frequency range (Section 2.4.2). At higher frequencies, the broadband S-wave does not contribute significantly to the motions, which are dominated by scattered energy (Fig. 4.8). The envelope functions for the differential phases become equal to $\pm\pi$ (bottom subplots of Fig. 4.10), the phase spectra (Fig. 4.11) vary randomly between $[-\pi, +\pi)$, and coherency should tend to zero. As indicated earlier (Section 2.4.2), however, the lagged coherency evaluated from recorded data (Fig. 4.12) can never reach zero values in the higher frequency range due to the presence of noise in the data, whereas the plane-wave coherency (Section 2.4.4 and Figs. 3.19(a), 3.20(b) and 3.21(b)) will tend to zero. Hence, this alternative approach for the investigation of the spatial variability of the seismic ground motions attributes the decay of the coherency to the dominant frequency range of the broadband S-wave, its wavelength, in relation to the separation distance between stations, and to the presence of scattered energy in the data.

4.2.9 ADDITIONAL APPLICATIONS

The aforementioned methodology was applied to additional data at the SMART-1 array [298], [608], [613]: the vertical and east-west components of the motions of Event 5, the three components of the motions of Event 39, and the horizontal components of the motions during Events 40 and 43. The common component characteristics were, obviously, different for each event and each direction of motion analyzed, but the trend of the amplitude and phase variability of the motions around their corresponding common component characteristics was similar to the trend described herein for the data in the north-south direction of Event 5 (Figs. 4.7–4.10). In order to investigate whether this pattern of behavior can be replicated at sites other than the "soft" soil site of the SMART-1 array, Liao and Zerva [298], [300] utilized the approach with data recorded at four additional arrays. The arrays and recorded events considered in their evaluation were: a subarray of the SMART-2 array [102] recording the Chi-Chi earthquake, the Pinyon Flat array [46] recording a small magnitude ($M_L = 3.7$) event, six close-by stations at Leona Valley [91] recording the Northridge earthquake, and the ZAYA array [568] recording a chemical explosion. Again, the examination of this diverse set of data indicated that the trend of the amplitude and phase variability of the motions around their corresponding common component characteristics, identified for each direction of motion and event analyzed, followed the pattern observed in the data presented herein. These investigations suggest that the approach provides a useful tool for obtaining insight into the physical causes underlying the spatial variability of the seismic ground motions.

5 Seismic Ground-Surface Strains

The transient deformation of the ground surface during an earthquake is closely related to the spatial variability of the seismic motions. The dynamic strain field caused by the seismic waves can have a significant effect on earthquake triggering, e.g., [175], [226], [502], ground failure, e.g., [276], [509] and damage to man-made structures, e.g., [36], [86], [523]. Generally, the seismic design of buried structures (pipelines and tunnels) relies directly on the characterization of the seismic ground deformations and strains, e.g., [217], [369], [388], [503], [542], which are imposed on the structure due to its interaction with the surrounding soil. In addition, the torsional response of above-ground building structures, can be significantly affected by the rotational components of the strain field, e.g., [213], [367], [369]. This chapter illustrates semi-empirical and analytical approaches for the estimation of seismic ground strains.

Seismic ground strains have been analyzed since the late 1960's, when Newmark [366], [367], based on the "traveling-wave" assumption, proposed strain estimates to be used in the seismic design of above-ground and buried structures. Consider, e.g., a single, monochromatic wave that propagates on the ground surface with a constant velocity c without any significant changes in its shape and without any interference from other waves. The expression for the displacement time history of such a wave can be written as:

$$u(x, t) = f\left(t - \frac{x}{c}\right) \tag{5.1}$$

The relationship between the strain, $\epsilon(x, t)$, and the (particle) velocity, $v(x, t)$, for the wave of Eq. 5.1 along its direction of propagation can be easily shown to be:

$$\epsilon(x, t) = \frac{\partial u(x, t)}{\partial x} = -\frac{1}{c} \frac{\partial u(x, t)}{\partial t} = -\frac{1}{c} v(x, t) \tag{5.2}$$

Equation 5.2 simply relates the space and time derivatives of the displacement field through a proportionality constant, the inverse of the apparent propagation velocity of the waves on the ground surface. With the consideration that the displaced configuration of a straight, buried pipeline is similar to that of the ground [369], [384], the maximum axial strain induced in the structure was then approximated by the maximum ground strain, ϵ_{max}, which, from Eq. 5.2, is given by [366], [369]:

$$\epsilon_{max} = \frac{(v_L)_{max}}{c} \tag{5.3}$$

where $(v_L)_{max}$ is the maximum horizontal (particle) velocity in the longitudinal direction of the pipeline and c is the corresponding component of the apparent velocity of the waves with respect to the ground surface. The rotational ground deformation for the evaluation of the torsional response in structures was also approximated by a similar expression. In this case, the maximum angle of rotation of the ground about the vertical axis, θ_{max}, is estimated from Eq. 5.2 as [367], [369]:

$$\theta_{max} = \frac{(v_T)_{max}}{c} \tag{5.4}$$

where $(v_T)_{max}$ is now the maximum horizontal (particle) velocity in the transverse direction.

Observational data and direct measurements of the transient seismic strain field are limited. Recent evaluations of strains still rely on the traveling-wave assumption of Eqs. 5.1–5.4, and consider that the seismic ground motions are composed mainly of body or surface waves for the estimation of the apparent propagation velocity. Strain estimates based on these expressions, or similar ones illustrated later in this chapter, are commonly referred to as "single-station" strain estimates, and their evaluation requires the knowledge of the particle velocity and a representative propagation velocity of the motions. The particle velocity may be available from recorded data at a single station and/or attenuation relations. As illustrated in Chapters 2 and 4, appropriate estimates for the apparent propagation velocity of the motions are difficult to obtain. At the sites of dense instrument arrays, frequency-wavenumber techniques (Section 4.1) can be utilized for the estimation of the propagation characteristics of the contributing waves during different time windows and at different frequencies for each component of the motions and each event. However, in the absence of such information, the apparent propagation velocity of the motions needs to be evaluated analytically from the source-site geometry and the characteristics of the soil profile. It can be clearly seen from Eqs. 5.3 and 5.4 that uncertainties in the estimated apparent propagation velocity of the waveforms will directly affect the value of the seismic ground strain. Additionally, any strain estimate based on the traveling-wave assumption will only be an approximation for the actual strains, as Eq. 5.2 does not consider the effect of the change in the shape of the motions (dispersion, amplitude variability and loss of coherency) on the seismic ground deformation.

This chapter presents an overview of the modeling and estimation of transient seismic ground-surface strains, and illustrates the effect of the issues addressed above, namely, (i) the assessment of the apparent propagation velocity of the motions, and (ii) the consideration of the various causes underlying the ground deformation. It begins, in Section 5.1, with a description of initial efforts to determine the apparent propagation velocity of the motions at a given soil profile considering that the motions are comprised mostly of either body waves or surface waves. The sites were modeled as horizontally layered media, which is an approximation commonly used in numerical estimations of seismic strains. It is followed by Section 5.2 that describes briefly initial efforts for the analytical/numerical estimation of the seismic strain field (longitudinal and shear strains, and rotational and torsional components), as well as its approximation from array data. The accuracy of the single-station strain estimate is examined in Section 5.3 through its comparison with strainmeter data; the difficulty in assessing the value of the apparent propagation velocity of the motions and its effect on the final value of the strain estimate are also described. Section 5.4 illustrates the relative effects of amplitude variability, loss of coherency and apparent propagation of the waveforms on the seismic strain field. Displacement gradients, estimated with an approach introduced in seismology from the geodetic community, are presented in Section 5.5. The comparison of the single-station strain estimates of Eq. 5.2 with the "seismo-geodetic" ones at low and higher frequencies is presented first, and followed by the evaluation of displacement gradients on the ground surface and at depth. A brief reference on the significance of downhole array data for the 1-D characterization of the site is also included in Section 5.5. Finally, some observations on the accuracy of strain estimates conclude the chapter.

5.1   SEMI-EMPIRICAL ESTIMATION OF THE PROPAGATION VELOCITY

Early studies addressing the identification of the apparent propagation velocity of the seismic ground motions were geared towards the estimation of seismic strains (Eq. 5.3) for the seismic design of buried pipelines and tunnels. Axial deformations in these structures are induced by the seismic wave components with motions parallel to the axis of the structure, and bending deformations are caused by the seismic wave components with motions perpendicular to the axis of the structure. Axial strains are considered to be more important in the seismic design of straight, buried, small diameter pipelines [384]. As the radius of the pipeline increases, however, axial stresses decrease, and the structure becomes sensitive to bending stresses as well; an example illustration of the effects of the loss of coherency on the axial and transverse response of a buried pipeline [221] will be presented in Section 6.3.1. In addition, for tunnels, shear waves propagating in a direction normal or nearly normal to the structure's axis can lead to deformations of the cross section of the tunnel lining (ovaling or racking deformations) [217].

In a series of publications, O'Rourke *et al.* [384], [385], [386], [387] proposed semi-empirical approaches for the evaluation of the propagation velocities of body (shear) and surface (Rayleigh) waves. Sections 5.1.1 and 5.1.2 present their proposed methodologies for the estimation of the apparent velocity of shear waves and the evaluation of the dispersive phase velocity of Rayleigh waves, respectively.

5.1.1   ESTIMATION OF THE APPARENT VELOCITY OF BODY WAVES

The approach developed by O'Rourke *et al.* [384] considers a model site consisting of horizontal layers overlying a half-space (rock), as illustrated in Fig. 5.1. It is assumed that both P- and S-waves are incident from the half-space to the bottom layer of the site with the same angle γ_r. As the waves travel through the softer layers, their path changes direction, and they impinge the ground surface with angles γ_p and γ_s for the P- and S-wave, respectively. The evaluation of the apparent propagation velocity of the shear wave, during which strong shaking occurs, was the focus of the study. If c_s indicates the S-wave velocity of the top soil layer, the apparent propagation velocity of the motions on the ground surface is $c = c_s / \sin \gamma_s$ (Fig. 5.1).

To identify the value of γ_s, O'Rourke *et al.* [384] presented the following procedure. The principal directions of the ground motions were first evaluated using a

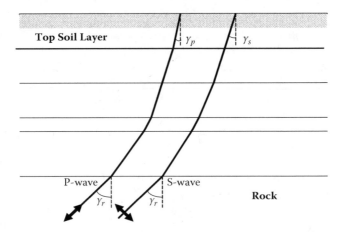

FIGURE 5.1 Illustration of the propagation pattern of body waves at a horizontally layered site. The incident angle, γ_r, of P- and SV-waves from the half-space (rock) to the bottom layer is assumed to be the same. The thick arrows indicate the direction of particle motions. The rays shift direction as they propagate through the layers to a more vertical path, and impinge the ground surface at angles γ_p and γ_s for the P- and SV-waves, respectively. (After M.J. O'Rourke, M.C. Bloom and R. Dobry, "Apparent propagation velocity of body waves," *Earthquake Engineering and Structural Dynamics*, Vol. 10, pp. 283–294, Copyright © 1982 John Wiley & Sons Limited; reproduced with permission.)

moving time window intensity tensor, $\mathbf{G}(t)$, with elements defined as:

$$g_{ij}(t) = \int_{t-0.5\delta t}^{t+0.5\delta t} a_i(\tau) a_j(\tau) d\tau \tag{5.5}$$

where $a_i(\tau)$ indicates the ground acceleration in the three mutually perpendicular directions, $i = X$ for east, Y for north and Z for vertical (up), and δt is the width of the moving window. As suggested by Arias [35] and Penzien and Watabe [400], described earlier in Section 3.3.2, the eigenvalues of $\mathbf{G}(t)$ correspond to the principal values of the moving time window intensity, and the components of the ground motions along the three principal directions are uncorrelated. The eigenvector corresponding to the largest eigenvalue indicates the predominant direction of the motions in the time window considered, and is illustrated by the vector $\vec{e}_G(t) = \{x(t), y(t), z(t)\}^T$ in Fig. 5.2. The figure also presents the longitudinal angle $\theta(t)$ and the collateral angle $\phi(t) = |\tan^{-1}\frac{x(t)}{z(t)\cos\theta(t)}|$ associated with the major principal eigenvector of $\mathbf{G}(t), \vec{e}_G(t)$. During the P-wave window, the motion is predominately in the vertical direction, yielding small values for the angle $\phi(t) \simeq \phi_p$, whereas at the onset of the S-wave, the motion becomes predominately horizontal, yielding values of $\phi(t)$ close to 90°. Additionally, at the beginning of the record, the motions contain the dominant P-wave, as well as reflected P- and SV-waves, whereas, during the strong motion S-wave window, incident SV- and SH-waves, as well as incident P-waves and reflected P-, SV- and SH- waves are present in the record. O'Rourke *et al.* [384] suggested that the examination of the S-wave window makes the direct evaluation of the relationship between $\phi(t)$ and the incident angle of the S-waves to the ground

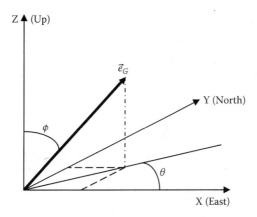

FIGURE 5.2 Illustration of the predominant direction of the surface ground motions indicated by $\vec{e}_G = \{x, y, z\}^T$. θ and $\phi = |\tan^{-1}\left(\frac{x}{z\cos\theta}\right)|$ are the longitudinal and collateral angles of $\vec{e}_G$, respectively. The dependence of the parameters on time t, utilized in the text, has been dropped in the figure. (After M.J. O'Rourke, M.C. Bloom and R. Dobry, "Apparent propagation velocity of body waves," *Earthquake Engineering and Structural Dynamics*, Vol. 10, pp. 283–294, Copyright ©1982 John Wiley & Sons Limited; reproduced with permission.)

surface, γ_s, difficult. Therefore, they first established the relationship between the incident angle of the P-wave to the ground surface, γ_p, and the collateral angle, ϕ_p, during the P-wave window, and then estimated γ_s from γ_p.

Because the compressional wave window incorporates not only the effect of the incident P-waves, but, also, the effect of the reflected P- and SV-waves on the ground surface, the predominant direction of the seismic motion ($\phi \simeq \phi_p$ in Fig. 5.2) does not coincide with the angle of incidence of the P-waves on the ground surface (γ_p in Fig. 5.1). For small values of ϕ_p and γ_p, O'Rourke *et al.* [384] suggested the approximate expression:

$$\gamma_p = \frac{\phi_p \beta}{2}; \quad \beta = \sqrt{\frac{2(1-\hat{v})}{1-2\hat{v}}} \tag{5.6}$$

with $\hat{v}$ indicating the Poisson's ratio of the medium. They further showed that, for $\hat{v} < 0.35$, the predominant direction of ground motion is slightly larger than the angle of incidence, whereas the opposite occurs for $\hat{v} > 0.35$.

The relationship between γ_p and γ_s for P- and S-waves, respectively, was then established by applying Snell's law at the rock-soil and soil-soil interfaces of the site (Fig. 5.1), as [384]:

$$\sin\gamma_s = \sqrt{\frac{(1-\hat{v}_r)(1-2\hat{v}_s)}{(1-2\hat{v}_r)(1-\hat{v}_s)}} \sin\gamma_p \tag{5.7}$$

where $\hat{v}_r$ and $\hat{v}_s$ are the Poisson's ratios for the rock and the top soil layer, respectively. Equations 5.6 and 5.7, and the assumptions that $\hat{v}_r = 0.25$ and that γ_p, γ_s and ϕ_p are small (less than 0.44 rad $= 25°$) led to the following approximate expression for the apparent propagation velocity of S-waves on the ground surface:

$$c = \frac{c_s}{\sin \gamma_s} \simeq \frac{c_s}{0.87 \phi_p} \tag{5.8}$$

It is noted that, for a site that can be approximated by horizontal layers (Fig. 5.1), the incident waves impinging from the bedrock shift direction, and their path tends to become vertical, so that the assumption that γ_p, γ_s and ϕ_p are small is valid [384].

Based on the aforementioned methodology, O'Rourke et $al.$ [384] estimated the apparent propagation velocity of the shear waves radiated from the 1971 San Fernando and the 1979 Imperial Valley earthquakes. In their evaluation, they utilized as δt the duration of the entire P-wave window, i.e., the window from the onset of the record to the first S-wave arrival. For the 1971 San Fernando earthquake and for recording locations at which information on the S-wave velocity of the top soil layer, c_s, was available, the apparent propagation velocity of the motions was estimated in the range of 1.26 – 9.33 km/sec, with a median value of 2.12 km/sec. For sites where the value of c_s was inferred from near-by soil conditions, the apparent propagation velocity of the motions ranged between 0.54 km/sec and 7.18 km/sec, with a median value of 2.22 km/sec. For the 1979 Imperial Valley earthquake and for a value of $c_s \approx 150$ m/sec obtained from preliminary information, the values for the apparent propagation velocity of the motions were estimated in the range of 1.34 – 20.9 km/sec, with a median value of 3.76 km/sec.

5.1.2 ESTIMATION OF THE APPARENT VELOCITY OF SURFACE WAVES

Surface waves are dispersive, i.e., their propagation velocities are functions of wavelength. Hence, a suitable value for the propagation velocity needs to be selected such that the strain estimates neither underestimate nor overestimate the actual ground strains. In addition, Love and Rayleigh waves yield different ground deformation patterns: Love waves cause horizontal motions perpendicular to the direction of propagation and induce bending strains in a buried pipeline, whereas Rayleigh waves follow an elliptical trajectory on a vertical plane in the direction of propagation, and the horizontal component of the motions causes axial strains. Focusing their attention on the seismic response of small diameter, buried pipelines, which is dominated by axial stresses [384], O'Rourke et $al.$ [387] evaluated a semi-empirical approach for the estimation of the dispersive propagation velocity of Rayleigh waves.

The peak ground strains induced by Rayleigh waves were also approximated by Eq. 5.3, rewritten as [387]:

$$\epsilon_{max} = \frac{v_{max}}{c_{ph}} \tag{5.9}$$

where c_{ph}, the phase velocity of the wave, is given by $c_{ph} = \lambda f$, with λ indicating the wavelength and f the frequency. The following semi-empirical dispersion curve for the phase velocity of Rayleigh waves at a site of a single layer of thickness H

overlying a half-space was then proposed [387]:

$$
c_{ph} = \begin{cases} 0.875\,c_H & Hf/c_L \leq 0.25 \\ 0.875\,c_H - (0.875\,c_H - c_L)(Hf/c_L - 0.25)/0.25 & 0.25 \leq Hf/c_L \leq 0.5 \\ c_L & Hf/c_L \geq 0.5 \end{cases}
$$

$$(5.10)$$

where c_L and c_H are the S-wave velocities of the layer and the half-space, respectively. The rationale behind Eq. 5.10 follows from the general properties of surface waves. For low frequencies ($Hf/c_L \leq 0.25$), i.e., long wavelengths, the waves are not significantly affected by the layer and propagate with a velocity that is, essentially, the S-wave velocity of the half-space. For large frequencies ($Hf/c_L \geq 0.5$), i.e., short wavelengths, the wave energy propagates inside the layer with the S-wave velocity of the layer. In-between these two extremes ($0.25 \leq Hf/c_L \leq 0.5$), Eq. 5.10 approximates the dispersion curves with a linear transition region.

The above approximation can be extended to more than one layer, as illustrated by O'Rourke et al. [387]: Consider, e.g., that the site consists of two layers with thicknesses of 9.1 m and 15.2 m and S-wave velocities of 122 m/sec and 228 m/sec, respectively, overlying a "half-space" with an S-wave velocity of 386 m/sec, as shown in Fig. 5.3(a). For high frequencies (short wavelengths), the incoming waves do not "see" the half-space. Instead the second layer acts as the half-space in Eq. 5.10, the soil profile is approximated by the model of Fig. 5.3(c), and the resulting approximate dispersion curve is presented in Fig. 5.3(d); the solid part of the line is valid for these higher frequency components. On the other hand, for low frequencies (long wavelengths), the incoming waves "see" a single layer, equivalent to the two actual ones ($c_L = (c_1 H_1 + c_2 H_2)/(H_1 + H_2) = 188$ m/sec), and the half-space, as shown in Fig. 5.3(e). The approximate dispersion curve in this case is the solid part of the line in Fig. 5.3(f). The approximate dispersion curve for the site is the combination of the solid lines of the dispersion curves of the high- and low-frequency models of Figs. 5.3(d) and (f), respectively, and is presented in Fig. 5.3(b) next to the actual soil profile (Fig. 5.3(a)). O'Rourke et al. [387] compared this approximate curve with the exact one (for Poisson's ratio and material density of 0.30 and 2.0 g/cm^3, respectively), and found that the two curves are in good agreement.

The strain estimates described by Eqs. 5.3, 5.8, 5.9 and 5.10, or similar derivations [503], are still being utilized in the evaluation of seismic strains for the design of buried pipelines when the site can be approximated by a horizontally layered medium [217], [389], provided that there is no slippage between the pipeline and the surrounding soil [388]. Katayama [261], from his evaluation of pipeline and soil strains during 45 earthquakes recorded at the Chiba experimental station of the Institute of Industrial Science, University of Tokyo, Japan, also reported that the axial pipeline strain was almost equal to the soil strain along the axis of the pipeline. He further noted that the ratio of the maximum axial pipeline strain (in 10^{-6}) to the corresponding peak ground acceleration (in cm/sec^2) was of the order of 0.15–0.30 when the major contributing waves were shear waves, but that the ratio was approximately equal to 4 when the strain was caused by surface waves, which can be attributed to the frequency content of the different wave types. The Naganoken-Seibu earthquake

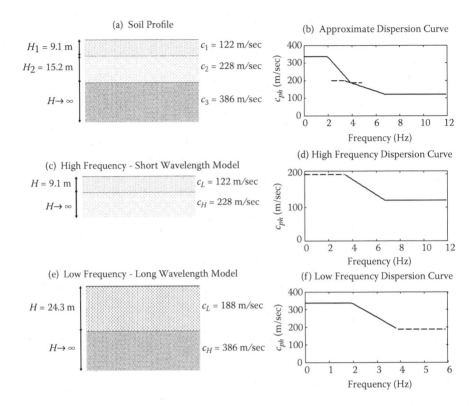

FIGURE 5.3 Approximate dispersion curve for the phase velocity, c_{ph}, of a Rayleigh wave in a horizontally layered site. Part (a) presents the horizontally layered soil profile. Part (b) shows the approximate dispersion curve for the entire site developed through the steps displayed in parts (c)-(f): Part (c) illustrates the soil profile for the high-frequency (short-wavelength) wave components and part (d) the approximate dispersion curve for the soil profile in (c). Parts (e) and (f) present the soil profile and the corresponding approximate dispersion curve for the low-frequency (long-wavelength) wave components. The solid lines in parts (d) and (f) represent the valid frequency range of the curves for the actual soil profile of part (a) and are combined to yield the approximate dispersion curve of part (b). (After M.J. O'Rourke, G. Castro and I. Hossain, "Horizontal soil strain due to seismic waves," *Journal of Geotechnical Engineering*, ASCE, Vol. 110, pp. 1173–1187, 1984; reproduced with permission from ASCE).

data ($M = 6.8$, epicentral distance of 232 km and focal depth 2 km) were dominated by surface waves, and, even though the peak acceleration was only 4.4 cm/sec^2, the peak strain reached 18.5×10^{-6}. On the other hand, for seismic data dominated by body waves, as, e.g., those of the Chibaken-Toho-Oki earthquake ($M = 6.7$, epicentral distance of 46 km and focal depth 58 km), the peak ground acceleration was 326.1 cm/sec^2 and the maximum strain 54.2×10^{-6}. However, Katayama [261] observed that the soil strain time history showed similarity to the particle velocity waveform and that the magnitude of the strain agreed roughly with the ratio of the particle velocity to the apparent propagation velocity of the motions only in the case of surface wave propagation. This suggests that causes in addition to the wave passage effect contribute to seismic strains, as will be illustrated in Sections 5.3.1, 5.4.1 and 5.5.1.

5.2 ESTIMATION OF THE SURFACE STRAIN FIELD

The previous section dealt with the estimation of longitudinal ground strains evaluated along the axis of the buried structure. These ground deformations, in addition to causing damage in buried pipelines [36], could also be the cause of the collapse of bridges during the 1971 San Fernando and the 1978 Miyaki-Ken-Oki earthquakes [86]. However, as indicated earlier (Eq. 5.4), the rotational components of the seismic excitation can have a significant effect on the structural response as well [367]. For example, Hart *et al.* [213] attributed a large part of the torsional response of high-rise buildings during the 1971 San Fernando earthquake to the rotational components of the motions.

These observations prompted the semi-empirical and analytical investigation of the rotational characteristics of the seismic wavefield: Nathan and MacKenzie [361], Tso and Hsu [544] and Rutenberg and Heidebrecht [433], among others, developed rotational response spectra based on the traveling-wave assumption. Castellani and Boffi [96], [97] presented a semi-empirical approach that takes into consideration the contribution of the various wave types to the rotational components of the motions: A recorded time history was first decomposed into its participating body and surface waves, deconvolved through a viscoelastic medium so that the amplitudes of the incoming waves from the bedrock were identified, and then propagated forward to yield spatially variable ground motions, from which the rotational components were inferred [97]. A stochastic-domain methodology based on the aforementioned approach was more recently presented by Zembaty *et al.* [583].

Bouchon and Aki [78] and Bouchon *et al.* [79] evaluated analytically the strain field components in the vicinity of faults. Trifunac [530], Wong and Trifunac [567] and Lee and Trifunac [289], [290] synthesized translational, rotational, strain and curvature time histories based on empirical response spectra and estimates for the duration of the motions. Teng and Qu [519] calculated long-period translations and rotations in the Los Angeles basin caused by a large earthquake at the San Andreas fault using the surface-wave Gaussian beam method for a 3-D Southern California crustal structure. Takeo and Ito [517] suggested that not only dislocations at the fault, but, also, additional "defects" contribute to the generation of waves. In particular, they noted that rotational components at the source generate directly rotational components in seismic waves, and derived expressions for the description of the rotational motions of the seismic wavefield.

The lack of observations of the seismic strain field, so that the analytical results can be compared and validated, was noted by the early analytical investigations. Spatially recorded array data provide a means for obtaining estimates of the rotational components of the seismic ground motions and have been utilized for such evaluations. Niazi [371] inferred rotational components of the data recorded at the El Centro differential array during the 1979 Imperial Valley earthquake, and Oliveira and Bolt [381] determined the rotational characteristics of the seismic motions recorded during five events at the SMART-1 array. Hao [196] proposed a coherency model and evaluated the torsional response spectrum of two events recorded at the SMART-1 array. Huang [229] inferred large rotational motions from a dense array near the northern end of the 1999 Chi-Chi earthquake rupture zone. Takeo [516] also reported large rotational components recorded, however, by one of few existing six-degree-of-freedom observational systems in the vicinity of a fault. The last two authors also pointed out that the analyses of the rotational ground motions may lead to useful constraints for the identification of the rupture mechanism at the source.

As illustration of the aforementioned developments, Section 5.2.1 presents the analytical estimates of the surface strain field in the vicinity of faults reported by Bouchon and Aki [78]. The observations by Takeo [516] from rotational components recorded in the vicinity of a fault are also presented in the section for comparison purposes. Section 5.2.2 briefly describes the simulation scheme developed by Trifunac *et al.* [525], [527], [567] and presents synthetic translational, rocking and torsional time histories resulting from the methodology. The torsional and rocking components of the seismic motions evaluated by Niazi [371] and by Oliveira and Bolt [381] from spatially recorded data are highlighted in Section 5.2.3. Hao's approach [196] for the development of the stochastic description of the torsional components of SMART-1 data is presented later in Section 5.4.2.

5.2.1 STRAIN COMPONENTS IN THE VICINITY OF FAULTS

Bouchon and Aki [78] presented one of the first analytical evaluations of seismic ground strains, tilts (rocking components) and rotations (torsional components) in the vicinity of a strike-slip and a dip-slip fault buried in layered media utilizing the discrete wavenumber representation method presented earlier by Bouchon [76], [77]. Compressional, shear and surface waves contributed to the motions over a frequency range between 0 and 2.5 Hz, although Bouchon and Aki [78] noted that the extension of the frequency range to 5 Hz did not significantly affect the amplitude of their results.

The surface view of the strike-slip fault model considered in their analysis is presented in Fig. 5.4(a). The 30 km long, vertical fault was buried at a depth of 1.5 km below the surface and extended to a depth of 10 km. Rupture at the fault initiated at its left end (point A in the figure) and propagated horizontally with a velocity of 2.2 km/sec. The dislocation at the source was described by a ramp function with rise time of 0.3 sec and final amplitude of 1 m. The medium was considered to be a layered half-space with the top layer having a thickness of 1.5 km, P- and S-wave velocities of 3 km/sec and 1.75 km/sec, respectively, and a density of 2.3 g/cm^3. The half-space had P- and S-wave velocities of 6 km/sec and 3.5 km/sec, respectively, and its density was 2.8 g/cm^3. Bouchon and Aki [78] based the configuration of this strike-slip fault model, in part, on the data of the 1966 Parkfield earthquake, and, hence, the results emulate translations and rotations of a moderate, strike-slip California earthquake. Components of the strain field were then evaluated at four stations located on a line perpendicular to the fault at distances of 1, 5, 10 and 20 km (stations 1-4 in Fig. 5.4(a), respectively), as well as four additional stations (not shown in the figure) located on a line parallel to the fault. Figure 5.5 presents the decay of the peak values of the strain components with distance from the source. In the figure, x_1

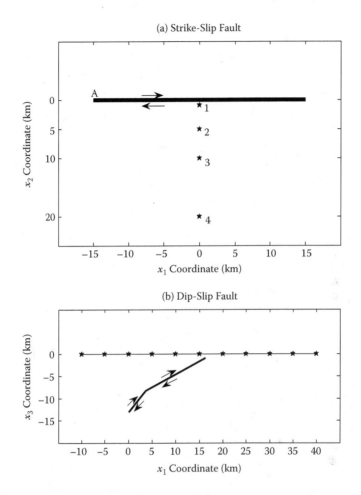

FIGURE 5.4 Source-station geometry of the strike-slip and dip-slip faults analyzed by Bouchon and Aki [78]. Stars indicate the location of the stations at which the ground motion characteristics were evaluated. Part (a) presents the plan view of the strike-slip model and the stations are referred to by the numbers 1-4; the letter A in the figure indicates the epicenter location. Part (b) presents the cross section of the dip-slip fault; the surface stations are referred to by their x_1 coordinate. (Reproduced from M. Bouchon and K. Aki, "Strain, tilt, and rotation associated with strong ground motion in the vicinity of earthquake faults," *Bulletin of the Seismological Society of America*, Vol. 72, pp. 1717–1738, Copyright © 1982 Seismological Society of America.)

and u_1 indicate the coordinate and displacement along the fault, x_2 and u_2 those normal to the fault on the ground surface (Fig. 5.4(a)), and x_3 and u_3 the vertical ones. The longitudinal ($\partial u_1/\partial x_1$ and $\partial u_2/\partial x_2$), shear ($\partial u_1/\partial x_2$ and $\partial u_2/\partial x_1$), rocking ($\partial u_3/\partial x_1$ and $\partial u_3/\partial x_1$) and torsional ($[\partial u_2/\partial x_1 - \partial u_1/\partial x_2]/2$) components of the strain field are presented in parts (a), (b), (c) and (d) of Fig. 5.5, respectively. It can be seen from the figure that the strain components decay rapidly as the distance from the fault

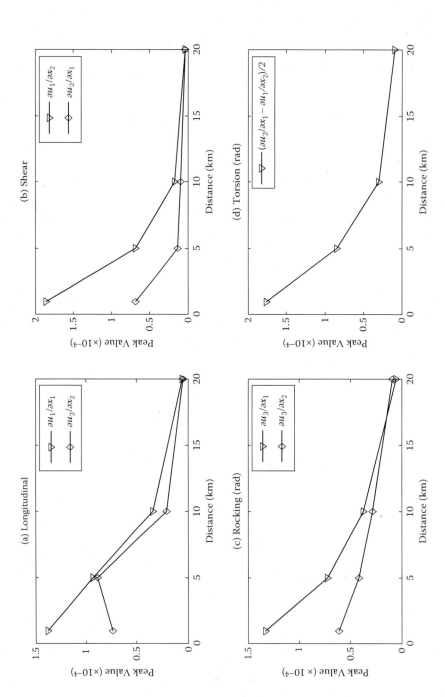

FIGURE 5.5 Variation of peak strains and rotations with distance obtained from the strike-slip fault model of Fig. 5.4(a). Part (a) represents the longitudinal strain components, part (b) the shear strain components, part (c) the rocking components (in rad) and part (d) indicates torsion (in rad) (adapted from Bouchon and Aki [78]).

increases. The peak values for the longitudinal and rocking (in rad) components are close to 1.5×10^{-4}, whereas those for shear and torsion (in rad) close to 2×10^{-4}.

For the dip-slip fault (Fig. 5.4(b)), Bouchon and Aki [78] utilized a source geometry inferred for the 1971 San Fernando earthquake except that, in the model, the fault stopped 1 km below the surface. The two fault segments had a dip of 52° and 30°, respectively, and the fault break was at a depth of 8.25 km. The rupture started at a depth of 13 km and propagated upward, with the width of the fractured zone being 12 km. The medium was approximated by a 2 km thick sedimentary layer with P- and S-wave velocities of 2.8 km/sec and 1.6 km/sec, respectively, overlying a homogeneous half-space with corresponding velocities of 5.6 km/sec and 3.2 km/sec. The density of the sediments and the basement rock were 2.3 g/cm³ and 2.8 g/cm³, respectively. The velocity of rupture propagation was considered to be 2 km/sec in the bedrock and 1 km/sec in the sediments with a fault slip of 1 m. Bouchon and Aki [78] computed strain field components at 11 locations on the ground surface distributed at 5 km intervals along the trace of the plane of symmetry of the source (Fig. 5.4(b)). Figure 5.6 presents the distribution of the peak longitudinal and rocking strain components along the fault trace; because the stations are located

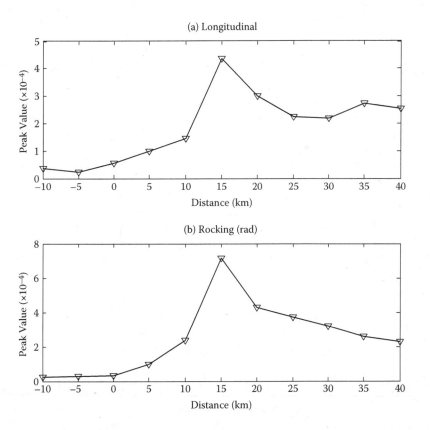

FIGURE 5.6 Variation of peak longitudinal strain, in part (a), and rocking component (in rad), in part (b), with distance obtained from the dip-slip fault model of Fig. 5.4(b) (adapted from Bouchon and Aki [78]).

along the trace of the fault (Fig. 5.4(b)), they are not subjected to shear strains. The peak strains occur at the tip of the fault and assume values close to 4×10^{-4} for the longitudinal strain, and close to 8×10^{-4} rad for the rocking component (Fig. 5.6). Bouchon and Aki [78] further evaluated translations and rotations for the same fault geometry depicted in Fig. 5.4(b), but for a softer surface layer of 1 km thickness, P- and S-wave velocities of 2.0 km/sec and 0.8 km/sec, respectively, and same values for the remaining parameters. They observed that the low-velocity sediments amplified the longitudinal strain but decreased the tilt; the corresponding peak values at the tip of the fault were 8×10^{-4} for the strain and 7×10^{-4} rad for the rocking component.

In examining the validity of Eq. 5.3, Bouchon and Aki [78] noted that the apparent propagation velocity in the equation should be related to the velocity in the bedrock and the rupture velocity at the source rather than the velocity of the uppermost layer, a result also previously obtained by Luco and Sotiropoulos [323]. It is noted that this conclusion does not contradict the rationale of Eqs. 5.3 and 5.8, where c (Eq. 5.8) is evaluated from the propagation velocity of the top layer, divided, however, by a factor that increases its value. They also noted that their assumption of uniform slip and rupture velocity over the entire fault plane and the approximation of the medium by a horizontally layered one resulted in low estimates of strains and rotations, and suggested that the consideration of nonuniform rupture at the fault and irregular subsurface topography would amplify their results.

The latter observation by Bouchon and Aki [78] was later confirmed by Takeo [516], who reported one of the very few instances of recorded rotations in the vicinity of a fault. The observation system consisted of a triaxial translational and a triaxial rotational sensor, similar to the six-degree-of-freedom recording system designed by Nigbor [374], that recorded a swarm of earthquakes offshore Ito, in the Izu peninsula, Japan, during March 1997. Takeo [516] reported that, even though the seismic moments of the two largest events in the swarm were two orders of magnitude lower than the seismic moment of the analytical strike-slip fault problem considered by Bouchon and Aki [78] (Fig. 5.4(a)), the recorded torsional velocity was approximately 3.3×10^{-3} rad/sec and 8.1×10^{-3} rad/sec for the largest and second largest events, respectively, at a source-site distance of 3.3 km, whereas the maximum torsional velocity evaluated by Bouchon and Aki [78] was approximately 1.2×10^{-3} rad/sec, 1 km away from the fault. Takeo [516], based on previous work by Takeo and Ito [517], attributed these differences to the spatial variation of the slip velocity at the fault and, also, to rotational strains at the seismic source that directly generate rotational components in seismic waves.

5.2.2 SYNTHETIC TRANSLATIONAL AND ROTATIONAL COMPONENTS

Trifunac et al. [289], [290], [525], [527], and Refs. [286], [287] and [288] for recent reviews of the evolution of the work, developed a methodology for the generation of synthetic translational and rotational (rocking and torsional) accelerations, as well as surface ground strains and curvatures. The steps of the approach are briefly summarized herein as an illustration for the generation of synthetic ground motions; alternative techniques based on seismological spectra are described in Chapter 8.

In the approach, the site profile is approximated by N horizontal, elastic layers, with the N-th layer denoting the bedrock. The layers are described by their depth, with the depth of the bedrock being equal to infinity, their P- and S-wave propagation velocities and their density. The site profile is used to evaluate the phase and group velocities of Rayleigh and Love waves. Compressional and shear wave contributions are approximately incorporated in the approach by including "higher-order modes" to the dispersion curves of the surface waves, which, however, have essentially constant velocities over a significant frequency range [567]. The ground motions are generated such that they have a prescribed Fourier amplitude spectrum, $FS(\omega)$, and duration, $T_d(\omega)$. The Fourier amplitude spectrum, $FS(\omega)$, is estimated from empirical scaling relations as function of earthquake magnitude, site classification, confidence level and depth of layers [286], [525], [528], [567], and the duration of the time history is evaluated from derived empirical expressions and the first and last arrival times of the wave components [287], [533], [537].

For each frequency band $\omega_n \pm \Delta\omega_n$, with ω_n indicating the discrete frequency and $\Delta\omega_n$ the frequency step, the synthetic acceleration time history component is generated according to the expression [527], [567]:

$$f_n(R, t) = \alpha_n \sum_{m=1}^{M} A_{mn} \frac{\sin\left[\Delta\omega_n\left(t - t_{mn}^*(R)\right)\right]}{t - t_{mn}^*(R)} \cos(\omega_n t + \phi_n) \tag{5.11}$$

in which, ϕ_n is a random phase angle uniformly distributed between $[-\pi, +\pi)$ and R is the source-site distance;

$$t_{mn}^*(R) = \frac{R}{v_m(\omega_n)} \tag{5.12}$$

is the arrival time at the station of the m-th mode of the surface wave propagating with velocity $v_m(\omega_n)$; the relative amplitudes of the different modes of the surface waves, A_{mn}, are empirically estimated as [527]:

$$A_{mn} = A_1(m)A_2(\omega_n) \tag{5.13}$$

$$A_1(m) = \left|\exp\left[-\frac{(m - m_0)^2}{2C_0^2}\right] + C_R X_m\right| \tag{5.14}$$

$$A_2(\omega_n) = \left|B_0 \exp\left[-\frac{(\omega_n - \omega_p)^2}{2\omega_B^2}\right] + B_R X_n\right| \tag{5.15}$$

where X_m and X_n are random numbers uniformly distributed in the range $[-1, +1)$ and the coefficients in Eqs. 5.14 and 5.15 are provided in Table 5.1; and

$$\alpha_n = \frac{2\Delta\omega_n FS(\omega)}{\frac{\pi}{2}\int_{\omega_n - \Delta\omega_n}^{\omega_n + \Delta\omega_n}\left|\sum_{m=1}^{M} A_{mn}\exp\{-i[(\omega - \omega_n)t_{mn}^*(R) - \phi_n]\}\right|d\omega} \tag{5.16}$$

is a scaling factor that makes the amplitude of the synthetic time history compatible with the target Fourier amplitude spectrum, $FS(\omega)$. The total time history is then the superposition of the contributions of each frequency band, i.e.,

$$f(R, t) = \sum_{n=1}^{N} f_n(R, t) \tag{5.17}$$

TABLE 5.1
Empirical coefficients of Eqs. 5.14 and 5.15. (Reproduced from *Soil Dynamics and Earthquake Engineering*, Vol. 4, V.W. Lee and M.D. Trifunac, "Torsional accelerograms," pp. 132–139, Copyright ©1985, with permission from Elsevier.)

Mode	C_0	m_0	C_R	B_0	ω_p	ω_B	B_R
1	3	5	0.2	1.5	10	5	0.1
2	3	5	0.2	1.5	10	5	0.1
3	3	5	0.2	1.5	10	5	0.1
4	3	5	0.2	2.0	25	15	0.1
5	3	5	0.2	2.0	25	15	0.1
6	3	6	0.2	3.0	30	10	0.3
7	3	7	0.2	1.5	30	5	0.25

where N is the total number of frequency bands and $f_n(R, t)$ is given by Eq. 5.11.

An example of the application of the approach to the Westmoreland site in the Imperial Valley, California, [288] is presented in the following. The site is approximated by five layers overlying the bedrock, with the total depth of the five top layers being 5.58 km, and P- and S-wave velocities and densities ranging from 1.70 km/sec, 0.98 km/sec and 1.28 g/cm³ for the uppermost layer to 6.40 km/sec, 3.70 km/sec and 2.71 g/cm³ for the half-space. Figure 5.7 presents the phase and group velocities (in parts (a) and (b), respectively) of the first five modes of the Rayleigh (solid lines) and Love waves (dashed lines) developed for the site. The solid and dashed lines farthest to the left model approximately the arrival of the components associated with the direct P- and S-waves. At low periods (high frequencies), all phase and group velocities approach the minimum S-wave velocity at the site, i.e., that of the top layer. The radial component of the synthetic acceleration, velocity and displacement time histories generated for a 6.5 magnitude earthquake at a source-site distance of 10 km is presented in Fig. 5.8.

Trifunac [531], based on wave propagation analyses, proceeded then with the evaluation of the rotational components of the motions for incident body waves, and Lee and Trifunac [289], [290] extended the methodology for the generation of synthetic time histories of the rocking component of the wavefield caused by Rayleigh waves and the torsional component caused by Love waves. The rocking and torsional acceleration, velocity and displacement time histories, corresponding to the characteristics of the earthquake used to generate the translational motions of Fig. 5.8, are presented in Figs. 5.9(a) and (b), respectively. Using asymptotic expansion of the expressions for the rocking and torsional components of the motions at low and high frequencies, Lee and Trifunac [289], [290] further noted that these components are indeed inversely proportional to the propagation velocity, which, however, varies: At low frequencies, the velocity approaches the maximum S-wave velocity at the site, i.e., the bedrock S-wave velocity, and, at high frequencies, the minimum S-wave velocity at the site, i.e., the S-wave velocity of the top layer. These conclusions are also in agreement with the approximate expression for the phase velocity (Eq. 5.10) proposed by O'Rourke *et al.* [387].

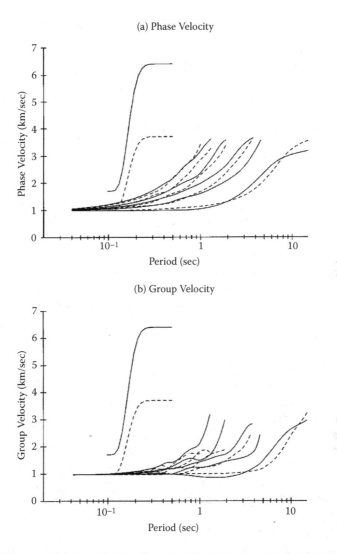

FIGURE 5.7 Phase and group velocities (in parts (a) and (b), respectively) of Rayleigh and Love waves developed for the Westmoreland site at Imperial Valley, California. The solid curves correspond to Rayleigh waves and the dashed curves to Love waves. The solid and dashed lines farthest to the left model approximately the arrival of the components associated with the direct P- and S-waves, whereas those farthest to the right the first mode of the Rayleigh and Love waves, respectively; the curves in-between represent modes two through five in increasing order from right to left. (Reprinted from V.W. Lee, "Empirical scaling of strong earthquake ground motion – Part III: Synthetic strong motion," *ISET Journal of Earthquake Technology*, Vol. 39, pp. 273–310, Copyright © 2002, with permission from the Indian Society of Earthquake Technology.)

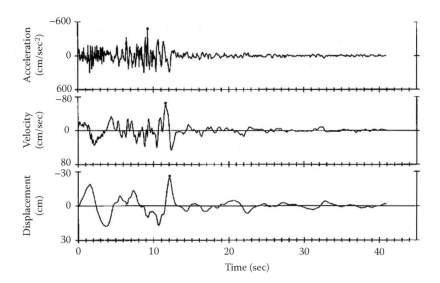

FIGURE 5.8 Synthetic acceleration, velocity and displacement time histories of the radial component of the seismic motions generated at the Westmoreland site, Imperial Valley, California, for a 6.5 magnitude earthquake with an epicentral distance of 10 km. (Reprinted from V.W. Lee, "Empirical scaling of strong earthquake ground motion – Part III: Synthetic strong motion," *ISET Journal of Earthquake Technology*, Vol. 39, pp. 273–310, Copyright © 2002, with permission from the Indian Society of Earthquake Technology.)

Ground-surface strains and curvatures were also analytically evaluated and their time histories synthesized [288], [530], [532]. Based on this methodology, Trifunac and Lee [536] presented attenuation relationships for ground strains evaluated as functions of peak ground velocity. Later, Todorovska and Trifunac [523] developed an approach for the peak strain microzonation of large areas, and, as an example application, conducted the microzonation of the Los Angeles metropolitan area. They also noted that larger strains ought to be expected in soft and deep soils, due to the smaller S-wave velocity of the layer, and smaller strains ought to be expected in the basement rock [523].

5.2.3 ROTATIONAL CHARACTERISTICS FROM ARRAY DATA

Niazi [371] reported perhaps the first evaluation of rotational components of the seismic wavefield based on spatially recorded data, the data recorded at the El Centro differential array (Fig. 1.1) during the 1979 Imperial Valley earthquake. To evaluate the rotational components of the motions, Niazi [371] assumed that a long, rigid, unembedded foundation extended over the length of the station spacing, i.e., the length of the foundation varied between 18 m, which was the shortest station separation distance of the array (between stations DA1 and DA2), and 213 m, i.e., the distance between stations DA1 and DA5 (Fig. 1.1).

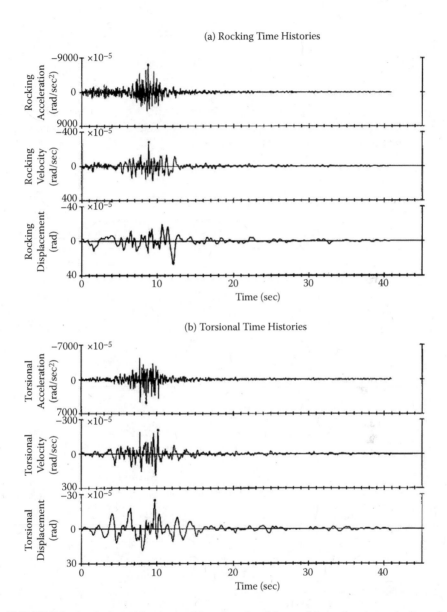

FIGURE 5.9 Synthetic rocking, in part (a), and torsional, in part (b), acceleration, velocity and displacement time histories generated for the earthquake whose translational, radial component of the motions is presented in Fig. 5.8. (Reprinted from V.W. Lee, "Empirical scaling of strong earthquake ground motion – Part III: Synthetic strong motion," *ISET Journal of Earthquake Technology*, Vol. 39, pp. 273–310, Copyright ©2002, with permission from the Indian Society of Earthquake Technology.)

Table 5.2 presents the results of his evaluation for the rigid translational, torsional (rotation about the vertical axis) and rocking (rotation about the east-west direction) components of the motions. The values of the ground motion parameters for the longest spacing in Table 5.2 are in parentheses: Niazi [371], based on an earlier study by Spudich and Cranswick [499], suggested that, due to the different noise characteristics at station DA5 in comparison to the other stations, the data from this station may not be reliable for differential motion evaluation[1].

Niazi [371] indicated that the scatter of the data in Table 5.2 was too large to lead to concrete conclusions, except that the largest rotation estimates occurred at the shortest station separation distance and dropped very significantly from a station spacing of 18 m to 37 m. He also noted that the largest rotations coincided with the arrival of the S-wave.

TABLE 5.2

Peak values of the rigid translational and rotational ground motion parameters evaluated from the data of the 1979 Imperial Valley earthquake at the location of the El Centro differential array as functions of the length of a rigid foundation or, equivalently, station separation distance. (After M. Niazi, "Inferred displacements, velocities and rotations of a long rigid foundation located at El Centro differential array site during the 1979 Imperial Valley, California, earthquake," *Earthquake Engineering and Structural Dynamics*, Vol. 14, pp. 531–542, Copyright © 1986 John Wiley & Sons Limited; reproduced with permission.)

	Stations DA1, DA2 $\xi = 18$ m	Stations DA2, DA3 $\xi = 37$ m	Stations DA1, DA3 $\xi = 55$ m	Stations DA1, DA4 $\xi = 128$ m	Stations DA1, DA5 $\xi = 213$ m
Displacement (cm)					
Vertical	12.2	12.5	12.3	—	(18.7)
East-West	49.4	49.6	49.6	49.1	—
North-South	16.0	15.9	16.0	15.9	—
Velocity (cm/sec)					
Vertical	20.0	20.7	20.3	—	(23.7)
East-West	72.9	73.0	72.9	72.5	—
North-South	44.6	42.6	43.9	43.4	—
Rotation (μrad)					
Torsion	1131.0	170.0	328.0	217.0	—
Rocking	1017.0	103.0	275.0	—	(851.0)
Rotation rate (μrad/sec)					
Torsion	1894.0	681.0	812.0	602.0	—
Rocking	1092.0	911.0	699.0	—	(903.0)

---

[1] This characteristic of the records at station DA5 was also observed in the evaluation of the spatial covariance of the data by Smith *et al.* [480] and King and Tucker [270], presented in Figs. 3.2 and 3.3. These authors, however, suggested that, even though this peculiarity increased the uncertainty in some of their estimates, the trend of the results was consistent (Figs. 3.2 and 3.3), and they felt justified in using the data at station DA5 in their analyses.

Oliveira and Bolt [381] evaluated rotational components of five earthquakes (Events 5, 24, 39, 43 and 45) recorded at the SMART-1 array. The rocking components along the horizontal axes, θ_1 and θ_2, were estimated as:

$$\theta_1 = \frac{\Delta u_3}{\Delta x_2}; \quad \theta_2 = \frac{\Delta u_3}{\Delta x_1} \tag{5.18}$$

where axis 1 indicates east, axis 2 north and axis 3 the vertical direction, and the torsional strain component, θ_3, as:

$$\theta_3 = \frac{1}{2}\left(\frac{\Delta u_2}{\Delta x_1} - \frac{\Delta u_1}{\Delta x_2}\right) \tag{5.19}$$

In Eqs. 5.18 and 5.19, $\Delta u_i, i = 1, 2, 3$, indicates the difference in the displacements between two stations in the i-th direction, and Δx_j, $j = 1, 2$, the component of the station separation distance along the j-th horizontal direction. The estimates of Eqs. 5.18 and 5.19 were evaluated from the data at the inner ring stations (Fig. 1.2) and averaged using stacking techniques. Table 5.3 presents the Peak Ground Acceleration (PGA) and Peak Ground Displacement (PGD) of the analyzed five events at the SMART-1 array, and the associated peak torsional and rocking components.

TABLE 5.3

Peak values of ground acceleration (PGA), ground displacement (PGD), and torsional and rocking components reported by Oliveira and Bolt [381] for five events at the SMART-1 array. (After C.S. Oliveira and B.A. Bolt, "Rotational components of surface strong ground motion," *Earthquake Engineering and Structural Dynamics*, Vol. 18, pp. 517–526, Copyright © 1989 John Wiley & Sons Limited; reproduced with permission.)

	PGA [cm/sec^2]	PGD [cm]	Torsion [$\times 10^{-6}$ rad]	Rocking [$\times 10^{-6}$ rad]
Event 5				
Vertical	64.0	0.5	—	4.4
Horizontal	244.0	2.5	8.5	—
Event 24				
Vertical	15.0	0.5	—	5.7
Horizontal	65.0	2.0	6.8	—
Event 39				
Vertical	314.0	0.6	—	5.1
Horizontal	375.0	10.0	14.6	—
Event 43				
Vertical	224.0	0.6	—	5.7
Horizontal	283.0	6.0	7.4	—
Event 45				
Vertical	104.0	2.5	—	12.8
Horizontal	238.0	8.5	39.3	—

Oliveira and Bolt [381] noted that there is a trend in the relation between the peak vertical displacement and the peak rocking component, i.e., small and comparative values of the peak vertical displacements correspond to small and comparative values of the peak rocking components, and an increase in the PGD leads to an increase in the peak rocking component (Table 5.3). However, this is not the case for the relation between the peak horizontal displacement and the peak torsional component of the motions. The horizontal PGD and the peak torsional component show significant scatter and no trend, which Oliveira and Bolt [381] attributed to the possible effect of the low frequency content of the horizontal displacements. They further reported that the torsional component for the events analyzed was larger than the rocking component (Table 5.3) and exhibited smaller error dispersion, that rocking in the radial direction was larger than rocking in the transverse direction, and that the highest values of the rotational components were observed at the onset of the S-wave and/or during the surface-wave windows of the motions.

5.3   ACCURACY OF SINGLE-STATION STRAIN ESTIMATION

The single-station strain estimates of Eqs. 5.2 and 5.9, as well as comparative ones for radial, tangential and shear strains (presented in the following subsection), rely on the following assumptions: (i) the seismic energy travels as plane waves with known azimuth and horizontal velocity; (ii) the surface deformation is caused solely by the propagation of the motions; and (iii) the medium is laterally homogeneous.

This section examines the adequacy of assumption (i): Section 5.3.1 presents a comparative evaluation of surface strains estimated from seismometer data and elastic wave theory with strains measured on a three-component long-base strainmeter system located at Pinyon (Piñon) Flat in California [174]. Section 5.3.2 illustrates the difficulty in assessing the value of the apparent propagation velocity for the evaluation of single-station strain estimates by comparing surface strains derived from the data recorded during the 1994 Northridge earthquake by two independent research groups. The contribution of causes other than the wave passage effect to the seismic strain field (assumption (ii)) is presented in Section 5.4. The third assumption considers that the site can be approximated by horizontal layers extending to infinity, as utilized in the analytical approaches described in Sections 5.1, 5.2.1 and 5.2.2. Lateral variations in the site's subsurface topography, however, can significantly amplify seismic ground strains, e.g., [297], as will be briefly illustrated in Section 5.6. In addition, assumption (iii) implies that the soil properties within each layer can be considered "deterministic", when, in reality, they are uncertain, e.g., [159], a fact that will invariably affect the value of the strain estimate. It is noted that the spatial variation of the seismic ground motions incorporates the effect of the variability in the soil properties at a site in the local "scattering" effect (Figs. 1.4 and 3.8).

5.3.1   COMPARISON OF STRAINS FROM STRAINMETER AND SEISMOMETER DATA

Using elastic plane-wave theory, Gomberg and Agnew [174] derived the following relationships for radial, ϵ_{rr}, tangential, $\epsilon_{\phi\phi}$, and shear, $\epsilon_{r\phi}$, strains for a single mode of Rayleigh and Love waves as:

$$\epsilon_{rr} \simeq -\frac{1}{c_R}\frac{\partial u_r}{\partial t}; \qquad \epsilon_{\phi\phi} = 0; \qquad \epsilon_{r\phi} \simeq -\frac{1}{2c_L}\frac{\partial u_\phi}{\partial t} \tag{5.20}$$

where c_R and c_L are the Rayleigh and Love wave phase velocities, respectively, u_r and u_ϕ indicate displacements in the radial and tangential directions, and the dependence of the functions in Eq. 5.20 on time, t, and frequency, f, has been dropped for convenience. Equation 5.20 indicates that the tangential strain, $\epsilon_{\phi\phi}$, should be zero, a consequence of plane-wave propagation away from the source. The equation also suggests that, for a single surface wave mode or multiple modes with phase velocities of similar functional forms, the strain estimates ϵ_{rr} and $\epsilon_{r\phi}$ can be determined from the particle velocity time histories, provided that the functional forms of the phase velocities are known at each frequency [174].

Gomberg and Agnew [174] then compared strain estimates resulting from the application of Eq. 5.20 to seismometer data recorded at the Pinyon Flat Observatory with seismic strains measured directly by strainmeters. They considered three regional earthquakes: St. George, Utah, at a distance of 470 km; Little Skull Mountain, Nevada, at a distance of 345 km; and the 1994 Northridge, California, earthquake at a distance of 206 km. The range of frequencies considered was $f = 0.05 - 0.5$ Hz, the data were rotated to the radial and tangential direction of each event, and the phase velocities were estimated from the soil profile underneath the site. Figure 5.10 presents the comparison between the single-station strain estimates obtained from the recorded translational motions and the strainmeter data. The traces of the estimated and recorded radial (ϵ_{rr}) and shear ($\epsilon_{r\phi}$) strains appear to have the same trend, but, also, exhibit significant differences, more in the amplitudes than in the phases. Additionally, the tangential strain, $\epsilon_{\phi\phi}$, obtained from the strainmeter data has finite values, when its theoretical estimate is zero (Eq. 5.20). The root-mean-square (rms) value of the recorded tangential strain, in the worst case, exceeded the rms value of the other nonzero components. Gomberg and Agnew [174] reported that the ratio of the rms tangential strain to the radial one was $0.82, 0.29$ and 0.48 for the Northridge, Little Skull Mountain and St. George earthquakes, respectively, and the corresponding ratios of the rms tangential strain to the shear strain were $1.11, 0.61$ and 0.62. In examining the possible causes for the differences between the analytical and recorded tangential strain estimates, Gomberg and Agnew [174] re-evaluated their assumptions on the backazimuthal direction of the events, as well as the assumption of neglecting the curvature of the cylindrical wave fronts, but concluded that these factors did not contribute to the discrepancy in the tangential strain estimates. Instead, they attributed the existence of the recorded tangential strain to the distortion and scattering of the wavefield by lateral material heterogeneities and topographic effects.

To quantify the differences between the single-station estimates and the strainmeter data for the radial and shear strains (Fig. 5.10), Gomberg and Agnew [174] conducted a cross spectral estimation. Let the strainmeter (or true strain), $\bar{\epsilon}_{ij}(f)$, be related to the approximate strain estimate, ϵ_{ij} of Eq. 5.20, at a frequency f as:

$$\bar{\epsilon}_{ij}(f) = \Delta s(f)\epsilon_{ij}(f) \tag{5.21}$$

where $\Delta s(f)$ is a "correction factor" for the frequency-dependent slowness, $s(f) = c^{-1}(f)$, with $i = j = r$ and $c(f) = c_R(f)$ for the radial strain, and $i = r$, $j = \phi$

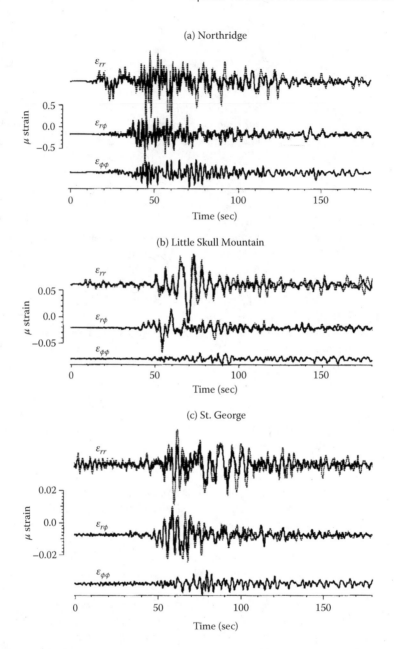

FIGURE 5.10 Comparison of single-station strain estimates from seismometers (dashed lines) and strainmeter data (solid lines) for the Northridge (part (a)), Little Skull Mountain (part (b)) and St. George (part (c)) earthquakes recorded at the Pinyon (Piñon) Flat Observatory. ϵ_{rr} is the radial, $\epsilon_{r\phi}$ the shear, and $\epsilon_{\phi\phi}$ the tangential strain; for the single-station estimates $\epsilon_{\phi\phi} = 0$. (Reprinted from J. Gomberg and D. Agnew, "The accuracy of seismic estimates of dynamic strains: An evaluation using strainmeter and seismometer data from Piñon Flat Observatory, California," *Bulletin of the Seismological Society of America*, Vol. 86, pp. 212–220, Copyright ©1996 Seismological Society of America.)

and $c(f) = c_L(f)$ for the shear strain. The correction factor was then estimated from the ratio of the smoothed cross spectral density between the approximate and the true strain estimates over the smoothed power spectral density of the approximate strains as [55]:

$$\Delta s(f) = \frac{\widehat{\epsilon_{ij}(f)\tilde{\epsilon}^*_{ij}(f)}}{\widehat{\epsilon_{ij}(f)\epsilon^*_{ij}(f)}} \tag{5.22}$$

where $*$ denotes complex conjugate, and $\widehat{\,\ldots\,}$ indicates smoothed estimators, for which a smoothing window of 0.051 Hz width was utilized. In the estimation of $\Delta s(f)$ of Eq. 5.22, Gomberg and Agnew [174] noted that, since the phase velocities, $c(f)$, are real-valued, the phase of $\Delta s(f)$ should be zero, and, since $c(f)$ is a smooth function, $|\Delta s(f)|$ should also be a smooth function.

Figure 5.11 presents the amplitude and phase estimates of $\Delta s(f)$ together with the 95% confidence intervals for the three earthquakes. It can be seen from the figure that the phases of the strain estimates appear to be more closely related to those of the actual strains below approximately 0.15 Hz for the Little Skull Mountain and St. George earthquakes (top subplots of parts (b) and (c), respectively); in this frequency range, phases are within ±10% of a cycle (at the 95% confidence level), but the uncertainty increases at higher frequencies. The uncertainty in the phases of $\Delta s(f)$ for the Northridge earthquake data (top subplots of part (a)) is higher than that for the other two events. The strain amplitudes for the Little Skull Mountain and St. George earthquakes (bottom subplots of parts (b) and (c), respectively) are at approximately ±20% in the frequency range between 0.04 − 0.1 Hz, but their values fluctuate significantly, and the uncertainty of the estimates can reach up to ±50%. For the Northridge data (part (a)), the "true" strain characteristics can be predicted by the single-station strain estimates only over narrow frequency bands with a ±25% uncertainty. Gomberg and Agnew [174] attributed the overall differences between the single-station strain estimates and the strainmeter data to their assumed site profile and to the contribution of higher surface-wave modes in the ground motions, which propagate with velocities that vary differently with frequency for each mode (see, e.g., Fig. 5.7(a)) and, hence, their effect is not captured by Eq. 5.20. They also suggested that the accuracy of the phase velocities limits the accuracy of the single-station strain estimates.

5.3.2 EFFECT OF APPARENT VELOCITY IN SINGLE-STATION STRAIN ESTIMATION

To illustrate how the uncertainty in the apparent propagation velocity of the waveforms is reflected on single-station strain estimates, this section presents strains obtained from data recorded during the Northridge earthquake by two different research groups.

Using a single-station strain estimate approach and assuming that mostly surface waves contribute to the motions, Trifunac *et al.* [541] estimated peak velocities and peak ground strains during the 1994 Northridge earthquake. Peak seismic ground strain factors were evaluated from the expression v_{max}/c_s for the horizontal (radial, ϵ_{rr}, and shear, $\epsilon_{r\phi}$) strains, and $v_{max}/1.73c_s$ for the vertical (ϵ_{zz}) ones, with v_{max} indicating, again, the maximum particle velocity in the corresponding direction, and

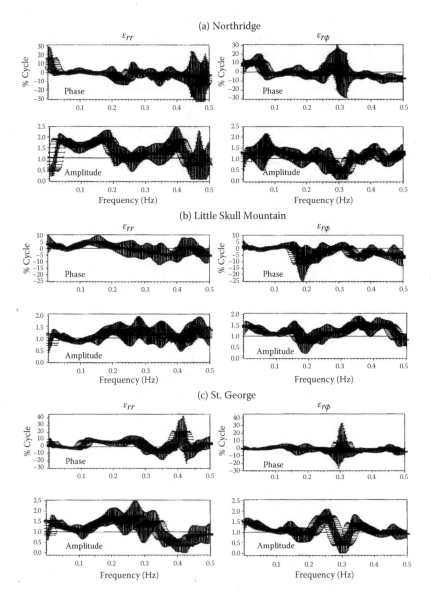

FIGURE 5.11 The phase (in the top subplots) and the amplitude (in the bottom subplots) of $\Delta s(f)$ (Eq. 5.22) for ϵ_{rr} (left subplots) and $\epsilon_{r\phi}$ (right subplots) evaluated from the single-station strain estimates and the strainmeter data at the Pinyon (Piñon) Flat Observatory. Part (a) presents the results for the Northridge, part (b) for the Little Skull Mountain, and part (c) for the St. George earthquake. The vertical bars indicate the 95% confidence intervals, and the horizontal bars are equal to the width of the spectral smoothing window (0.051 Hz). (Reprinted from J. Gomberg and D. Agnew, "The accuracy of seismic estimates of dynamic strains: An evaluation using strainmeter and seismometer data from Piñon Flat Observatory, California," *Bulletin of the Seismological Society of America*, Vol. 86, pp. 212–220, Copyright ©1996 Seismological Society of America.)

c_s the propagation velocity of the S-waves in the top soil layer. The velocity time histories were obtained from 169 records of the Northridge earthquake, and the average propagation velocity, c_s, was estimated as 275 m/sec for sites over quaternary deposits and 475 m/sec over geological rock. Figure 5.12 presents the contours of the seismic ground strain factor estimates for the horizontal and vertical strains, in parts (a) and (b), respectively, in the epicentral area of the event. The maximum values of the peak strain factors over the examined area were determined as approximately $10^{-2.2}(= 6.31 \times 10^{-3})$ for the horizontal strains (at the Rinaldi Receiving Station - RRS), and as approximately 10^{-3} for the vertical ones [541]. Trifunac et al. [541] suggested that, as a first approximation for an estimate of the actual peak strains, one may multiply the radial strain factors by $A = 0.36$, the shear ones by $A = 0.22$ and the vertical ones by $A = 1.7$, which yields peak strains of $\epsilon_{rr} = 2.27 \times 10^{-3}$ and $\epsilon_{zz} = 1.7 \times 10^{-3}$. Trifunac et al. [541], however, cautioned that these preliminary multiplication factors were evaluated for a site near Westmoreland in Imperial Valley, California, and their values may not apply to all locations of the contour maps of Fig. 5.12. They also suggested that, given the approximation of strains by the expression $\epsilon_{max} \simeq A v_{max}/c_s$ and the uncertainty in the value of c_s, their strain estimates should be accurate to within a factor of 3.

Gomberg [173] also used the single-station strain estimation technique to evaluate principal strains in the epicentral area of the Northridge earthquake, considering, however, that mostly body (S-) waves contribute to the motions; these strain estimates are presented in Fig. 5.13. In the approach, the velocity time histories were obtained from recorded accelerograms bandpassed in the frequency range $0.3 - 3$ Hz. The apparent propagation of the waveforms was evaluated from the standard Southern California model assuming a first-arriving S-wave. The estimated apparent propagation velocity was assumed to be $3 - 5$ km/sec for the large amplitude S-wave pulses that dominated many of the traces [173], as was also reported by Zhang and Papageorgiou [609]. To account for source finiteness, Gomberg [173] considered that the entire energy of the Northridge earthquake was emitted from one of three subevents, according to the source model of Wald et al. [557]; the locations of these subevents are indicated by stars labeled as A, B, and C, respectively, in Fig. 5.13. Three different analyses were conducted, each utilizing a different subevent. Figure 5.13 presents the principal horizontal strain traces (positive for extension and negative for compression) resulting from the consideration of subevent A; the vertical strain time histories are shown underneath the horizontal traces in the figure. The three numbers next to each trace indicate the peak values obtained from the analysis of each subevent. Gomberg [173] noted that the shape of the strain traces evaluated from the different subevent analyses varied significantly only for stations close to the location of each subevent. For illustration purposes, the strain traces at station U55 resulting from each of the three subevents are also shown in the figure. Figure 5.13 indicates that, throughout the analyzed region, the horizontal strains exceeded the vertical ones by more than a factor of two. The peak horizontal strain of amplitude 2.29×10^{-4} was observed at the RRS station for subevent B. Gomberg [173] also noted that the later-arriving energy in the seismograms, which travels more horizontally with lower apparent propagation velocities, can result in the seismic ground strains of Fig. 5.13 being underestimated by a factor of 3–4.

(a)

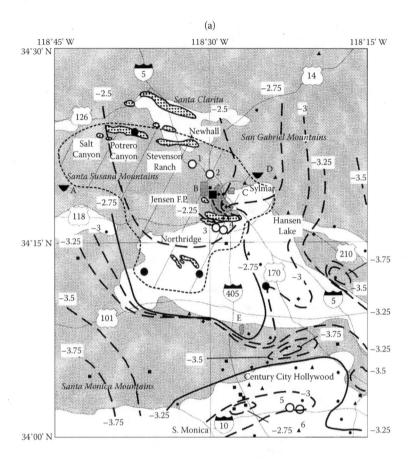

FIGURE 5.12 Part (a): The epicentral area of the Northridge earthquake with the short dashed line indicating the surface projection of the fault. The light gray areas denote the bedrock and the white areas the quaternary sediments. The triangular, circular, diamond and square symbols show the locations of the recording stations used in the analysis. The shaded regions outlined by heavy solid lines indicate zones of concentrated ground breakage. The letter A denotes the location of the Sinai Valley Tailings Dam, B the Jansen filtration plant, C the San Fernando Valley Juvenile Hall, D the Pacoima Dam site, and E the intersection of Mulhollad Drive and Beverly Glen Boulevard. The open circles, denoted by the numbers 1-6, indicate locations of extensive damage or collapse of freeway structures. Heavy full and dashed lines present the contours of the horizontal peak strain factors $\log_{10} \epsilon$ ($\epsilon = v_{max}/c_s$). Part (b), shown on the next page, is the same as part (a), but presents the contours of the vertical peak strain factors $\log_{10} \epsilon$ ($\epsilon = v_{max}/1.73c_s$). (Reprinted from *Soil Dynamics and Earthquake Engineering*, Vol. 15, M.D. Trifunac, M.I. Todorovska and S.S. Ivanovic, "Peak velocities and peak surface strains during Northridge, California, earthquake of 17 January 1994," pp. 301–310, Copyright ©1996, with permission from Elsevier; courtesy of M.D. Trifunac.)

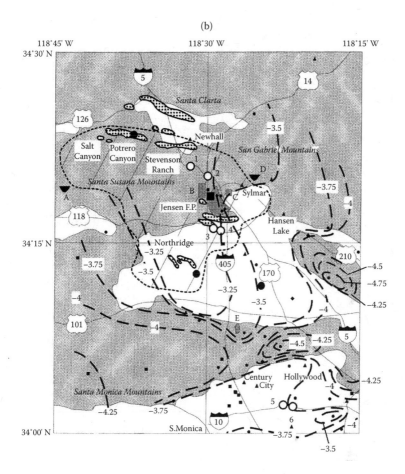

FIGURE 5.12 (Continued).

The results of the two independent investigations differ by approximately an order of magnitude[2]. The major differences between the peak values reported by the two studies (Figs. 5.12 and 5.13) are caused by the types of waves assumed to dominate the motions and the estimates of their apparent velocities: One study [541] considers that the dominant waves are mostly surface waves propagating with a velocity controlled by the velocity of the surface layer divided by a factor less than one for horizontal strains, whereas the other [173] that they are shear waves propagating with a velocity of 3 – 5 km/sec. The differences in the vertical strains may also be attributed to the same cause. It is noted, however, that with the error bounds provided by the independent investigations, the strain estimates of both approaches have an overlapping region. The difficulty in obtaining solid estimates for the apparent propagation velocity of

---

[2] It is noted that Zhang and Papageorgiou [609] reported strains of the order of 10^{-4} in the major epicentral area, whereas Hutchings and Jarpe [231] estimated that strains were of the order of 10^{-3} at the Interchange between Highways SR14 and I-5, where several of the interchange structures collapsed.

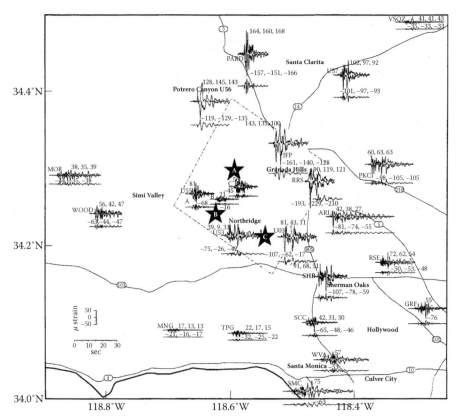

FIGURE 5.13 The epicentral area of the Northridge earthquake with the short dashed line indicating the surface projection of the fault. The stars labeled A, B and C represent the subevent locations used in the evaluation. The time histories in the figure are estimates of the horizontal principal strains and the vertical strains at the recording stations considered in the study. For the horizontal strains, positive traces correspond to extension and negative traces to compression. The vertical strains are plotted underneath the horizontal traces. Three strain time histories are plotted for station U55; they correspond to the three different subevents, A, B and C. For the remaining stations only the plots corresponding to subevent A are presented; the numbers next to the peaks indicate maximum values for each subevent. (Reprinted from *Soil Dynamics and Earthquake Engineering*, Vol. 16, J. Gomberg, "Dynamic deformation and the M6.7, Northridge, California, earthquake," pp. 471–494, Copyright©1997, with permission from Elsevier.)

the motions to be used in seismic ground strain evaluations will be illustrated again in Section 5.5.2.

5.4 INCOHERENCE VS. PROPAGATION EFFECTS IN SURFACE STRAINS

The single-station strain estimation assumes that ground deformations are caused solely by the wave passage of a single plane wave that propagates with constant velocity on the ground surface. When surface waves dominate, harmonic components propagating with different phase velocities (Fig. 5.7(a)) contribute to the motions, and their omission will lead to discrepancies between the actual strains and the single-station strain estimates. When body waves dominate, the variability in the amplitude and the phase of the wavefield (Fig. 4.8), the latter in excess to the phase caused by the wave passage effect and reflecting the loss of coherency in the motions (Section 2.4.2), will constitute an additional cause of ground deformation. Dispersion of surface waves and amplitude and (random) phase variability of body waves result in changes in the shape of the motions as they propagate on the ground surface.

Section 5.4.1 describes the effects of the apparent propagation and the change in the shape of the seismic motions on longitudinal strain estimates obtained from recorded data during the 1971 San Fernando earthquake [386]. The trend of these estimates is compared in the same section with strains generated through simulations [593] from a spatially variable random field. The contribution of wave passage and loss of coherency effects to the torsional characteristics of the wavefield estimated from data recorded at the SMART-1 array [196] is presented in Section 5.4.2.

5.4.1 ESTIMATED AND SIMULATED LONGITUDINAL STRAINS

O'Rourke *et al.* [386] evaluated and compared three seismic strain estimates from data recorded during the 1971 San Fernando earthquake. Twenty-two station pairs were utilized in the evaluation that were oriented approximately along the radial direction of the event. The radial component of the acceleration time histories was integrated twice to yield displacement time series, which were subsequently aligned (Sections 2.3.2 and 2.4.1) to remove the effect of the independent triggering of the accelerograms.

The first two strain estimates in the evaluation were maximum average strains obtained from the differences in the displacements at two stations i and j, denoted by $u_i(t)$ and $u_j(t)$, respectively, divided by the station separation distance L. As estimates of derivatives obtained from finite differences, these average maximum strains provide a lower bound for the actual peak strains. The first strain estimate was the maximum value of the average ground strain between the two stations due to apparent propagation effects only. In this case, the ground motions propagate on the ground surface without any change in shape, and the time history at the further-away station, j, is identical in shape to that at the close-by station, i, except for a time delay. The first strain estimate, $\epsilon_1(\tau)$, becomes then [386]:

$$\epsilon_1(\tau) = \frac{|u_i(t) - u_i(t - \tau)|_{\max}}{L} \tag{5.23}$$

where τ is an effective time lag to account for wave propagation and given by:

$$\tau = L/c \tag{5.24}$$

Different selections for τ reflect the effect of variable propagation velocity, c, on the strain estimate. The second strain estimate took into consideration the change in the shape of the motions at the two recording stations i and j, and, hence, utilized the displacements at both stations. The estimate is given by [386]:

$$\epsilon_2(\tau) = \frac{|u_i(t) - u_j(t - \tau)|_{max}}{L} \tag{5.25}$$

The third estimate was the single-station strain estimate of Eqs. 5.3 or 5.9:

$$\epsilon_3(\tau) = \frac{v_{max}\,\tau}{L} = \frac{v_{max}}{c} \tag{5.26}$$

with v_{max} indicating, again, the maximum particle velocity. Equation 5.26 is a "point" estimate of the strains, i.e., the expression represents the actual values of the peak strains caused solely by the wave passage effects.

Figure 5.14 presents the three ground strain estimates as functions of the time lag τ obtained from the data at two recording stations (R253 and F089) at a separation distance of 760 m. At $\tau = 0.1$ sec (or, equivalently, from Eq. 5.24, $c = 7600$ m/sec), the first and third strain estimates (Eqs. 5.23 and 5.26, respectively) coincide, i.e., $\epsilon_1(0.1\ \text{sec}) = \epsilon_3(0.1\ \text{sec})$. This is a consequence of the fact that $\tau = 0.1$ sec was the time step used in the evaluation, and, hence, at this value, $\epsilon_1(\tau)$ (Eq. 5.23) and $\epsilon_3(\tau)$ (Eq. 5.26) are identical. As indicated earlier, both $\epsilon_1(\tau)$ and $\epsilon_3(\tau)$ are strains caused solely by the apparent propagation of the waveforms. At low values of τ, the two estimates produce similar values, but as τ increases, $\epsilon_3(\tau)$ increases linearly with τ (Eq. 5.26) and becomes greater than $\epsilon_1(\tau)$. Because $\epsilon_3(\tau)$ is a point estimate of the

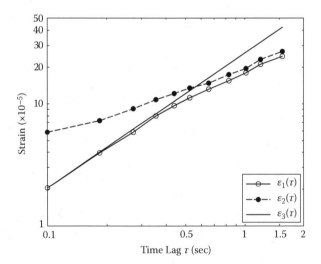

FIGURE 5.14 Strain estimates as functions of the time lag τ from data at a station pair (R253 and F089) with separation distance of 760 m recording the 1971 San Fernando earthquake. $\epsilon_1(\tau)$ is an average strain estimate incorporating wave passage effects; $\epsilon_2(\tau)$ is also an average strain estimate incorporating the effects of both the wave passage and the change in the shape of the motions at the two stations; and $\epsilon_3(\tau)$ is the single-station strain estimate. (After M.J. O'Rourke, G. Castro and N. Centola, "Effects of seismic wave propagation upon buried pipelines," *Earthquake Engineering and Structural Dynamics*, Vol. 8, pp. 455–467, Copyright ©1980 John Wiley & Sons Limited; reproduced with permission.)

motions, whereas $\epsilon_1(\tau)$ is an average estimate, $\epsilon_3(\tau)$ will always be an upper bound for $\epsilon_1(\tau)$. Additionally, past approximately 0.40 sec (or, equivalently, $c = 1900$ m/sec), $\epsilon_1(\tau)$ starts tapering off to a constant value, because it is bounded by the expression [386]:

$$\epsilon_1(\tau) \leq \frac{2|u_i(t)|_{\max}}{L} \qquad (5.27)$$

On the other hand, $\epsilon_2(\tau)$ (Eq. 5.25) incorporates the effects of both the change in the shape of the motions between the two stations and their apparent propagation on the ground surface. For low values of τ (large apparent propagation velocities), this estimate is approximately 2.5 times the values of the estimates caused by the apparent propagation effects only. For the range of τ less than approximately 0.6 sec ($c > 1270$ m/sec), $\epsilon_2(\tau)$ provides the highest values for the ground strain. Past this value, $\epsilon_2(\tau)$ becomes lower than $\epsilon_3(\tau)$, approaches $\epsilon_1(\tau)$ from above, and becomes, eventually, parallel to it. It is noted that $\epsilon_2(\tau)$ is also bounded [386], since:

$$\epsilon_2(\tau) \leq \frac{|u_i(t)|_{\max} + |u_j(t)|_{\max}}{L} \qquad (5.28)$$

The effective time lag $\tau \approx 0.6$ sec ($c \approx 1270$ m/sec) was termed the "cross-over" time lag by O'Rourke et al. [386]. They suggested that for $\tau < 0.6$ sec ($c > 1270$ m/sec) both the change in the shape of the motions and the apparent propagation effects contribute to the value of the seismic ground strains, and expressions like Eq. 5.25 should be used in the design of buried pipelines. On the other hand, when $\tau > 0.6$ sec ($c < 1270$ m/sec), the assumption that seismic motions propagate unchanged on the ground surface is a valid approximation for the evaluation of the seismic strains. O'Rourke et al. [386] also evaluated the three strain estimates (Eqs. 5.23, 5.25 and 5.26) from the data recorded at the 22 station pairs (with separation distances ranging between 50 m and 1000 m), and averaged the results. These average strains from the 22 station pairs followed the exact same pattern as those of the single station pair in Fig. 5.14. The single station pair results, however, are presented herein, as they render a one-to-one correspondence between the effective time lag and the propagation velocity through Eq. 5.24; this correspondence is no longer valid for the average strains from multiple station pairs, because their variable separation distance (L in the equation) yields multiple values for c for the same effective time lag τ.

Zerva [593] reached similar conclusions regarding the effect of the change in the shape of the motions and their apparent propagation on seismic ground strains by simulating strains from stationary and homogeneous random fields; the simulation scheme is described in Section 7.2.1. In the approach, the power spectral density of the motions was described by the Clough-Penzien spectrum (Eq. 3.7) with parameters $\omega_g = 15.46$ rad/sec, $\omega_f = 1.636$ rad/sec, $\zeta_g = 0.623$ and $\zeta_f = 0.619$ [221]. The coherency model utilized was the model of Luco and Wong [324], presented in Eq. 3.27, with three values for the coherency drop parameter, α, namely 1×10^{-3}, 5×10^{-4} and 2×10^{-4} sec/m. The first value of α corresponds to strongly incoherent motions, the last value to strongly coherent motions at short separation distances, and the second one represents an intermediate case. The exponential decay of the model with frequency for the three values of α at a short separation distance ($\xi = 10$ m) is presented in Fig. 5.15(a); simulated displacement time histories for $\alpha = 2 \times 10^{-4}$ sec/m and

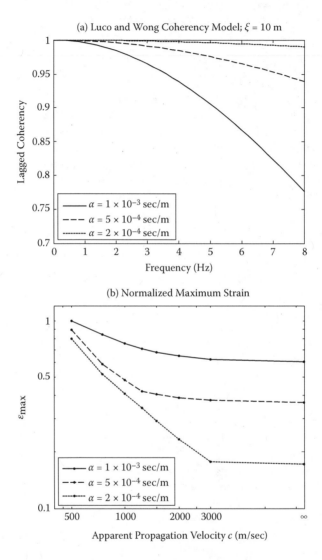

(a) Luco and Wong Coherency Model; $\xi = 10$ m

(b) Normalized Maximum Strain

Apparent Propagation Velocity c (m/sec)

FIGURE 5.15 Part (a): Exponential decay of the coherency model of Luco and Wong [324] for three values of the coherency drop parameter, α, at a separation distance of $\xi = 10$ m. Part (b): Peak values of normalized simulated longitudinal strains for the three values of α illustrated in part (a) as functions of the apparent propagation velocity of the motions; the simulated strains incorporate the effects of both the wave passage and the change in the shape of the motions. (Part (b) reproduced from *Probabilistic Engineering Mechanics*, Vol. 7, A. Zerva, "Spatial incoherence effects on seismic ground strains," pp. 217–226, Copyright ©1992, with permission from Elsevier.)

$\alpha = 1 \times 10^{-3}$ sec/m are presented later in Figs. 7.10(b) and (c), respectively. The apparent propagation velocity of the motions was considered to range from 500 m/sec to infinity, the latter value reflecting vertical incidence of the waves at the site. It is noted that the simulation scheme used in the evaluation preserves both the amplitude and the phase variability in the generated motions (Section 7.3.1).

Peak strains obtained from the simulations and normalized with respect to the maximum peak strain value of all simulations are presented in Fig. 5.15(b). The peak strains in the figure are point estimates and incorporate the effects of both the change in the shape in the motions and the propagation of the waveforms on the ground surface. The variability of the peak strain, ϵ_{max}, with the apparent velocity c shows the same pattern for all values of the coherency drop parameter α. Above $c = 3000$ m/sec, ϵ_{max} is constant for all values of α, suggesting that, in this propagation velocity range, the wave passage effects do not contribute significantly to the strains. Peak strains increase as the motions become less coherent and as the apparent propagation velocity decreases. For the low value of α (2×10^{-4} sec/m), as the apparent propagation velocity decreases below 3000 m/sec, the peak strains start increasing (essentially proportionally to $1/c$ in the figure), i.e., the apparent propagation of the motions becomes the dominant contributor to the strains. For the intermediate value of the coherency ($\alpha = 5 \times 10^{-4}$ sec/m), the peak strains are fairly constant for c greater than approximately 1250 m/sec, suggesting that the change in the shape of the motions is still the dominant factor; as the propagation velocity decreases further, the peak strain for $\alpha = 5 \times 10^{-4}$ sec/m also starts increasing, and, in the range $500 < c < 700$ m/sec, it becomes parallel to ϵ_{max} for $\alpha = 2 \times 10^{-4}$ sec/m. For these motions, there is a transition region, where the wave passage effect starts contributing to the strains until, at the lower propagation velocities, it becomes the dominant contributor to the deformation. Finally, the peak strains for the strongly incoherent motions ($\alpha = 1 \times 10^{-3}$ sec/m) in Fig. 5.15(b) increase very gradually with decreasing apparent propagation velocity and, even for the low values of c, they tend to be but are not yet proportional to $1/c$.

The trend of the $\epsilon_2(\tau)$ strain estimate in Fig. 5.14, which incorporates the effects of both the change in the shape of the motions and the wave passage, is similar to the strain estimates of Fig. 5.15(b). For low values of the time lag (large apparent propagation velocities), the strains appear to be almost constant and increase as the time lag increases (the propagation velocity decreases). $\epsilon_2(\tau)$, however, as a bounded, average estimate, tapers off to a constant value at large τ (Fig. 5.14), whereas the peak simulated strains, as point estimates, keep increasing with decreasing velocity (Fig. 5.15(b)). The difference between the characteristics of average and point strain estimates is also the reason why Fig. 5.15(b) does not support the existence of a "cross-over" apparent propagation velocity corresponding to the cross-over time lag observed by O'Rourke et al. [386] in Fig. 5.14. The change in the shape of the motions and their apparent propagation on the ground surface contribute constructively to the point estimates of the strains, and, therefore, their combined effect on seismic strains is greater than their individual contribution [593]. This can be seen, e.g., in the comparison of ϵ_{max} in the range $500 < c < 700$ m/sec (Fig. 5.15(b)), where the wave passage effect dominates: The less coherent motions yield higher peak strains than the more coherent ones. Hence, a cross-over apparent propagation velocity cannot exist for point estimates of the strains. For these estimates, the regions of dominance of the two contributing factors to the surface strains are separated by a "transition region" of apparent propagation velocities, that becomes wider as the motions become less coherent (Fig. 5.15(b)). Figures 5.14 and 5.15(b) also confirm, again, that higher values of ground strains are obtained at lower apparent propagation velocities.

5.4.2 TORSIONAL CHARACTERISTICS OF ARRAY DATA

Hao [196] determined the torsional power spectral density of the seismic field from the spatially variable translational motions of Events 24 and 45 recorded at the SMART-1 array, and conducted sensitivity analyses of the effects of the loss of coherency and the wave passage on the torsional characteristics of the motions.

A coherency expression of the following form was first proposed [196]:

$$|\gamma(\xi_l, \xi_t, \omega)| = \exp\left\{-\left[\alpha_1(\omega)\xi_l^2 + \alpha_2(\omega)\xi_t^2\right]\omega\right\} \tag{5.29}$$

where, as previously in Eq. 3.14, ξ_l and ξ_t indicate the projected station separation distances along and normal to the direction of propagation of the waves. The dependence of the coherency function in Eq. 5.29 on the square separation distances is necessary for the subsequent evaluation of the torsional power spectral density. The functions $\alpha_1(\omega)$ and $\alpha_2(\omega)$ were assumed to be of the form:

$$\alpha_j(\omega) = \frac{a_j}{\ln(\omega) + b_j} \quad j = 1, 2 \tag{5.30}$$

with a_j and b_j being parameters estimated from regression analyses of the data, $\omega \geq 0.314$ rad/sec and $b_j > |\ln(0.314)| = 1.1584$. The apparent propagation of the waveforms was also considered in the derivation of the torsional power spectral density, i.e., the spatial variability of the motions was given by (Eqs. 3.3 and 3.4):

$$\gamma(\xi_l, \xi_t, \omega) = |\gamma(\xi_l, \xi_t, \omega)| \exp\left(-\frac{i\omega\xi_l}{c}\right) \tag{5.31}$$

in which c indicates, again, the apparent propagation velocity of the motions. Table 5.4 presents the model parameters identified by Hao [196] for the two events at the SMART-1 array, and Fig. 5.16(a) illustrates the exponential decay of the model (Eqs. 5.29 and 5.30 and Table 5.4) at a separation distance of $\xi = 200$ m for Event 45, labeled "intermediate coherency"; the curves labeled "high coherency" and "low coherency" in the figure are used later in the sensitivity analyses. The data of Event 45 were more highly correlated than the data of Event 24 [196]. It is also noted, from the comparison of Fig. 5.16(a) with Fig. 3.18, that this model produces a comparable exponential decay with frequency with the model of Hao et al. [200] and Oliveira et al. [382] (Eqs. 3.14 and 3.15 and Table 3.5) that was based on the same data.

With the spatial variability model of Eqs. 5.29–5.31, the power spectral density of the torsional acceleration, $S_{\ddot{\theta}_3\ddot{\theta}_3}(\omega)$, with θ_3 indicating the rotation angle around the

TABLE 5.4
Parameters of the Hao [196] coherency model (Eqs. 5.29–5.31) for Events 24 and 45 at the SMART-1 array. (After H. Hao, "Characteristics of torsional ground motions," *Earthquake Engineering and Structural Dynamics*, Vol. 25, pp. 599–610, Copyright ©1996 John Wiley & Sons Limited; reproduced with permission.)

Event	a_1 (sec/m^2)	b_1	a_2 (sec/m^2)	b_2	c (m/sec)
24	1.68934×10^{-6}	1.42983	1.74054×10^{-6}	1.57179	770
45	1.1167×10^{-6}	1.66726	1.1593×10^{-6}	1.72263	875

vertical axis, takes the form [196]:

$$S_{\ddot{\theta}_3\ddot{\theta}_3}(\omega) = S(\omega)\,\omega\left[\alpha_1(\omega) + \alpha_2(\omega) + \frac{\omega}{2c^2}\right] \tag{5.32}$$

where $S(\omega)$ is the power spectral density of the translational acceleration, which was assumed to be the same in the two horizontal directions. Hao [196] then conducted a sensitivity analysis of the effect of loss of coherency and apparent propagation of the motions on the torsional spectra considering three different degrees for the exponential decay of the coherency and three values for the apparent propagation velocity. A "high" coherency was obtained for $a_j = 1.0 \times 10^{-7}$ sec/m^2 and $b_j = 1.16$, $j = 1, 2$, and a "low" coherency for $a_j = 5.0 \times 10^{-6}$ sec/m^2 and $b_j = 2.0$, $j = 1, 2$ (Eqs. 5.29 and 5.30), and the coherency of Event 45 was considered as an "intermediate" coherency; the resulting coherencies are shown in Fig. 5.16(a). The values for the apparent propagation velocity were $c = 500, 1000$ and 2000 m/sec. For illustration purposes, Fig. 5.16(b) presents the torsional acceleration power spectral density (Eq. 5.32) based on the Clough-Penzien spectrum (Eq. 3.7 normalized by S_o) for medium soil conditions (Table 3.1(b)), the high, intermediate and low coherency models of Fig. 5.16(a), and an apparent propagation velocity of 1000 m/sec; the lower the value of the coherency, the higher the peak value of the torsional spectrum. Figures 5.16(c) and (d) present, respectively, the effect of loss of coherency only (i.e., $c \rightarrow \infty$) and the effect of wave passage only (i.e., propagating translational motions identical in shape) on the torsional spectrum. Significantly higher values for the peak of the spectra are obtained as the propagation velocity decreases (Fig. 5.16(d)), which Hao [196] attributed, in part, to the assumed coherency model and its effect on the torsional response spectrum expression (Eq. 5.32). However, when the motions are only weakly correlated, the consideration of the apparent propagation effects only underestimates the torsional spectra [196], as can be seen from the comparison of Fig. 5.16(b) with the torsional spectra of motions propagating unchanged with a velocity of $c = 1000$ m/sec in Fig. 5.16(d). It is also noted that the trend of the relative effects of loss of coherency and wave passage on the torsional spectra is similar to their effect on the longitudinal strains of Fig. 5.15(b). As the apparent propagation velocity decreases, the contribution of the wave passage effect becomes more dominant than the effect of the loss of coherency in the motions. Laouami *et al.* [282], [283] reached similar conclusions regarding the effect of loss of coherency on the torsional characteristics of the motions evaluated from data recorded at the LSST array.

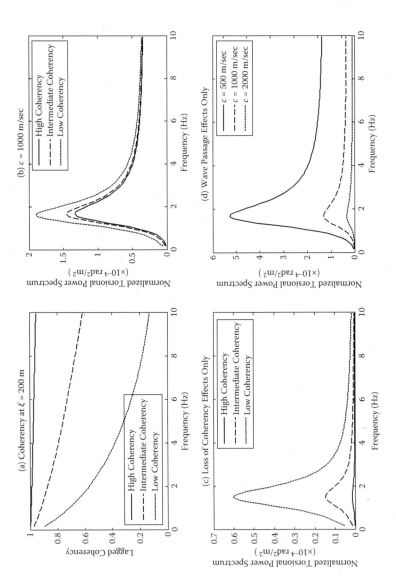

FIGURE 5.16 Effect of loss of coherency and wave passage on torsional spectra according to the ground motion model reported by Hao [196]. Part (a) presents the exponential decay of the coherency model proposed in this evaluation for the data of Event 45 recorded at the SMART-1 array and for low, intermediate (coinciding with the coherency of Event 45) and high values of the coherency. Part (b) indicates the effect of loss of coherency and wave passage (for an apparent velocity of $c = 1000$ m/sec) on torsional spectra. Part (c) presents the effect of loss of coherency only, and part (d) the effect of wave passage only on the torsional spectra (adapted from Hao [196]).

5.5   DISPLACEMENT GRADIENT ESTIMATION FROM ARRAY DATA

An alternative methodology for the estimation of seismic ground strains, that requires array data, was introduced in seismology from the geodetic community by Spudich *et al.* [502]. The seismo-geodetic approach waives some of the single-station method restrictions but introduces some new ones.

In the approach, the displacement gradient tensor, $\mathbf{D}$, with elements $\partial u_i / \partial x_j$, u_i indicating displacement in the i-th direction and x_j distance in the j-th direction, is evaluated from an array of N sensors using the first terms of a Taylor series expansion as [502]:

$$\Delta u_i^{(n)} = u_i^{(n)} - u_i^{(0)} = \sum_{j=1}^{3} \frac{\partial u_i}{\partial x_j} \Delta x_j^{(n)}; \quad i = 1, 2, 3; \quad n = 1, \dots, N-1 \quad (5.33)$$

where $\Delta u_i^{(n)}$ is the differential displacement in the i-th direction between station (n) and the reference station (0), $\Delta x_j^{(n)}$ is the separation distance between the stations in the j-th direction, and the dependence of the displacement field on time has been dropped in the notation.

The displacement gradients, $\partial u_i / \partial x_j$, fully describe strains and rotations, i.e.,

$$\epsilon_{ij} = \frac{1}{2} \left(\frac{\partial u_i}{\partial x_j} + \frac{\partial u_j}{\partial x_i} \right); \quad \theta_{ij} = \frac{1}{2} \left(\frac{\partial u_i}{\partial x_j} - \frac{\partial u_j}{\partial x_i} \right) \quad (5.34)$$

respectively. At the free surface, the stress-free boundary conditions impose the following constraints on the displacement gradients:

$$\frac{\partial u_1}{\partial x_3} = -\frac{\partial u_3}{\partial x_1}; \quad \frac{\partial u_2}{\partial x_3} = -\frac{\partial u_3}{\partial x_2};$$

$$\frac{\partial u_3}{\partial x_3} = -\frac{\hat{\lambda}}{\hat{\lambda} + 2\hat{\mu}} \left(\frac{\partial u_1}{\partial x_1} + \frac{\partial u_2}{\partial x_2} \right) \quad (5.35)$$

where $\hat{\lambda}$ and $\hat{\mu}$ are the Lamé constants, and the subscripts for direction are specified as 1 for east, 2 for north and 3 for vertical. The remaining six independent displacement gradients can then be evaluated by solving Eq. 5.33 at $x_3 = 0$. If more than three stations are used in the identification, a least-squares minimization scheme should be utilized [502].

Limitations of the seismo-geodetic strain estimates obtained from Eq. 5.33 include the following: The derivations consider that deformation is spatially uniform and, since the approach is based on the first terms of a Taylor series expansion, displacements should vary linearly within the array. Additionally, the accuracy of the displacement gradients is constrained by the aperture of the array. The ratio of the estimated over the true displacement gradient is given by [63], [173]:

$$\frac{[\partial u_i / \partial x_j]_{\text{geodetic}}}{[\partial u_i / \partial x_j]_{\text{exact}}} = \frac{sin[\pi L/\lambda]}{[\pi L/\lambda]} = \frac{sin[\pi L f/c]}{[\pi L f/c]} \quad (5.36)$$

where L is the distance between stations, λ indicates wavelength and c is the velocity of the contributing wave. Equation 5.36 suggests that, for approximately 90% accuracy in the displacement gradient estimates, L should be less than approximately $1/4$ of the wavelength of the dominant energy in the signal. An additional consideration that affects the accuracy of the seismo-geodetic method is that it utilizes differences of displacements (Eq. 5.33): When taking differences of measured displacements, the coherent signal subtracts out, but noise does not [176]. The method also requires that three-component data from, at least, three stations with synchronous timing are available. On the other hand, the advantages of the approach are that it does not require any *a priori* information regarding the backazimuth of the impinging waves and their apparent propagation velocities, and, also, is not limited by any relationship between strain and particle velocity.

Section 5.5.1 presents a comparison between seismic ground strains estimated by means of the seismo-geodetic approach and the single-station approximation utilizing data recorded at the Geyokcha array in Turkmenistan [176]. It is followed by Section 5.5.2 that illustrates the derivation of surface displacement gradients and strains as well as the variation of strain estimates with depth obtained from data recorded at a 3-D array in the lake-bed region of Mexico City [63].

5.5.1 SEISMO-GEODETIC VS. SINGLE-STATION STRAIN ESTIMATES

Gomberg *et al.* [176] compared single-station and seismo-geodetic displacement gradient estimates using the data of the Geyokcha, Turkmenistan, array. The array, shown schematically in Fig. 5.17, consisted of 12 broadband STS-2 and 36 strings of L28 short-period sensors located on the diameters of a circle with radius of 423.5 m. Gomberg *et al.* [176] noted that, if subarrays of stations from each quadrant are used in the evaluation, then the maximum frequency utilized should be less than approximately 0.6 V_h/km, where V_h is the identified propagation velocity of the incoming waves, for an accuracy of approximately 90% in the estimation of the displacement gradients (Eq. 5.36). The signal-to-noise ratio (SNR) for many of the events recorded at the array was SNR $\geq$ 100, and the epicentral distance of the events analyzed ranged between 64 km and 904 km.

The single-station estimates were determined from the following approximations:

$$\frac{\partial u_i}{\partial x} = -\frac{\partial u_i}{\partial t}\frac{\sin\phi}{V_h}; \quad \frac{\partial u_i}{\partial y} = -\frac{\partial u_i}{\partial t}\frac{\cos\phi}{V_h} \quad i = x, y, z \tag{5.37}$$

where ϕ indicates the backazimuth. The single-station estimates $\partial u_i/\partial z, i = x, y, z$, were obtained from Eq. 5.35, and the seismo-geodetic estimates from Eqs. 5.33 and 5.35 for $x \equiv x_1, y \equiv x_2$ and $z \equiv x_3$.

Figure 5.18 presents the comparison of single-station and seismo-geodetic estimates of the displacement gradients in the frequency range of 0.5 – 1.0 Hz for an earthquake with epicentral distance of 64 km and backazimuth of 325°; the apparent propagation velocity of the waveforms was estimated as 2.72 km/sec. The STS-2 data of the stations of the eastern quadrant subarray (Fig. 5.17) were used in the evaluation; the maximum station separation distance was 400 m. Four single-station estimates and one seismo-geodetic estimate using the data at all four stations are shown in the figure. The agreement between the estimates in Fig. 5.18 is satisfactory, suggesting that the plane-wave assumption is valid in this frequency range, and the

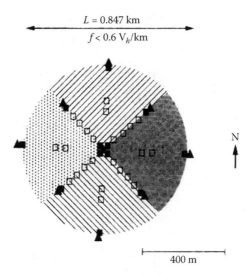

FIGURE 5.17 Schematic diagram of the Geyokcha, Turkmenistan, array configuration. The solid squares indicate the L28 stations used in the study by Gomberg *et al.* [176] and the open squares other L28 stations. The triangles indicate broadband STS-2 stations, all of which were used in the evaluation. For approximately 90% accuracy in the estimation of the displacement gradients (Eq. 5.36) from subarrays of stations in each quadrant, indicated by different shading in the figure, the maximum frequency utilized should be less than approximately 0.6 V_h/km, where V_h is the identified propagation velocity of the incoming waves. (Reprinted from J. Gomberg, G. Pavlis and P. Bodin, "The strain in the array is mainly in the plane (waves below $\sim$ 1Hz)," *Bulletin of the Seismological Society of America*, Vol. 89, pp. 1428–1438, Copyright ©1999 Seismological Society of America.)

single-station estimate can be used as a proxy for the actual displacement gradient. Gomberg *et al.* [176] noted, however, that for events with longer epicentral distances, more significant differences were observed in the estimates a few seconds after the onset of the S-wave, possibly due to forward scattering and contributions of dispersive surface waves.

Gomberg *et al.* [176] then proceeded with the comparison of the displacement gradients estimated by the two techniques at higher ($4.0 - 8.0$ Hz) frequencies using the stations close to the center of the array with a maximum separation distance of approximately 50 m; this station separation distance yields a frequency range of $f \leq 5$ V_h /km for approximately 90% accuracy in the displacement gradient estimates (Eq. 5.36). The epicentral distance of the earthquake in this evaluation was 136 km with a backazimuth of 320°, and the apparent propagation velocity of the motions was 3.84 km/sec. Gomberg *et al.* [176] first evaluated single-station estimates at the four L28 array stations located close to the center of the array and the STS-2 station at the array center (Fig. 5.17). The resulting displacement gradients, shown in Fig. 5.19(a), are very similar for all stations considered, which suggests that the single-station estimate is robust [176]. This is not the case, however, when single-station and seismo-geodetic estimates are compared in this higher frequency range:

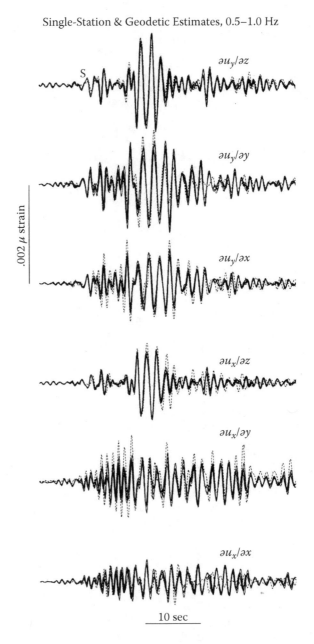

FIGURE 5.18 Comparison of four single-station estimates (solid lines) and one seismo-geodetic estimate (dashed lines) of displacement gradients in the frequency range of 0.5 – 1.0 Hz evaluated from the STS-2 data at four stations of the eastern quadrant of the Geyokcha array (Fig. 5.17); the data were recorded during a 64 km distant earthquake with backazimuth of 325°. The letter S indicates the arrival of the S-wave. (Reprinted from J. Gomberg, G. Pavlis and P. Bodin, "The strain in the array is mainly in the plane (waves below ∼ 1Hz)," *Bulletin of the Seismological Society of America*, Vol. 89, pp. 1428–1438, Copyright ©1999 Seismological Society of America; courtesy of J. Gomberg.)

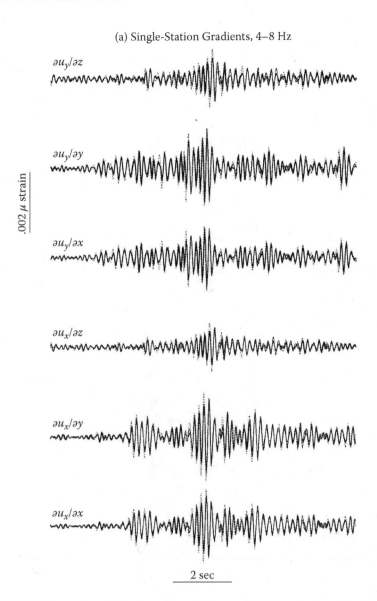

FIGURE 5.19 Comparison of single-station and seismo-geodetic estimates of displacement gradients in the frequency range of 4.0 – 8.0 Hz evaluated from data recorded at the center stations of the Geyokcha array (Fig. 5.17) during an earthquake with epicentral distance of 136 km and backazimuth of 320° (maximum separation distance between stations was approximately 50 m). Part (a): Comparison of single-station estimates of the displacement gradients from data recorded at the four center L28 stations (solid lines) and the center STS-2 station (dashed lines); the estimates of the gradients from the L28 data are undistinguishable. Part (b), shown on the next page: Comparison of a single-station estimate of the displacement gradients from one of the four center L28 stations shown in part (a) (solid lines) with the corresponding seismo-geodetic estimate (dashed lines) evaluated from the same four center L28 stations. (Reprinted from J. Gomberg, G. Pavlis and P. Bodin, "The strain in the array is mainly in the plane (waves below ~ 1Hz)," *Bulletin of the Seismological Society of America*, Vol. 89, pp. 1428–1438, Copyright ©1999 Seismological Society of America; courtesy of J. Gomberg.)

(b) Single-Station & Geodetic Gradients, 4–8 Hz

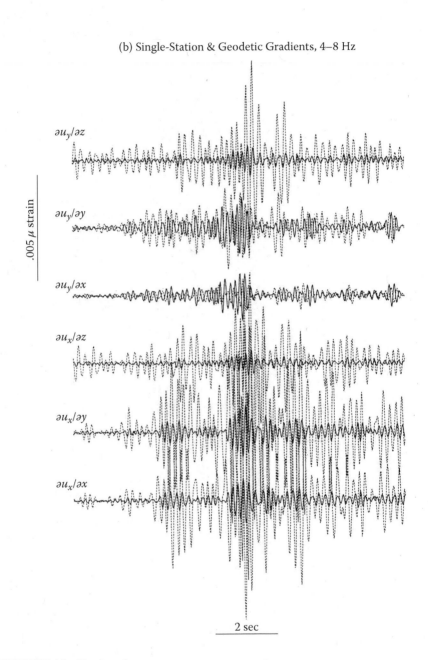

FIGURE 5.19 (Continued).

Figure 5.19(b) presents single-station displacement gradients of the data from one representative L28 station of the ones shown in Fig. 5.19(a) and the seismo-geodetic estimates determined from the use of all four L28 stations at the center of the array. The differences between the two estimates are very significant, and, furthermore, Gomberg et al. [176] reported that the seismo-geodetic estimates were unstable, i.e., they depended on the selection of the reference station in Eq. 5.33. The authors concluded that, in this higher frequency range, displacements vary nonlinearly over distances as short as 50 m, local scattering underneath the array becomes important, and the assumption of plane-wave propagation can no longer be valid. The conclusions of Gomberg et al. [176], however, also suggest that the traditional single-station estimate cannot be used as a proxy for the actual strains in this higher frequency range, as the variability in the shape of the motions contributes significantly to the ground deformations.

More recently, Kamiyama and Sato [252] used the seismo-geodetic approach to estimate displacement gradients, and, subsequently, strains and rotations at the KASSEM array observation system in the prefectures of Miyaki and Fukushima, Japan. The seven-station array was installed on a shallow basin (in the north-south direction) that is dipping from the west towards the east. In the same study, Kamiyama and Sato [252] conducted a 2-D analysis of the anti-plane (SH-waves) and plane-strain (P- and SV-waves) wave propagation problem at the irregular site using the pseudo-spectral method. They suggested that the numerical results agreed reasonably well with the strain estimates obtained from the recorded data, and, also, demonstrated numerically the significance of the contributions of the generated surface waves in the basin that led to quite different waveforms (accelerations and strains) at stations with separation distances as close as 100 m.

5.5.2 DISPLACEMENT GRADIENTS FROM 3-D ARRAYS

Bodin et al. [63] and Singh et al. [479] conducted an extensive seismic ground strain evaluation from data recorded in the shallow sediment valley of Mexico. After the 1985 Michoacan earthquake, a significant number (approximately 100) of surface accelerographs were deployed in the valley of Mexico, and four sites in the lake-bed zone included downhole accelerographs as well. Bodin et al. [63] based their investigation on one of these 3-D arrays, the Roma array, illustrated in Fig. 5.20. The array consisted of three surface accelerographs, Roma A, Roma B and Roma C, with minimum and maximum station separation distances of 91.97 m and 139.92 m, respectively (Fig. 5.20); two additional instruments were deployed underneath Roma C at depths of 30 m and 102 m. Bodin et al. [63] noted that the clay of the lake-bed zone exhibits certain unusual characteristics: The ratio of water to solid often exceeds 4:1, the S-wave velocity is very low (between approximately 50 m/sec and 100 m/sec), and, also, laboratory tests suggested a linear behavior even at high shear strains (exceeding 0.1%). An additional characteristic of the lake-bed zone is that it can amplify (by a factor of $10 - 50$ at certain frequencies) the seismic motions in comparison to those recorded in the surrounding hill-zone sites [477], [478].

Bodin et al. [63] analyzed displacement gradients, strains and rotations obtained through seismo-geodetic evaluation of the data recorded during the October 24, 1993,

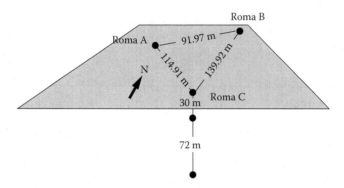

FIGURE 5.20 Schematic diagram of the 3-D configuration of the Roma array in Mexico City. Stations Roma A, Roma B and Roma C are surface stations; underneath Roma C there are two additional stations at depths of 30 m and 102 m. (Reprinted from P. Bodin, J. Gomberg, S.K. Singh and M. Santoyo, "Dynamic deformations of shallow sediments in the valley of Mexico. Part I: Three-dimensional strains and rotations recorded on a seismic array," *Bulletin of the Seismological Society of America*, Vol. 87, pp. 528–539, Copyright ©1997 Seismological Society of America.)

earthquake ($M_w = 6.7$, depth 35 km and distance from the array 311 km). Translational accelerations, velocities and displacements at the Roma C station are presented in Fig. 5.21 in the east-west, north-south and vertical directions, denoted by the letters E, N and Z, respectively; the data were bandpassed in the frequency range of $0.1 - 2.0$ Hz. Theoretical dispersion curves for Love and Rayleigh waves for a velocity model of the Roma site suggested a strong site resonance for both types of waves at a period of $2\frac{1}{4}$ sec and a dominant phase velocity of approximately 1.5 km/sec, which corresponds to the S-wave velocity of the medium right below the lake bed [63]. For this value of the period, for phase velocities in the range of $0.5 - 1.5$ km/sec and for $L = 140$ m (Fig. 5.20), the seismo-geodetic estimates are $97.5\% - 99.7\%$ accurate (Eq. 5.36).

Figure 5.22 presents the surface displacement gradients for the October 1993 earthquake (in part (a)) and the corresponding rotations and strains (in parts (b) and (c), respectively). Bodin *et al.* [63] noted that the largest gradients were associated with surface and coda waves, that they were dominated by 2.5 sec period waveforms, and that the horizontal gradients exceeded by a factor of 2 the vertical ones. Bodin *et al.* [63] also examined the validity of the single-station strain estimation approach by comparing single-station and seismo-geodetic strain estimates. They noted that the single-station strain estimate is a valid approximation, provided, however, that the apparent propagation velocity is reduced from 1.5 km/sec to $0.5 - 0.7$ km/sec. They attributed this discrepancy to a possible overestimation of the S-wave velocity in the layers underneath the lake bed and the lack of consideration of the single-station estimate for causes of ground deformation other than that of plane-wave propagation. From their evaluation of the Roma array data, as well as additional data recorded in the lake-bed zone of the Valley of Mexico, Bodin *et al.* [63] and Singh *et al.* [479] suggested that the single-station approximation may be uncertain to within a factor

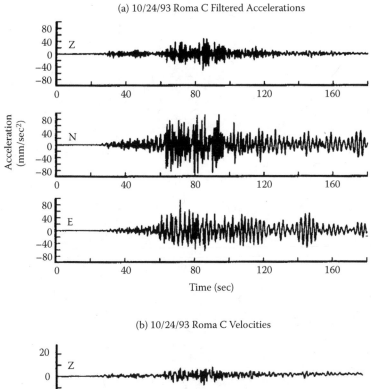

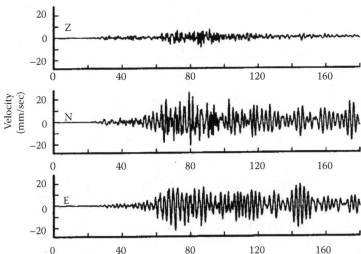

FIGURE 5.21 Translational acceleration (part (a)), velocity (part (b)) and displacement (part (c)), shown on the next page) time histories evaluated from the records at Roma C during the October 24, 1993, earthquake with $M_w = 6.7$, depth 35 km and distance from the array 311 km. The letters Z, N and E indicate direction: Z for vertical, N for North and E for East. The acceleration time histories were bandpassed in the frequency range of $0.1 - 2.0$ Hz. (Reprinted from P. Bodin, J. Gomberg, S.K. Singh and M. Santoyo, "Dynamic deformations of shallow sediments in the valley of Mexico. Part I: Three-dimensional strains and rotations recorded on a seismic array," *Bulletin of the Seismological Society of America*, Vol. 87, pp. 528–539, Copyright © 1997 Seismological Society of America.)

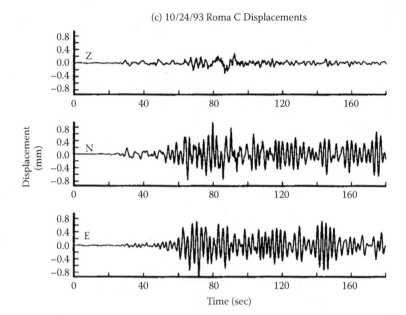

FIGURE 5.21 (Continued).

of 2 to 3. It is noted that this range of uncertainty in the estimate was also reported by Trifunac *et al.* [541] and Gomberg [173] in their evaluation of seismic ground strains from the Northridge earthquake (Section 5.3.2).

Using the downhole data from the stations underneath Roma C, Bodin *et al.* [63] also evaluated vertical displacement gradients, presented in Fig. 5.23. These vertical estimates were evaluated as the difference of the displacement time histories at two stations divided by the distance between the stations (i.e., $\Delta u_i / \Delta x_Z$, $i = $ E, N, and Z, rather than $\partial u_i / \partial x_Z$), and, hence, they represent only a lower bound of the actual maximum displacement gradients. Part (a) of Fig. 5.23 presents the vertical displacement gradients between the data at Roma C and the station at 30 m depth, and part (b) the corresponding values between the two downhole stations at a separation distance of 72 m. The vertical gradients between the surface and the 30 m depth exceed by at least a factor of 10 the corresponding values at the surface, which can be recognized from the comparison of the surface displacement gradients, $\partial u_Z / \partial x_E = -\partial u_E / \partial x_Z$ and $\partial u_Z / \partial x_N = -\partial u_N / \partial x_Z$ (Eq. 5.35) indicated as ZE and ZN in Fig. 5.22(a), with the lower bounds of the vertical displacement gradients (indicated by EZ and NZ) in Fig. 5.23(a) [63]. As depth increases, the lower bounds of the vertical displacement gradient estimates assume values comparable to those at the surface (Figs. 5.22(a) and 5.23(b)). Bodin *et al.* [63] also analyzed the data of three additional events recorded at the Roma array and reported that the results for the October 24, 1993, earthquake (both for the ground surface data and those at depth) were representative of all events considered.

Figure 5.23 illustrates the significance of downhole arrays in providing insight into the vertical wave propagation pattern. Downhole arrays are deployed around the

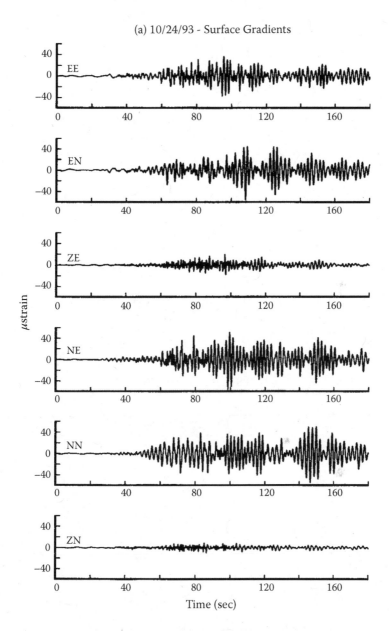

(a) 10/24/93 - Surface Gradients

FIGURE 5.22 Seismo-geodetic surface displacement gradients (in part (a)), and rotations and strains (in parts (b) and (c), respectively, shown on the next page) obtained from the data recorded at the Roma array (Fig. 5.20) during the October 24, 1993, earthquake. A geographic coordinate system is used in the figures, i.e., ZE stands for the gradient $\partial u_Z/\partial x_E$, EN for $\partial u_E/\partial x_N$, etc., whereas, for rotations, the single letter (Z, N or E) indicates rotation about the respective axis. (Reprinted from P. Bodin, J. Gomberg, S.K. Singh and M. Santoyo, "Dynamic deformations of shallow sediments in the valley of Mexico. Part I: Three-dimensional strains and rotations recorded on a seismic array," *Bulletin of the Seismological Society of America*, Vol. 87, pp. 528–539, Copyright ©1997 Seismological Society of America.)

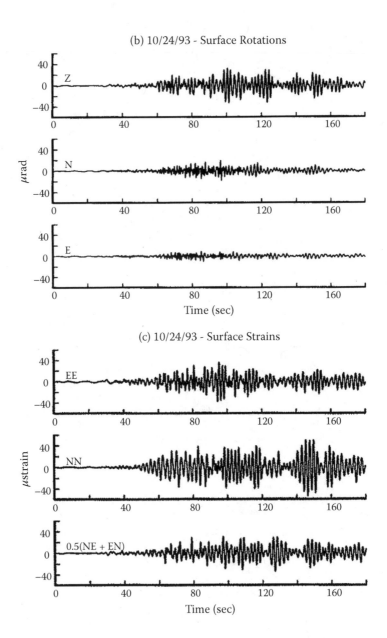

FIGURE 5.22 (Continued).

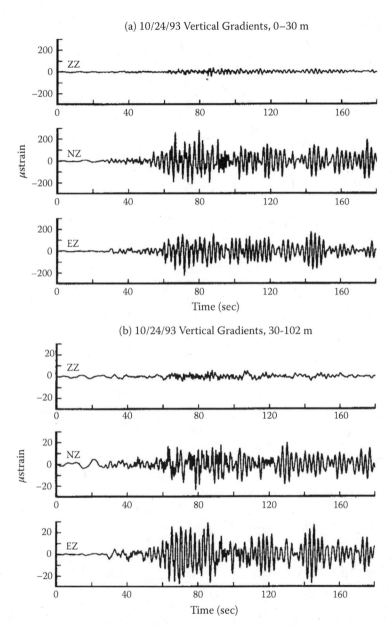

FIGURE 5.23 Vertical displacement gradients obtained from the difference of displacement time histories at the vertical subarray at Roma C divided by the separation distance between the stations (Fig. 5.20). Part (a) of the figure presents the vertical displacement gradient between the surface station and the station at 30 m depth; part (b) of the figure presents the corresponding values from the two downhole stations of the array at a distance of 72 m. In the figure, the same geographic coordinate system and conventions are used as in Figure 5.22. (Reprinted from P. Bodin, J. Gomberg, S.K. Singh and M. Santoyo, "Dynamic deformations of shallow sediments in the valley of Mexico. Part I: Three-dimensional strains and rotations recorded on a seismic array," *Bulletin of the Seismological Society of America*, Vol. 87, pp. 528–539, Copyright ©1997 Seismological Society of America.)

world [504], in some cases together with a surface strong motion array, as, e.g., the LSST array (Fig. 1.3) or the EUROSEIS array, near Thessaloniki, Greece [330]. Data from these arrays facilitate the evaluation of the 1-D site response characteristics that cannot be readily estimated *in-situ* or in the laboratory [34], [40]. Downhole array data are used for the identification of the S-wave velocity at the site, the resonant 1-D site response characteristics, the shear stress-strain time history, the local shear modulus and damping coefficient, and the global site dynamic properties, as described by Elgamal *et al.* [145] in a recent, comprehensive review article of the work that has been conducted in this subject area; the interested reader is referred to this publication and the references therein for information about the uses and applications of downhole array data.

5.6   CONSIDERATIONS IN THE ESTIMATION OF SEISMIC GROUND STRAINS

Equations 5.3 and 5.4 constitute an attractive approximation for seismic ground deformations, but the accuracy of their prediction may be within a factor of 2 – 4 [63], [173], [479], [541]. This is an anticipated consequence of the limiting assumptions on which the equations are based, namely that a single plane wave, that propagates on the ground surface with constant velocity without any changes in its shape and without interference from other waves, constitutes the seismic ground motion. The seismic time series, however, are comprised of different wave components that propagate with different and, in the case of dispersive surface waves, frequency-dependent velocities. The first consideration in the application of the traveling-wave assumption for the estimation of strains is whether the seismic records at the site of interest are dominated by body or surface waves. For the same value of the peak particle velocity, surface waves with lower phase velocities will lead to higher values for the strains than inclined body waves that propagate with higher apparent velocities. The differences in the strain estimates obtained from the data of the Northridge earthquake by the two investigating groups [173], [541], presented in Section 5.3.2, can be attributed, to a large extent, to this consideration. For engineering applications, empirical "rules of thumb" for the identification of regions of dominance of body and surface waves have been proposed, e.g., [251], [387]. The second consideration is whether the assumption of the similarity in the time and space variation of the displacement field, on which Eqs. 5.3 and 5.4 are based, is valid or not. This is illustrated in Figs. 5.18 and 5.19, which compare single-station estimates, i.e., the time derivative of the displacement time series, with seismo-geodetic estimates, i.e., the space derivative (gradient) of the displacements. In the low frequency range (0.5 – 1.0 Hz in Fig. 5.18), where the signals are highly correlated, the two estimates essentially coincide. Similarly, in the higher frequency range (4 – 8 Hz in Fig. 5.19(a)), the time derivatives of the displacement time series at close-by stations are robust. This is not, however, the case for the space derivatives in this higher frequency range, which vary nonlinearly over distances of 50 m, and which the seismo-geodetic approach fails to capture. Hence, the time and space variations of the displacement field are no longer similar, and the single-station strain estimate may not be used as a proxy for the seismic ground strains. The reasons for this may be summarized as follows: When a broad-band body (S-) wave controls the motions, the approximation that it propagates unchanged on the ground surface is not valid, especially as frequency increases, due to the amplitude and (random) phase variability of the time series (Fig. 4.8). Since each of the causes of the spatial variation of the seismic ground motions, i.e., wave passage, amplitude variability and random phase variability (loss of coherency), gives rise to ground deformation, their combined effect will, generally, result in seismic strains higher than the strain estimates evaluated from the consideration of the wave passage effect only (Fig. 5.15(b)). Hence, Eqs. 5.3 and 5.4 will underestimate the actual ground-surface deformation field. On the other hand, when dispersive surface waves dominate the motions, the assumption of a single, constant phase velocity for the estimation of strains will underestimate the effect of the higher frequency wave components and overestimate the effect of the lower frequency components, as, for sites where the S-wave velocity increases with depth, the phase velocity decreases with increasing frequency (Figs. 5.3 and 5.7(a)). In addition, the seismic ground motion is comprised, at any time instant, of many harmonic waves that propagate with different frequency-dependent phase velocities, as illustrated herein in the description of the methodology for the generation of synthetic time histories at the Westmoreland site, Imperial Valley, California (Eqs. 5.11–5.17 and Figs. 5.7–5.9). The contributions of these harmonic components are not reflected in Eqs. 5.3 and 5.4, and, hence, the discrepancies between the actual strains and rotations and their estimates, in this case, can be attributed to this reason. An additional significant consideration is the accuracy of the estimates of the apparent propagation velocity of the motions. The study by O'Rourke et al. [384] presented a wide range of possible apparent propagation velocities for the incident body waves at the examined sites (Section 5.1.1). The differences between the single-station strain estimates and the strainmeter data (Figs. 5.10 and 5.11) and the observations on the discrepancies between the single-station and the seismo-geodetic estimates at the Roma array by Bodin et al. [63] can be attributed, among other reasons, to limited geotechnical information on the soil profile at the site for the evaluation of the dispersive surface wave velocities. The assumption of plane-wave propagation, the contribution of scattering effects at the local topography, the variability in the soil properties, as well as the interference of different wave components, that can arrive at the site during the same time window, are additional causes for the limitation of the single-station estimates in describing the seismic ground deformations.

   The aforementioned considerations were also recently addressed by the International Organization for Standardization (ISO) in its ISO 23469:2005 standard [235]. The standard explicitly addresses the design of geotechnical works[3] "with" and "without" spatial variability considerations. The standard suggests that the wave passage effects are predominant at low frequencies, whereas the effects of the loss of coherency in the motions are predominant at high frequencies (see, e.g., Section 5.5.1).

---

[3] The standard defines "geotechnical works" as "those comprised of soil or rock, including buried structures (e.g., buried tunnels, box culverts, pipelines, and underground storage facilities), foundations (e.g., shallow and deep foundations, and underground diaphragm walls), retaining walls (e.g., soil retaining and quay walls), pile-supported wharves and piers, earth structures (e.g., earth and rockfill dams and embankments), gravity dams, landfill and waste sites" [235].

In an informative annex, ISO 23469:2005 [235] evaluates the seismic ground strains, $\epsilon(\omega)$, due to the wave passage effect in a manner similar to Eqs. 5.3 and 5.9, i.e.,

$$\epsilon(\omega) = v(\omega)/c(\omega) \tag{5.38}$$

where $v(\omega)$ is the particle velocity and $c(\omega)$ the apparent propagation velocity of the seismic waves, both, however, functions of the frequency ω. The standard notes that the assumption of seismic waves with a single frequency dominating the seismic motions is an oversimplification, but also recognizes that the decomposition of the recorded time series into its participating components for the proper estimation of seismic strains is a difficult task. Since the phase velocity of surface waves is, generally, lower than the velocity of shear waves, the informative annex of the ISO 23469:2005 standard [235] proposes the following approach for the conservative estimation of the strains in Eq. 5.38: A site specific evaluation of the phase velocity of surface waves be conducted from information on the site's soil profile; an illustration of such an evaluation for the Westmoreland site at Imperial Valley in California [288] was presented in Fig. 5.7(a). The phase velocities of the first mode of Love or Rayleigh waves be then used in Eq. 5.38, as, at each frequency, they correspond to the lowest values for the propagation velocities (Fig. 5.7(a)). For sites, for which the phase velocity of surface waves can be reliably estimated, the recommendation of the ISO 23469:2005 standard [235] is an upper-bound approximation for the values of seismic ground strains, that are caused, however, solely by the wave passage effect. Alternatively, Trifunac and Gicev [534] suggested the establishment of an "equivalent," frequency-dependent propagation velocity of surface waves to account for the effects of the lower and higher phase velocities of the dispersive surface waves at each frequency (Fig. 5.7(a)) in the estimation of the ground deformation.

The validity of the aforementioned approximations needs to be verified with direct measurements of seismic ground strains. As indicated in the introductory remarks of this chapter, direct measurements of seismic ground deformations for validation of the estimated seismic strain field and its rotational components are limited, but, clearly, needed. Hashash et al. [217] also identified as a research need the instrumentation of tunnels, so that a better understanding of the effects of the spatial variation and the directivity of the motions on these structures can be obtained. Currently, the estimation of the rotational components of the seismic wave field is being investigated by the recently formed International Working Group on Rotational Seismology [236].

The strain estimates described in this chapter dealt with sites that can be approximated by horizontally layered media (Fig. 5.1). It should be noted, however, that significant strain concentrations occur at sites with lateral variations in their subsurface topography and soil conditions. Figure 5.24, reproduced from the ISO 23469:2005 standard [235], illustrates this effect: The intensity of the seismic ground motions can vary significantly across the boundary of a rock outcrop and the softer local site deposit, and induce significant strains in a buried structure crossing the boundary (Fig. 5.24(a)); similar phenomena can occur when the structure is placed above a buried lateral heterogeneity in the site conditions (Fig. 5.24(b)). This issue was addressed in an extensive review article by Liang and Sun [297], who presented field observations and analytical and experimental studies of site effects on the seismic

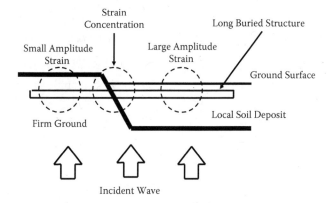

(a) Buried Structure crossing the Boundary
between Rock Outcrop and Local Soil Deposits

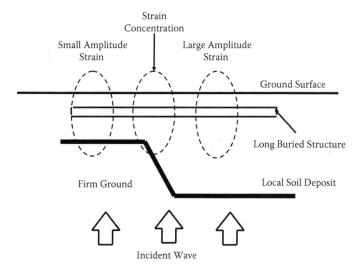

(b) Buried Structure extending above the Boundary
between Rock Outcrop and Local Soil Deposits

FIGURE 5.24 Illustration of the regions of high strain concentration in long structures buried at sites with irregular subsurface topography according to ISO 23469:2005 [235]. Part (a) of the figure presents a buried pipeline/tunnel crossing the boundary between a rock outcrop and the softer local site deposits. Part (b) presents a similar effect when the structure is placed above a buried lateral heterogeneity in the site conditions. (The terms and definitions taken from ISO 23469:2005 Basis for Design of Structures - Seismic Actions for Designing Geotechnical Works, Figure F.1, are reproduced with permission of the International Organization for Standardization, ISO[4]. Copyright remains with ISO.)

---

[4] This standard can be obtained from any ISO member and from the Web site of ISO Central Secretariat at the following address: www.iso.org

response of pipelines. They reported that significant strain amplification occurs at such discontinuities: When the interface between the two media was considered to extend to infinity, the amplification was of the order of two, but was considerably higher, when the pipeline crossed finite interfaces between the rock outcrop and the sediment site. These large amplifications of the incident excitation at the edge of a basin result from the constructive interference of diffracted waves in the basin and surface waves that are generated at its edges by incident body waves [266]. These effects cannot be captured by the 1-D modeling of the site, but require 2-D or 3-D models of the basin [15]. The limitations of 1-D wave propagation approaches at sites with complex subsurface topography will be further elaborated upon in Section 9.5.

6 Random Vibrations for Multi-Support Excitations

The random field of spatially variable seismic ground motions (described in Chapters 2 and 3) can be readily applied as the input excitation at the supports of lifelines in a random vibration approach. Random vibrations have been utilized extensively in the examination of the effects of loss of coherency and apparent propagation of seismic motions on a variety of lifeline systems, including, e.g., beams [195], [209], [210], [315], [590], [591], [600], foundations [192], [203], [324], [391], pipelines [221], [285], [600], arches [193], [194], suspension bridges [2], [306], [571], cable-stayed bridges [296], [307], [362], [511], and highway bridges and viaducts [306], [348], [379], [401]. Initial efforts on this broad topic have been reported by Sandi [438] in 1970, who conducted a stochastic analysis of lifeline response to nonsynchronous seismic excitations, by Hindy and Novak [221], [376] in 1979–1980, who presented the first stochastic analysis of a lifeline system (buried pipeline) subjected to seismic motions exhibiting loss of coherency, and by Abdel-Ghaffar and Rubin [2], [3] in 1982–1983, who evaluated the random vibration response of suspension bridges subjected to spatially recorded seismic data.

This chapter illustrates the concept of linear, stationary random vibrations as applied to multiply-supported structures. For a system subjected to a zero-mean seismic excitation, as is the case for acceleration time histories (Section 2.2.1), the response will have a zero mean and be described, through a random vibration analysis, by its power spectral density (PSD). The PSD can then be utilized for the evaluation of the root-mean-square (rms) response of the system, which provides information about its maximum response [125]. This is the major advantage of the random vibration approach: A single evaluation that utilizes the stochastic characterization of the excitation suffices for the evaluation of the probabilistic properties of the maximum response of the structure. Its disadvantage, however, as will be briefly illustrated later herein, is that, at this point, it is limited to linear structural problems.

The chapter begins, in Section 6.1, with a brief introduction to random vibrations utilizing a single-degree-of-freedom oscillator subjected to an external force or a base excitation; the steps for the evaluation of the derivatives of random processes and fields are also described in this section. The approach is then extended to discrete, multi-degree-of-freedom systems: Section 6.2.1 presents the response of a two-degree-of-freedom system subjected to spatially variable excitations at its two supports; this model was introduced by Nelson and Weidlinger [363] for the investigation of the seismic response behavior of a pipeline with soft joints. Section 6.2.2 describes the random vibration approach for the evaluation of the response of a discrete, multi-degree-of-freedom, multiply-supported system. The random vibration response of distributed-parameter systems on multiple supports is derived in Section 6.3: The first subsection (Section 6.3.1) illustrates the response of continuous systems, for which the soil-structure interaction is modeled by Wrinkler foundations.

This model was introduced by Hindy and Novak [221] for the evaluation of the effect of loss of coherency on the axial and lateral response of a continuous, buried pipeline; their results are also presented in this subsection. Section 6.3.2 highlights the evaluation of the response of a "generic" lifeline model, a continuous beam on multiple supports; due to its simplicity, this model has been extensively used by researchers (e.g., [210], [315], [590]) for insight into the effects of spatial variability on the response of lifelines. The physical interpretation of the various terms contributing to the PSD of the response is presented next: Section 6.4.1 illustrates the ground motion characteristics used in the evaluation, and Sections 6.4.2, 6.4.3 and 6.4.4 present, respectively, the contributions of the pseudo-static response, the dynamic response, and the cross terms of the multipart PSD expression to the total response of the structure. The differences in the seismic loads predicted by different coherency models for the response evaluation of lifeline systems are also illustrated in these sections. Section 6.4.5 presents the effect of strongly incoherent and fully coherent excitations on the linear response of a generic lifeline model. Although this evaluation utilizes a simple model, namely a continuous, three-span beam, its results serve as a reference for the differences in the lifeline response induced by uniform and spatially variable seismic excitations. The last section of this chapter describes some additional random vibration considerations: Section 6.5.1 transforms the PSD of the seismic excitation to the more acceptable response spectrum, and Section 6.5.2 illustrates efforts in analyzing the nonlinear structural response by means of random vibrations.

An important assumption that is made throughout this chapter is that the excitation is stationary, and described by its power spectral density and a coherency model. In the time domain, this assumption implies that the seismic ground motion time histories possess the same properties over time, and have no beginning and no end (Section 2.2.1). The resulting response of the structural systems is also stationary, which, in a time-domain evaluation, would correspond to its steady-state response. Seismic ground motions, however, exhibit non-stationarity, both in their intensity and their frequency content, as will be illustrated in Sections 7.1.2 and 7.1.3, and the non-stationary characteristics of the seismic excitations can have a significant effect on the structural response. The assumption of stationarity is made in the random vibration formulation presented herein, so that the effect of the spatial coherency on the response of lifelines can be clearly identified; it has been shown that the assumed shape of the intensity modulating function, that gives the stationary ground motions a non-stationary character (Section 7.1.2), can have a significant effect on the structural response [19], [240]. Additionally, it is considered that the excitations in the three orthogonal directions (along the axis of the structures, transverse horizontal and vertical) are uncorrelated (Section 3.3.2), and, hence, the structural response in each direction can be evaluated independently. This assumption was further validated by Zerva [595], who analyzed the random vibration response of continuous buried pipelines subjected to directionally and spatially correlated seismic ground motions, and concluded that the effect of the cross correlation of the excitations in the two horizontal directions on the response of the structures was insignificant. It is also noted that the derivations in this chapter are valid for any structure with deterministic properties and for seismic excitations exhibiting loss of coherency and propagating on the ground surface. For simplicity in presentation, however, the example applications consider symmetric structures and support motions exhibiting loss of coherency only.

It should be emphasized at this point that the material presented herein is not intended to cover the broad topic of random vibrations. An extensive bibliography is available on the subject including a large number of books, as, e.g., by Benaroya [50], Crandall and Mark [120], Lin [308], Lin and Cai [309], Lutes and Sarkani [326], Newland [365], Nigam and Narayanan [373], Roberts and Spanos [426], Solnes [483], Soong and Grigoriu [492], To [522] and Wirsching et al. [566]. The interested reader is referred to these publications and references therein for additional information on concepts and derivations.

6.1 INTRODUCTION TO RANDOM VIBRATIONS

The random vibration approach follows directly from the evaluation of the deterministic dynamic response of structures, only in this case, the excitation is described by a random process (or field) rather than a time series. This section illustrates the basic steps that lead from the derivation of the deterministic response of a Single-Degree-of-Freedom (SDOF) oscillator to the evaluation of its mean value and PSD caused by a random excitation.

6.1.1 DETERMINISTIC INPUT

Consider a SDOF oscillator subjected to either an external force (Fig. 6.1(a)) or a seismic excitation at its support (Fig. 6.1(b)). The equations of motion for the two systems in Figs. 6.1(a) and (b) become, respectively (e.g., [107]):

$$M\ddot{w}(t) + C\dot{w}(t) + Kw(t) = F(t) \tag{6.1}$$

$$M\ddot{w}_t(t) + C\dot{w}_t(t) + Kw_t(t) = C\dot{u}(t) + Ku(t) \tag{6.2}$$

where, in both equations, M, C and K are the mass, damping and stiffness of the system, and overdot and double overdot indicate first and second derivatives with respect to time, respectively. $F(t)$ in Eq. 6.1 indicates the forcing function and $w(t)$ the system's response (Fig. 6.1(a)). In Eq. 6.2, $w_t(t)$ is the (total) response of the oscillator, and $u(t)$ the support displacement, which, in Fig. 6.1(b), is illustrated by its acceleration. The consideration of the relative response of the system in Fig. 6.1(b), i.e.,

$$w(t) = w_t(t) - u(t) \tag{6.3}$$

simplifies the equation of motion (Eq. 6.2), which becomes:

$$M\ddot{w}(t) + C\dot{w}(t) + Kw(t) = -M\ddot{u}(t) \tag{6.4}$$

Equations 6.1 and 6.4 can be expressed in a single form as:

$$\ddot{w}(t) + 2\zeta_0\omega_0\dot{w}(t) + \omega_0^2 w(t) = \left\{ \begin{array}{c} F(t)/M \\ -\ddot{u}(t) \end{array} \right\} = f(t) \tag{6.5}$$

(a) Single-Degree-of-Freedom System with Force Excitation

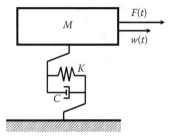

(b) Single-Degree-of-Freedom System with Base Excitation

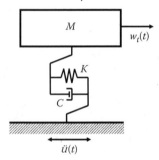

FIGURE 6.1 Illustration of SDOF systems with mass M, stiffness K and damping C. A system subjected to an external force, $F(t)$, is illustrated in part (a); $w(t)$ indicates the displacement. A system subjected to a base excitation, $\ddot{u}(t)$, is presented in part (b); the total displacement, consisting of the pseudo-static and dynamic components, is indicated by $w_t(t)$.

where $\omega_0 = \sqrt{K/M}$ is the natural frequency of the oscillator and $\zeta_0 = C/(2\sqrt{MK})$ its damping coefficient.

For a linear system starting from rest (i.e., $w(0) = 0$ and $\dot{w}(0) = 0$), Eq. 6.5 can be solved by means of Duhamel's (or superposition) integral as:

$$w(t) = \int_0^t h(t - \tau) f(\tau) \, d\tau \tag{6.6}$$

where the impulse response function, $h(t)$, is given by:

$$h(t) = \frac{1}{\omega_0 \sqrt{1 - \zeta_0^2}} e^{-\zeta_0 \omega_0 t} \sin\left(\omega_0 \sqrt{1 - \zeta_0^2} \, t\right) \tag{6.7}$$

6.1.2 RANDOM EXCITATION

If the excitation $f(t)$ is a random process, the response of the system, $w(t)$, will also be a random process, with mean value (Eq. 2.14):

$$\mu_w(t) = \mathrm{E}[w(t)] = \mathrm{E}\left[\int_0^t h(t - \tau) f(\tau) \, d\tau\right] = \int_0^t h(t - \tau) \mu_f(\tau) \, d\tau \tag{6.8}$$

where the operations of expectation and integration have been interchanged, and $\mu_f(\tau)$ indicates the mean value of the excitation. When the mean value of the excitation is equal to zero, i.e., $\mu_f(\tau) = 0$, then $\mu_w(t) = 0$. Since, for all practical purposes (Section 2.2.1), the mean value of the ground acceleration ($f(t) = -\ddot{u}(t)$ in Eq. 6.5) can be considered to be equal to zero, it will be assumed hereafter that $\mu_w(t) = 0$. The autocovariance function of the response becomes then (Eq. 2.15):

$$R_{ww}(t_1, t_2) = E[w(t_1)w(t_2)]$$

$$= E\left[\int_0^{t_1} h(t_1 - \tau_1) f(\tau_1) \, d\tau_1 \int_0^{t_2} h(t_2 - \tau_2) f(\tau_2) \, d\tau_2 \right]$$

$$= \int_0^{t_1} \int_0^{t_2} h(t_1 - \tau_1) h(t_2 - \tau_2) R_{ff}(\tau_1, \tau_2) \, d\tau_1 \, d\tau_2 \qquad (6.9)$$

where, again, the operations of expectation and integration have been interchanged, and $R_{ff}(\tau_1, \tau_2)$ indicates the autocovariance function of the excitation.

If, in addition, the random process $f(t)$ is stationary (or can be reasonably approximated by a stationary process or only the stationary response is of interest), it is customary to work in the frequency rather than the time domain (Section 2.2.1). The response of the system, given by Eq. 6.6, is then rewritten as:

$$w(t) = \int_0^t h(t - \tau) f(\tau) \, d\tau = \int_{-\infty}^{+\infty} h(t - \tau) f(\tau) \, d\tau = \int_{-\infty}^{+\infty} h(\tau) f(t - \tau) \, d\tau$$

$$(6.10)$$

The change of the lower limit in the second integral of the above expression from zero to $-\infty$ can be justified either by considering that the transient response has already died out or that the stationary excitation, since it has no beginning and no end, started at time equal to $-\infty$. The change of the upper limit in the second integral of Eq. 6.10 from t to $+\infty$ is due to the fact that the contribution of $h(t - \tau)$ to the integral for $\tau > t$ is equal to zero. The last expression in the equation results from the second integral with a change of variables of integration, i.e., $(t - \tau) \to \tau$. The autocovariance function of the stationary response of the system (Eqs. 2.22 and 6.9) becomes then:

$$R_{ww}(\tau) = R_{ww}(t, t + \tau) = E[w(t)w(t + \tau)]$$

$$= E\left[\int_{-\infty}^{+\infty} f(t - \tau_1) h(\tau_1) \, d\tau_1 \int_{-\infty}^{+\infty} f(t + \tau - \tau_2) h(\tau_2) \, d\tau_2 \right]$$

$$= \int_{-\infty}^{+\infty} \int_{-\infty}^{+\infty} h(\tau_1) h(\tau_2) R_{ff}(\tau + \tau_1 - \tau_2) \, d\tau_1 \, d\tau_2 \qquad (6.11)$$

Taking the Fourier transform (Eq. 2.31) of both sides of the above equation leads to the PSD of the response (Eq. 2.33) as:

$$S_{ww}(\omega) = \frac{1}{2\pi} \int_{-\infty}^{+\infty} \int_{-\infty}^{+\infty} \int_{-\infty}^{+\infty} h(\tau_1) h(\tau_2) R_{ff}(\tau + \tau_1 - \tau_2) e^{-i\omega\tau} \, d\tau_1 \, d\tau_2 \, d\tau$$

$$(6.12)$$

The transformation of variables: $u = \tau + \tau_1 - \tau_2$, $v_1 = \tau_1$ and $v_2 = \tau_2$ simplifies the triple integral in Eq. 6.12 to:

$$S_{ww}(\omega) = \int_{-\infty}^{+\infty} h(v_1)e^{+i\omega v_1}dv_1 \int_{-\infty}^{+\infty} h(v_2)e^{-i\omega v_2}dv_2 \left[\frac{1}{2\pi}\int_{-\infty}^{+\infty} R_{ff}(u)e^{-i\omega u}du\right]$$

(6.13)

where the expression in the brackets is the PSD of the excitation (Eq. 2.33), i.e.,

$$S_{ff}(\omega) = \frac{1}{2\pi}\int_{-\infty}^{+\infty} R_{ff}(\tau)e^{-i\omega\tau}\,d\tau$$

(6.14)

The second integral in Eq. 6.13 is the frequency transfer function of the system, $H(\omega)$, and the first integral its complex conjugate. The frequency transfer function and the impulse response function are Fourier transform pairs:

$$H(\omega) = \int_{-\infty}^{+\infty} h(t)e^{-i\omega t}dt$$

(6.15)

$$\Longleftrightarrow h(t) = \frac{1}{2\pi}\int_{-\infty}^{+\infty} H(\omega)e^{+i\omega t}dt$$

(6.16)

and $H(\omega)$ is given by:

$$H(\omega) = \left[(\omega_0^2 - \omega^2) + 2i\zeta_0\omega_0\omega\right]^{-1}$$

(6.17)

It is noted that the Fourier transforms of Eqs. 6.15 and 6.16 are not consistent with the Fourier transform pair definition used herein (Eqs. 2.30 and 2.31), since the factor 2π appears in the denominator of the right-hand side of Eq. 6.16 instead of Eq. 6.15. This inconsistency is a result of necessity rather than choice, as was the case for the selection of the Fourier transform pair of Eqs. 2.30 and 2.31: Equation 6.13 states the derived relation between the input excitation and the system response that yields the frequency transfer function, $H(\omega)$, as in Eq. 6.15 and not Eq. 2.31. The Fourier transform pair of Eq. 6.15, i.e., Eq. 6.16, has then to include the factor 2π in its denominator for consistency in the transformation, as was discussed in Section 2.2.2. Indeed, if it is assumed that the excitation is sinusoidal, as, e.g., $f(t) = F_0\exp(i\omega t)$, the steady-state system response becomes $w(t) = F_0 H(\omega)\exp(i\omega t)$, with $H(\omega)$ being, again, given by Eqs. 6.15 and 6.17. The latter description of the input excitation is utilized in the pseudo-excitation method of random vibration analyses [306], [307].

With the aforementioned considerations, the PSD of the response becomes:

$$S_{ww}(\omega) = H^*(\omega)H(\omega)S_{ff}(\omega) = |H(\omega)|^2 S_{ff}(\omega)$$

(6.18)

with:

$$|H(\omega)|^2 = \left\{\left[(\omega_0^2 - \omega^2)\right]^2 + 4\zeta_0^2\omega_0^2\omega^2\right\}^{-1}$$

(6.19)

Equation 6.18 is the well-known input-output relationship in random vibrations. It is recalled that the seismic excitation is, generally, considered to be a zero-mean,

Gaussian random process or field (Section 2.2.1). One important property of linear random vibrations is that the transformation of Eq. 6.18 results in a zero-mean, Gaussian distribution for the response. Hence, the first two moments (mean value and PSD) of the resulting process fully describe the stochastic characteristics of the system's response.

6.1.3 EXAMPLE APPLICATION

As an illustration of the above derivations, Fig. 6.2 presents the relative displacement of the SDOF oscillator of Fig. 6.1(b) subjected to a ground acceleration described by the Clough-Penzien PSD (Eq. 3.7) for soft soil conditions (Table 3.1(b)), i.e., $\omega_g = 5$ rad/sec, $\omega_f = 0.5$ rad/sec, $\zeta_g = 0.2$ and $\zeta_f = 0.6$ (Fig. 6.2(a))[1]. The oscillator has a natural frequency of $f_0 = \omega_0/(2\pi) = 1.5$ Hz and a damping coefficient of $\zeta_0 = 5\%$; the square of the modulus of the frequency transfer function (Eq. 6.19) is illustrated in Fig. 6.2(b). As per Eq. 6.18, the multiplication of the input excitation (Fig. 6.2(a)) with the squared modulus of the frequency transfer function (Fig. 6.2(b)) yields the PSD of the relative displacement of the system, shown in Fig. 6.2(c). It can be seen from the figure that the PSD of the relative system response contains two peaks: one at the dominant frequency of the excitation and one at the natural frequency of the oscillator.

6.1.4 DERIVATIVES OF RANDOM PROCESSES

In random vibration analyses, the power and cross spectral densities of derived random processes are broadly utilized, as has been already illustrated in Eq. 3.6 and will also be needed in the derivation of subsequent equations in this chapter. The evaluation of these derivatives (e.g., [365], [551]) is based on the following approach.

Consider a stationary random process, $w(t)$, with zero mean and autocovariance function given by Eq. 2.22, i.e., $R_{ww}(t_1, t_2) = R_{ww}(t_2 - t_1) = R_{ww}(\tau)$. The cross covariance function between $w(t)$ and its first derivative, $\dot{w}(t)$, provided that the process is mean-square differentiable, is expressed as:

$$R_{w\dot{w}}(t_1, t_2) = \mathrm{E}\left[w(t_1)\dot{w}(t_2)\right] = \frac{\partial R_{ww}(t_1, t_2)}{\partial t_2}$$

The above equation and Eq. 2.32 lead to the cross spectral density between the process and its derivative by means of the following simple steps:

$$R_{ww}(t_1, t_2) = \int_{-\infty}^{+\infty} S_{ww}(\omega)e^{i\omega(t_2-t_1)}\, d\omega$$

---

[1] In random vibration analyses, it is customary to work with the frequency ω (measured in [rad/sec]) rather than the cyclic frequency f (in units of [Hz]). This results in the differences between the units of the power spectral densities in the figures of this chapter and the units of the PSDs in the figures of the previous and following chapters.

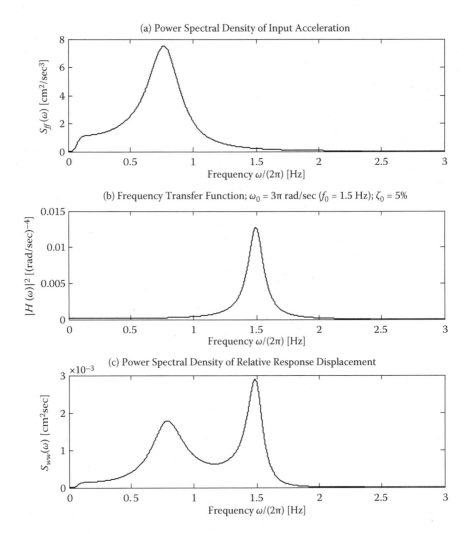

FIGURE 6.2 Illustration of the random vibration approach: The PSD of the excitation, $S_{ff}(\omega)$, (part (a)) is multiplied by the squared modulus of the frequency transfer function, $|H(\omega)|^2$, of the SDOF system (part (b)) to yield the PSD of the relative response displacement, $S_{ww}(\omega)$, in part (c). The PSD of the excitation is based on the Clough-Penzien spectrum for soft soil conditions (Table 3.1(b)), and the system's characteristics are $\omega_0 = 3\pi$ rad/sec and $\zeta_0 = 5\%$. The peaks of $S_{ff}(\omega)$ and $|H(\omega)|^2$ at distinct frequencies, in parts (a) and (b), respectively, are clearly reflected in the PSD of the response (part (c)).

$$\Longrightarrow R_{w\dot{w}}(t_1, t_2) = \frac{\partial}{\partial t_2} \int_{-\infty}^{+\infty} S_{ww}(\omega) e^{i\omega(t_2 - t_1)}\, d\omega$$

$$\Longrightarrow \int_{-\infty}^{+\infty} S_{w\dot{w}}(\omega) e^{i\omega(t_2 - t_1)}\, d\omega = \int_{-\infty}^{+\infty} i\omega\, S_{ww}(\omega) e^{i\omega(t_2 - t_1)}\, d\omega$$

$$\Longrightarrow S_{w\dot{w}}(\omega) = i\omega\, S_{ww}(\omega) \tag{6.20}$$

Power and cross spectral densities of higher derivatives of a random process or a random field can be determined in a similar manner.

6.1.5 TOTAL RESPONSE DUE TO BASE EXCITATION

For a system subjected to a ground excitation (e.g., Fig. 6.1(b)), the PSD of the total response, $w_t(t) = u(t) + w(t)$ (Eq. 6.3), can be derived as follows. The autocovariance function of the total response is:

$$R_{w_t w_t}(\tau) = \mathrm{E}[w_t(t)w_t(t+\tau)] = \mathrm{E}[\{u(t)+w(t)\}\{u(t+\tau)+w(t+\tau)\}]$$

$$\Longleftrightarrow R_{w_t w_t}(\tau) = R_{uu}(\tau) + R_{uw}(\tau) + R_{wu}(\tau) + R_{ww}(\tau) \tag{6.21}$$

The Fourier transformation of both sides of the above equation and derivations similar to those presented in Eqs. 6.11–6.14 result in the following expression for the PSD of the total response:

$$S_{w_t w_t}(\omega) = S_{uu}(\omega) + S_{uw}(\omega) + S_{wu}(\omega) + S_{ww}(\omega)$$

$$= S_{uu}(\omega) - H(\omega)S_{u\ddot{u}}(\omega) - H^*(\omega)S_{\ddot{u}u}(\omega) + |H(\omega)|^2 S_{\ddot{u}\ddot{u}}(\omega)$$

$$= S_{uu}(\omega) - 2\Re[H(\omega)]S_{u\ddot{u}}(\omega) + |H(\omega)|^2 S_{\ddot{u}\ddot{u}}(\omega) \tag{6.22}$$

where $\Re[.]$ indicates the real part. Equations 6.17 and 6.22 and expressions for the derivatives of the random process, which can be evaluated by means of the procedure presented in Eq. 6.20, lead to:

$$S_{w_t w_t}(\omega) = \frac{1}{\omega^4} \frac{\omega_0^4 + 4\zeta_0^2 \omega_0^2 \omega^2}{\left(\omega_0^2 - \omega^2\right)^2 + 4\zeta_0^2 \omega_0^2 \omega^2} S_{\ddot{u}\ddot{u}}(\omega) \tag{6.23}$$

or

$$S_{\ddot{w}_t \ddot{w}_t}(\omega) = \frac{\omega_0^4 + 4\zeta_0^2 \omega_0^2 \omega^2}{\left(\omega_0^2 - \omega^2\right)^2 + 4\zeta_0^2 \omega_0^2 \omega^2} S_{\ddot{u}\ddot{u}}(\omega) \tag{6.24}$$

Equation 6.23 presents the relation between the input base acceleration and the total response displacement of the oscillator, and Eq. 6.24 the relation between the input base acceleration and the total response acceleration. It is noted that Eq. 6.24 and the expression for the Kanai-Tajimi spectrum (Eq. 3.5) are identical. Indeed, the Kanai-Tajimi spectrum [253], [514] was derived by passing a white-noise bedrock excitation through a linear soil filter characterized by its natural frequency ($\omega_0 \equiv \omega_g$) and its damping coefficient ($\zeta_0 \equiv \zeta_g$).

6.2 DISCRETE-PARAMETER SYSTEMS

The SDOF random vibration approach can be readily extended to multi-degree-of-freedom systems subjected to multi-support excitations. This section evaluates first the response of a two-degree-of-freedom system subjected to spatially variable ground motions; this model was used to reproduce the behavior of buried pipelines with soft joints [363]. The general approach for the derivation of the response of a multi-degree-of-freedom system subjected to spatially variable seismic excitations at its supports is presented next.

6.2.1 PIPELINES WITH SOFT JOINTS

In 1977, Nelson and Weidlinger [363] introduced the "Interference Response Spectrum" in order to examine the effects of the apparent propagation of seismic motions on the axial response of segmented pipelines with soft joints. The approach is based on the differential motion of a two-degree-of-freedom pipeline, in which the two pipe segments behave as rigid bodies supported by springs, with stiffness K_g, and dashpots, with damping C_g, as shown in Fig. 6.3; the two pipe segments are connected with a soft joint of stiffness K_p and damping C_p. For the development of the Interference Response Spectrum, it was assumed that the motion at the left support (Fig. 6.3) is a displacement time history obtained through the double integration of a recorded accelerogram, and the input at the right support is the same displacement time history delayed by the travel time of the motions between the supports, i.e., equal to L/c, where L is the distance between the centers of gravity of the two pipe segments (Fig. 6.3) and c is the apparent propagation velocity of the predominant wave in the motions. This model was later extended by Zerva *et al.* [589], [600] to evaluate, by means of random vibrations, the effect of spatially variable seismic ground motions on the axial, transverse horizontal and vertical response of segmented pipelines. The effect of the loss of coherency in the seismic excitations on the axial response of the pipeline is illustrated in the following.

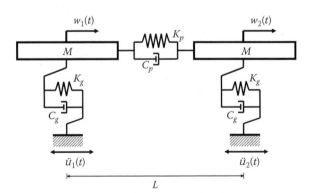

FIGURE 6.3 The model of a segmented pipeline with soft joints introduced by Nelson and Weidlinger [363] for the evaluation of the Interference Response Spectrum. The two rigid pipe segments, each of mass M, are connected by a spring of stiffness K_p and a dashpot of damping C_p. K_g and C_g are the stiffness and damping of the soil-structure interaction, and L is the distance between the mass centers of the two pipe segments. $w_1(t)$ and $w_2(t)$ indicate the displacements at the two structural degrees of freedom, and $\ddot{u}_1(t)$ and $\ddot{u}_2(t)$ the spatially variable excitations at the two supports.

The equations of motion for the structural system of Fig. 6.3 are given by [363], [600]:

$$M\ddot{w}_1(t) + C_p[\dot{w}_1(t) - \dot{w}_2(t)] + K_p[w_1(t) - w_2(t)]$$
$$+ C_g[\dot{w}_1(t) - \dot{u}_1(t)] + K_g[w_1(t) - u_1(t)] = 0 \qquad (6.25)$$

$$M\ddot{w}_2(t) + C_p[\dot{w}_2(t) - \dot{w}_1(t)] + K_p[w_2(t) - w_1(t)]$$
$$+ C_g[\dot{w}_2(t) - \dot{u}_2(t)] + K_g[w_2(t) - u_2(t)] = 0 \qquad (6.26)$$

where M is the mass of each pipe segment, $w_1(t)$ and $w_2(t)$ are the displacements at the degrees of freedom of the system, and $u_1(t)$ and $u_2(t)$ the displacement excitations at its supports. The addition of Eqs. 6.25 and 6.26 provides the equation of motion of the rigid body mode, whereas the subtraction of Eq. 6.26 from Eq. 6.25 yields the equation of differential motion between the two pipe segments. With the assumption that $K_p/K_g = C_p/C_g$, the equation of differential motion becomes:

$$\Delta\ddot{w}(t) + 2\zeta_0\omega_0\Delta\dot{w}(t) + \omega_0^2\Delta w(t) = \beta\left[2\zeta_0\omega_0\Delta\dot{u}(t) + \omega_0^2\Delta u(t)\right] \qquad (6.27)$$

where $\Delta w(t) = w_1(t) - w_2(t)$ is the differential displacement between the pipe segments, $\omega_0 = \sqrt{(K_g + 2K_p)/M}$ the natural frequency of the modified system, $2\zeta_0\omega_0 = (C_g + 2C_p)/M$, $\beta = K_g/(K_g + 2K_p) = C_g/(C_g + 2C_p)$, and $\Delta u(t) = u_1(t) - u_2(t)$ the differential ground displacement. The differential displacement between the pipe segments, $\Delta w(t)$, is then evaluated by means of Duhamel's integral (Eq. 6.6), and its PSD has the form [589]:

$$S_{\Delta w \Delta w}(\omega) = \frac{\beta^2}{\omega^4}\left(4\zeta_0^2\omega_0^2\omega^2 + \omega_0^4\right)|H(\omega)|^2 S_{\Delta\ddot{u}\Delta\ddot{u}}(\omega) \qquad (6.28)$$

where $|H(\omega)|^2$ is given by Eq. 6.19, and the PSD of the differential acceleration is given by:

$$S_{\Delta\ddot{u}\Delta\ddot{u}}(\omega) = S_{\ddot{u}_1\ddot{u}_1}(\omega) + S_{\ddot{u}_2\ddot{u}_2}(\omega) - 2\Re\left[S_{\ddot{u}_1\ddot{u}_2}(\omega)\right]$$
$$= 2S_{\ddot{u}\ddot{u}}(\omega)(1 - \Re[\gamma(L, \omega)]) \qquad (6.29)$$

where $\gamma(L, \omega)$ is the coherency function at frequency ω and the separation distance L between the supports of the two pipe segments (Fig. 6.3), and the latter expression in Eq. 6.29 is valid if the seismic ground motions at the two supports are described by the same PSD, $S_{\ddot{u}\ddot{u}}(\omega)$.

As an illustration of the response of pipelines with soft joints, the system of Fig. 6.3 is subjected to spatially variable motions described by the Clough-Penzien spectrum (Eq. 3.7) and the lagged coherency model of Harichandran and Vanmarcke (Eq. 3.12). The parameters of the Clough-Penzien spectrum are $\omega_g = 5$ rad/sec, $\omega_f = 0.5$ rad/sec, $\zeta_g = 0.2$ and $\zeta_f = 0.6$ (Table 3.1(b) and Fig. 6.2(a)). The parameters of the lagged coherency model are those obtained from the time histories of Event 20 in the radial direction (Section 3.4.1), namely, $A = 0.736$, $a = 0.147$, $k = 5210$ m, $f_o = 1.09$ Hz and $b = 2.78$. The structural model parameters are assumed to be [600]: $K_p/K_g = 1/10$ and $\zeta_0 = 50\%$, the latter value reflecting the effect of the soil-structure interaction for a buried structure. Figure 6.4(a) presents the exponential decay of the lagged coherency model for distances between the mass centers of the two segments of $L = 5$, 10 and 20 m; it can be observed from the figure that the correlation of the motions for all values of L is significant over the frequency range of the excitation (Fig. 6.2(a)). Figure 6.4(b) presents the PSD of

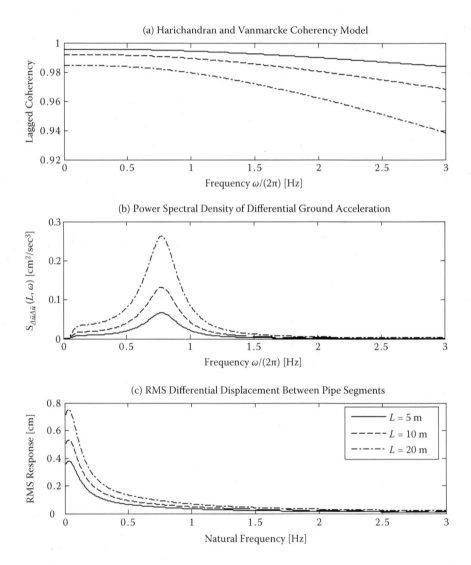

FIGURE 6.4 Part (a): The lagged coherency model of Harichandran and Vanmarcke [208] at separation distances of $L = 5, 10$ and 20 m between the supports of the pipe segments of Fig. 6.3. Part (b): The PSD of the differential ground acceleration resulting from the Clough-Penzien spectrum [107] at soft soil conditions (Fig. 6.2(a)) and the coherency model of part (a). Part (c): The rms differential displacement between the pipe segments for the excitation of part (b) as function of the natural frequency of the system of Eq. 6.27.

the differential ground accelerations (Eq. 6.29), and Fig. 6.4(c) the rms differential displacement between the pipe segments, i.e., the relative displacement experienced by the joint. The response assumes its higher values at the very low natural frequencies, consistent with the PSD of the ground displacement ($S_{\ddot{u}\ddot{u}}(\omega)/\omega^4$) and the frequency transfer function of the modified system, as indicated by Eq. 6.28.

6.2.2 MULTI-DEGREE-OF-FREEDOM, MULTIPLY-SUPPORTED SYSTEMS

Consider a discretized system with K degrees of freedom resting on N supports. The equation of motion for the system in matrix notation takes the form, e.g., [107]:

$$\begin{bmatrix} [\mathbf{M}_{ff}] & [\mathbf{M}_{fs}] \\ [\mathbf{M}_{fs}]^T & [\mathbf{M}_{ss}] \end{bmatrix} \begin{Bmatrix} \{\ddot{w}_t(t)\} \\ \{\ddot{u}(t)\} \end{Bmatrix} + \begin{bmatrix} [\mathbf{C}_{ff}] & [\mathbf{C}_{fs}] \\ [\mathbf{C}_{fs}]^T & [\mathbf{C}_{ss}] \end{bmatrix} \begin{Bmatrix} \{\dot{w}_t(t)\} \\ \{\dot{u}(t)\} \end{Bmatrix}$$

$$+ \begin{bmatrix} [\mathbf{K}_{ff}] & [\mathbf{K}_{fs}] \\ [\mathbf{K}_{fs}]^T & [\mathbf{K}_{ss}] \end{bmatrix} \begin{Bmatrix} \{w_t(t)\} \\ \{u(t)\} \end{Bmatrix} = \begin{Bmatrix} \{0\} \\ \{P_s(t)\} \end{Bmatrix} \tag{6.30}$$

where $\{w_t(t)\}$ is the $K \times 1$ vector of the total displacements at the free (unconstrained) degrees of freedom of the structure; $\{u(t)\}$ is the $N \times 1$ vector of restrained (support) displacements; $[\mathbf{M}_{ff}]$, $[\mathbf{C}_{ff}]$ and $[\mathbf{K}_{ff}]$ are the $K \times K$ mass, damping and stiffness matrices associated with the free degrees of freedom; $[\mathbf{M}_{ss}]$, $[\mathbf{C}_{ss}]$ and $[\mathbf{K}_{ss}]$ are the $N \times N$ mass, damping and stiffness matrices associated with the degrees of freedom at the supports; $[\mathbf{M}_{fs}]$, $[\mathbf{C}_{fs}]$ and $[\mathbf{K}_{fs}]$ are the $K \times N$ coupling mass, damping and stiffness matrices associated with both sets of degrees of freedom; $\{0\}$ indicates a $K \times 1$ vector of zeros; and $\{P_s(t)\}$ is the $N \times 1$ vector of the reaction forces at the supports. For a lumped-mass model $[\mathbf{M}_{fs}] = 0$, which will be considered to be the case in the following derivations.

The total displacement is commonly decomposed into its pseudo-static, $\{w_{ps}(t)\}$, and dynamic, $\{w(t)\}$, components, namely [107]:

$$\{w_t(t)\} = \{w_{ps}(t)\} + \{w(t)\} \tag{6.31}$$

The pseudo-static component is evaluated from the static equilibrium equations obtained by omitting the dynamic forces in Eq. 6.30 as:

$$\{w_{ps}(t)\} = -[\mathbf{K}_{ff}]^{-1} [\mathbf{K}_{fs}] \{u(t)\} = [\mathbf{R}] \{u(t)\} \tag{6.32}$$

$$[\mathbf{R}] = -[\mathbf{K}_{ff}]^{-1} [\mathbf{K}_{fs}] \tag{6.33}$$

with $[\mathbf{R}]$ denoting the matrix of influence coefficients. Substituting Eqs. 6.31–6.33 into Eq. 6.30, and assuming that, for low values of damping, the terms involving pseudo-static velocities are much smaller than the terms involving pseudo-static accelerations, the differential equation of motion for the dynamic response becomes:

$$[\mathbf{M}_{ff}]\{\ddot{w}(t)\} + [\mathbf{C}_{ff}]\{\dot{w}(t)\} + [\mathbf{K}_{ff}]\{w(t)\} \simeq -[\mathbf{M}_{ff}][\mathbf{R}]\{\ddot{u}(t)\} \tag{6.34}$$

The use of the normal-mode approach, i.e., expressing the dynamic response in terms of the mode-shape matrix, $[\boldsymbol{\Phi}]$, the columns of which, $\{\Phi_k\}$, represent the mode shape for each mode $k = 1, \ldots, K$, and the vector of the generalized coordinates, $\{q(t)\}$, as:

$$\{w(t)\} = [\boldsymbol{\Phi}]\{q(t)\} \tag{6.35}$$

and the pre-multiplication of Eq. 6.34 by $[\boldsymbol{\Phi}]^T$ uncouples the differential equation of motion to:

$$\ddot{q}_k(t) + 2\zeta_k\omega_k\dot{q}_k(t) + \omega_k^2 q_k(t) = \sum_{J=1}^{N}\Gamma_{kJ}\ddot{u}_J(t); \quad k = 1, \ldots, K \tag{6.36}$$

where ω_k and ζ_k are the natural frequency and damping coefficient, respectively, for mode k, and the participation factors, Γ_{kJ}, are given by:

$$\Gamma_{kJ} = -\frac{\{\Phi_k\}^T[\mathbf{M}_{ff}]\{R_J\}}{\{\Phi_k\}^T[\mathbf{M}_{ff}]\{\Phi_k\}} \tag{6.37}$$

with $\{R_J\}$ being the J-th column of $[\mathbf{R}]$ (Eq. 6.33). Alternatively, the generalized coordinate, $q_k(t)$ in Eq. 6.36, can be expressed as:

$$q_k(t) = \sum_{J=1}^{N}\Gamma_{kJ}s_{kJ}(t) \tag{6.38}$$

where $s_{kJ}(t)$, evaluated from:

$$\ddot{s}_{kJ}(t) + 2\zeta_k\omega_k\dot{s}_{kJ}(t) + \omega_k^2 s_{kJ}(t) = \ddot{u}_J(t) \tag{6.39}$$

is the normalized modal response, i.e., $s_{kJ}(t)$ reflects the response of a SDOF oscillator with frequency ω_k and damping ζ_k subjected to the excitation at support J of the multiply-supported structure. Equations 6.36 or 6.39 can be readily solved by means of Duhamel's integral (Eq. 6.6) to yield the dynamic response of the system.

Any response quantity of interest (displacement, force, stress or strain), $z(t)$, can be expressed as a linear function of the nodal displacements, as, e.g., [138], [285]:

$$z(t) = \{b\}^T\{w_t(t)\} \tag{6.40}$$

where $\{b\}$ is a vector depending on the geometry and material properties of the structure. The decomposition of the total displacement into its pseudo-static and dynamic components (Eq. 6.31) and the consideration of Eqs. 6.32–6.39 lead to the following expression for $z(t)$ (Eq. 6.40):

$$z(t) = \sum_{J=1}^{N}\beta_J u_J(t) + \sum_{J=1}^{N}\sum_{k=1}^{K}\eta_{kJ}s_{kJ}(t) \tag{6.41}$$

in which,

$$\beta_J = \{b\}^T\{R_J\}; \qquad J = 1, .., N \tag{6.42}$$

$$\eta_{kJ} = \{b\}^T\{\Phi_k\}\Gamma_{kJ}; \qquad J = 1, .., N; \quad k = 1, .., K \tag{6.43}$$

The PSD of $z(t)$ can then be evaluated as:

$$S_{zz}(\omega) = \sum_{J=1}^{N} \sum_{L=1}^{N} \beta_J \beta_L S_{u_J u_L}(\omega)$$

$$+ \sum_{J=1}^{N} \sum_{L=1}^{N} \sum_{k=1}^{K} \left[\beta_J \eta_{kL} H_k(\omega) S_{u_J \ddot{u}_L}(\omega) + \beta_L \eta_{kJ} H_k^*(\omega) S_{\ddot{u}_J u_L}(\omega) \right]$$

$$+ \sum_{J=1}^{N} \sum_{L=1}^{N} \sum_{k=1}^{K} \sum_{m=1}^{K} \eta_{kJ} \eta_{mL} H_k^*(\omega) H_{\dot{m}}(\omega) S_{\ddot{u}_J \ddot{u}_L}(\omega) \qquad (6.44)$$

where the frequency transfer function, $H_k(\omega)$, is now defined, from Eq. 6.17, for each mode k as:

$$H_k(\omega) = \left[(\omega_k^2 - \omega^2) + 2i\zeta_k \omega_k \omega \right]^{-1} \qquad (6.45)$$

and the input excitation cross spectral densities can be obtained through derivations similar to those presented for Eq. 6.20 as:

$$S_{\ddot{u}_J \ddot{u}_L}(\omega) = S_{\ddot{u}\ddot{u}}(\omega) \gamma(\xi_{JL}, \omega); \quad S_{u_J u_L}(\omega) = S_{\ddot{u}\ddot{u}}(\omega) \gamma(\xi_{JL}, \omega)/\omega^4$$

$$S_{u_J \ddot{u}_L}(\omega) = S_{\ddot{u}_J u_L}(\omega) = -S_{\ddot{u}\ddot{u}}(\omega) \gamma(\xi_{JL}, \omega)/\omega^2 \qquad (6.46)$$

with $\gamma(\xi_{JL}, \omega)$ indicating the coherency function evaluated at the separation distance ξ_{JL} between the two supports J and L.

6.3 DISTRIBUTED-PARAMETER SYSTEMS

As examples of random vibration analyses of distributed-parameter systems, this section describes the derivations for the evaluation of the response of continuous, buried pipelines (Section 6.3.1) and continuous beams on multiple supports (Section 6.3.2) subjected to spatially variable seismic excitations.

6.3.1 CONTINUOUS PIPELINES

Hindy and Novak [221] presented a detailed analysis of the seismic response of buried, continuous pipelines in the axial and lateral (transverse horizontal) directions. The basic steps of their approach are presented next and followed by the description of their results, which illustrate the effects of the ground motion characteristics and the structural properties on the response of these long structures.

Axial Response

The response in the axial direction of the pipeline can be evaluated from the model shown in Fig. 6.5(a). The pipeline rests on elastic foundations and its end conditions (roller supports) allow a rigid body motion in the axial direction. If the absolute pipeline displacement in the axial direction is indicated by $v(y, t)$, and $v_g(y, t)$ is

(a) Continuous Pipeline Model - Axial Direction

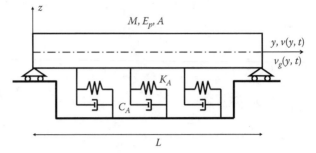

(b) Continuous Pipeline Model - Transverse Direction

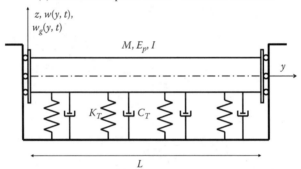

FIGURE 6.5 The continuous, buried pipeline model of Hindy and Novak [221] for the axial (part (a)) and the transverse (part (b)) response of the structure. M, E_p, A and I indicate mass per unit length, modulus of elasticity, cross-sectional area and second moment of area, respectively; y is the coordinate along the axis of the structure. For the axial direction (part (a)), $v(y, t)$ is the system's response, $v_g(y, t)$ the excitation, and K_A and C_A are the stiffness and damping of the soil-structure interaction, respectively. Correspondingly, for the transverse direction (part (b)), $w(y, t)$ is the system's response, $w_g(y, t)$ the excitation, and K_T and C_T are the stiffness and damping of the soil-structure interaction, respectively.

the ground excitation in the same direction, the equation of motion of the pipeline becomes:

$$M\frac{\partial^2 v(y, t)}{\partial t^2} + C_A\frac{\partial v(y, t)}{\partial t} + K_A v(y, t) - E_p A\frac{\partial^2 v(y, t)}{\partial y^2}$$
$$= C_A\frac{\partial v_g(y, t)}{\partial t} + K_A v_g(y, t) \qquad (6.47)$$

where M is the mass per unit length of the pipeline, C_A and K_A are the damping and stiffness of the soil-structure interaction, and E_P and A are the modulus of elasticity and cross sectional area of the pipeline, respectively. In the derivation, the internal damping in the pipeline is neglected as it is much smaller than the damping derived from the soil [221], and buckling of the pipeline is not considered. The stress-free boundary conditions of the pipeline model in Fig. 6.5(a) lead to the natural frequencies,

ω_j, and mode shapes, $\Phi_j(y)$, of the structure, which are given by:

$$\omega_j = \sqrt{\frac{K_A}{M}\left[1 + \frac{E_p A}{K_A}\frac{(j-1)^2\pi^2}{L^2}\right]} \qquad (6.48)$$

$$\Phi_j(y) = \cos\frac{(j-1)\pi y}{L} \qquad (6.49)$$

for $j = 1, \ldots, K$, with K being the number of modes used in the analysis. It is noted that the first mode ($j = 1$ in Eq. 6.49) represents the rigid body mode, i.e., $\Phi_1(y) \equiv 1$.

After some lengthy, but otherwise straightforward derivations, the PSD of the axial displacement at location y along the pipeline axis becomes [600]:

$$S_{vv}(y,\omega) = \frac{1}{M^2}\left[\frac{C_A^2}{\omega^2} + \frac{K_A^2}{\omega^4}\right]\sum_{j=1}^{K}\sum_{k=1}^{K}\frac{\Phi_j(y)\Phi_k(y)}{L_j L_k}H_j^*(\omega)H_k(\omega)$$

$$\times \int_0^L\int_0^L \Phi_j(y_1)\Phi_k(y_2)S_{\ddot{v}_g(y_1)\ddot{v}_g(y_2)}(\omega)\,dy_1 dy_2 \qquad (6.50)$$

where $H_k(\omega)$ is given by Eq. 6.45,

$$L_j = \begin{cases} L & \text{if } j = 1 \\ L/2 & \text{if } j = 2, \ldots, K \end{cases} \qquad (6.51)$$

and $S_{\ddot{v}_g(y_1)\ddot{v}_g(y_2)}(\omega)$ is the cross spectral density of the ground acceleration between two locations y_1 and y_2 along the pipeline axis. Similarly, the PSD of the axial stress, $\sigma_A(y)$, is given by:

$$S_{\sigma_A\sigma_A}(y,\omega) = \frac{E_p^2\pi^2}{M^2 L^2}\left[\frac{C_A^2}{\omega^2} + \frac{K_A^2}{\omega^4}\right]\sum_{j=1}^{K}\sum_{k=1}^{K}\frac{(j-1)(k-1)}{L_j L_k}$$

$$\times \sin\frac{(j-1)\pi y}{L}\sin\frac{(k-1)\pi y}{L}H_j^*(\omega)H_k(\omega)$$

$$\times \int_0^L\int_0^L \Phi_j(y_1)\Phi_k(y_2)S_{\ddot{v}_g(y_1)\ddot{v}_g(y_2)}(\omega)\,dy_1 dy_2 \qquad (6.52)$$

Lateral Response

The response in the direction normal to the pipeline axis (i.e., the transverse horizontal or vertical direction) can be evaluated from the model shown in Fig. 6.5(b). Again, the pipeline rests on elastic foundations and its end conditions (guided ends) allow a rigid body motion in the transverse direction. Hindy and Novak [221] noted that the selection of the boundary conditions in the models of Figs. 6.5(a) and (b) simplifies the analysis and does not affect the value of the stresses in the central part of the structures, provided that the pipeline is sufficiently long so that the boundary effects die out. The equation of motion of the pipeline in the transverse direction becomes then:

$$
M\frac{\partial^2 w(y,t)}{\partial t^2} + C_T\frac{\partial w(y,t)}{\partial t} + K_T w(y,t) + E_p I\frac{\partial^4 w(y,t)}{\partial y^4}
$$

$$
= C_T\frac{\partial w_g(y,t)}{\partial t} + K_T w_g(y,t) \tag{6.53}
$$

where $w(y,t)$ is the absolute displacement of the pipeline, $w_g(y,t)$ the ground displacement, C_T and K_T are the damping and stiffness of the soil-structure interaction, respectively, I is the second moment of area, and, again, the internal damping in the pipeline is neglected [221]. The boundary conditions of the model in Fig. 6.5(b), i.e., that the slope and shear force at the two ends are equal to zero, lead to the following natural frequencies for the pipeline:

$$
\omega_j = \sqrt{\frac{K_T}{M}\left[1 + \frac{E_p I}{K_T}\frac{(j-1)^4\pi^4}{L^4}\right]} \tag{6.54}
$$

with $j = 1,\ldots,K$, and the mode shapes in the transverse direction are also given by Eq. 6.49.

The PSD of the transverse displacement at location y along the pipeline axis is derived as [600]:

$$
S_{ww}(y,\omega) = \frac{1}{M^2}\left[\frac{C_T^2}{\omega^2} + \frac{K_T^2}{\omega^4}\right]\sum_{j=1}^{K}\sum_{k=1}^{K}\frac{\Phi_j(y)\Phi_k(y)}{L_j L_k}H_j^*(\omega)H_k(\omega)
$$

$$
\times \int_0^L\int_0^L \Phi_j(y_1)\Phi_k(y_2)S_{\ddot{w}_g(y_1)\ddot{w}_g(y_2)}(\omega)\,dy_1 dy_2 \tag{6.55}
$$

and the PSD of the maximum bending stress, $\sigma_B(y)$, as:

$$
S_{\sigma_B\sigma_B}(y,\omega) = \frac{E_p^2 R^2\pi^4}{M^2 L^4}\left[\frac{C_T^2}{\omega^2} + \frac{K_T^2}{\omega^4}\right]\sum_{j=1}^{K}\sum_{k=1}^{K}\frac{(j-1)^2(k-1)^2}{L_j L_k}\Phi_j(y)\Phi_k(y)
$$

$$
\times H_j^*(\omega)H_k(\omega)\int_0^L\int_0^L \Phi_j(y_1)\Phi_k(y_2)S_{\ddot{w}_g(y_1)\ddot{w}_g(y_2)}(\omega)\,dy_1 dy_2
$$

$$
\tag{6.56}
$$

in which, R is the pipeline radius and $S_{\ddot{w}_g(y_1)\ddot{w}_g(y_2)}(\omega)$ the cross spectral density of the transverse excitation.

Example Application

For the evaluation of the seismic response of buried pipelines, Hindy and Novak [221] introduced the Clough-Penzien spectrum parameters presented in Table 3.1(a), and, perhaps, the first spatial coherency model, described by Eq. 3.9, and repeated here for convenience:

$$
|\gamma(\xi,\omega)| = \exp\left[-\kappa\left(\frac{\omega\xi}{V_s}\right)^\nu\right] \tag{6.57}
$$

In the model, κ is a parameter controlling its exponential decay, V_s the S-wave velocity of the medium, and v was assumed to be equal to unity. Hindy and Novak [221]
modeled the soil-structure interaction by a complex dynamic stiffness, the real part
of which represents the effect of the springs and the imaginary part the effect of the
dashpots. The complex dynamic stiffness was reported to be a function of frequency,
the outer radius of the pipeline, and the shear modulus of the soil, its mass density, Poisson's ratio and material damping [377]. The response of the structures, for
frequency-independent soil-structure interaction, is provided by Eqs. 6.50, 6.52, 6.55
and 6.56. Regarding the effects of the ground motion characteristics and the structural
properties on the pipeline response, Hindy and Novak [221] noted the following:

1. The S-wave velocity, V_s, in the coherency expression (Eq. 6.57) does not
   affect the stresses in the pipeline, except when its value is very low. The
   lack of dependence of the pipeline stresses on V_s was attributed to two
   counteracting effects: As V_s increases, the correlation of the excitation
   increases (Eq. 6.57), which reduces the pipeline stresses, but an increase in
   V_s is also associated with an increase in the soil stiffness, which increases
   the stresses.
2. The coherency parameter κ (Eq. 6.57) strongly affects the resulting stresses:
   Stresses increase significantly as the parameter increases from $\kappa = 0.1$ to
   $\kappa \approx 30$, but decrease rapidly for higher values of κ.
3. Stresses increase proportionally to the ratio $\sigma_{v_g}/\sigma_{\ddot{v}_g}$ for the axial direction
   and $\sigma_{w_g}/\sigma_{\ddot{w}_g}$ for the transverse direction, where σ_{v_g} and σ_{w_g} are the rms
   ground displacements, and $\sigma_{\ddot{v}_g}$ and $\sigma_{\ddot{w}_g}$ the rms ground accelerations in the
   respective directions.
4. For longer pipelines, the response is minimally affected by the length
   of the structure. However, when the pipeline length becomes shorter than
   approximately $100\,R$, the seismic stresses start decreasing due to the higher
   values of the coherency at the shorter separation distances (Eq. 6.57).
5. Axial stresses decrease with increasing radius, because the axial stiffness of
   the pipeline increases with increasing radius. On the other hand, bending
   stresses are obtained from the product of the radius of the pipeline and
   its curvature. As the radius decreases, the curvature increases and, hence,
   bending stresses remain fairly constant.
6. Both axial and bending stresses decrease with increasing pipeline
   thickness.

Figure 6.6 presents an example application of their evaluation: Figures 6.6(a) and
(b) present the axial response in terms of rms relative displacements and stresses,
respectively, and Figs. 6.6(c) and (d) the corresponding results in the transverse horizontal direction. The seismic excitation was the same in both directions: The PSD
was described by the characteristics of "Spectrum 1" in Table 3.1(a), i.e., soft soil
conditions, with $\sigma_{\ddot{v}_g} = \sigma_{\ddot{w}_g} = 61$ cm/sec^2 (= 2 ft/sec^2). The pipeline was long enough
so that the results were independent of its actual length, indicated by L in the fig-
ures, as per item 4 in the aforementioned description of Hindy and Novak's [221]
conclusions. The value of V_s was such that the results were independent of its actual
value (item 1), and κ was assumed to be equal to 0.5. The remaining parameters were

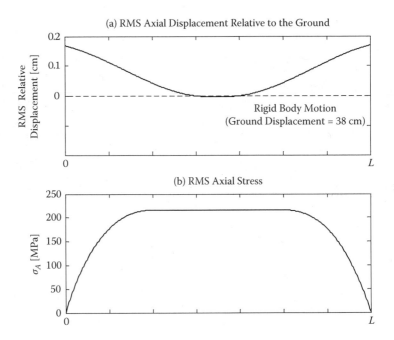

FIGURE 6.6 The seismic response characteristics of the continuous pipeline models of Fig. 6.5. Parts (a) and (b) present the axial response in terms of rms relative displacements and stresses, and parts (c) and (d), shown on the next page, the corresponding results in the transverse horizontal direction; the same spatially variable excitation was used in both directions. The response is symmetric about the midpoint of the structure. The rms relative displacements (parts (a) and (c)) indicate that the pipeline deformation relative to the rigid body motion (presented by the dashed line in the figures) is small. Axial stresses (part (b)) are higher than bending stresses (part (d)), and, in both cases, stresses are affected by the boundary conditions only close to the boundaries, and are constant along the center part of the structure. (After A. Hindy and M. Novak, "Pipeline response to random ground motion," *Journal of the Engineering Mechanics Division, ASCE,* Vol. 106, pp. 339–360, 1980; reproduced with permission from ASCE).

specified as: $R = 68.625 \, \text{cm} \, (= 2.25 \, \text{ft}), t/R = 1.622 \times 10^{-2}$ and $E_p/G = 2.74 \times 10^4$, where G indicates the shear modulus of the soil.

The pipeline models in Figs. 6.5(a) and (b) are symmetric about the midpoint of the structures, and their response to stationary spatially variable excitations exhibiting loss of coherency only is also symmetric (Fig. 6.6) for the following reason: Realizations of spatially variable ground motions will lead to time series that vary at different locations on the ground surface, as illustrated, through simulations, in Figs. 7.10(a)–(c) for motions exhibiting loss of coherency only and in Fig. 7.10(d) for motions exhibiting loss of coherency and propagating on the ground surface. The application of such time histories as input excitations at the supports of a symmetric structure will lead to a nonsymmetric response. However, for an infinite number of realizations and for seismic ground motions exhibiting loss of coherency only, another realization of the time series will exist that will look like the exact "reverse" of the first

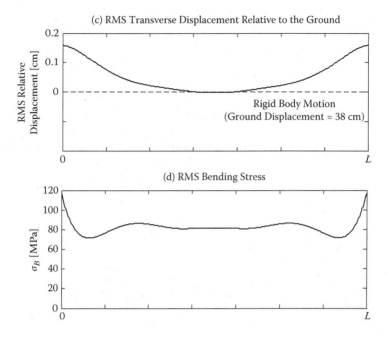

(c) RMS Transverse Displacement Relative to the Ground

Rigid Body Motion
(Ground Displacement = 38 cm)

(d) RMS Bending Stress

FIGURE 6.6 (Continued).

realization, i.e., the time series at location $x = 0$ m in Figs. 7.10(a)–(c) will appear at location $x = 400$ m of the "reverse" realization, the time series at location $x = 100$ m in Figs. 7.10(a)–(c) will appear at location $x = 300$ m of the "reverse" realization, and so on. This is a consequence of the fact that the random field of seismic ground motions exhibiting loss of coherency only is direction-independent (Section 3.3.3). The response of the symmetric structure due to the "reverse" realization will also be nonsymmetric, but it will be the "mirror image" about the structure's midpoint of the response due to the first ground motion realization. The combined effect of the responses due to the two excitations will then lead to a symmetric response. Since, for every realization of a direction-independent random field, a "reverse" realization will always exist, the stochastic response of the symmetric structure, determined by means of a random vibration approach that encompasses the results of all possible simulations, will be symmetric. On the other hand, when the wave passage effect is included in the evaluation, the response of symmetric structures ceases to be symmetric, as the input random field is now direction-dependent (Fig. 7.10(d)).

Figures 6.6(a) and (c) present the rms pipeline displacements relative to the ground displacement in the axial and transverse directions, respectively. The dashed lines in the figures indicate the rigid body displacement, the value of which is 38 cm; the relative rms displacements in either direction are less than 0.2 cm. It can be clearly seen from the figures that the pipeline follows the motion of the ground, which validates the assumption that strains in these buried structures can be approximated by the seismic ground strains (Chapter 5). Figures 6.6(b) and (d) indicate that the stresses in the pipeline are affected by the boundary conditions only close to the end supports of the structure, and are, essentially, constant in its central part, thus justifying the use of the simple boundary conditions in the approach. The significance of axial vs. bending strains in small-diameter, buried pipelines was briefly described in Section 5.1, and is also illustrated in Figs. 6.6(b) and (d): Axial and bending stresses are of the same order of magnitude, but the bending stresses are significantly lower than the axial stresses. It should be noted, however, as suggested earlier in item 5 of Hindy and Novak's [221] conclusions, that the value of axial stresses decreases as the pipeline radius increases, whereas bending stresses remain fairly constant. Sakurai and Takahashi [437], based on measurements of strains in pipelines during earthquakes, suggested that, for small diameter pipelines, axial stresses dominate, but, for large diameter pipelines, bending stresses should be superimposed to the axial ones. Random vibration analyses of a larger radius ($R = 1$ m) pipeline [600] further confirmed the latter observation.

6.3.2 CONTINUOUS BEAMS ON MULTIPLE SUPPORTS

The structural model for a continuous beam on multiple supports is presented in Fig. 6.7. The model, due to its simplicity, has been utilized by various researchers for the investigation of the effects of spatial variability on the response of long structures (e.g., [210], [315], [590]); such an example application will be presented later in Section 6.4.5.

The equation of motion of the system in the transverse direction is given by:

$$M\frac{\partial^2 w_t(y, t)}{\partial t^2} + C\frac{\partial w_t(y, t)}{\partial t} + E_p I\frac{\partial^4 w_t(y, t)}{\partial y^4} = 0 \qquad (6.58)$$

where $w_t(y, t)$ indicates the total displacement, M the mass per unit length, C the damping, E_p the modulus of elasticity, I the second moment of area, and y the coordinate along the axis of the beam. As described in the evaluation of the response of multi-degree-of-freedom, multiply-supported discrete systems (Eq. 6.31 in

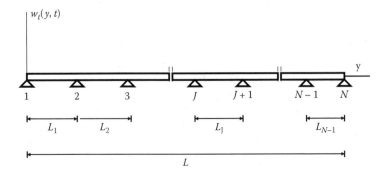

FIGURE 6.7 The model of continuous beams on multiple (N) supports. y is the coordinate along the axis of the structure and $w_t(y, t)$ the total displacement in the lateral direction. The length of each span is indicated by L_J, $J = 1, \ldots, N - 1$, and the total length of the structure is L.

Section 6.2.2), the total displacement is, generally, decomposed into its pseudo-static and dynamic components:

$$w_t(y, t) = w_{ps}(y, t) + w(y, t) \tag{6.59}$$

The pseudo-static response is expressed as:

$$w_{ps}(y, t) = \sum_{J=1}^{N} R_J(y) u_J(t) \tag{6.60}$$

in which, $R_J(y)$ is the shape function for unit displacement at support J, $u_J(t)$ is the ground displacement at support J, and N is the number of supports. The substitution of Eqs. 6.59 and 6.60 in Eq. 6.58 and the assumption of low damping yield the following equation of motion in terms of the structure's dynamic response:

$$M\ddot{w}(y, t) + C\dot{w}(y, t) + E_p I w^{\mathrm{IV}}(y, t) \simeq -M \sum_{J=1}^{N} R_J(y) \ddot{u}_J(t) \tag{6.61}$$

where the superscript IV indicates the fourth derivative with respect to y. The consideration of a normal mode approach leads to the following expression for the dynamic displacement:

$$w(y, t) = \sum_{k=1}^{K} \Phi_k(y) q_k(t) \tag{6.62}$$

where, as in Eq. 6.35, $\Phi_k(y)$ and $q_k(t)$ indicate the mode shape and the generalized coordinate, respectively, for mode k, and K is the number of modes used in the analysis. The natural frequencies, ω_k, and mode shapes, $\Phi_k(y)$, $k = 1, \ldots, K$, are evaluated from the boundary conditions of the continuous beam of Fig. 6.7, namely, zero displacements at the supports, continuity of slopes and equality of bending moments from left and right at the inner supports, and zero moments at the end supports. Finally, using the orthogonality property of the mode shapes, the equation of motion for mode k becomes:

$$\ddot{q}_k(t) + 2\zeta_k \omega_k \dot{q}_k(t) + \omega_k^2 q_k(t) = \sum_{J=1}^{N} \Gamma_{kJ} \ddot{u}_J(t) \tag{6.63}$$

where the participation factor Γ_{kJ} for mode k and the excitation at support J is given by:

$$\Gamma_{kJ} = -\frac{\int_0^L R_J(y) \Phi_k(y) \, dy}{\int_0^L [\Phi_k(y)]^2 \, dy} \tag{6.64}$$

and $\zeta_k = \zeta_1 \omega_1 / \omega_k = C/(2M\omega_k)$. Equation 6.63 can be readily solved by means of Duhamel's integral (Eqs. 6.6 and 6.7) leading to the solution for the total response as (Eqs. 6.59, 6.60, 6.62 and 6.63):

$$w_t(y, t) = \sum_{J=1}^{N} R_J(y) u_J(t) + \sum_{k=1}^{K} \Phi_k(y) q_k(t) \tag{6.65}$$

After some lengthy, but otherwise straightforward manipulations, the PSD of the total displacement response and its space derivatives at a location y along the structure's axis (Fig. 6.7) is found to be [590], [594]:

$$S_{w_t^{(n)}w_t^{(n)}}(y,\omega) = \sum_{J=1}^{N}\sum_{L=1}^{N} R_J^{(n)}(y)R_L^{(n)}(y)S_{u_J u_L}(\omega)$$

$$+ \sum_{J=1}^{N}\sum_{L=1}^{N}\sum_{k=1}^{K}\left[\Phi_k^{(n)}(y)\left\{R_J^{(n)}(y)\Gamma_{kL}H_k(\omega)S_{u_J\ddot{u}_L}(\omega)\right.\right.$$

$$\left.\left.+ R_L^{(n)}(y)\Gamma_{kJ}H_k^*(\omega)S_{\ddot{u}_J u_L}(\omega)\right\}\right]$$

$$+ \sum_{J=1}^{N}\sum_{L=1}^{N}\sum_{k=1}^{K}\sum_{m=1}^{K}\Phi_k^{(n)}(y)\Phi_m^{(n)}(y)\Gamma_{kJ}\Gamma_{mL}H_k^*(\omega)H_m(\omega)S_{\ddot{u}_J\ddot{u}_L}(\omega)$$

$$(6.66)$$

where the superscript (n) indicates the n-th derivative of the quantity with respect to y. For example, the power spectral density of bending moments, $\mathcal{M}(y,t)$, and shear forces, $\mathcal{V}(y,t)$, along the structure can be obtained from:

$$S_{\mathcal{M}\mathcal{M}}(y,\omega) = (EI)^2 S_{w_t'' w_t''}(y,\omega)$$

$$S_{\mathcal{V}\mathcal{V}}(y,\omega) = (EI)^2 S_{w_t''' w_t'''}(y,\omega) \qquad (6.67)$$

in which, $w_t''(y,t)$ and $w_t'''(y,t)$ indicate the second ($n=2$ in Eq. 6.66) and third ($n=3$ in Eq. 6.66) derivatives with respect to y, respectively. The input cross spectral densities in Eq. 6.66 are provided by Eq. 6.46.

6.4   ANALYSIS OF RMS LIFELINE RESPONSE

The PSDs of the response of discrete- and distributed-parameter systems (Eqs. 6.44 and 6.66) consist of a significant number of terms: The double summation in the equations corresponds to the pseudo-static response, the quadruple summation to the dynamic response, and the triple summation to the cross terms between the pseudo-static and the dynamic response. This section analyzes the relative significance of these terms to the lifeline response and provides insight into the effect of spatial variability on the response of multiply-supported structures. Section 6.4.1 describes the ground motion characteristics, which, unless otherwise noted, will be used in the evaluations of this section. Sections 6.4.2 and 6.4.3 describe the dominant pseudo-static and dynamic terms of the response, respectively, as reported by Zerva [594], [596], and Section 6.4.4 examines the contribution of the cross terms to the PSD by means of the cross correlation coefficients defined by Der Kiureghian and Neuenhofer [138]. Finally, Section 6.4.5 illustrates, with a simple example, the differences in the total response of a lifeline system induced by uniform and spatially variable excitations at its supports.

6.4.1   GROUND MOTION CHARACTERISTICS

The PSD of the ground motions is described by the Clough-Penzien spectrum [107] of Eq. 3.7 for firm soil conditions with filter parameters as specified by Hindy and Novak [221] in "Spectrum 2" of Table 3.1(a), i.e., $\omega_g = 5\pi$ rad/sec, $\omega_f = 0.5\pi$ rad/sec and $\zeta_g = \zeta_f = 0.6$; the amplitude of the white-noise excitation spectrum is assumed to be $S_o = 1$ cm^2/sec^3. The acceleration PSD is presented in Fig. 6.8(a); velocity and displacement power spectral densities for these soil conditions are presented in Fig. 7.1.

Two coherency models, the model of Harichandran and Vanmarcke [208] and the model of Luco and Wong [324], are utilized: The model of Harichandran and Vanmarcke (Eq. 3.12 in Section 3.4.1) is used herein with parameters, as before in Section 6.2.1, $A = 0.736$, $a = 0.147$, $k = 5210$ m, $f_o = 1.09$ Hz and $b = 2.78$. The model of Luco and Wong (Eq. 3.27 in Section 3.4.2) is used with two values for the coherency drop parameter, $\alpha = 2 \times 10^{-4}$ sec/m and $\alpha = 1 \times 10^{-3}$ sec/m, the former representing a slow exponential decay with frequency and separation distance and the latter a significantly sharper one. The three models are compared in Figs. 6.8(b), (c) and (d), which present, respectively, their decay with frequency at separation distances of $\xi = 100, 300$ and 500 m. As discussed earlier in Sections 3.4.1 and 3.4.2, the model of Harichandran and Vanmarcke is only partially correlated at low frequencies and decreases slowly with separation distance and frequency over the frequency range presented in the figures, whereas the model of Luco and Wong is fully correlated in the low frequency range irrespective of the value of the coherency drop parameter, but approaches zero faster than the model of Harichandran and Vanmarcke. The significance of the differences in the exponential decay of these widely-used coherency models on the lifeline response is illustrated in the following three subsections.

6.4.2   PSEUDO-STATIC RESPONSE

The PSD of the pseudo-static response of the structure is given by the double summation in Eq. 6.66, i.e.,

$$S_{PS}(y, \omega) = \sum_{J=1}^{N} \sum_{L=1}^{N} R_J^{(n)}(y) R_L^{(n)}(y) S_{u_J u_L}(\omega) \tag{6.68}$$

The mean-square pseudo-static response can be shown to be [594], [596]:

$$\sigma_{PS}^2(y) = \int_{-\infty}^{+\infty} S_{PS}(y, \omega) \, d\omega$$

$$= \sum_{J=1}^{N} \left[R_J^{(n)}(y) \right]^2 \sigma_u^2 + \sum_{J=1}^{N-1} \sum_{L=J+1}^{N} R_J^{(n)}(y) R_L^{(n)}(y) D^2(\xi_{JL}) \tag{6.69}$$

where,

$$\sigma_u^2 = \int_{-\infty}^{+\infty} S_{uu}(\omega) \, d\omega; \quad D^2(\xi_{JL}) = \int_{-\infty}^{+\infty} [S_{u_J u_L}(\omega) + S_{u_L u_J}(\omega)] \, d\omega \tag{6.70}$$

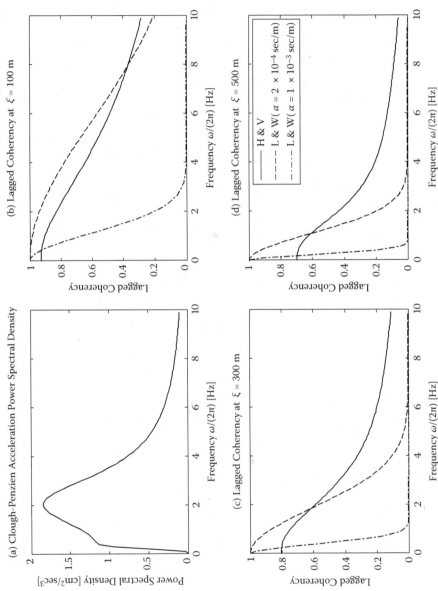

FIGURE 6.8

with the cross spectral densities of the ground displacements, $S_{u_J u_L}(\omega)$ and $S_{u_L u_J}(\omega)$, being defined in Eq. 6.46.

Because of the property of the shape functions:

$$\sum_{J=1}^{N} R_J(y) = 1 \Rightarrow \sum_{J=1}^{N} R_J^{(n)}(y) = 0 \quad \text{for} \quad n > 0 \tag{6.71}$$

the mean-square pseudo-static response (Eq. 6.69) of the internal forces in the structure, as, e.g., bending moments and shear forces (Eq. 6.67), can be rewritten as [594], [596]:

$$\sigma_{PS}^2(y) = -\sum_{J=1}^{N-1} \sum_{L=J+1}^{N} R_J^{(n)}(y) R_L^{(n)}(y) \left[2\sigma_u^2 - D^2(\xi_{JL}) \right] \quad \text{for} \quad n > 0 \tag{6.72}$$

The expression in the brackets depends only on the seismic ground motion characteristics, and, from Eqs. 6.29 and 6.70, it can be seen that it represents the mean-square differential ground displacement, $\sigma_{\Delta u}^2(\xi_{JL})$, between the motions at the two supports J and L, i.e.,

$$\sigma_{\Delta u}^2(\xi_{JL}) = 2\sigma_u^2 - D^2(\xi_{JL}) \tag{6.73}$$

Equations 6.72 and 6.73 then suggest that the pseudo-static internal forces in the structure are controlled by the differential displacements between the supports. Figure 6.9(a) presents the rms differential displacements resulting from the ground motions described in the previous subsection (Fig. 6.8) for separation distances ranging between 0 and 500 m. The coherency model of Luco and Wong with the slow exponential decay ($\alpha = 2 \times 10^{-4}$ sec/m) produces the lowest contribution to the pseudo-static internal forces in the structure, because it is fully correlated in the low frequency range (< 1 Hz) that controls the displacements (Figs. 6.8(b)–(d) and 7.1(c)). On the other hand, the model of Harichandran and Vanmarcke is partially correlated even at zero frequency (Figs. 6.8(b)–(d)), and yields large values for the rms differential displacements even at short separation distances. The rms differential displacements resulting from the model of Luco and Wong with the sharp exponential decay ($\alpha = 1 \times 10^{-3}$ sec/m in Figs. 6.8(b)–(d)) fall in-between the rms values produced by the other two models: Because the model for this value of the coherency drop parameter is still fully correlated in the low frequency range, it produces lower values for the rms differential displacements than those induced by the model of Harichandran and Vanmarcke. As separation distance increases, however, the decay

---

FIGURE 6.8 Part (a): The PSD of the Clough-Penzien spectrum [107] for firm soil conditions ("Spectrum 2" in Table 3.1(a)). Part (b): The exponential decay of the lagged coherency models of Harichandran and Vanmarcke [208] (indicated by "H & V" in the figure caption) and Luco and Wong [324] (indicated by "L & W" in the figure caption) for two values of the coherency drop parameter ($\alpha = 2 \times 10^{-4}$ sec/m and $\alpha = 1 \times 10^{-3}$ sec/m) at a separation distance of $\xi = 100$ m. Parts (c) and (d): Same as part (b) but for separation distances of $\xi = 300$ m and $\xi = 500$ m, respectively. The legend in subplot (d) refers to parts (b)–(d) of the figure.

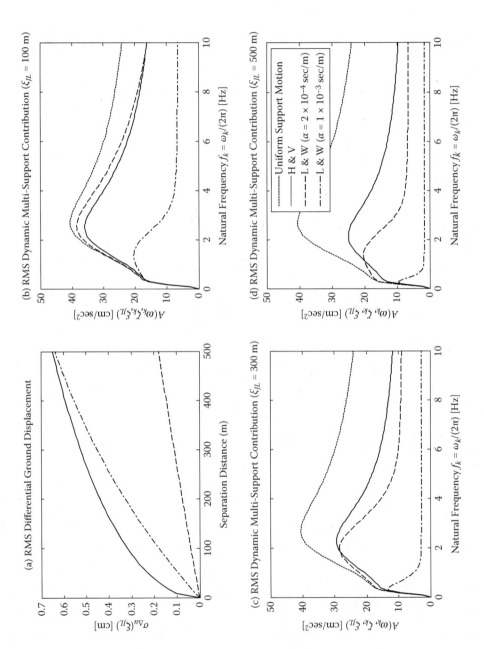

FIGURE 6.9

of the model becomes sharper, and its contribution to the pseudo-static response of the structure increases steeply. For the longer separation distances in Fig. 6.9(a), the models of Harichandran and Vanmarcke and Luco and Wong for $\alpha = 1 \times 10^{-3}$ sec/m yield comparable results, with the rms differential displacements of the model of Harichandran and Vanmarcke tending to a constant value and those of the model of Luco and Wong for $\alpha = 1 \times 10^{-3}$ sec/m still increasing. It is noted that, considering that the rms absolute displacement of the input excitation is 0.83 cm, the differential motions at the longer separation distances induced by the models of Harichandran and Vanmarcke and Luco and Wong for $\alpha = 1 \times 10^{-3}$ sec/m assume significantly high values. Obviously, when uniform motions are applied as input excitations at the lifeline supports, the rms differential motions are equal to zero, and the seismic excitation does not contribute to the pseudo-static response of the structure.

6.4.3 DYNAMIC RESPONSE

The dynamic response of the structure is given by the quadruple summation in Eq. 6.66, i.e.,

$$S_D(y, \omega) = \sum_{J=1}^{N} \sum_{L=1}^{N} \sum_{k=1}^{K} \sum_{m=1}^{K} \Phi_k^{(n)}(y) \Phi_m^{(n)}(y) \Gamma_{kJ} \Gamma_{mL} H_k^*(\omega) H_m(\omega) S_{\ddot{u}_J \ddot{u}_L}(\omega) \quad (6.74)$$

When $J = L$, the summation terms in the above expression reflect the contribution of the seismic motion at support J to the dynamic response of the structure. When $J \neq L$, the summation terms represent the contribution of the multi-support excitations in the response. The terms with $k = m$ correspond to the contribution of mode k to the response, whereas the terms with $k \neq m$ are the modal cross terms. Insight into the contribution of the spatially variable ground motions to the dynamic response of the structure can be obtained by examining the contribution of the multi-support excitation to each individual mode of the structure, i.e., by considering the terms with $k = m$ in Eq. 6.74. The equation then takes the form [594], [596]:

$$S_{D|k=m}(y, \omega) = \sum_{J=1}^{N} \sum_{L=1}^{N} \sum_{k=1}^{K} \left[\Phi_k^{(n)}(y) \right]^2 \Gamma_{kJ} \Gamma_{kL} |H_k(\omega)|^2 S_{\ddot{u}_J \ddot{u}_L}(\omega) \quad (6.75)$$

For firm soil conditions and for well-separated frequencies, the terms in Eq. 6.75 dominate the dynamic response of the structure [138], and are illustrated in the following. The modal cross terms ($k \neq m$) in Eq. 6.74 are examined in the next subsection.

FIGURE 6.9 Contribution of the pseudo-static and dynamic terms to the total response of multiply-supported structures. The effect is shown for the ground motion characteristics (*firm soil conditions*) illustrated in Fig. 6.8, as well as uniform motions at the structure's supports. Part (a): rms differential ground displacement, contributing to the pseudo-static response, as function of support separation distance; the contribution of uniform support motions is identically equal to zero. Part (b): rms multi-support contribution to the dynamic response of the structure as function of its k-th natural frequency (damping is 5% of critical); the support separation distance is $\xi_{JL} = 100$ m. Parts (c) and (d): Same as in part (b), but for support separation distances of $\xi_{JL} = 300$ m and $\xi_{JL} = 500$ m, respectively (after Zerva [594], [596]).

With the aforementioned considerations, the mean-square dynamic response of the structure for the excitation of the individual modes (Eq. 6.75) can be rewritten as [594], [596]:

$$
\sigma^2_{D|k=m}(y) = \int_{-\infty}^{+\infty} S_{D|k=m}(y, \omega) \, d\omega
$$

$$
= \sum_{k=1}^{K} \frac{[\Phi_k^{(n)}(y)]^2}{\omega_k^4} \left\{ \sum_{J=1}^{N} \Gamma_{kJ}^2 A^2(\omega_k, \zeta_k) \right.
$$

$$
\left. + \sum_{J=1}^{N-1} \sum_{L=J+1}^{N} \Gamma_{kJ} \Gamma_{kL} A^2(\omega_k, \zeta_k, \xi_{JL}) \right\} \quad (6.76)
$$

in which, $A(\omega_k, \zeta_k)$ is the rms contribution of the acceleration time history at each support to the k-th mode of the structure, i.e.,

$$
A^2(\omega_k, \zeta_k) = \omega_k^4 \int_{-\infty}^{+\infty} |H_k(\omega)|^2 S_{\ddot{u}\ddot{u}}(\omega) \, d\omega \quad (6.77)
$$

$A(\omega_k, \zeta_k)$ is proportional to the mean-maximum response of a SDOF oscillator with frequency ω_k and damping ζ_k subjected to a seismic excitation described by its power spectrum, $S_{\ddot{u}\ddot{u}}(\omega)$. It depends only on the PSD of the motions and the structural properties of the oscillator, and, hence, it is comparable to the concept of the response spectrum [594]. On the other hand, $A(\omega_k, \zeta_k, \xi_{JL})$ in Eq. 6.76 is the rms contribution of the cross spectral densities between the motions at the two supports J and L to the k-th mode of the structure, i.e.,

$$
A^2(\omega_k, \zeta_k, \xi_{JL}) = \omega_k^4 \int_{-\infty}^{+\infty} |H_k(\omega)|^2 (S_{\ddot{u}_J \ddot{u}_L}(\omega) + S_{\ddot{u}_L \ddot{u}_J}(\omega)) \, d\omega
$$

$$
= 2\omega_k^4 \int_{-\infty}^{+\infty} |H_k(\omega)|^2 S_{\ddot{u}\ddot{u}}(\omega) \Re[\gamma(\xi_{JL}, \omega)] \, d\omega \quad (6.78)
$$

The rms multi-support contribution to the dynamic response of the structure, $A(\omega_k, \zeta_k, \xi_{JL})$, filters the cross spectra of the spatially variable seismic excitations at the two supports through the frequency transfer function of a SDOF oscillator with frequency ω_k and damping ζ_k. Because of its similarity to Eq. 6.77, Eq. 6.78 may also be viewed as the rms response of a SDOF oscillator, that is subjected, however, to the cross spectral densities of the spatially variable motions between the two supports [594].

Figures 6.9(b), (c) and (d) present the multi-support contribution of the spatially variable ground motions of Fig. 6.8 to the excitation of the individual modes of the lifeline system (Eq. 6.78) for support separation distances of $\xi_{JL} = 100, 300$, and 500 m, respectively. For comparison purposes, the corresponding results caused by uniform motions at the two supports are also shown in the figures. The plots present the values of $A(\omega_k, \zeta_k, \xi_{JL})$ as functions of the natural frequency of the SDOF oscillator, which corresponds to the k-th natural frequency of the structure; in all cases, damping is assumed to be 5% of critical. It is noted that, because $|\gamma(\xi_{JL}, \omega)| \equiv 1$ for uniform motions, the rms values for uniform support excitations in Figs. 6.9(b)–(d) are the same. Furthermore, uniform support excitations produce consistently the highest multi-support contribution to the dynamic response of the system. For $\xi_{JL} = 100$ m in Fig. 6.9(b), the contribution of the uniform excitations to the response is only slightly higher than that of the models of Harichandran and Vanmarcke and Luco and Wong for $\alpha = 2 \times 10^{-4}$ sec/m at the lower frequencies, but the differences in the response induced by uniform and spatially variable excitations increase with increasing frequency. For this separation distance (Fig. 6.9(b)), the models of Harichandran and Vanmarcke and Luco and Wong for $\alpha = 2 \times 10^{-4}$ sec/m produce similar ordinates, with the model of Luco and Wong yielding slightly higher results in the frequency range presented in the figure. The model of Luco and Wong for $\alpha = 1 \times 10^{-3}$ sec/m produces the lowest contribution to the response (Fig. 6.9(b)). As separation distance increases (Figs. 6.9(c) and (d)), the rms multi-support contribution of the spatially variable ground motions becomes significantly lower than that of the uniform motions at the supports. For the separation distance of $\xi_{JL} = 300$ m (Fig. 6.9(c)), the models of Harichandran and Vanmarcke and Luco and Wong for $\alpha = 2 \times 10^{-4}$ sec/m produce similar results, with the model of Luco and Wong producing slightly higher ordinates for oscillator frequencies lower than approximately 2 Hz, but lower ordinates past this frequency. The model of Luco and Wong for the higher value of α produces the lowest contribution: It follows the increase of the other two models at the very low frequencies, but decreases quickly to a constant value from a frequency lower than 2 Hz (Fig. 6.9(c)). For the longest separation distance considered, $\xi_{JL} = 500$ m in Fig. 6.9(d), the model of Luco and Wong for $\alpha = 2 \times 10^{-4}$ sec/m still produces slightly higher ordinates than the model of Harichandran and Vanmarcke at low oscillator frequencies, but, past approximately 1 Hz, its contribution becomes lower and tends to a constant value at lower frequencies than the contribution of the model of Harichandran and Vanmarcke. At the longest separation distance (Fig. 6.9(d)), the model of Luco and Wong for the higher value of the coherency drop parameter ($\alpha = 1 \times 10^{-3}$ sec/m) produces a peak ordinate at the very low frequencies and, at approximately 1 Hz, has reached a low, constant value.

The rms multi-support contribution to the dynamic response of the structure (Figs. 6.9(b)–(d)) is proportional to the mean-maximum response of a SDOF oscillator subjected to stationary ground motions described by twice their cross spectral density (Eq. 6.78), and, hence, physically, it is comparable to the concept of the conventional response spectrum for the aforementioned excitation [594]. When the modal frequency falls within the range of dominant frequencies of the seismic excitation, the major contribution to the response results from excitation frequencies in the vicinity of the modal frequency under consideration. The seismic excitation for the evaluation of the rms response induced by uniform support motions is equal to twice the PSD of the ground acceleration (Eq. 6.78 with $|\gamma(\xi_{JL}, \omega)| \equiv 1$). On the other hand, when spatially variable ground motions are utilized, the input excitation is twice the cross spectral density, i.e., the product of the PSD of Fig. 6.8(a) with the coherency models of Figs. 6.8(b)–(d), for which $|\gamma(\xi_{JL}, \omega)| \leq 1$. Consequently, uniform support excitations will always produce the highest multi-support contribution to the dynamic response of the system. Furthermore, the coherency model that produces higher correlation in the vicinity of the natural frequency of the oscillator (i.e., the modal frequency of the system) will also contribute more to the dynamic response of the structure. Indeed, the "cross-over" oscillator frequencies, that separate the regions where the coherency models of Harichandran and Vanmarcke and Luco and Wong for $\alpha = 2 \times 10^{-4}$ sec/m produce the higher response (Figs. 6.9(b)–(d)), correspond, basically, to the frequencies where these two coherency models intersect (Figs. 6.8(b)–(d)). For the longer support separation distances, the cross-over frequencies of these two models (Figs. 6.8(c) and (d)) and of their induced response (Figs. 6.9(c) and (d)) essentially coincide due to the narrower bandwidth of the oscillator frequency transfer functions at these lower frequencies. Additionally, response spectra tend to constant values when the modal frequencies increase past the dominant frequencies of the seismic excitation, as, in this case, the excitation is viewed as pseudo-static. This effect can be clearly observed in the rms multi-support contribution to the dynamic response of the spatially variable motions described by the model of Luco and Wong for $\alpha = 1 \times 10^{-3}$ sec/m: When the coherency model tends to zero in Figs. 6.8(b)–(d), the rms response in Figs. 6.9(b)–(d) tends to constant values. Similarly, the rms response induced by the spatially variable ground motions described by the model of Luco and Wong for $\alpha = 2 \times 10^{-4}$ sec/m at the longer separation distances (Figs. 6.9(c) and (d)) tends to constant values at the frequencies where the coherency model, and, hence, the system's excitation (Eq. 6.78), tends to zero. On the other hand, the model of Harichandran and Vanmarcke decays slower than the other two models (Figs. 6.8(b)–(d)), and, thus, its rms response tends slowly to constant values at the higher frequencies.

Figures 6.9(b)–(d) clearly indicate that the loss of coherency in the seismic motions contributes beneficially to the excitation of the individual modes of the structure as compared to uniform excitations at the structure's supports. The terms $A(\omega_k, \zeta_k, \xi_{JL})$ of Eq. 6.78, however, are embedded in the expression for the total response of the system (Eq. 6.66), which contains additional dynamic contributions (Eq. 6.74), and, for nonuniform support motions, the pseudo-static contribution (Section 6.4.2), as well as the cross terms of the pseudo-static response and between the pseudo-static and the dynamic response. Furthermore, uniform and spatially variable ground motions do not excite the same structural modes. For example, uniform excitations induce only the symmetric modes of a symmetric structure, whereas spatially variable excitations induce also its antisymmetric modes (Section 6.4.5). Hence, it cannot be determined *a-priori* which type of excitation will lead to the highest structural response, as will be further elaborated upon in Chapter 9.

6.4.4 CROSS TERMS

The significance of the contributions of the cross terms of the pseudo-static response, between the pseudo-static and the dynamic response, and the modal cross terms to the total rms response of multiply-supported systems was noted by Lee and Penzien [285]; their example application dealt with a piping system supported within a nuclear reactor building. The analysis of cross terms has also been reported, e.g., among others, in Refs. [56], [138], [142], [220], [296], [315] and [571]. This section examines their effect on the system's response by means of the cross correlation coefficients defined by Der Kiureghian and Neuenhofer [138]; this approach has also led to the convenient incorporation of the response spectrum in random vibration evaluations, which will be presented later in Section 6.5.1.

The rms total response of the generic quantity, $z(t)$, of a multi-degree-of-freedom system (Eq. 6.44) can be expressed as [138], [285]:

$$\sigma_z^2 = \sum_{J=1}^{N}\sum_{L=1}^{N} \beta_J \beta_L \rho_{u_J u_L} \sigma_{u_J} \sigma_{u_L} + 2\sum_{J=1}^{N}\sum_{L=1}^{N}\sum_{m=1}^{K} \beta_J \eta_{mL} \rho_{u_J s_{mL}} \sigma_{u_J} \sigma_{s_{mL}}$$

$$+ \sum_{J=1}^{N}\sum_{L=1}^{N}\sum_{k=1}^{K}\sum_{m=1}^{K} \eta_{kJ} \eta_{mL} \rho_{s_{kJ} s_{mL}} \sigma_{s_{kJ}} \sigma_{s_{mL}} \tag{6.79}$$

In the above expression, σ_{u_J} is the rms displacement at support J, and $s_{mL}(t)$ the rms response of a SDOF oscillator with frequency ω_m subjected to the excitation of support L, which was defined in Eq. 6.39. These quantities are given by:

$$\sigma_{u_J}^2 = \int_{-\infty}^{+\infty} S_{u_J u_J}(\omega)\, d\omega \tag{6.80}$$

$$\sigma_{s_{mL}}^2 = \int_{-\infty}^{+\infty} |H_m(\omega)|^2 S_{\ddot{u}_L \ddot{u}_L}(\omega)\, d\omega \tag{6.81}$$

The cross correlation coefficients in Eq. 6.79, $\rho_{u_J u_L}$, $\rho_{u_J s_{mL}}$ and $\rho_{s_{kJ} s_{mL}}$, are evaluated at $\tau = 0$, i.e., they reflect the correlation of the respective quantities at the same time instant. They can be evaluated, with an approach similar to the one used for the derivation of Eqs. 2.98 and 2.99, as:

$$\rho_{u_J u_L} = \rho_{u_J u_L}(\tau = 0) = \frac{1}{\sigma_{u_J}\sigma_{u_L}} \int_{-\infty}^{+\infty} S_{u_J u_L}(\omega)\, d\omega \tag{6.82}$$

$$\rho_{u_J s_{mL}} = \rho_{u_J s_{mL}}(\tau = 0) = \frac{1}{\sigma_{u_J}\sigma_{s_{mL}}} \int_{-\infty}^{+\infty} H_m(\omega) S_{u_J \ddot{u}_L}(\omega)\, d\omega \tag{6.83}$$

$$\rho_{s_{kJ} s_{mL}} = \rho_{s_{kJ} s_{mL}}(\tau = 0) = \frac{1}{\sigma_{s_{kJ}}\sigma_{s_{mL}}} \int_{-\infty}^{+\infty} H_k^*(\omega) H_m(\omega) S_{\ddot{u}_J \ddot{u}_L}(\omega)\, d\omega \tag{6.84}$$

Equation 6.82 represents the cross correlation between the displacements at supports J and L, Eq. 6.83 the cross correlation between the displacement at support J and the response, $s_{mL}(t)$, of a SDOF oscillator with characteristics ω_m and ζ_m subjected to the ground motions at support L, and Eq. 6.84 the cross correlation between $s_{mL}(t)$ and the response, $s_{kJ}(t)$, of a SDOF oscillator with characteristics ω_k and ζ_k subjected to the ground motions at support J. The cross spectral densities in Eqs. 6.82–6.84 are given by Eq. 6.46. The integrals in these equations are real-valued, as the imaginary part of the complex integrands cancels out in the integration from $-\infty$ to $+\infty$ due to antisymmetry. Additionally, the integration of the two terms in the triple summation of Eq. 6.44 yields the same real-valued results, which appear as a single term in Eq. 6.79 multiplied by a factor of 2.

The contribution of the cross terms to the PSD of the total response (Eq. 6.44 or 6.66) can then be evaluated by means of the cross correlation coefficients (Eqs. 6.82–6.84), which are functions of the ground motion characteristics and the natural frequencies and damping coefficients of the system. Der Kiureghian and Neuenhofer [138] presented cross correlation coefficients for uniform and variable site conditions, the coherency model of Luco and Wong (Eq. 3.27), as well as various values for the apparent propagation velocity of the motions. For consistency with previous derivations in this section, cross correlation coefficients are presented in the following for ground motions at firm soil conditions, the coherency models of Harichandran and Vanmarcke and Luco and Wong, presented in Fig. 6.8, as well as uniform seismic ground motions at all supports of the structure; these results are illustrated in Figs. 6.10–6.12. Because the modal cross correlation coefficients vary significantly depending on the soil conditions [138], their evaluation at soft soil conditions is presented in Fig. 6.13.

Cross Correlation Coefficients of the Pseudo-Static Response

The cross correlation coefficients $\rho_{u_J u_L}$ at separation distances ranging between 0 and 500 m for the ground motions described in Fig. 6.8 and for uniform ground motions are presented in Fig. 6.10(a). It can be seen from the figure that the cross correlation coefficients assume high values for all ground motion scenarios. Obviously, for uniform support excitations, the coefficient is identically equal to unity. For spatially variable ground motions, the value of $\rho_{u_J u_L}$ is controlled by the slow exponential decay of the coherency models in the low frequency range (< 1 Hz) that dominates the displacements (Figs. 6.8(b)–(d) and 7.1(c)). The model of Luco and Wong for $\alpha = 2 \times 10^{-4}$ sec/m yields values that are essentially equal to unity for all support separation distances. For the other two models, the coefficient decays slowly, with the model of Harichandran and Vanmarcke producing the lowest values due to its partial correlation even at $\omega = 0$. Der Kiureghian and Neuenhofer [138] noted that these cross correlation coefficients assume consistently high values for all types of ground excitations considered in their evaluation.

Cross Correlation Coefficients between the Pseudo-Static and Dynamic Response

These cross correlation coefficients, $\rho_{u_J s_{mL}}$, are presented in Figs. 6.10(b)–(d) for the ground motion characteristics of Fig. 6.8 and support separation distances of $\xi_{JL} = 100, 300$ and 500 m, as well as uniform ground motions; damping for the oscillator is assumed to be 5% of critical. Der Kiureghian and Neuenhofer [138]

---→

FIGURE 6.10 Cross correlation coefficients, as defined by Der Kiureghian and Neuenhofer [138], of the pseudo-static terms and those between the pseudo-static and the dynamic terms for the ground motion characteristics (*firm soil conditions*) illustrated in Fig. 6.8, as well as uniform motions at the structure's supports. Part (a): Cross correlation coefficients between the pseudo-static terms, i.e., between the ground displacements $u_J(t)$ and $u_L(t)$ at supports J and L, as functions of support separation distance; the contribution of uniform support motions is identical to unity. Part (b): Cross correlation coefficients between the pseudo-static and dynamic terms, i.e., between the displacement, $u_J(t)$, at support J and the response of a SDOF, $s_{mL}(t)$, with frequency f_m subjected to the excitation of support L, as functions of the natural frequency of the oscillator for a support separation distance of $\xi_{JL} = 100$ m. Parts (c) and (d): Same as in part (b), but for support separation distances of $\xi_{JL} = 300$ m and $\xi_{JL} = 500$ m, respectively.

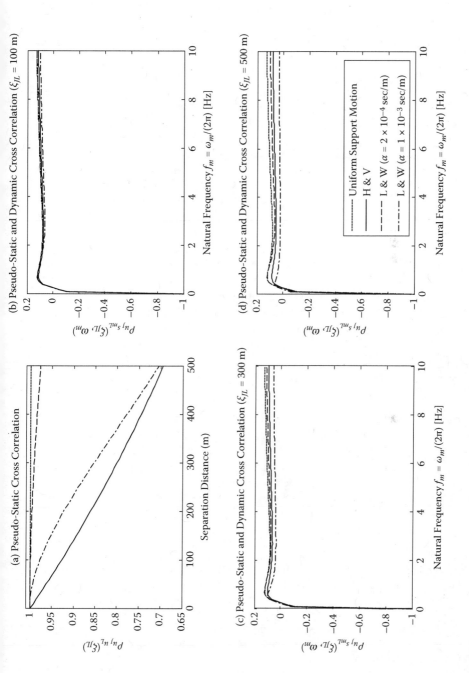

FIGURE 6.10

indicated that, when the oscillator is extremely flexible (i.e., $\omega_m \to 0$), then $s_{mL}(t) \to -u_L(t)$, since the oscillator has zero absolute displacement; this observation implies that $\rho_{u_J s_{mL}} = -\rho_{u_J u_L}$ as $\omega_m \to 0$. Additionally, for fully coherent motions, $\rho_{u_J u_L} \equiv 1$. Hence, the cross correlation coefficient assumes values equal to (-1) at zero frequency for uniform support excitations, and significantly low (negative) values for spatially variable ground motions (Figs. 6.10(b)–(d)); these low negative values, however, are concentrated in the frequency range close to $\omega_m = 0$. On the other hand, as $\omega_m \to \infty$, i.e., for an infinitely stiff oscillator, $s_{mL}(t) \to -\ddot{u}_L(t)/\omega_m^2$, because the inertia and damping forces become negligible compared to the restoring force [138]. Hence, as ω_m increases, $\rho_{u_J s_{mL}} \to -\rho_{u_J \ddot{u}_L}$, which is independent of the oscillator's frequency. Since the cross correlation between the ground displacement and the ground acceleration is small and negative (as can be deduced from the expression for the cross spectral density in the second row of Eq. 6.46 at large frequencies), the cross correlation coefficient $\rho_{u_J s_{mL}}$ assumes low, positive values as ω_m increases. Figures 6.10(b)–(d) illustrate this effect: For uniform support excitations as well as spatially variable ones, these cross correlation coefficients tend to a low, positive value as the frequency of the oscillator increases. For the short separation distance ($\xi_{JL} = 100$ m in Fig. 6.10(b)), the results for uniform motions and spatially variable ones for all coherency models are, essentially, the same, and their differences remain small even for the longest separation distance ($\xi_{JL} = 500$ m in Fig. 6.10(d)). It is noted that the cross correlation coefficients for uniform support excitations are the same irrespective of the separation distance. Der Kiureghian and Neuenhofer [138] reported that higher values for these cross correlation coefficients result for variable soil conditions at the supports, and if the apparent propagation of the motions is taken into consideration.

Modal Cross Correlation Coefficients

Figures 6.11 and 6.12 present the modal cross correlation coefficients, $\rho_{s_{kJ} s_{mL}}$, when both oscillators are located on firm soil conditions; uniform seismic ground motions as well as the spatially variable ones presented in Fig. 6.8 are used in the evaluation. In Figs. 6.11 and 6.12, the natural frequency of one oscillator is considered fixed and the natural frequency of the other oscillator is allowed to vary. Specifically, Fig. 6.11 presents the cross correlation coefficients when $\omega_m = 2\pi$ rad/sec ($f_m = 1$ Hz) and ω_k varies, and Fig. 6.12 the corresponding coefficients for $\omega_m = 10\pi$ rad/sec

---→

FIGURE 6.11 Modal cross correlation coefficients, as defined by Der Kiureghian and Neuenhofer [138], for the ground motion characteristics (*firm* soil conditions) illustrated in Fig. 6.8, as well as uniform motions at the structure's supports. Part (a): The square modulus of the frequency transfer function of an oscillator with frequency $f_m = 1$ Hz. Part (b): Cross correlation coefficients between the response of a SDOF oscillator, $s_{kJ}(t)$, with variable frequency f_k subjected to the excitation of support J, and that of a SDOF oscillator, $s_{mL}(t)$, with a fixed natural frequency $f_m = 1$ Hz (part (a)) subjected to the excitation of support L, as functions of the natural frequency of the first oscillator, f_k, for a support separation distance of $\xi_{JL} = 100$ m. Parts (c) and (d): Same as in part (b), but for support separation distances of $\xi_{JL} = 300$ m and $\xi_{JL} = 500$ m, respectively. The legend in subplot (d) refers to parts (b)–(d) of the figure.

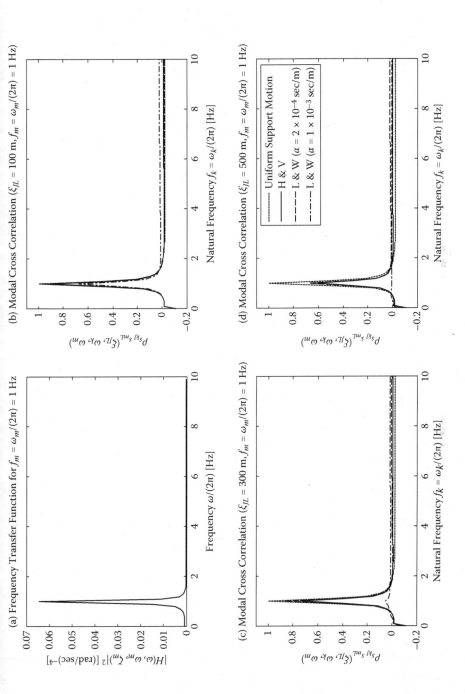

FIGURE 6.11

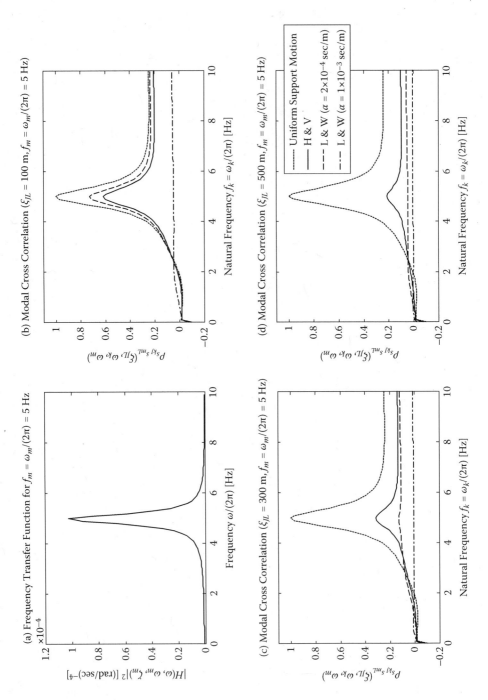

FIGURE 6.12

($f_m = 5$ Hz) with ω_k again varying; for both oscillators, damping is assumed to be 5% of critical. Figures 6.11(a) and 6.12(a) present the square modulus of the frequency transfer function of the oscillator with the fixed frequency. The behavior of the cross correlation coefficients as functions of the natural frequency of the first oscillator for the three separation distances considered, i.e., $\xi_{JL} = 100, 300$ and 500 m, are presented in parts (b), (c) and (d), respectively, of the figures. Again, for uniform support excitations, the results in parts (b)-(d) of each figure are the same, as $|\gamma(\xi_{JL}, \omega)| \equiv 1$. The cross correlation coefficients in all cases are small and negative at the very low frequencies. The natural frequencies of the oscillators with the fixed properties ($f_m = 1$ Hz and 5 Hz in Figs. 6.11 and 6.12, respectively) fall within the dominant frequency range of the excitation (Fig. 6.8(a)). Hence, the contribution of the ground motions to the response of the oscillators with the fixed natural frequency results from excitation frequencies in the vicinity of the natural frequency of each system, and the frequency range of the peak values of the coefficients coincides, basically, with the dominant frequency range of the transfer function of the oscillator with the fixed properties.

When the second oscillator has its frequency fixed at 1 Hz and the separation distance is short (Fig. 6.11(b)), the modal cross correlations are sharply peaked for all types of excitations. As separation distance increases (Figs. 6.11(c) and (d)), the modal cross correlation coefficients for the models of Harichandran and Vanmarcke and Luco and Wong with $\alpha = 2 \times 10^{-4}$ sec/m still exhibit sharp peak values, though smaller than the peak values for the shorter separation distance. On the other hand, the cross correlation coefficient for the model of Luco and Wong with $\alpha = 1 \times 10^{-3}$ sec/m has a very small peak when the separation distance is equal to 300 m (Fig. 6.11(c)), and is, essentially, equal to zero past the very low frequencies when $\xi_{JL} = 500$ m (Fig. 6.11(d)). This is due to the fact that at the frequency of 1 Hz, the first two coherency models yield highly correlated ground motions, whereas the coherency of the third model has already reached zero values (Figs. 6.8(c) and (d)). Since uncorrelated excitations lead to uncorrelated response for the two oscillators, the modal cross correlation coefficient for the coherency model of Luco and Wong with $\alpha = 1 \times 10^{-3}$ sec/m at the longer separation distances assumes values that tend to zero.

When the second oscillator has its frequency fixed at 5 Hz (Fig. 6.12), the peak values of the coefficients for the spatially variable motions are significantly smaller than the values presented in Fig. 6.11 due to the lower coherency of the models at this higher frequency: The model of Luco and Wong with $\alpha = 1 \times 10^{-3}$ sec/m produces essentially zero values for the coefficient, as the correlation between the support excitations for this model has already reached zero values, even for the shorter separation distance considered (Fig. 6.8(b)). The models of Harichandran and Vanmarcke and Luco and Wong for $\alpha = 2 \times 10^{-4}$ sec/m lead to cross correlation coefficients with significant peaks for the shorter separation distance ($\xi_{JL} = 100$ m in Fig. 6.12(b)). The peak values, however, decrease significantly with increasing separation distance

FIGURE 6.12 Same as in Fig. 6.11, only the frequency, f_m, of the SDOF oscillator $s_{mL}(t)$ is fixed at 5 Hz. The legend in subplot (d) refers to parts (b)–(d) of the figure.

(Figs. 6.12(c) and (d)), with the model of Harichandran and Vanmarcke, due to its slower exponential decay (Figs. 6.8(c) and (d)), producing higher peaks for the coefficient than the model of Luco and Wong for $\alpha = 2 \times 10^{-4}$ sec/m.

Figure 6.13 presents the modal cross correlation coefficients when the two systems are located on soft soil conditions, the PSD of which was presented in Fig. 6.2(a). The coherency models and separation distances considered are those described in Figs. 6.8(b)–(d), and the frequency of the second oscillator is fixed at 5 Hz. The variation of the modal cross correlation coefficients in this case differs from the variation of the coefficients presented in Figs. 6.11 and 6.12: Past the low frequency range of the natural frequency of the first oscillator, the modal cross correlation coefficients assume high, essentially constant values, that decrease, at a different rate, for the various coherency models as the separation distance increases. In this evaluation, the frequency content of the excitation (Fig. 6.2(a)) is lower than the fixed frequency of the second oscillator, which views the excitation as pseudo-static. Der Kiureghian [134] noted that, for a multi-degree-of-freedom system, the higher modes, which view the excitation as pseudo-static, are highly correlated. The behavior of the cross correlation coefficients at soft soil conditions and higher oscillator frequencies can be attributed to the same rationale [138]. Hence, for the shorter separation distance (Fig. 6.13(b)), all coherency models yield highly correlated excitations (Fig. 6.8(b)) leading to significant values for the cross correlation coefficient. The models of Harichandran and Vanmarcke and Luco and Wong for $\alpha = 2 \times 10^{-4}$ sec/m produce significant values for the cross correlation coefficients at the longer separation distances (Figs. 6.13(c) and (d)) due to their high correlation in the low frequency range (Figs. 6.2(a), and 6.8(c) and (d)); in this case, the model of Luco and Wong produces higher coefficients than the model of Harichandran and Vanmarcke because of its higher correlation at the lower frequencies. As the separation distance increases, the model of Luco and Wong with $\alpha = 1 \times 10^{-3}$ sec/m produces, again, low values for the coefficient (Figs. 6.13(c) and (d)) due to its sharp exponential decay even at low frequencies (Figs. 6.2(a), and 6.8(c) and (d)).

The models of Harichandran and Vanmarcke [208] and Luco and Wong [324] are commonly used to describe the loss of coherency in the seismic ground motions for the evaluation of the seismic response of lifelines (Chapter 9). Figures 6.9–6.13 suggest that the differences in the exponential decay of the models result in different seismic loads for the structures, an observation that needs to be taken into consideration when the effects of the spatial variability of the seismic excitations on the response of lifeline systems are analyzed; this issue will be further discussed in Chapter 9. It needs to be emphasized, however, that the way the individual terms presented in Figs. 6.9–6.13 interact to yield the total response of a multiply-supported system (Eq. 6.44 or 6.66) depends on the structural properties of the system; an example illustration of this effect is presented in the following section.

FIGURE 6.13 Same as in Fig. 6.12, only the supports of the two oscillators rest on *soft* soil conditions, the PSD of which is illustrated in Fig. 6.2(a). The legend in subplot (b) refers to parts (b)–(d) of the figure.

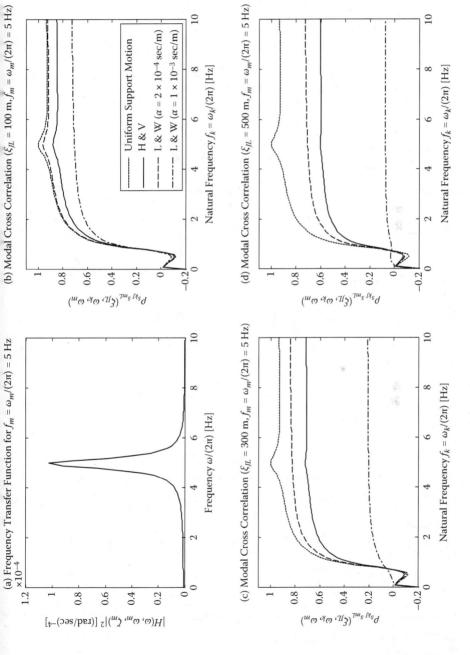

FIGURE 6.13

6.4.5 RESPONSE TO COHERENT AND INCOHERENT EXCITATIONS

Zerva [590], [591] conducted a comparative analysis of the response of "generic" lifeline models, i.e., two- and three-span continuous beams of various span lengths, subjected to a wide range of spatially variable excitations including loss of coherency and wave passage effects. This section presents the lateral response of a three-span beam subjected to two extreme scenarios of seismic support excitations from the ones reported in Ref. [590]. In the first scenario, the seismic motions are essentially incoherent, whereas, in the second case, they are fully coherent, i.e., the excitations applied at the structure's supports are identical.

The model of the three-span beam is presented in Fig. 6.14(a). The end spans of the beam are 20 m long, and the length of the center span is 40 m. The characteristics of the beam are: $E_p = 2 \times 10^{11}$ N/m^2, $M = 2.45 \times 10^3$ kg/m and $I = 6 \times 10^{-2}$ m^4, and represent average values of the characteristics of superstructures of highway bridges. The first twelve frequencies of the beam were evaluated as: 3.39, 8.69, 11.00, 13.58, 22.95, 34.76, 39.24, 44.00, 59.89, 78.22, 84.87 and 91.80 Hz; the odd-numbered modes are symmetric and the even-numbered modes are antisymmetric. For both ground motion scenarios, the beam supports were considered to rest on firm soil conditions described by the Clough-Penzien spectrum of Fig. 6.8(a). The coherent motions are represented by $|\gamma(\xi, \omega)| \equiv 1$ for any support separation distance and frequency. The characteristics of the incoherent motions were captured by increasing the value of the coherency drop parameter of the Luco and Wong model to $\alpha = 5 \times 10^{-3}$ sec/m, which leads to sharp drops for the coherency at the short separation distances between the beam supports in this example application [590].

Figures 6.14(b)–(d) presents the results of the evaluation. Because the structure is symmetric and the excitations exhibit loss of coherency only, the random vibration results are also symmetric about the beam's midpoint (Section 6.3.1), and are presented in the figures over the first half of the beam. Additionally, because the structure is symmetric, the fully coherent motions excite only the symmetric modes of the structure. The introduction of spatial variability in the evaluation induces the pseudo-static response and excites both the symmetric and the antisymmetric modes. Figure 6.14(b) presents the envelope functions of the rms total displacements for the two ground motion scenarios normalized with respect to the rms ground displacement. The displacements are larger for the incoherent motions than the fully coherent ones, especially over the first span of the beam, but the differences are insignificant. Figures 6.14(c) and (d) present, respectively, the envelope functions of the rms bending moments and shear forces along the structure (Eq. 6.67) normalized with respect to the maximum bending moment (in Fig. 6.14(c)) and the maximum shear force (in Fig. 6.14(d)) that occurs along the structure for the uniform support excitations. Also shown in the figures is the pseudo-static response (double summation in Eq. 6.66), which is induced by the partially correlated motions only. Clearly, the bending moments induced by the incoherent motions are dominated by the pseudo-static terms, which can be, by themselves, higher than the purely dynamic terms excited by the uniform motions over a significant part of the structure. Around the midpoint of the structure ($L_2/2$ in Fig. 6.14(c)), the dynamic terms caused by the incoherent excitations become very significant as well. At this location, only the symmetric modes

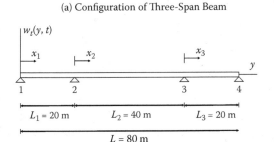

(a) Configuration of Three-Span Beam

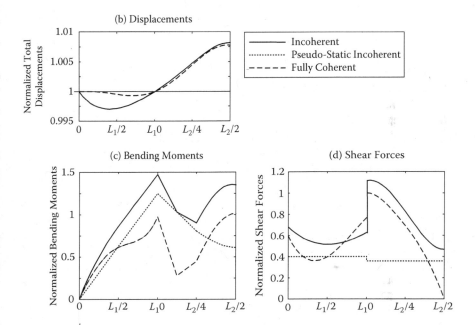

FIGURE 6.14 Part (a): The configuration of the three-span beam subjected to fully coherent and strongly incoherent motions at its supports; the two end spans are 20 m long, and the length of the middle span is 40 m. Part (b) presents the envelopes of rms total displacements normalized with respect to the rms ground displacement. Parts (c) and (d) present the envelopes of rms bending moments and the envelopes of rms shear forces, respectively; the results are normalized with respect to the maximum response that occurs along the beam for uniform support excitations. The response is symmetric about the midpoint of the beam and is presented in parts (b)-(d) only along the first half of the structure. In parts (c) and (d), the dotted lines indicate the pseudo-static response induced by the incoherent motions; fully coherent excitations do not induce any pseudo-static internal forces (after Zerva [590]).

contribute to the dynamic response. The first mode of the beam is symmetric and its natural frequency (at 3.39 Hz) is well within the dominant frequency range of the seismic excitation (Fig. 6.8(a)). This mode controls the dynamic response at the midpoint of the structure yielding significant contributions to the moments caused by the incoherent seismic motions and values for the moments induced by the uniform motions that surpass the large pseudo-static moments of the incoherent excitation case. For shear forces (Fig. 6.14(d)) induced by the spatially variable excitations, the pseudo-static response is significant but not to the same degree as for the bending moments, as shear forces are one order higher derivatives of displacements than bending moments [594]. For the same reason, the values of the shear forces resulting from the uniform support excitations are closer to the values of the shear forces caused by the incoherent motions. At the inner support of the first span (at L_1 in Fig. 6.14(d)), the shear forces due to uniform support excitations surpass the values of the shear forces induced by the spatially variable ground motions. This is caused by the contribution of the third mode of the structure (at a natural frequency of 11.0 Hz), that is fully excited by the fully coherent motions. For the incoherent excitations, although this mode still contributes to the shear forces, its effect is not so significant, because the correlation of the input excitations at the frequency of the mode is already very low (Eq. 3.27 for $\alpha = 5 \times 10^{-3}$ sec/m). At the midpoint of the beam ($L_2/2$ in Fig. 6.14(d)), shear forces for the uniform input motion scenario are zero due to the symmetry of the participating mode shapes.

Zerva [591] further evaluated the relative effects of loss of coherency and wave passage in the seismic excitations on the response of the multiply-supported, generic lifeline models by utilizing the two coherency scenarios described herein and a wide range of apparent propagation velocities. The analysis suggested that, for the strongly incoherent motions, the wave passage effects can be neglected, but, for fully coherent, propagating ground motions, the time delays in the propagation of the excitation may produce higher or lower response than the uniform support motions depending on the structural configuration and the response quantity examined (shear force or bending moment).

Even though the lifeline model analyzed in this section is simple (Fig. 6.14(a)) and the analysis is linear, the results in Figs. 6.14(c) and (d) demonstrate the differences in the response of symmetric lifeline systems induced by uniform and spatially variable seismic ground motions: Spatially variable ground motions excite the pseudo-static response of the structure and all its dynamic modes, whereas the uniform support motions induce the symmetric dynamic response only. This is, perhaps, the major reason why it is difficult to establish equivalent, uniform support motions that would have the same effect on the response of the structures as spatially variable excitations. Further evaluations of the effect of spatially variable ground motions on the response of long structures are presented in Chapter 9.

6.5   ADDITIONAL RANDOM VIBRATION CONSIDERATIONS

As indicated in the beginning of this chapter, the advantage of the random vibration approach is that a single analysis suffices for the estimation of the probabilistic properties of any response quantity of the considered structure. The approach becomes more attractive in engineering applications if the PSD of the excitation is replaced by the more widely-acceptable response spectrum, and if the evaluations can be extended to the nonlinear range of the structural response. Section 6.5.1 describes a response-spectrum compatible methodology for random vibrations, and Section 6.5.2 briefly illustrates a commonly utilized concept in nonlinear random vibrations.

6.5.1 RESPONSE SPECTRUM APPROACH FOR MULTI-SUPPORT EXCITATION

Efforts to incorporate the response spectrum in random vibrations of multiply-supported structures have been reported, e.g., among others, in Refs. [18], [56], [138], [142], [296], [315], [512], [571] and [584]. The Multiple Support Response Spectrum (MSRS) method, presented by Der Kiureghian and Neuenhofer [138], for the evaluation of the mean maximum response of a multiply-supported structure is summarized in the following.

The expected maximum value of a zero-mean, Gaussian random process can be expressed in terms of its rms value and a peak factor [125], [550], as will be illustrated briefly in Section 7.1.4. Let $RS_J(\omega_k, \zeta_k) = E[\max |s_{kJ}(t)|]$ indicate the value of the response spectrum of a SDOF oscillator with frequency ω_k and damping ζ_k subjected to the excitation at support J (Eq. 6.39). It follows then that $RS_J(\omega_k, \zeta_k) = p_{s_{kJ}} \sigma_{s_{kJ}}$, where $p_{s_{kJ}}$ is the aforementioned peak factor, and $\sigma_{s_{kJ}}$ is given by Eq. 6.81. Similarly, the expected peak value of the ground displacement at support J can be expressed as $u_{J,\max} = p_{u_J} \sigma_{u_J}$. With these considerations, the mean maximum response of the generic quantity $z(t)$ of Eq. 6.79 can be expressed as $E[\max |z(t)|] = p_z \sigma_z$, which leads to:

$$E[\max |z(t)|] = \left[\sum_{J=1}^{N} \sum_{L=1}^{N} \beta_J \beta_L \rho_{u_J u_L} \frac{p_z^2}{p_{u_J} p_{u_L}} u_{J,\max} u_{L,\max} \right.$$
$$+ 2 \sum_{J=1}^{N} \sum_{L=1}^{N} \sum_{k=1}^{K} \beta_J \eta_{kL} \rho_{u_J s_{kL}} \frac{p_z^2}{p_{u_J} p_{s_{kL}}} u_{J,\max} RS_L(\omega_k, \zeta_k)$$
$$\left. + \sum_{J=1}^{N} \sum_{L=1}^{N} \sum_{k=1}^{K} \sum_{m=1}^{K} \eta_{kJ} \eta_{mL} \rho_{s_{kJ} s_{mL}} \frac{p_z^2}{p_{s_{kJ}} p_{s_{mL}}} RS_J(\omega_k, \zeta_k) RS_L(\omega_m, \zeta_m) \right]^{1/2}$$

$$(6.85)$$

Der Kiureghian and Neuenhofer [138] suggested that the peak factors in the above equation depend only mildly on the characteristics of the each process, and simplified Eq. 6.85 to:

$$E[\max |z(t)|] = \left[\sum_{J=1}^{N} \sum_{L=1}^{N} \beta_J \beta_L \rho_{u_J u_L} u_{J,\max} u_{L,\max} \right.$$
$$+ 2 \sum_{J=1}^{N} \sum_{L=1}^{N} \sum_{k=1}^{K} \beta_J \eta_{kL} \rho_{u_J s_{kL}} u_{J,\max} RS_L(\omega_k, \zeta_k)$$
$$\left. + \sum_{J=1}^{N} \sum_{L=1}^{N} \sum_{k=1}^{K} \sum_{m=1}^{K} \eta_{kJ} \eta_{mL} \rho_{s_{kJ} s_{mL}} RS_J(\omega_k, \zeta_k) RS_L(\omega_m, \zeta_m) \right]^{1/2}$$

$$(6.86)$$

The cross correlation coefficients in the above equation, which are described by Eqs. 6.82–6.84 and were illustrated in Figs. 6.10–6.13 for the Clough-Penzien PSD, need now be evaluated in terms of a PSD compatible with the utilized response spectrum. Der Kiureghian and Neuenhofer [138] suggested the following first-order approximation for the PSD of the ground acceleration:

$$S^a_{\ddot{u}_J\ddot{u}_J}(\omega) = \frac{\omega^{p+2}}{\omega^p + \omega^p_J}\left(\frac{2\zeta\omega}{\pi} + \frac{4}{\pi T_e}\right)\left[\frac{RS_J(\omega,\zeta)}{p_s}\right]^2 ; \quad \omega \geq 0 \qquad (6.87)$$

where p and ω_J are parameters selected such that the PSD at low frequencies is consistent with the peak ground displacement, $u_{J,\max}$, T_e is the duration of the strong motion, ζ a reference damping ratio, and p_s a peak factor associated with the response of an oscillator (as, e.g., in Eq. 7.25). The approximate PSD of Eq. 6.87 can then be used in Eqs. 6.82–6.84 for the evaluation of the cross correlation coefficients. Der Kiureghian and Neuenhofer [138] noted that, since the cross correlation coefficients (Eqs. 6.82–6.84) are ratios, they are not overly sensitive to the amplitude of the individual power spectral densities, and further indicated that they are also not overly sensitive to the parameters ζ, ω_J, p and T_e. An alternative approach for converting a response spectrum to an equivalent PSD is presented in Section 7.1.4, and an iterative scheme for the generation of response-spectrum compatible time series in Section 7.3.2.

6.5.2 NONLINEAR RANDOM VIBRATIONS

The random vibration formulation, as described herein in the applications for multi-support excitations, is linear, as it is based on the principle of superposition (Duhamel's integral) and the Fourier transform. Hence, its extension to the nonlinear range of the structural response is not straightforward. The topic of nonlinear random vibrations has attracted significant interest, and extensive work has been and is currently being conducted on the topic, as, e.g., among many others, in Refs. [38], [39], [58], [95], [98], [139], [151], [255], [308], [309], [326], [358], [397], [426], [450], [482], [491], [522] and [559]. The interested reader is referred to these publications and references therein for insight into nonlinear random vibrations. The following briefly highlights the concept of the Equivalent Linearization (EQL) method, which is the most commonly used random vibration approach for the evaluation of the hysteretic-type nonlinear response of structures in earthquake engineering [450].

The basic idea underlying the EQL approach is to replace a given system of nonlinear equations with a system of "statistically equivalent" ones, so that the mean-square error between the actual nonlinear and the equivalent linearized systems is minimized in a statistical sense [151]. In its simplest form, EQL involves the estimation of the coefficients of a linear function that approximates a given nonlinear one through the minimization of the expectation of the mean-square error, e.g., [326]: For example, if $g(z(t))$ indicates the nonlinear function of the random process $z(t)$, the approach determines the coefficients c_0 and c_1 of an equivalent linear function, $c_0 + c_1[z(t) - \mu_z(t)]$, through the minimization of the expectation of the mean-square error, i.e.,

$$\text{minimize } \left\{\text{E}\left[\{c_0 + c_1[z(t) - \mu_z(t)] - g(z(t))\}^2\right]\right\} \qquad (6.88)$$

which can be achieved by setting the derivatives of the above expression with respect to c_0 and c_1 equal to zero. Considering further that the process is zero-mean, i.e., $\mu_z(t) = 0$, as is the case, for all practical purposes, in earthquake engineering applications

(Section 2.2.1), the solution of the resulting two equations leads to [326]:

$$c_0 = \mathrm{E}[g(z(t))]$$

$$c_1 = \frac{\mathrm{E}[z(t)g(z(t))]}{\mathrm{E}[z^2(t)]} \tag{6.89}$$

If the probability distribution of $z(t)$ is known, then, in principle, the estimates of c_0 and c_1 can be obtained. The probability distribution of $z(t)$, however, is not necessarily known: Under the assumption of Gaussianity for the excitation, the response of a linear system will also be Gaussian, whereas the probability distribution of the response of a nonlinear system will be unknown. Hence, if $z(t)$ is the response of a nonlinear system, in which $g(z(t))$ is the nonlinear restoring force, the probability distribution of $z(t)$ will be unknown and the coefficients in Eq. 6.89 cannot be determined. The traditional approach, e.g., [38], [98], is to use the rationale that, since the response of the linearized system subjected to a Gaussian excitation will be Gaussian, the assumption of Gaussianity can be invoked for the derivations in Eq. 6.89. However, even if the mean and the variance of the response can be sufficiently approximated by the EQL approach, their use in reliability analyses is limited, as such derivations require the realistic modeling of the tails of the distribution functions [450]. An additional limitation of the procedure is that, even if the coefficients c_0 and c_1 can be determined exactly, the linearization process has matched only the second moment of the error term (Eq. 6.88) without consideration of other measures of error [326]. In spite of its shortcomings, however, EQL has been effectively implemented in nonlinear random vibration applications [39], [151], [397], [450], [491], [559].

In concluding this chapter it should be emphasized that, because the random vibration formulation is based on the stochastic description of the excitation, its results do not depend on the particular features of a single time series, nor does it necessitate the large number of time history evaluations required in Monte Carlo simulations. There are, however, associated limitations that counteract the strengths of the random vibration approach: As indicated earlier, its extension to the nonlinear range of the structural response is not a straightforward task, and, hence, at this point, the general applicability of random vibrations is still limited to the linear response of the structures. A much more versatile approach for the solution of nonlinear, complex dynamic problems is, still, through simulations. Chapter 7 introduces simulations of stationary and non-stationary random processes and fields based on the description of the point estimates of the seismic ground motions by either power spectral densities or response spectra. Chapter 8 presents conditional simulations of spatially variable ground motions based on predefined accelerograms that can be either recorded or synthetically generated. The effect of spatially variable excitations on the response of lifelines evaluated by means of random vibrations as well as linear and nonlinear time history analyses will be further illustrated in Chapter 9.

7 Simulations of Spatially Variable Ground Motions

The spatially variable seismic ground motion models described in Chapter 3 can be used directly as input motions at the supports of lifelines in random vibration analyses (Chapter 6). The nonlinear response evaluation of the structural systems, however, requires time history analyses. Such analyses, in turn, necessitate the generation of spatially variable time histories to be used as input excitations at the structures' supports in a Monte Carlo simulation framework. Spatially variable seismic ground motions can be generated either from the description of the random field through a power spectral density (or a response spectrum) and a spatial variability model or from a predefined seismic ground motion time history (e.g., a recorded accelerogram) and a spatial variability model. The former is referred to herein as "simulations" of spatially variable seismic ground motions and is described in this chapter, whereas the latter is referred to as "conditional simulations" of spatially variable seismic ground motions and is presented in the following chapter.

An extensive list of publications addressing the topic of simulations of random processes and fields has appeared in the literature. In 1955, Housner [227] stated that "It is assumed that an accelerogram is formed by the superposition of a large number of elemental [one-cycle sine-wave] acceleration pulses random in time. It is shown that this agrees with recorded accelerograms, and an accelerogram composed in this fashion is shown to have the characteristics of actual recorded accelerograms." This statement is, essentially, the basis for the generation (or synthesis) of artificial accelerograms. Housner's concept was utilized by, among others, Bycroft [85], Hudson [230] and Rosenblueth [428], who modeled seismic ground motions by a series of pulses distributed randomly in time. Housner and Jennings [228] further generated Gaussian random time series based on a power spectral density compatible with the average (undamped) velocity spectra of recorded accelerograms. Bogdanoff *et al.* [64] generated non-stationary acceleration time histories as the sum of cosine functions with random phase angles uniformly distributed between $[0, 2\pi)$ modulated with an exponential time function. Amin and Ang [26], [27], Cornell [117], Goto and Toki [177], Jennings *et al.* [244], and Shinozuka and Sato [467] also generated non-stationary time series.

One of the techniques that has been widely applied in engineering problems is the spectral representation method [423], [460]. Simulations based on AR (auto-regressive), MA (moving-average) and ARMA (auto-regressive-moving-average) techniques have also been used extensively (e.g., [115], [149], [275], [344], [494]). Simulations by means of the local average subdivision method [160], the turning bands method [333], and based on wavelet [187], [440], [582] and Hilbert-Huang [560] transforms have been generated. Simulations of non-stationary time series by means of processes modulated by time varying functions have been further reported (e.g., [136], [184], [577]). Spatially variable seismic ground motions have also been simulated by means of a wide variety of techniques. Examples include, but are not limited to: covariance matrix decomposition [127], [200], [259], [260]; spectral representation method [592], [596]; envelope functions containing random phase variability [6]; coherency function approximation by a Fourier series [419], [420]; ARMA approximation [496]; and FFT [294] and hybrid DFT and digital filtering approach [295] for non-stationary random processes. The topic still attracts significant research interest and notable developments are being reported (e.g., among others, [129], [141], [181], [182], [183], [212], [274], [280], [335], [403], [407], [411], [416], [610]).

Generally, random processes and fields in structural engineering applications are simulated as Gaussian or non-Gaussian, stationary or non-stationary, and homogeneous or non-homogeneous (e.g., [146], [180], [275], [451], [465], [492], [497] and references therein). The appropriate simulation technique for the particular problem at hand depends on the characteristics of the problem itself [158]. The main objectives are that the characteristics of the simulated motions match those of the target field, and that the computational cost for the simulations is not excessive. The material in this chapter is not intended to exhaust the topic of simulations of random processes and fields, but, instead, to introduce the concept from simulations of stationary random processes to simulations of non-stationary, non-homogeneous, response-spectrum-compatible spatially variable random fields. It concentrates on the spectral representation method utilizing Gaussian processes and fields.

The concept of representations of Gaussian random processes was introduced by Rice in 1944 [423] (reprinted in [424]), who presented two analytical representations for a stationary Gaussian process. The first representation scheme is:

$$f(t) = \sum_{k=1}^{K} [A_k \cos(\omega_k t) + B_k \sin(\omega_k t)] \qquad (7.1)$$

where ω_k is the discrete frequency, and A_k and B_k are independent Gaussian random variables with zero mean and standard deviation equal to $\sqrt{2S(\omega_k)\Delta\omega}$, with $S(\omega_k)$ indicating the power spectral density and $\Delta\omega$ the frequency step, i.e., Eq. 7.1 is the sum of trigonometric functions with random amplitudes. Written, alternatively, as:

$$f(t) = \sum_{k=1}^{K} \sqrt{(A_k^2 + B_k^2)} \cos\left(\omega_k t + \tan^{-1}\left(-\frac{B_k}{A_k}\right)\right) \qquad (7.2)$$

it represents a trigonometric series with random amplitudes and phases. The second representation is given by:

$$f(t) = \sqrt{2} \sum_{k=1}^{K} C_k \cos(\omega_k t + \phi_k) \qquad (7.3)$$

where $C_k = \sqrt{2S(\omega_k)\Delta\omega}$, and ϕ_k, $k = 1, \ldots, K$, are independent random phase angles uniformly distributed over the range $[0, 2\pi)$, i.e., Eq. 7.3 represents a trigonometric series with random phases. It was Shinozuka [459], [460], [466], however, who extended the use of the approach, particularly its second formulation due to its ergodicity properties, for the generation of simulations of random processes and fields. The methodology was presented in a comprehensive article by Shinozuka [462], and was further elaborated upon in Shinozuka and Deodatis [463], [464], [465] and references therein. In its initial formulation, the spectral representation method dealt with the summation of large numbers of weighted trigonometric functions (Eq. 7.3), which is, computationally, not efficient. Yang [573] introduced the Fast Fourier Transform (FFT) technique in simulating envelopes of random processes, and Shinozuka [461] extended the approach to random processes and fields. The use of the FFT dramatically reduces the computational requirements for simulations. Improvements on the approach and evaluations of its properties and capabilities have been reported by various researchers (e.g., [179], [258], [327], [343], [592]). Mignolet and Harish [343] compared the performance of the spectral representation algorithm, the randomized spectral representation scheme [466] and the random frequencies algorithm [459], and concluded that, irrespectively of computational effort, the latter performs, generally, better than the other two in terms of first and second order distributions. As computational efficiency is a significant consideration, the simulation schemes described in this chapter utilize the spectral representation method at fixed frequencies.

Simulations of random processes are highlighted in the following section as an introduction to the simulation of random fields and vector processes, which are described in Sections 7.2 and 7.3, respectively. Section 7.1.1 first illustrates the concept of simulating stationary random processes by means of the trigonometric series of Rice's [424] second representation approach (Eq. 7.3), and then introduces the FFT in the simulation scheme. Intensity modulation and intensity and frequency modulation of the stationary time series are described in Sections 7.1.2 and 7.1.3, respectively, along with examples that illustrate the transformation of simulations from stationary to non-stationary, exhibiting finite duration (Section 7.1.2) or becoming compatible to a recorded accelerogram (Section 7.1.3). The simulations in Sections 7.1.1 and 7.1.2 are based on the description of the point estimates of the motions through power spectral densities. However, a more widely-accepted description of the characteristics of seismic ground motions is the response spectrum. For illustration purposes, Section 7.1.4 converts, in a non-iterative scheme, a response spectrum to a power spectral density; an iterative scheme for such a conversion is briefly highlighted in Section 7.3.2. Section 7.2 simulates stationary, spatially variable time series by means of the rigorous frequency-wavenumber representation of the random field. Section 7.2.1 describes the approach, and Section 7.2.2 compares the time-domain characteristics of simulations generated from different coherency models. Section 7.3 introduces the more versatile generation of spatially variable ground motions as realizations of vector processes. Section 7.3.1 describes briefly two simulation schemes, and compares the characteristics of the stationary time series resulting from these approaches with the characteristics of the target random field as well as the trend of spatially recorded data. Section 7.3.2 presents a methodology for the generation of non-stationary, non-homogeneous and response-spectrum-compatible spatially variable simulations and illustrates the approach with an example application. The section also includes a brief discussion on the assumptions and approximations utilized in the generation of spatially variable seismic ground motions for engineering applications; some additional remarks on these issues are presented in Section 9.5.

7.1   SIMULATION OF RANDOM PROCESSES

This section presents the concept of simulation of random processes based on the spectral representation method. In Section 7.1.1, the time histories are generated as stationary, based on a widely used description for the power spectral density. Common formulations that introduce finite duration in the generated motions through intensity modulation are then presented in Section 7.1.2, and followed in Section 7.1.3 by an illustration of simulations compatible with recorded accelerograms through modulation of both their intensity and frequency content. A non-iterative scheme that converts response spectra into power spectral densities is then described in Section 7.1.4.

7.1.1   GENERATION OF STATIONARY PROCESSES

Consider a zero-mean Gaussian random process with autocovariance function $R(\tau)$ and power spectral density (PSD) $S(\omega)$, in which τ and ω indicate time lag and frequency (in rad/sec), respectively. $S(\omega)$ and $R(\tau)$ are even functions of their respective arguments and obey the Wiener-Khinchine transformation (Eqs. 2.32 and 2.33). According to Shinozuka [460], time history simulations can be generated by means of the following expression:

$$f(t) = \sqrt{2} \sum_{k=k_0}^{K} [2S(\omega_k)\Delta\omega]^{1/2} \cos(\omega_k t + \phi_k) \qquad (7.4)$$

in which $k_0 = 1$, the discrete frequencies are described by a centroidal rule, i.e., $\omega_k = (k - 1/2)\Delta\omega$, with $\Delta\omega$ indicating, again, the frequency step, and ϕ_k, $k = k_0, \ldots, K$, are realizations of random phase angles uniformly distributed between $[0, 2\pi)$. Equation 7.4 is valid if there exists an upper cut-off frequency ω_u, above which the contribution of the power spectral density to the simulations is negligible for practical purposes. The simulations have the following inherent characteristics: (i) they are asymptotically Gaussian as $K \to \infty$ due to the central limit theorem; (ii) they are periodic with period $T_0 = 4\pi/\Delta\omega$; (iii) they are ergodic, at least up to second moment, over an infinite time domain or over their period; and (iv) for $K \to \infty$, the ensemble mean, autocovariance and power spectral density functions of the simulations become identical to those of the process itself, due to the orthogonality property of the cosine functions in Eq. 7.4. For finite K, the autocovariance function of the simulations,

$$R_{ff}(\tau) = 2 \sum_{k=k_0}^{K} S(\omega_k) \cos(\omega_k \tau) \Delta\omega \qquad (7.5)$$

approximates the autocovariance function of the process, namely,

$$R(\tau) = 2 \int_{0}^{+\infty} S(\omega) \cos(\omega\tau) \, d\omega \qquad (7.6)$$

within a numerical integration error. The accuracy of the integration, which is equivalent to the convergence rate of the stochastic characteristics of the simulations to those of the random field, depends on the choices of the cut-off frequency ω_u (truncation error) and the frequency step $\Delta\omega$ (integration error). The truncation error decreases as ω_u increases; ω_u has to be large enough so that the contribution of the PSD for $\omega > \omega_u$ in Eqs. 7.5 and 7.6 is insignificant. Shinozuka and Deodatis [463] suggested that the cut-off frequency ω_u be estimated from the criterion:

$$\int_0^{\omega_u} S(\omega)\,d\omega \geq (1 - \epsilon) \int_0^{+\infty} S(\omega)\,d\omega \tag{7.7}$$

Since $\int_0^{+\infty} S(\omega)\,d\omega$ equals one half the variance of the process, ϵ can be viewed as an acceptable portion of the variance that can be ignored. The integration error of Eq. 7.5 relative to Eq. 7.6 is of the order of $(\Delta\omega)^3$, provided that there exists a Taylor series expansion of the PSD around ω_k, and the discrete frequency is given by $\omega_k = (k - 1/2)\Delta\omega$.

It is noted that the centroidal rule is a subcase of a more general discretization scheme, as, e.g., presented in Katafygiotis *et al.* [258]: The interval $[0, \omega_u]$ is divided into K equal parts, each having length $\Delta\omega = \omega_u/K$, and ω_k is selected as some point in the interval $[(k - 1)\Delta\omega, k\Delta\omega]$ so that for any $k \geq 1$, $\omega_{k+1} - \omega_k = \Delta\omega$. This general discretization scheme provides simulations with period equal to $T_0 = 2\pi q/\Delta\omega$, assuming that the ratio $\omega_1/\Delta\omega$ is equal to a rational number p/q, with p and q being mutually prime numbers, and, also, requires that either $\omega_1 \neq 0$ or $S(\omega_1 = 0) = 0$ for the simulations to be ergodic in the mean.

Utilization of FFT in Simulations

As a large sum of trigonometric functions, Eq. 7.4 requires an extensive computational effort. Yang [573] introduced the Fast Fourier Transform (FFT) approach in simulating envelopes of random processes and Shinozuka [461], [462] extended the approach to simulations of random processes and random fields. The introduction of the FFT technique has the advantage of dramatically reducing the computational time in the simulations.

Any frequency discretization scheme that does not contain the origin $\omega_1 = 0$, as is the case for Eq. 7.4, requires a shifted FFT for the generation of the simulations. Zerva [592] utilized the centroidal rule, and discretized the frequency domain as $\omega_k = (1/2)\Delta\omega + k\Delta\omega$ with $k = 0, \ldots, K - 1$, i.e., the discrete frequencies are shifted by $(1/2)\Delta\omega$ to accomodate the fact that the FFT starts at $k = 0$. Equation 7.4 is then rewritten as [592]:

$$f_r = \sqrt{2}\,\Re\left[e^{i\pi r/M} \left\{ \sum_{k=0}^{M-1} ([2\,S(\omega_k)\Delta\omega]^{1/2} e^{i\phi_k})\, e^{i2\pi rk/M} \right\} \right] \tag{7.8}$$

in which, $f_r = f(r\Delta t)$, $r = 0, \ldots, M - 1$, $\Delta t = 2\pi/(M\Delta\omega)$, and $\Re[.]$ indicates the real part. M is a power of 2 ($M = 2^m$, $m =$ integer) and $M \geq 2K$ so that aliasing effects can be avoided, i.e., the PSD is zero-padded past the cut-off frequency, ω_u, with at least K additional discrete frequencies. The FFT is then applied to the expression in the braces. Equation 7.8 provides simulations over a time length that is equal to one half the period of the original approach (Eq. 7.4), i.e., $T_0' = T_0/2 = 2\pi/\Delta\omega$.

This restriction is easily overcome due to the periodicity of the simulations: Let $0 \leq t \leq T_0/2$, $T_0/2 \leq t' \leq T_0$, and $t' = t + T_0/2$. The substitution of t' in Eqs. 7.4 and 7.8 results in

$$f(t') = -f(t) \Leftrightarrow f_{r+(M-1)} = -f_r \tag{7.9}$$

which implies that the duration of the simulations can be extended to their original period of $T_0 = 4\pi/\Delta\omega$ by means of Eq. 7.9. Additionally, the simulations can be extended to any arbitrary time (longer than T_0) by "placing" at the end of each period a ·repetition of the simulation over the initial period, since Eq. 7.4 is periodic. Hence, the simulations obtained through Eqs. 7.8 and 7.9 are completely equivalent to Eq. 7.4, but, since they make use of the FFT technique, the computational effort is reduced to a fraction of the original trigonometric approach.

Example Applications

An illustration of the aforementioned simulation scheme is presented in Figs. 7.1–7.3. Figure 7.1 presents the Clough-Penzien [107] PSD for accelerations, $S(\omega)$, velocities, $S_v(\omega)$, and displacements, $S_d(\omega)$, given by Eqs. 3.6 and 3.7 and repeated here for convenience:

$$S(\omega) = S_\circ \frac{1 + 4\zeta_g^2 \left(\frac{\omega}{\omega_g}\right)^2}{\left[1 - \left(\frac{\omega}{\omega_g}\right)^2\right]^2 + 4\zeta_g^2 \left(\frac{\omega}{\omega_g}\right)^2} \frac{\left(\frac{\omega}{\omega_f}\right)^4}{\left[1 - \left(\frac{\omega}{\omega_f}\right)^2\right]^2 + 4\zeta_f^2 \left(\frac{\omega}{\omega_f}\right)^2} \tag{7.10}$$

$$S_v(\omega) = \frac{1}{\omega^2} S(\omega); \quad S_d(\omega) = \frac{1}{\omega^4} S(\omega) \tag{7.11}$$

The parameters of the Clough-Penzien spectrum are: $\omega_g = 5\pi$ rad/sec, $\omega_f = 0.5\pi$ rad/sec, $\zeta_g = \zeta_f = 0.6$ (i.e., the firm soil conditions of "Spectrum 2" [221] in Table 3.1(a)), and $S_\circ$ is assumed to be equal to 1 cm^2/sec^3. The dominant frequency range of accelerations, velocities and displacements can be clearly identified from Fig. 7.1. Simulations are then generated according to Eq. 7.8 for $\Delta t = 0.01$ sec and $T_0 = 20.48$ sec. Figure 7.2 presents a set of simulated acceleration, velocity and displacement time histories, each generated directly from Eq. 7.8 with the use of the appropriate PSD and the same seed for the generation of the random phase angles. The simulations are presented over two periods, so that the periodicity in the resulting time series can be clearly observed. As indicated earlier, an infinite number of repetitions of the simulations over one period can be "placed" before $t = 0$ and after $t = 40.96$ sec in Fig. 7.2, i.e., the time series have no beginning and no end. The stationarity of the resulting motions is an inherent characteristic of simulations generated by means of trigonometric series, and is also in accordance with the basic assumption for the development of the power spectral density from recorded data (Section 2.2.3), which considers that the strong motion S-wave window of the motions is a segment of a stationary process. Figure 7.3 presents a different set of simulations generated from the same PSDs of Fig. 7.1 again over two periods. The simulated time series in Figs. 7.2 and 7.3 have the same stochastic properties even though their shape differs. A vast number of such realizations of the random process, resulting from different seeds for the random phase angles, can then be generated quickly and efficiently, which illustrates the versatility of the use of Monte Carlo simulations in

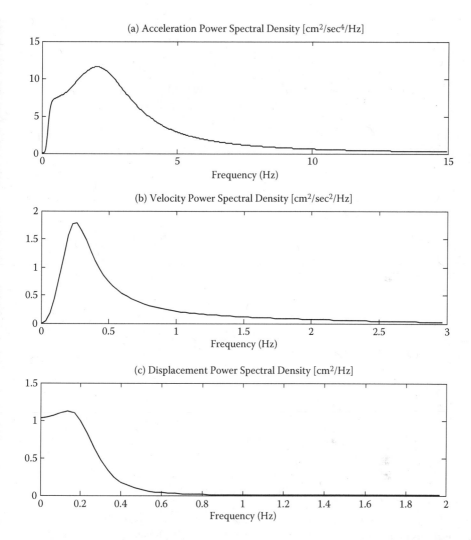

FIGURE 7.1 Illustration of acceleration (part (a)), velocity (part (b)) and displacement (part (c)) power spectral densities resulting from the Clough-Penzien [107] spectrum of Eqs. 7.10 and 7.11. The parameters of the spectra are as suggested by Hindy and Novak [221] for the firm soil conditions of "Spectrum 2" in Table 3.1(a). The frequency scale in the three subplots of the figure is not the same.

engineering problems. As stationary time series, however, the generated motions are not directly applicable in time history response evaluations of structures. Intensity and frequency modulation of the stationary time series is then performed, so that the generated motions emulate the characteristics of recorded accelerograms, as described in the following two subsections.

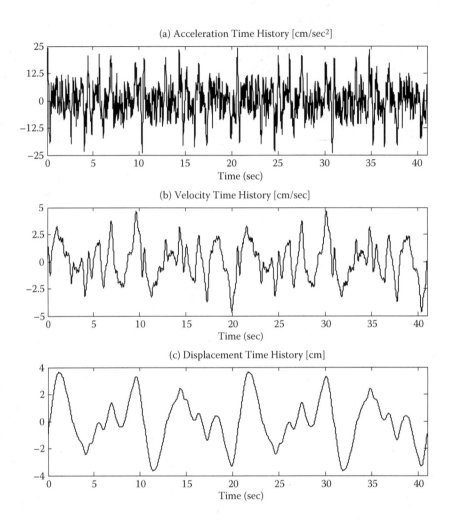

FIGURE 7.2 One set of stationary acceleration (part (a)), velocity (part (b)) and displacement (part (c)) time series generated from the corresponding Clough-Penzien spectra of Fig. 7.1; the same seed for the generation of the random phase angles was used in the simulation of the three time series. The simulations are presented over a duration of twice their period, which is equal to $T_0 = 20.48$ sec.

7.1.2 INTENSITY MODULATION OF SIMULATED TIME SERIES

A simple approach for the introduction of finite duration in the generated stationary motions, such as the ones shown in Figs. 7.2 and 7.3, is through their multiplication with an intensity modulating (envelope) function, $\tilde{I}(t)$, also commonly referred to as amplitude modulating function. A number of such intensity modulating functions have been introduced since the 1960's, e.g., Refs. [27], [64], [237], [244], [467], [514]. The most straightforward intensity modulating function is a box-car type of

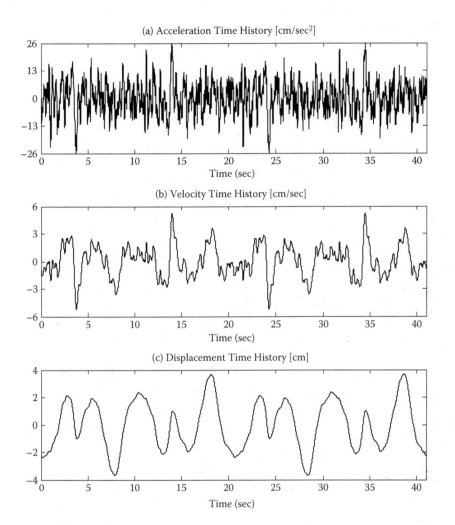

(a) Acceleration Time History [cm/sec²]

(b) Velocity Time History [cm/sec]

(c) Displacement Time History [cm]

FIGURE 7.3 Another set of simulations as in Fig. 7.2. Again, the same seed for the generation of the random phase angles was used for the acceleration (part (a)), velocity (part (b)) and displacement (part (c)) time series in the figure. The seed used for the simulations in this figure is different than the seed used for the simulations in Fig. 7.2.

duration t_1 [514]:

$$\tilde{I}(t) = \begin{cases} I_0 & 0 \le t \le t_1 \\ 0 & t > t_1 \end{cases} \tag{7.12}$$

Single-exponential intensity modulating functions, e.g., [64]:

$$\tilde{I}(t) = a_1 t\, e^{-a_2 t} \qquad t \ge 0 \tag{7.13}$$

double-exponential functions, e.g., [467]:

$$\tilde{I}(t) = I_0 (e^{-b_1 t} - e^{-b_2 t}) \qquad b_2 > b_1 \tag{7.14}$$

and piecewise-continuous functions, e.g., [27], [244]:

$$\tilde{I}(t) = \begin{cases} I_0(t/t_1)^2 & 0 \le t \le t_1 \\ I_0 & t_1 \le t \le t_2 \\ I_0 e^{-b(t-t_2)} & t \ge t_2 \end{cases} \qquad (7.15)$$

have also been suggested. The stationary simulation, e.g., $f(t)$ from Eq. 7.4, is modified as:

$$\tilde{f}(t) = \tilde{I}(t)f(t) \qquad (7.16)$$

and the modulated time series, $\tilde{f}(t)$, acquires finite duration. The evolutionary (time varying) power spectrum of the non-stationary process becomes then [414]:

$$\tilde{S}(t, \omega) = [\tilde{I}(t)]^2 S(\omega) \qquad (7.17)$$

where $S(\omega)$ is the power spectral density of the stationary process. The parameters in Eqs. 7.12–7.15 can be selected such that the modulated simulations exhibit the desired characteristics (e.g., strong motion duration), or, alternatively, can be estimated from recorded data, as will be illustrated in Section 7.1.3.

An example application of the procedure is presented in Fig. 7.4: The stationary acceleration time history of Fig. 7.2(a), also presented in Fig. 7.4(a), is multiplied with the piecewise-continuous modulation function of Eq. 7.15 (Fig. 7.4(b)) to yield a non-stationary time history (Fig. 7.4(c)), that resembles more an actual accelerogram than Fig. 7.4(a). The parameters of the modulating function (Fig. 7.4(b)) for unit energy are: $I_0 = 0.4123, t_1 = 1.47$ sec, $t_2 = 6.47$ sec (so that the flat part of the modulating function, or, equivalently, the strong motion duration of the series, is $t_2 - t_1 = 5$ sec) and $b = 0.85/$sec [240].

It needs to be noted, however, that the use of different intensity modulating functions may induce different structural response. Jangid [240] examined the linear, non-stationary random vibration response of single-degree-of-freedom oscillators subjected to seismic excitations resulting from various intensity modulating functions, including those presented in Eqs. 7.12, 7.14 and 7.15, as well as a variation of Eq. 7.13 and two triangular ones, and concluded that the shape of the modulating function has a significant effect on the response of the systems.

7.1.3 NON-STATIONARY SIMULATED TIME SERIES COMPATIBLE WITH RECORDED DATA

Even though the time-dependent intensity modulating functions (Eqs. 7.12–7.15) introduce non-stationarity in the simulated motions (Fig. 7.4), they cannot capture the time-varying frequency content of recorded data, that result from the different characteristics of the various wave components that constitute the seismic motions (e.g., body and surface waves propagate with different velocities, dominate in different time windows of the series, and can vary significantly in frequency content). More realistic simulations can be obtained if both their intensity and frequency content are modulated based on information extracted from recorded data. Indeed, Eq. 7.16 is a special case of Priestley's [414] evolutionary spectrum, in which the modulating function depends on both time and frequency. Significant effort has been placed in

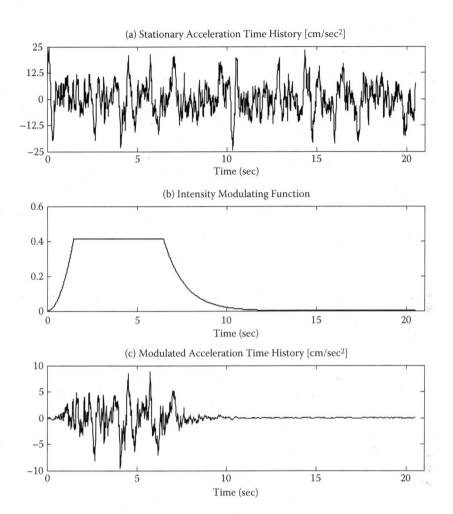

FIGURE 7.4 Illustration of transformation of stationary simulations to non-stationary time series through intensity modulating functions: The stationary simulation in part (a), which is the same as the acceleration simulation of Fig. 7.2(a), is multiplied with the piecewise-continuous intensity modulating function of part (b) to yield the non-stationary series in part (c) that acquires finite duration and a strong motion part.

identifying the evolutionary nature of seismic records, as, e.g., in Refs. [114], [115], [131], [136], [155], [184], [248], [277], [310], [311], [406], [439], [446], [577].

Additionally, the simple multiplication of the stationary time series with the intensity modulating function (Eq. 7.16) does not take into consideration the relationship between the envelope function and the group delay time of a recorded accelerogram. As indicated in Section 2.2.4, the probability distribution of the group delay time, defined as the derivative of the Fourier phase of an accelerogram with respect to the frequency ω (or, equivalently, the probability distribution of the Fourier phase differences of the accelerogram), resembles the shape of the accelerogram envelope function [262], [380], i.e., it contains information about the duration and the evolutionary character of the motions. Simulation schemes that also incorporate the phase properties of accelerograms have been reported, among others, in Refs. [71], [346], [440], [471], [472], [520], [521], [527] and [567].

As an illustration of simulations compatible with the non-stationary (in intensity and frequency content) characteristics of a recorded accelerogram, the approach developed by Yeh and Wen [577] is presented herein. The methodology, which is a modification of the work of Grigoriu *et al.* [184] with a more rigorous interpretation of the instantaneous power spectrum, identifies intensity and frequency modulating functions from recorded data, scales the non-stationary data with the identified modulating functions to make them approximately stationary, and fits stationary power spectral densities to the scaled stationary data.

Let $f(t)$ be a stationary random process, and let $\psi(t)$ be defined as $\psi(t) = f(y(t))$, with $y(t)$ being a smooth, strictly increasing function of t. If $y(t)$ is a linear function of t, $\psi(t)$ can be easily shown to be stationary but with a different time scale, whereas, if $y(t)$ is a nonlinear function of t, $\psi(t)$ is not stationary due to the continuously changing time-scale effect [577]. The instantaneous power spectral density of $\psi(t)$ becomes then:

$$S_{\psi\psi}(t, \omega) = \frac{1}{\dot{y}(t)} S_{ff}\left(\frac{\omega}{\dot{y}(t)}\right) \tag{7.18}$$

where overdot indicates derivative with respect to time, and $S_{ff}(\omega)$ is the power spectral density of $f(t)$. The intensity and frequency modulated process is then described by:

$$\tilde{f}(t) = \tilde{I}(t) f(y(t)) \tag{7.19}$$

where, as before in Eq. 7.16, $\tilde{I}(t)$ is a deterministic envelope function.

To estimate the intensity and frequency modulating functions from a recorded accelerogram, Yeh and Wen [577] proposed the following approach. The intensity modulating function, $\tilde{I}(t)$, is estimated so that the expected energy of the model fits the energy of a recorded accelerogram in the least-squares sense. Yeh and Wen [577] suggested that the following form for the intensity modulation function be used:

$$[\tilde{I}(\tau)]^2 = A \frac{t^B}{D + t^E} e^{-Ct} \tag{7.20}$$

with A, B, C, D and E being parameters estimated from the record. The frequency modulating function is estimated from the zero crossing rate of the seismogram, as suggested earlier by Saragoni and Hart [439]. Let $\mu_0(t)$ indicate the mean value of the total zero crossings up to time t and be described by a continuous, differentiable and non-decreasing function of t. The functional form suggested for $\mu_0(t)$ was [577]:

$$\mu_0(t) = r_1 t + r_2 t^2 + r_3 t^3 \tag{7.21}$$

with r_1, r_2 and r_3 being estimated from the data, again, in a least-squares sense, and the frequency modulating function was determined from the expression:

$$y(t) = \mu_0(t) / \dot{\mu}_0(t_0) \tag{7.22}$$

where t_0 indicates the starting time of significant excitation.

Once $\tilde{I}(t)$ and $y(t)$ were identified, the accelerogram was scaled in intensity and time so that it became approximately stationary. This stationary part was then described by a power spectral density function as, e.g., the Kanai-Tajimi (Eq. 3.5) or the Clough-Penzien (Eq. 7.10) spectrum, the parameters of which were determined through system identification. Simulated ground motions can be generated from the (stationary) power spectral density according to, e.g., Eq. 7.8, scaled through the frequency modulating function and then multiplied with the intensity modulating function (Eq. 7.19) to yield a non-stationary time history with characteristics similar to those of the recorded accelerogram. To account for the variable frequency content of the recorded seismic time histories in different time windows, Yeh and Wen [577] also suggested that the original accelerogram be decomposed into several components.

An example of the aforementioned approach is illustrated herein for the S16°E component of the Pacoima Dam accelerogram recorded during the 1971 San Fernando, California, earthquake. The parameters of the intensity and frequency modulating functions and the Clough-Penzien power spectral densities were identified in two frequency ranges, $0 - 2$ Hz and $2 - 25$ Hz, and are presented, for completeness, in Table 7.1 [577].

TABLE 7.1

Parameters of the intensity and frequency modulating functions and the (stationary) Clough-Penzien PSDs identified by Yeh and Wen [577] from the 1971 Pacoima Dam accelerogram in two frequency ranges. (Reproduced from *Structural Safety*, Vol. 8, C.-H. Yeh and Y.K. Wen, "Modeling of nonstationary ground motion and analysis of inelastic structural response," pp. 281–298, Copyright © 1990, with permission from Elsevier.)

Parameters	$0 - 2$ Hz	$2 - 25$ Hz
Intensity modulating function		
A	1.2285×10^5	6.4545×10^{32}
B	17.938	-2.8402×10^{-2}
C	1.0328×10^{-1}	-4.7086×10^{-1}
D	5.4141×10^7	1.4854×10^{30}
E	20.054	31.763
Frequency modulating function		
r_1	8.8169×10^{-1}	14.982
r_2	1.6904×10^{-1}	-3.7098×10^{-1}
r_3	-3.6761×10^{-3}	8.2505×10^{-3}
t_0	2.0	1.0
$\mu_0(t_0)$	1.5137	14.264
Clough-Penzien PSD		
ω_g [rad/sec]	5.0	32.0
ζ_g	0.4	0.4
ω_f [rad/sec]	0.5	17.0
ζ_f	0.6	0.6

Figure 7.5 presents the intensity and frequency modulating functions and the Clough-Penzien power spectral densities identified from the recorded accelerogram in the two frequency ranges (Eqs. 7.20, 7.22 and 7.10 with the parameters of Table 7.1). The actual Pacoima Dam accelerogram is presented in Fig. 7.6(a) and an artificial time history, resulting from the combination of two simulated time series generated from the identified power spectra and the modulating functions in the two frequency ranges, is shown in Fig. 7.6(b). It can be clearly seen from the comparison of Figs. 7.6(a) and 7.6(b) that the simulated time history captures the non-stationary characteristics of the recorded data. For lower frequencies, which dominate in the beginning of the

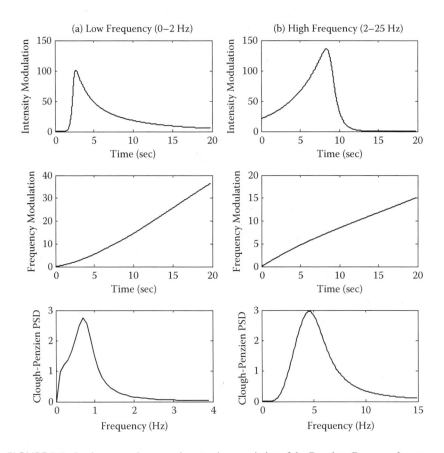

FIGURE 7.5 Stationary and non-stationary characteristics of the Pacoima Dam accelerogram recorded during the 1971 San Fernando earthquake in two frequency ranges: 0 – 2 Hz in part (a) and 2 – 25 Hz in part (b), as identified by Yeh and Wen [577]. The top subplots illustrate the identified intensity modulating functions, the middle subplots present the frequency modulating functions and the bottom subplots show the (stationary) Clough-Penzien power spectral densities; the parameters of the functions are given in Table 7.1. (Reproduced from *Structural Safety*, Vol. 8, C.-H. Yeh and Y.K. Wen, "Modeling of nonstationary ground motion and analysis of inelastic structural response," pp. 281–298, Copyright ©1990, with permission from Elsevier.)

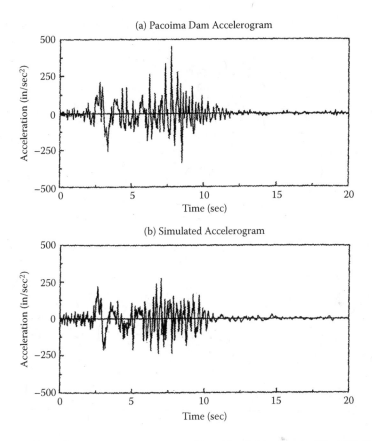

FIGURE 7.6 Illustration of simulations compatible with seismic records: The 1971 Pacoima Dam accelerogram is presented in part (a) of the figure. Part (b) presents a simulated accelerogram generated from the (stationary) Clough-Penzien spectra identified from the record in two frequency ranges (0 – 2 Hz and 2 – 25 Hz) and modified with the corresponding intensity and frequency modulating functions of Fig. 7.5. (Reprinted from *Structural Safety*, Vol. 8, C.-H. Yeh and Y.K. Wen, "Modeling of nonstationary ground motion and analysis of inelastic structural response," pp. 281–298, Copyright ©1990, with permission from Elsevier.)

record, the intensity and frequency modulating functions (top and middle subplot of Fig. 7.5(a)) transform the stationary simulations generated from the Clough-Penzien spectra (bottom subplot in Fig. 7.5(a)) to non-stationary time series peaked at approximately the same times as the recorded data. At later times, the frequency content changes (Fig. 7.5(b)), which the simulated time history closely reproduces, but there are some differences in the values of the peak amplitudes and the locations of the peaks. The differences between the record and the simulated motion should be anticipated, as simulations do not exactly replicate the original time history, but, rather, its stochastic characteristics. These differences, however, may also be attributed to the inherent assumptions for the parametric forms of the modulating functions and the subsequent identification schemes for the evaluation of the parameters, as well as in the reproduction of non-stationary processes through stationary (Fourier) transforms.

Recent developments for the description of non-stationary processes, including accelerograms, are based on wavelets, e.g., Ref. [493], and Hilbert-Huang Transforms (HHT), e.g., Ref. [560]. These approaches incorporate time and frequency domain features in their transforms, which preserve the non-stationary characteristics of the seismic ground motions. For illustration purposes, the potential of their use in earthquake engineering is presented in Fig. 7.7, which was developed by Wen and Gu [560]: The El Centro record of the 1940 Imperial Valley earthquake is shown at the top of the figure, and four simulations based on Hilbert spectra are shown below. It can be clearly seen from the figure that the characteristics of the acceleration record are closely reproduced by the simulated time series. Indeed, the main advantages of HHT based simulations are that they are truly non-stationary and model-free, and, hence, the problem of parameter estimation of the time-dependent power spectral density function is avoided. These approaches are not described herein: At the present time, coherency is defined based on the assumption of stationarity through Fourier transforms (Section 2.4), and simulations of spatially variable seismic ground motions are also based on this concept.

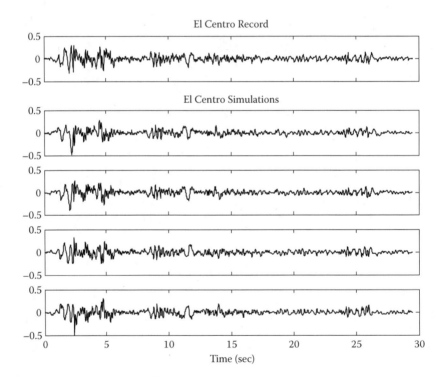

FIGURE 7.7 Illustration of simulations compatible with seismic records: The El Centro record is shown in the top part of the figure. Four simulations based on the record and utilizing Hilbert-Huang transforms are presented below the recorded time history. The transformation preserves the non-stationary characteristics of the record without relying on parametric modeling, as was the case in Figs. 7.5 and 7.6. (Reprinted from Y.K. Wen and P. Gu, "Description and simulation of nonstationary processes based on Hilbert spectra," *Journal of Engineering Mechanics, ASCE,* Vol. 130, pp. 942–951, 2004, with permission from ASCE; courtesy of Y.K. Wen.)

7.1.4 CONVERSION OF RESPONSE SPECTRUM TO POWER SPECTRAL DENSITY

The most widely acceptable description of the characteristics of seismic ground motions in engineering applications is the response spectrum. For motions to be simulated as response-spectrum-compatible, an equivalence between the response spectrum and the power spectral density of the motions needs to be established. Scanlan and Sachs [445] suggested that the phase angles of the simulated processes should be random, and the amplitudes iteratively modified until the differences between the target response spectrum and the response spectrum derived from the simulations become small. A similar concept was later utilized by Hao *et al.* [200] and Deodatis [127] in the generation of response-spectrum-compatible spatially variable simulations. The techniques evaluate the response spectrum of each generated time history, after the simulations have been modified to become non-stationary, and adjust the Fourier amplitudes of the series [200] or upgrade their (stationary) power spectral densities [127] in an iterative scheme; the approach by Deodatis [127] is presented in Section 7.3.2.

Alternatively, a power spectral density can be derived from the target response spectrum using extreme value theory, as, e.g., in Refs. [105], [166], [186], [263], [271], [396], [402], [429], [495], [548]. It is recalled that the response spectrum is the maximum response of a linear single-degree-of-freedom oscillator subjected to a given excitation. By using extreme value statistics, the maximum response of the oscillator to the (yet unknown) power spectral density can be estimated, and the parameters of the power spectral density can be identified from the equivalence between the estimated maximum system response and the target response spectrum. Pfaffinger [402] presented an iterative scheme based on the probability distribution of extremes and the peak value statistics derived by Davenport [125]. Park [396] extended Pfaffinger's method [402] and proposed a direct (non-iterative) technique for the conversion of a response spectrum to a power spectral density. Park's approach [396] is highlighted herein for illustration of the concept.

The unknown (equivalent) PSD, $S_R(\omega)$, that will be derived from the target response spectrum, is discretized at frequencies ω_j and expressed as the sum of discretized power components p_j as:

$$S_R(\omega) = \sum_{j=1}^{M} S(\omega_j)\Delta\omega_j\delta(\omega - \omega_j) = \sum_{j=1}^{M} p_j\delta(\omega - \omega_j) \qquad (7.23)$$

in which, $\delta(\omega)$ is the Dirac delta function and $\Delta\omega_j$ the incremental frequency. The target response spectrum, $RS_t(\omega_k, \zeta)$, at discrete frequencies, ω_k, $k = 1, \ldots, N$, is approximated by:

$$RS_t^2(\omega_k, \zeta) \cong \sum_{j=1}^{M} p_j R^2(\omega_k, \omega_j, \zeta); \quad k = 1, \ldots, N \qquad (7.24)$$

where $R(\omega_k, \omega_j, \zeta)$ is the peak acceleration response of a single-degree-of-freedom oscillator with natural frequency ω_k and damping ζ excited by the extremely narrowband process whose PSD is the Dirac delta function at frequency ω_j.

The relationship between the PSD and the response spectrum is established from extreme value theory. The expected maximum value of a zero-mean, Gaussian random process can be expressed, e.g., in terms of Davenport's peak factor approximation as [125]:

$$E[f_m] \cong \sigma_f \left(\sqrt{2\ln(\nu T_e)} + \frac{\gamma}{\sqrt{2\ln(\nu T_e)}} \right) \qquad (7.25)$$

$$\nu = \frac{1}{\pi} \left[\frac{\int_0^\infty \omega^2 S(\omega)\,d\omega}{\int_0^\infty S(\omega)\,d\omega} \right]^{1/2} \qquad (7.26)$$

with σ_f^2 indicating the variance of the process, T_e the considered time interval, and $\gamma = 0.577216$ being the Euler constant. The approximation of Eq. 7.25 is based on the Poisson model, which considers statistical independence of the crossings of a symmetric ($\pm$) barrier by the process. Alternative formulations for the peak factor in earthquake engineering applications have also been reported (see, e.g., Ref. [185] for a comparative evaluation). For example, Vanmarcke [550] considered the dependence between crossings through their clumping effect and developed the cumulative probability function of the maxima, and Der Kiureghian [133] fitted empirical functions to Vanmarcke's results [550] to provide alternative expressions for their mean and standard deviation. Der Kiureghian [133] reported that these modifications provide more accurate results, especially when the bandwidth of the underlying process decreases. Pfaffinger [402], based on the rationale that, given the overall number of approximations made in earthquake engineering problems, Eq. 7.25 can be considered sufficient, utilized Eq. 7.25, and Park [396] extended his approach and estimated $R(\omega_k, \omega_j, \zeta)$ (Eq. 7.24) as:

$$R(\omega_k, \omega_j, \zeta) \cong \sqrt{\frac{1 + 4\zeta^2(\omega_j/\omega_k)^2}{(1 - (\omega_j/\omega_k)^2)^2 + 4\zeta^2(\omega_j/\omega_k)^2}} \left\{ \sqrt{2\ln[\nu_j T_e]} + \frac{\gamma}{\sqrt{2\ln[\nu_j T_e]}} \right\} \qquad (7.27)$$

where now $\nu_j = \omega_j/\pi$. Park [396] further defined an effective duration of the time series, T_e, as:

$$T_e = \frac{\int_0^\infty \tilde{I}(t)\,dt}{\max[\tilde{I}(t)]} \qquad (7.28)$$

where $\tilde{I}(t)$ is a selected intensity modulating function. The power components, p_j, can then be obtained from the following least-squares problem:

$$\text{minimize} \quad \sum_{k=1}^{N} \left\{ RS_t^2(\omega_k, \zeta) - \sum_{j=1}^{M} p_j R^2(\omega_k, \omega_j, \zeta) \right\} \qquad (7.29)$$

$$\text{subject to} \quad p_j \geq 0; \quad j = 1, \dots, M$$

Figure 7.8 illustrates an example of such a conversion process. The UBC response spectrum [234] for soft soil conditions and 5% damping normalized to 1g is utilized as the target spectrum. The effective duration of the time series (Eq. 7.28), based on the intensity modulating function of Eq. 7.15 with parameters $I_0 = 1, b = 0.23$/sec, $t_1 = 3$ sec, $t_2 = 16$ sec and truncated past 20 sec, is evaluated as $T_e = 16.8$ sec [396]. The target spectrum is presented in Fig. 7.8(a) together with the response spectrum

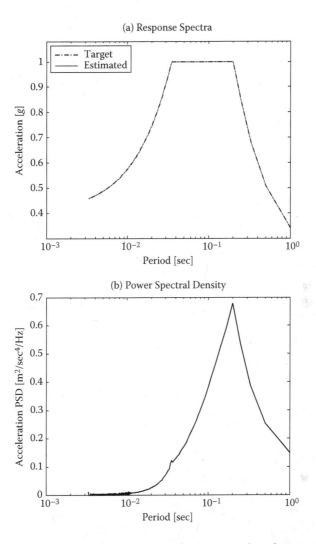

FIGURE 7.8 Illustration of a non-iterative process for the conversion of a response spectrum to a power spectral density based on Park's approach [396]: Part (a) presents the target UBC spectrum for soft soil conditions normalized to $1\,g$ and the response spectrum resulting from the estimated power spectral density. Part (b) shows the PSD estimated from the target spectrum of part (a); the estimated PSD is sharply peaked and, for comparison purposes, is presented as function of period (in [sec]) rather than the conventional frequency (in [Hz]).

estimated from the least-squares minimization process (Eq. 7.29). The identified PSD is presented in Fig. 7.8(b) as function of period (in [sec]), rather than the more conventional frequency (in [Hz]), for ease in comparison with the response spectrum (Fig. 7.8(a)). The response spectrum of the identified PSD is in close agreement with the target spectrum (Fig. 7.8(a)); the resulting PSD (Fig. 7.8(b)) is sharply peaked and, also, oscillates rapidly at lower periods (higher frequencies).

Inherent assumptions in the aforementioned conversion process are the peak factor approximation and the consideration that the single-degree-of-freedom oscillator has reached a steady-state (stationary) response, which may or may not be the case depending on the duration of the excitation and the damping value of the system [166], [271]. Additional issues that complicate the process of conversion are that one frequency component in a time history can affect the values of the response spectrum at other frequencies [271], and that there exist response spectra that cannot be realized [292].

7.2   SIMULATION OF RANDOM FIELDS

The concept of simulation of seismic ground motions at a single location on the ground surface based on random processes is extended in this section to the simulation of spatially variable motions generated from random fields. Section 7.2.1 presents the rigorous representation of the random field through its frequency-wavenumber spectrum, which, for recorded data, was defined in Section 4.1. In this approach, the space-time domain is transformed into the wavenumber-frequency domain, and the process of simulation is, essentially, a multi-domain extension of Eq. 7.4. Spatially variable seismic ground motions resulting from various spatial coherency models are then generated in Section 7.2.2., which also illustrates how the properties of the coherency models reflect into the characteristics of the simulated time series. The more versatile approach for the simulation of random fields as stochastic vector processes is then described in Section 7.3.

7.2.1   FREQUENCY-WAVENUMBER SPECTRAL REPRESENTATION

Consider a homogeneous space-time random field with zero mean, space-time covariance function $R(\xi, \tau)$, ξ being the separation distance and τ being the time lag, and frequency-wavenumber (F-K) spectrum $S(\kappa, \omega)$, in which κ indicates the wavenumber and ω indicates frequency. The frequency-wavenumber spectrum and the covariance function are Wiener-Khintchine transformation pairs and possess the same symmetries [551]. The F-K spectrum is obtained through the Fourier transform of the cross spectral density of the motions, $S(\xi, \omega)$, i.e., [365]:

$$S(\kappa, \omega) = \frac{1}{2\pi} \int_{-\infty}^{+\infty} S(\xi, \omega) e^{-i\kappa\xi} d\xi \tag{7.30}$$

Simulations of spatially variable seismic ground motions can then be generated by means of the spectral representation method [462]:

$$f(x, t) = \sqrt{2} \sum_{j=0}^{J-1} \sum_{n=0}^{N-1} \left([2S(\kappa_j, \omega_n)\Delta\kappa\,\Delta\omega]^{1/2} \cos\left(\kappa_j x + \omega_n t + \phi_{jn}^{(1)}\right) \right.$$
$$\left. + [2S(\kappa_j, -\omega_n)\Delta\kappa\,\Delta\omega]^{1/2} \cos\left(\kappa_j x - \omega_n t + \phi_{jn}^{(2)}\right) \right) \tag{7.31}$$

in which $\phi_{jn}^{(1)}$ and $\phi_{jn}^{(2)}$ are two sets of independent random phase angles uniformly distributed between $[0, 2\pi)$, $\kappa_j = (j + \frac{1}{2})\Delta\kappa$ and $\omega_j = (n + \frac{1}{2})\Delta\omega$ are the discrete wavenumber and frequency, respectively, and $\Delta\kappa$ and $\Delta\omega$ are the wavenumber and frequency steps. Simulations based on Eq. 7.31 can be obtained if there exists an upper cut-off wavenumber $\kappa_u = J \times \Delta\kappa$ and an upper cut-off frequency $\omega_u = N \times \Delta\omega$, above which the contribution of the F-K spectrum to the simulations is insignificant for practical purposes. As in the case of random processes, the introduction of the FFT algorithm in Eq. 7.31 dramatically improves the computational efficiency of the simulations [462]. A shifted Fourier transform applied to Eq. 7.31 results then in the following expression for the simulations [592]:

$$
f_{rs} = \sqrt{2}\,\Re \left[e^{\frac{i\pi r}{M}} e^{\frac{i\pi s}{L}} \left\{ \sum_{j=0}^{M-1} \sum_{n=0}^{L-1} \left([2S(\kappa_j, \omega_n)\,\Delta\kappa\,\Delta\omega]^{1/2} e^{i\phi_{jn}^{(1)}} \right) e^{\frac{i2\pi r j}{M}} e^{\frac{i2\pi s n}{L}} \right\} \right]
$$

$$
+ \sqrt{2}\,\Re \left[e^{\frac{i\pi r}{M}} e^{-\frac{i\pi s}{L}} \left\{ \sum_{j=0}^{M-1} \sum_{n=0}^{L-1} \left([2S(\kappa_j, -\omega_n)\Delta\kappa\Delta\omega]^{1/2} e^{i\phi_{jn}^{(2)}} \right) e^{\frac{i2\pi r j}{M}} e^{-\frac{i2\pi s n}{L}} \right\} \right]
$$

$$(7.32)$$

in which, $f_{rs} = f(x_r, t_s), x_r = r\Delta x, \Delta x = 2\pi/(M\Delta\kappa), r = 0, \ldots, M-1, t_s = s\Delta t$, $\Delta t = 2\pi/(L\Delta\omega), s = 0, \ldots, L-1$, and $\Re[.]$ indicates, again, the real part. M and L are powers of 2, and ought to satisfy the inequalities $M \geq 2J$ and $L \geq 2N$, so that aliasing effects can be avoided; i.e., for $J \leq j < M$ and $N \leq n < L$ the value of the F-K spectrum in Eq. 7.32 is zero. The FFT is applied to the two expressions in the braces in Eq. 7.32. The following characteristics are inherent in the simulations: (i) they are asymptotically Gaussian as $J, N \to \infty$ due to the central limit theorem; (ii) they are periodic with period $T_o = 4\pi/\Delta\omega$ and wavelength $\mathcal{L}_o = 4\pi/\Delta\kappa$; (iii) they are ergodic, at least up to second moment, over an infinite time and distance domain or over the period and wavelength of the simulations; and (iv) as $J, N \to \infty$, the ensemble mean, covariance function and frequency-wavenumber spectrum of the simulations are identical to those of the random field. The application of the FFT in Eq. 7.32 provides values for the simulations over half their period and half their wavelength, but the simulations can be easily extended to their full period and wavelength through the following expressions [592]:

$$
\begin{aligned}
f_{r+(M-1),s} &= -f_{rs} \\
f_{r,s+(L-1)} &= -f_{rs} \\
f_{r+(M-1),s+(L-1)} &= f_{rs}
\end{aligned}
$$

$$(7.33)$$

As was the case for simulations of random processes (Eq. 7.8), simulations of random fields can also be extended to distances longer than their wavelength, $\mathcal{L}_0$, and times longer than their period, T_0.

To illustrate the approach, seismic ground motions are generated at a sequence of five locations along a straight line on the ground surface with separation distance between consecutive stations of 100 m. The Clough-Penzien power spectral density [107], Eq. 7.10, is utilized with parameters as before in Fig. 7.1, namely, $\omega_g = 5\pi$ rad/sec, $\omega_f = 0.5\pi$ rad/sec and $\zeta_g = \zeta_f = 0.6$, i.e., the firm soil conditions of "Spectrum 2" [221] in Table 3.1(a), and S_o is assumed, again, to be equal to 1 cm^2/sec^3. The coherency model is that of Harichandran and Vanmarcke [208] (Eq. 3.12 repeated here for convenience):

$$|\gamma(\xi, \omega)| = A \exp\left(-\frac{2B\,|\xi|}{av(\omega)}\right) + (1 - A)\exp\left(-\frac{2B\,|\xi|}{v(\omega)}\right)$$

$$v(\omega) = k\left[1 + \left(\frac{\omega}{2\pi f_0}\right)^b\right]^{-1/2} \quad ; \quad B = (1 - A + aA) \tag{7.34}$$

with parameters $A = 0.736$, $a = 0.147$, $k = 5210$ m, $f_o = 1.09$ Hz and $b = 2.78$ [208]; the exponential decay of the coherency model with frequency at separation distances of $\xi = 100, 200, 300$ and 400 m is presented in Fig. 7.9(a). The F-K spectrum (Eq. 7.30) for the aforementioned PSD and coherency model becomes [592]:

$$S(\kappa, \omega) = \frac{S_d(\omega)}{\pi}\left\{\frac{A\rho_1(\omega)}{\rho_1^2(\omega) + \kappa^2} + \frac{(1 - A)\rho_2(\omega)}{\rho_2^2(\omega) + \kappa^2}\right\} \tag{7.35}$$

with $S_d(\omega)$ being the displacement power spectral density of the Clough-Penzien spectrum (Eqs. 7.10 and 7.11), $\rho_1(\omega) = 2B/(av(|\omega|))$ and $\rho_2(\omega) = 2B/v(|\omega|)$; the absolute value of the frequency in $v(|\omega|)$ is now necessary, because ω in Eq. 7.32 can assume both positive and negative values. The resulting space-time field is quadrant-symmetric [551], i.e., $S(\kappa, \omega) = S(-\kappa, \omega) = S(\kappa, -\omega) = S(-\kappa, -\omega)$.

The comparison of the waveforms at the various locations is made in terms of displacement rather than acceleration time series. Because of their low frequency content (Fig. 7.1), differences in displacements can be more clearly recognized than differences in accelerations (cf., e.g., the difficulty in the comparison of the acceleration time histories of Figs. 7.2(a) and 7.3(a), but the relative ease in the comparison of the displacement time histories of Figs. 7.2(c) and 7.3(c)). It is noted, however, that spatially variable displacement time histories reflect the loss of coherency in this lower frequency range only (Fig. 7.1(c) and 7.9(a)), whereas spatially variable acceleration time histories (Fig. 7.1(a) and 7.9(a)) reflect the loss of coherency at higher frequencies as well. In the simulations, the cut-off wavenumber and frequency were $\kappa_u = 0.0628$ rad/m and $\omega_u = 15.70$ rad/sec, respectively; above these values, the contribution of the F-K spectrum to the simulations was negligible. Figure 7.10(a) presents the displacement time histories at the five stations; the simulations are presented over half their period, but can be easily extended to their full period according to Eq. 7.33. The waveforms in Fig. 7.10(a) show significant variations due to the partial correlation of the coherency model in the low frequency range (Fig. 7.9(a)).

7.2.2 EFFECT OF COHERENCY MODEL SELECTION IN SIMULATIONS

Section 6.4 compared the effect of different coherency models on the random vibration response of generic lifeline systems. The coherency models considered were the models of Harichandran and Vanmarcke [208] (Eq. 7.34) and of Luco and Wong [324] (Eq. 3.27). It is interesting at this point to also compare the characteristics of simulated seismic waveforms resulting from these models. This section compares the characteristics of the spatially variable displacements simulated at the five locations on the ground surface with the model of Harichandran and Vanmarcke (Fig. 7.10(a)) with

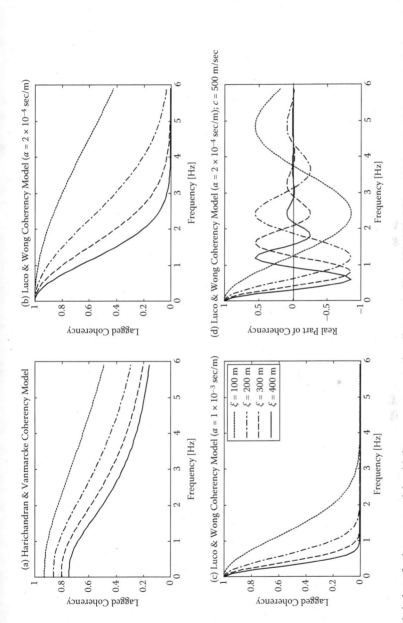

FIGURE 7.9 Variation of coherency models with frequency at separation distances of $\xi = 100, 200, 300$ and 400 m: Part (a) presents the lagged coherency of the model of Harichandran and Vanmarcke [208] based on the data of Event 20 at the SMART-1 array; part (b) the lagged coherency of the model of Luco and Wong [324] with a low value for the coherency drop parameter ($\alpha = 2 \times 10^{-4}$ sec/m); part (c) the lagged coherency of the model of Luco and Wong but with a higher value for the coherency drop parameter ($\alpha = 1 \times 10^{-3}$ sec/m); and part (d) presents the real part of the coherency derived from the model of Luco and Wong with the low value for the coherency drop parameter ($\alpha = 2 \times 10^{-4}$ sec/m) and an apparent propagation velocity of $c = 500$ m/sec.

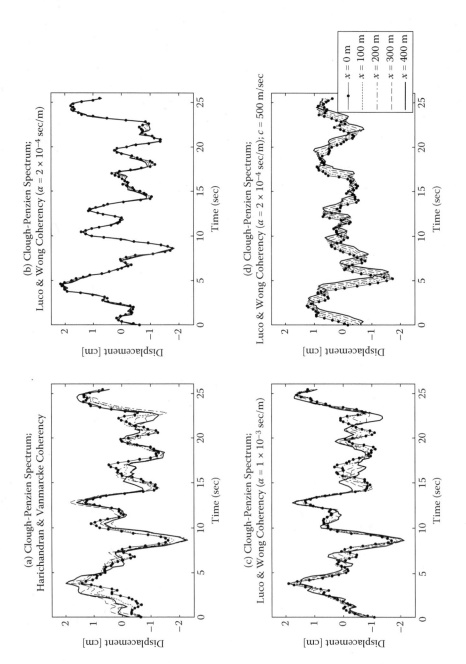

FIGURE 7.10

displacement time histories generated with the model of Luco and Wong (Eq. 3.27), the expression of which is repeated here for convenience:

$$|\gamma(\xi, \omega)| = e^{-\alpha^2 \omega^2 \xi^2} \tag{7.36}$$

where α indicates the coherency drop parameter. For the Clough-Penzien power spectral density (Eq. 7.10) and the coherency model of Luco and Wong (Eq. 7.36), the F-K spectrum (Eq. 7.30) takes the form [593]:

$$S(\kappa, \omega) = \frac{S_d(\omega)}{2\alpha |\omega| \sqrt{\pi}} \exp \left[-\frac{\kappa^2}{4\alpha^2 \omega^2} \right] \tag{7.37}$$

where $S_d(\omega)$, as in Eq. 7.35, indicates the displacement PSD. It can be recognized from Eq. 7.37 that this random field is also quadrant-symmetric.

The spectral representation method (Eq. 7.32) is used for the generation of the simulated displacement time histories. As in Section 6.4, two different values for the coherency drop parameter of the model of Luco and Wong [324] (Eq. 7.36) are used: a low value, $\alpha = 2 \times 10^{-4}$ sec/m, that reproduces slow exponential decay in the coherency as frequency and separation distance increase, and a high value, $\alpha = 1 \times 10^{-3}$ sec/m, that represents a sharp exponential decay. The exponential decay of the model with frequency at separation distances of $\xi = 100, 200, 300$ and 400 m is presented in Fig. 7.9(b) for $\alpha = 2 \times 10^{-4}$ sec/m and in Fig. 7.9(c) for $\alpha = 1 \times 10^{-3}$ sec/m. The parameters of the Clough-Penzien spectrum are the same as in the previous subsection. The displacement time histories based on the two variations of the model of Luco and Wong [324] are generated at the same five locations as before, i.e., along a straight line with separation distance between consecutive stations of 100 m.

The spatially variable displacement time histories simulated from the coherency model of Luco and Wong with the slow exponential decay ($\alpha = 2 \times 10^{-4}$ sec/m) are presented in Fig. 7.10(b), and the displacement simulations generated with the sharply decaying model ($\alpha = 1 \times 10^{-3}$ sec/m) in Fig. 7.10(c). For comparison

---

FIGURE 7.10 Simulated spatially variable displacement time histories generated by means of the frequency-wavenumber spectral representation method at five equidistant stations on the ground surface with separation distance between consecutive stations of 100 m: Part (a) illustrates simulations generated from the coherency model of Harichandran and Vanmarcke presented in Fig. 7.9(a), part (b) presents simulations resulting from the coherency model of Luco and Wong with a low value for coherency drop parameter ($\alpha = 2 \times 10^{-4}$ sec/m in Fig. 7.9(b)), and part (c) simulations generated from the coherency model of Luco and Wong with a high value for the coherency drop parameter ($\alpha = 1 \times 10^{-3}$ sec/m in Fig. 7.9(c)). The simulations in parts (a)-(c) represent standing waves. Part (d) presents simulations resulting from the coherency model of Luco and Wong with $\alpha = 2 \times 10^{-4}$ sec/m and an apparent propagation velocity of $c = 500$ m/sec (the real part of the coherency is illustrated in Fig. 7.9(d)); these motions propagate on the ground surface. In all simulations, the same Clough-Penzien PSD of Fig. 7.1(c), the same seed for the generation of the random phase angles and the same simulation parameters were utilized.

purposes, the same number of wavenumber and frequency points, the same discrete wavenumbers and frequencies, and the same seeds for the generation of the random phase angles were used in the simulations presented in Figs. 7.10(a)–(c); as a result, the simulations at $x = 0$ m for all cases have a similar, though not identical, trend. There are, however, significant differences as x increases: Figure 7.10(b) suggests that simulations based on the model of Luco and Wong with the low coherency drop parameter, $\alpha = 2 \times 10^{-4}$ sec/m, are essentially unchanged at all locations, whereas the simulations based on the other two models (Figs. 7.10(a) and (c)) vary with distance. It is also noted that the variability in the motions generated from the model of Luco and Wong with the high coherency drop parameter ($\alpha = 1 \times 10^{-3}$ sec/m) in Fig. 7.10(c) is dominated by higher frequency components (within the low frequencies that control the displacements). On the other hand, the displacement simulations resulting from the model of Harichandran and Vanmarcke in Fig. 7.10(a) exhibit significant low-frequency variability. These differences in the time histories are caused by the behavior of the models in the low frequency range (Figs. 7.9(a)–(c)): The model of Harichandran and Vanmarcke is only partially correlated even at zero frequency, yielding low-frequency variation between the time histories at the five stations. The model of Luco and Wong is fully correlated as $\omega \to 0$ for any value of the coherency drop parameter (Eq. 7.36 and Figs. 7.9(b) and (c)). For low values of α, it results in high values for the coherency in the lower frequency range (Fig. 7.9(b)), thus yielding essentially identical displacement simulations (Fig. 7.10(b)). For higher values of α, it decreases rapidly with frequency and separation distance (Fig. 7.9(c)), and the variability in the time histories shown in Fig. 7.10(c) is caused by the loss of coherency of the higher frequency components within the low frequency range that dominates the displacements (Fig. 7.1(c)). It needs to be emphasized, however, that seismic ground motions generated from the model of Luco and Wong with low values for the coherency drop parameter α show variability in space: Acceleration time histories are controlled by higher frequencies than displacements, as indicated in Fig. 7.1. As frequency increases, the model decreases with frequency and separation distance (Fig. 7.9(b)), leading to variability in the shape of the acceleration waveforms at the various locations. Displacement waveforms, obtained from recorded data at distances pertinent for engineering applications, often show close agreement, as observed, e.g., by Boore et al. [75] from the analysis of data recorded during the 1999 Hector Mine earthquake at a (fairly long) separation distance of 1.6 km. This observation suggests that seismic ground motions are highly correlated at low frequencies, and supports coherency models, such as the model of Luco and Wong [324], that reflect this effect; this issue will be further addressed in Section 9.5.

The comparison of the simulated time histories in Figs. 7.10(a)–(c) then indicates that the exponential decay of the selected coherency model is reflected in the characteristics of the generated ground motions, and that the simulation scheme of Eq. 7.31 or Eq. 7.32 reproduces the features of the target random field. The behavior of the spatially variable displacement time histories in Figs. 7.10(a)–(c) is also consistent with the seismic loads predicted by their respective coherency models (Section 6.4.2): It can be clearly seen from the time domain characteristics of the displacement time series generated from the model of Luco and Wong with the low coherency drop parameter (Fig. 7.10(b)) that differential displacements (Fig. 6.9(a)), and, consequently, the pseudo-static response of lifelines, will be lower than the corresponding results of the other two models. The model of Luco and Wong with the high coherency drop parameter results in differential displacements (Figs. 6.9(a) and 7.10(c)) that become more significant as the separation distance between stations increases, whereas the model of Harichandran and Vanmarcke results in significant differential displacements (Figs. 6.9(a) and 7.10(a)) even for close-by stations.

It is also noted from Figs. 7.10(a)–(c) that the simulated motions do not propagate on the ground surface. Indeed, it can be shown [592] that for quadrant-symmetric space-time fields, such as the random fields resulting from the models of Harichandran and Vanmarcke and of Luco and Wong (Eqs. 7.35 and 7.37, respectively), Eq. 7.31 can be rewritten as:

$$f(x, t) = 2\sqrt{2} \sum_{j=0}^{J-1} \sum_{n=0}^{N-1} [2S(\kappa_j, \omega_n)\Delta\kappa\,\Delta\omega]^{1/2}$$

$$\times \cos\left(\kappa_j x + \frac{\phi_{jn}^{(1)} + \phi_{jn}^{(2)}}{2}\right) \cos\left(\omega_n t + \frac{\phi_{jn}^{(1)} - \phi_{jn}^{(2)}}{2}\right) \quad (7.38)$$

which represents the superposition of standing waves, and reflects correctly the characteristics of the (direction-independent) lagged coherency. If the apparent propagation of the motions is included in the description of the random field, then it is reflected in the simulated motions. For example, for the Clough-Penzien spectrum, the coherency model of Luco and Wong and a unidirectional apparent propagation velocity c of the motions along the line connecting the stations, the F-K spectrum of the random field becomes [593]:

$$S(\kappa, \omega) = \frac{S_d(\omega)}{2\alpha|\omega|\sqrt{\pi}} \exp\left[-\frac{(\kappa + \omega/c)^2}{4\alpha^2\omega^2}\right] \quad (7.39)$$

The F-K spectrum of Eq. 7.39 is no longer quadrant-symmetric. Figure 7.9(d) presents the real part of the complex coherency consisting of the model of Luco and Wong with the low value for the coherency drop parameter ($\alpha = 2 \times 10^{-4}$ sec/m) and an apparent propagation velocity of $c = 500$ m/sec (Eqs. 2.73 and 2.74). Figure 7.10(d) presents displacement simulations based on this complex coherency, but for the same Clough-Penzien PSD and the same parameters for the simulations that were utilized in Figs. 7.10(a)–(c). It can be clearly observed from the figure that the seismic ground motions propagate along the x-direction with the specified apparent propagation velocity.

7.3   SIMULATION OF MULTI-VARIATE STOCHASTIC VECTOR PROCESSES

An alternative scheme to the F-K representation approach for the generation of spatially variable ground motions from random fields, presented in the previous section, is the use of multi-variate vector processes. These simulation schemes are, generally, based on the decomposition of the covariance or the cross spectral density matrix, through which the random field, viewed as a multi-variate process, is factorized into its subprocesses [414], [459] by either modal or Cholesky decomposition. Cholesky decomposition is more often utilized, because of its computational efficiency over modal decomposition[1]. The spectral decomposition approach, by eliminating the need for the evaluation of the F-K spectrum, becomes more versatile in engineering applications.

Hao *et al.* [200] first presented a spectral decomposition simulation scheme utilizing a sequential approach and random phase variability in the simulations, and generated spatially variable ground motions as non-stationary and response-spectrum-compatible, by adjusting, in an iterative process, the amplitudes of the simulations. Vanmarcke *et al.* [553] utilized covariance matrix decomposition in introducing the multivariate linear prediction method for conditionally simulating spatially variable ground motions (Section 8.1). Deodatis [127], based on an approach similar to the simulation scheme of Hao *et al.* [200], generated non-stationary, non-homogeneous and response-spectrum-compatible spatially variable seismic ground motions. Shrikhande and Gupta [471] presented a similar spectral decomposition scheme in a hybrid simulation/conditional simulation approach for the generation of spatially variable seismic ground motions that were consistent with the equivalent response spectrum of a recorded time history. Their approach also incorporated the duration and phase-difference spectra of the record in the simulation scheme, the latter by adding a random perturbation to the phase spectrum of the recorded time history. Hence, the resulting simulations exhibited amplitude and frequency content variation compatible with the characteristics of the record.

More recently, Katafygiotis *et al.* [259], [260] utilized the technique, which they termed Cholesky Decomposition with Random Amplitudes (CDRA), in a comparative evaluation of the characteristics of stationary, spatially variable ground motions resulting from four simulation methods. The first two techniques generated spatially variable time histories from modified F-K spectral representation schemes (Section 7.2.1). The modifications [258] entailed the explicit incorporation of the properties of homogeneity and partial isotropy (i.e., isotropy with respect to the space variables but not the time variable) in the random field that results from the use of the (direction-independent) lagged coherency in the simulation process. This consideration reduces significantly the computational effort in the generation of the time histories than when only the homogeneity of the random field is taken into account. It is noted that the wave passage effect cannot be used directly in these simulation schemes, because it would render the random field direction-dependent. However, it can be easily imposed, after the simulations have been generated, with appropriate delays in the arrival time of

---

[1] Cholesky decomposition requires that the matrix is Hermitian and positive definite. However, the cross spectral density matrix is Hermitian and positive semidefinite, i.e., its eigenvalues can be greater than or equal to zero, as, by definition, $0 \leq |\gamma(\xi, \omega)| \leq 1$ (Section 2.5). In simulations, ill-conditioning of the matrix may arise if, at low frequencies and/or short separation distances, the functional form of the selected coherency model yields values that tend to unity (i.e., $|\gamma(\xi, \omega)| \to 1$ as $\omega \to 0$ and/or $\xi \to 0$). In this case, the Cholesky decomposition of the matrix will lead to numerical problems.

the motions at the various locations on the ground surface. Katafygiotis *et al*. [258] termed the first of these two techniques as "Modified Rice," because its formulation considered random amplitudes and phases, as in Rice's [423] first representation approach (e.g., Eq. 7.1), and the second technique as "Modified Shinozuka," because its formulation considered random phases, as, e.g., in Eq. 7.31. The last two techniques compared in these studies [259], [260] generated spatially variable ground motions as vector processes: The third technique was the CDRA method, and the fourth the approach introduced by Hao *et al*. [200]. For each of the four simulation schemes, Katafygiotis *et al*. [259], [260] compared the behavior of the power and cross spectral densities and the Fourier amplitudes of the generated time series relative to the characteristics of the random field, and, also, examined the trend of the differences in the Fourier amplitudes and the Fourier phases of the generated motions between various locations. The first three techniques ("Modified Rice," "Modified Shinozuka" and CDRA) exhibited similar characteristics, whereas the stationary time series resulting from the approach of Hao *et al*. [200] displayed some interesting inherent features. The CDRA technique and the approach of Hao *et al*. [200] for the generation of stationary, spatially variable simulations are presented in the following section (Section 7.3.1) along with example applications that illustrate and compare their characteristics.

The comparisons in Section 7.3.1 consider stationary and homogeneous processes. This is consistent with the analysis of spatially recorded data at uniform site conditions, for which it is assumed that the strong motion window of the motions at the various stations is a segment of a stationary process (Section 2.2). As was the case for simulations of random processes (Section 7.1), however, the stationarity in the resulting time series is not a desirable feature. Spatially variable ground motions also need to be modified to exhibit non-stationarity. Additional desired features of simulated spatially variable seismic ground motions for engineering applications are that they are response-spectrum-compatible and, in cases, non-homogeneous. For example, the supports of an extended structure, such as a bridge, can be located on different ground types, i.e., rock outcrop for the abutments and sedimentary material for the central piers. In this case, the point estimates of the motions become functions of absolute location as well, and the multi-variate random process becomes non-homogeneous. The variations in the frequency content of the point estimates of the motions have been shown to have a significant effect on the seismic response of extended structures, as will be further elaborated in Section 9.4. For illustration purposes, non-stationary, non-homogeneous and response-spectrum-compatible spatially variable ground motions generated with the approach of Deodatis [127] are presented in Section 7.3.2. A brief discussion regarding assumptions and approximations made in the generation of spatially variable seismic ground motions concludes this chapter.

7.3.1 CHARACTERISTICS OF STATIONARY, SPATIALLY VARIABLE SIMULATIONS

This section describes briefly the CDRA technique [259], [260] and the part of the methodology introduced by Hao *et al*. [200] for the simulation of stationary, spatially variable time series. Example applications that illustrate the characteristics of the generated motions are also presented and discussed.

The Cholesky Decomposition with Random Amplitudes (CDRA) Technique

According to this technique, the random field at a point (location) x_i on the ground surface is simulated as a Fourier series with random space-dependent coefficients [259]:

$$f(x_i, t) = \sqrt{\Delta\omega} \sum_{k=1}^{K} \left[A_k^{(1)}(x_i) \cos(\omega_k t) + A_k^{(2)}(x_i) \sin(\omega_k t) \right] \qquad (7.40)$$

where, at each location x_i, $A_k^{(j)}(x_i)$, $k = 1, \ldots, K$, $j = 1, 2$, is a sequence of zero-mean, independent Fourier coefficients that follow a Gaussian distribution. For $i = 1$, i.e., when simulations are generated at a single location on the ground surface, Eq. 7.40 is, essentially, the first spectral representation scheme (Eq. 7.1) introduced by Rice [423], [424]. Let

$$\mathbf{A}_k^{(j)} = \left\{ A_k^{(j)}(x_1), \ldots, A_k^{(j)}(x_M) \right\}^T, \quad k = 1, \ldots, K, \quad j = 1, 2 \qquad (7.41)$$

define the vector of the random Fourier coefficients $A_k^{(1)}(x_i)$ and $A_k^{(2)}(x_i)$, $i = 1, \ldots, M$, to be generated at M locations on the ground surface. The elements of the covariance matrix, $\text{cov}(\mathbf{A}_k^{(j)})$, of the random vector $\mathbf{A}_k^{(j)}$ can be evaluated from the statistical relationships between the Fourier coefficients as [259]:

$$\left[\text{cov}(\mathbf{A}_k^{(j)}) \right]_{lm} = \text{E}\left[A_k^{(j)}(x_l) A_k^{(j)}(x_m) \right] = 2|\gamma(\xi_{lm}, \omega_k)| S(\omega_k) \qquad (7.42)$$

where E denotes expectation, ξ_{lm} is the distance between the two points (x_l and x_m), $|\gamma(\xi_{lm}, \omega_k)|$ is the lagged coherency, and $S(\omega_k)$ is the power spectral density of accelerations, velocities or displacements. It is noted that, by considering the lagged coherency in the evaluation of the covariance matrix in Eq. 7.42, the generated motions do not incorporate the wave passage effect (Eqs. 2.73 and 2.74). As indicated earlier, the wave passage effect can be included in the simulations, after they have been generated, with appropriate time delays in the arrival time of the motions at the various stations.

Let $\text{cov}(\mathbf{A}_k^{(j)}) = \mathbf{C}_k^T \mathbf{C}_k$ be the Cholesky decomposition of the covariance matrix of the random vector $\mathbf{A}_k^{(j)}$. $\mathbf{A}_k^{(j)}$ can then be simulated as:

$$\mathbf{A}_k^{(j)} = \mathbf{C}_k \mathbf{H}_j \qquad (7.43)$$

where the random vector $\mathbf{H}_j$ consists of M independent, normally distributed components. The generated $\mathbf{A}_k^{(j)}$ are then substituted into Eq. 7.40 to yield the simulated time series. The FFT can also be readily applied in the simulation process [258], [260].

As was the case for the simulations of random processes (Section 7.1.1), Eq. 7.40 is valid if there exists an upper cut-off frequency, ω_u, above which the contribution of the power spectral density to the simulations is negligible. If the centroidal rule is utilized for the discretization of the frequency domain (Sections 7.1.1 and 7.2.1), the following characteristics are inherent in the simulations [260]: (i) they are zero-mean Gaussian random processes; (ii) the period of the generated time histories is equal to $T_0 = 4\pi/\Delta\omega$; (iii) the simulations are ergodic of the first order, but asymptotically ergodic of the second order; and (iv) as $K \to \infty$, the ensemble autocovariance and cross covariance function of the simulations become identical to those of the random field. It is noted, as will also be elaborated upon later on, that the simulated motions at each station i are sums of trigonometric functions with random amplitudes and random phases. The random amplitudes at each frequency ω_k, $k = 1, \ldots, K$, are given by:

$$\Lambda_i(\omega_k) = \sqrt{\Delta\omega\left(\left[A_k^{(1)}(x_i)\right]^2 + \left[A_k^{(2)}(x_i)\right]^2\right)} \qquad (7.44)$$

To illustrate the stochastic characteristics of stationary simulations (accelerations) generated by the CDRA method, the Clough-Penzien spectrum (Eq. 7.10) and the coherency expression of Luco and Wong (Eq. 7.36) are utilized. The parameters of the Clough-Penzien spectrum are $S_\circ = 1$ cm^2/sec^3, $\omega_g = 5\pi$ rad/sec, $\omega_f = 0.5\pi$ rad/sec and $\zeta_g = \zeta_f = 0.6$, i.e., the same firm soil conditions as in the previous section, and the coherency drop parameter of the model of Luco and Wong is $\alpha = 2.5 \times 10^{-4}$ sec/m. One hundred simulations are generated at a sequence of three equidistant stations along a straight line with separation distance between consecutive stations of 500 m; this long distance is purposely selected so that the effect of the loss of coherency on the cross spectral density of the motions can be clearly identified. Figure 7.11 presents the ensemble averages of the unsmoothed power spectral densities, cross spectral densities, and Fourier amplitudes (Eq. 7.44) of the 100 realizations along with the target values. Specifically, Fig. 7.11(a) presents the PSD of the motions at the first two stations and Fig. 7.11(b) the cross spectral density between the motions at stations 1 and 2, and stations 2 and 3. The figures indicate that the CDRA technique reproduces well the prescribed spectral and cross spectral density functions as random functions with mean value equal to the corresponding target spectra and non-zero variance over the entire frequency range. Figure 7.11(c) presents the mean values of the amplitudes of the 100 realizations at the first two stations as generated by the CDRA method; the target amplitude of the Clough-Penzien spectrum, evaluated as $\Lambda(\omega_k) = 2\sqrt{S(\omega_k)\Delta\omega}$, is also presented in the figure. The ensemble averages of the amplitudes of the simulated motions behave as random variables, with mean value lower than the target amplitude, which follows from the χ^2 distribution with two degrees of freedom of the sum of squares of the two normal variates $A_k^{(1)}(x_i)$ and $A_k^{(2)}(x_i)$. It can be shown [301] that the mean value of the simulated Fourier amplitudes (Eq. 7.44) is approximately 88.6% of the target value at each frequency, as is also illustrated in Fig. 7.11(c). Figure 7.12 presents the ensemble lagged coherency of 100 simulations based on the CDRA method between stations 1 and 2, and stations 2 and 3 together with the target coherency (Eq. 7.36 for $\xi = 500$ m). For the evaluation of the lagged coherency of the simulations in Fig. 7.12, the power and cross spectral densities were smoothed with an 11-point Hamming spectral window (Eq. 2.47). It can be seen from the figure that the lagged coherency of the simulations follows closely the trend of the target coherency over the entire frequency range. Figures 7.11 and 7.12 then suggest that the CDRA technique produces time histories with characteristics that agree with the characteristics of the target random field and with similar attributes between stations, i.e., amplitudes that behave as random variables at all stations and at all frequencies. Parenthetically, it is noted that the smaller (larger) the number of simulations used in the evaluation of the ensemble averages, the larger (smaller) the variance of the average of the simulations relative to the characteristics of the target random field.

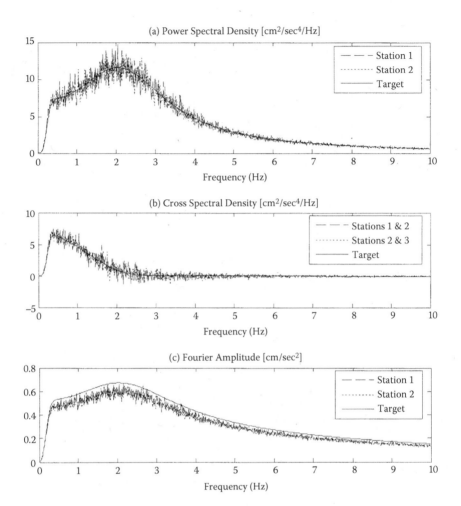

FIGURE 7.11 Comparison of the ensemble averages of the power and cross spectral densities, in parts (a) and (b), respectively, and the Fourier amplitudes, in part (c), of 100 simulations generated by the CDRA method [259], [260] with the target values. The distance between consecutive stations along a straight line is 500 m, the power spectral density is the Clough-Penzien spectrum for firm soil conditions illustrated in Fig. 7.1(a), and the coherency model is that of Luco and Wong with $\alpha = 2.5 \times 10^{-4}$ sec/m.

Katafygiotis *et al.* [259], [260] further examined the behavior of the normalized differential amplitudes and the differential phases of the stationary time series generated by the four simulation schemes described earlier herein. For this purpose, they introduced the "Difference Indexes" for the Fourier amplitudes and the Fourier phases of the simulated motions. The Difference Index (DI) for the Fourier amplitudes, $\Lambda_i(\omega_k)$ and $\Lambda_j(\omega_k)$, and the Fourier phases, $\phi_i(\omega_k)$ and $\phi_j(\omega_k)$, at two stations i and j and frequency ω_k were defined as:

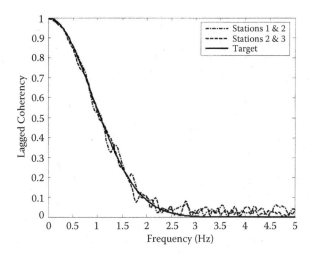

FIGURE 7.12 Comparison of the ensemble averages of the smoothed lagged coherency of 100 simulations generated by the CDRA method [259], [260] between stations 1 and 2, and stations 2 and 3, at a separation distance of $\xi = 500$ m, with the target values. The characteristics of the random field are the same as those in Fig. 7.11; for the evaluation of the lagged coherency of the simulations, the power and cross spectral densities were smoothed with an 11-point Hamming window (Fig. 2.4).

$$DI[\Lambda_i(\omega_k), \Lambda_j(\omega_k)] = \frac{\sqrt{E[\{\Lambda_i(\omega_k) - \Lambda_j(\omega_k)\}^2]}}{2\sqrt{S(\omega_k)\Delta\omega}} \qquad (7.45)$$

$$DI[\phi_i(\omega_k), \phi_j(\omega_k)] = \sqrt{E[\{\phi_i(\omega_k) - \phi_j(\omega_k)\}^2]} \qquad (7.46)$$

in which the denominator in the amplitude DI is evaluated from the target power spectral density. Considering that the Fourier amplitudes have the same mean value (88.6% of the target value [301]) and that the Fourier phases can be regarded as uniformly distributed in the range of $[0, 2\pi)$ [569], Eq. 7.45 reflects the normalized standard deviation of the differences of the Fourier amplitudes of the simulated motions at each frequency and Eq. 7.46 the standard deviation of the differences of the Fourier phases. It is noted that the DIs of Eqs. 7.45 and 7.46 are conceptually consistent with the normalized differential amplitudes and differential phases introduced for spatially recorded data in Eqs. 4.19 and 4.20, respectively, except that, for the recorded motions, the normalized differential amplitudes and differential phases were defined with respect to the common, coherent component in the data [604], [608].

Figure 7.13 presents the amplitude DI and phase DI (in parts (a) and (b), respectively) of the 100 simulations generated by the CDRA technique between stations 1 and 2, and stations 2 and 3. In the low frequency range, the DIs for both amplitudes and phases are small and gradually increase with frequency in a consistent manner. When the frequency exceeds approximately 2.0 Hz, at which value the coherency becomes low and starts tending to zero (Fig. 7.12), the DI values fluctuate around 0.66 for the amplitude DI and 2.57 rad for the phase DI, and remain fairly constant thereafter. Liao and Zerva [301] showed, from the statistical properties of the amplitudes and

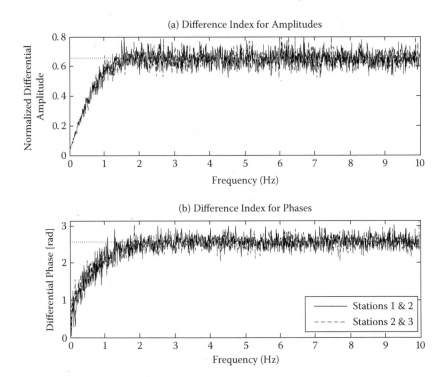

FIGURE 7.13 Comparison of the amplitude and phase difference indexes (DI in Eqs. 7.45 and 7.46) in parts (a) and (b), respectively, of 100 simulations generated by the CDRA method [259], [260] for the same characteristics of the random field as in Fig. 7.11. The DIs are shown for stations 1 and 2 at a separation distance of 500 m, and stations 2 and 3, also at a separation distance of 500 m. The dotted lines in the subplots indicate the expected values, to which the DIs tend to in the high frequency range, where the lagged coherency tends to zero.

phases of time series generated from expressions like Eq. 7.40, that the values of 0.66 and 2.57 rad (dotted lines in Fig. 7.13) are the expected values of the DIs in the high frequency range, where the coherency between the motions at the stations tends to zero, i.e., the motions become uncorrelated. It follows then that the amplitude and phase DIs of simulations based on the CDRA method will tend to these same constant values for any power spectral density as long as the value of the coherency for the separation distance between the two stations tends to zero.

Interestingly, the trend of the DIs for amplitudes and phases in Fig. 7.13 is compatible with the behavior of the normalized differential amplitudes and differential phases derived from recorded data[2] (Fig. 4.10): In the low frequency range, the

---

[2] It is noted that the normalized differential amplitudes and the differential phases derived from recorded data in Fig. 4.10 are differences (Eqs. 4.19 and 4.20), and, hence, assume positive and negative values contained within envelope functions. On the other hand, the DIs for amplitudes and phases in Fig. 7.13 are square roots of sums of squares (Eqs. 7.45 and 7.46), and, hence, assume only positive values exhibiting variability around their trend.

envelope functions for both the normalized differential amplitudes and differential phases derived from the recorded data increase with frequency (Fig. 4.10), as do the trends of the DIs for both amplitudes and phases (Fig. 7.13). As frequency increases within the dominant frequency range of the motions, the envelope functions for the normalized differential amplitudes and the differential phases of the recorded data start tapering off to constant values (Fig. 4.10), as do the DIs for the amplitudes and the phases of the simulations (Fig. 7.13) in the frequency range where the cross spectral density of the random field starts decreasing (Fig. 7.11(b)). At higher frequencies, the differential phases of the recorded data vary randomly between $[-\pi, +\pi)$ (bottom subplots of Fig. 4.10), and the DIs of the Fourier phases of the simulations fluctuate around the value of 2.57 rad (Fig. 7.13(b)). In this frequency range, the amplitude DIs also fluctuate around the constant value of 0.66, whereas the normalized differential amplitudes of the recorded data assume large values caused by the low amplitude of the common component in the frequency range where scattered energy dominates the motions (top subplots of Fig. 4.10 and Section 4.2.7). However, the behavior of the normalized standard deviation of the differences of the Fourier amplitudes of the simulations in Fig. 7.13(a) is consistent with the observed behavior of the standard deviation of the differences of the logarithms of the Fourier amplitudes of the spatially recorded data observed by Abrahamson et al. [14], which was illustrated in Fig. 3.19(b). Hence, the trend of the amplitude and phase variability of the stationary, spatially variable time series generated by the CDRA technique conforms with the behavior of spatially recorded seismic data.

Another interesting observation that can be deduced from Fig. 7.13 is that, although coherency describes the phase variability in the data (Section 2.4.2), the generated spatially variable ground motions exhibit both amplitude and phase variability. This can be attributed to the fact that, in the simulation process, the random amplitudes at the various stations are constrained to follow the cross spectral density of the motions (Fig. 7.11(b)), and, thus, indirectly, also obey the target coherency model. Hence, the CDRA technique simulates spatially variable time series, which, without a prescribed amplitude variation pattern, still exhibit amplitude variability with a trend similar to that of recorded data. The sequential approach of Hao et al. [200], on the other hand, leads to simulations with different properties, as illustrated in the following.

The Hao, Oliveira and Penzien [200] Method

Hao et al. [200], in their comprehensive article on the analysis of data at the SMART-1 array, parts of which were described in Sections 3.1 and 3.4.1, also proposed an approach for the generation of spatially variable, non-stationary and response-spectrum-compatible seismic time series. The stationary time series based on their approach are simulated according to the following expression:

$$f(x_i, t) = 2\sqrt{\Delta\omega} \sum_{m=1}^{i} \sum_{k=1}^{K} A_{im}(\omega_k)\cos(\omega_k t + \phi_{mk}); \quad i = 1, \ldots, M \quad (7.47)$$

where $A_{im}(\omega_k)$ are the amplitude coefficients of the sinusoids and ϕ_{mk} are the phases, which are independent random variables uniformly distributed in the range $[0, 2\pi)$.

The amplitudes at each frequency ω_k are evaluated, as for the CDRA technique, by Cholesky decomposition of the cross spectral density matrix, $\mathbf{S}(\omega_k)$ of Eq. 2.103, with elements, in this illustration, provided by Eq. 7.42.

Let $\mathbf{S}(\omega_k) = 2S(\omega_k)\mathbf{L}(\omega_k)\mathbf{L}^T(\omega_k)$, where $\mathbf{L}(\omega_k)$ is the lower triangular matrix, denote the Cholesky decomposition of the cross spectral density matrix. The amplitude coefficients of the cosine functions in Eq. 7.47 become then $A_{im}(\omega_k) = l_{im}(\omega_k)\sqrt{S(\omega_k)}$ with $l_{im}(\omega_k)$ being the elements of $\mathbf{L}(\omega_k)$. The technique (Eq. 7.47) simulates spatially variable seismic ground motions in sequence, rather than in parallel as the other techniques. This simulation scheme generates the time history at the first station with a single cosine function at each frequency, and adds an additional cosine term for each subsequent station. The phases at all locations vary randomly, but the amplitudes at each station are evaluated in a deterministic manner to ensure that the currently generated time history satisfies the necessary coherency relations with all previously generated ones. The process results in "semi-deterministic" power spectral densities [260], meaning that, for some frequency range, it produces random spectral values with mean value equal to the target spectra, but, for the remaining frequencies, the spectra are deterministic (zero variance) and equal to the target values.

To illustrate this point, Figs. 7.14(a) and (b) present, respectively, the ensemble averages of the PSDs of 100 simulations generated with the approach of Hao *et al.* [200] at station 1, i.e., the first simulation in the sequence with $i = 1$ in Eq. 7.47, and station 2, i.e., the second simulation in the sequence with $i = 2$ in Eq. 7.47, along with the target PSD. The characteristics of the random field used in this example are the same as those of the random field used for the simulations generated by the CDRA technique in Figs. 7.11–7.13. The PSD of each simulation at station 1 (Fig. 7.14(a)) is fully deterministic and coincides with the target PSD, whereas the PSDs at station 2 (Fig. 7.14(b)) behave as random variables only over a short frequency range (less than approximately 2 Hz) and become deterministic and identical with the target PSD thereafter. The behavior of the PSDs in Figs. 7.14(a) and (b) is directly related to the variability in the amplitudes of the simulations generated by this technique (Eq. 7.47), which, for the motions at stations 1 and 2, are presented in Figs. 7.14(c) and (d), respectively: The time series at station 1 consists of a single cosine term at each frequency ω_k with amplitude coefficient $A_{11}(\omega_k) = l_{11}(\omega_k)\sqrt{S(\omega_k)}$, where $l_{11}(\omega_k) = \gamma_{11}(\omega_k) = 1$. Hence, the Fourier amplitudes at this station are fully deterministic, coinciding with the target amplitude ($\Lambda(\omega_k) = 2\sqrt{S(\omega_k)\Delta\omega}$), and, consequently, the PSD of the motions at this station is also fully deterministic and coincides with the target PSD (Fig. 7.14(a)). The time series at station 2 consists of two cosine terms at each frequency ω_k with amplitude coefficients $A_{21}(\omega_k) = l_{21}(\omega_k)\sqrt{S(\omega_k)}$, and $A_{22}(\omega_k) = l_{22}(\omega_k)\sqrt{S(\omega_k)}$; in these expressions, $l_{21}(\omega_k) = \gamma_{21}(\omega_k)$ and $l_{22}(\omega_k) = \sqrt{1 - \gamma_{21}^2(\omega_k)}$. When the coherency between the two stations starts tending to zero, i.e., $\gamma_{21}(\omega_k) \to 0$, depending on the frequency, the station separation distance and the selected coherency model, the amplitude coefficient of the first cosine term, $A_{21}(\omega_k)$, also tends to zero, and the series is, again, comprised of a single term with amplitude coefficient $A_{22}(\omega_k) = \sqrt{S(\omega_k)}$. Hence, in this frequency range, the amplitude as well as the PSD of the motions are deterministic, whereas, in the remaining frequency range, they are random numbers with non-zero variance (Figs. 7.14(b) and (d)). A similar rationale can be used to explain the behavior of the amplitudes and PSDs of the

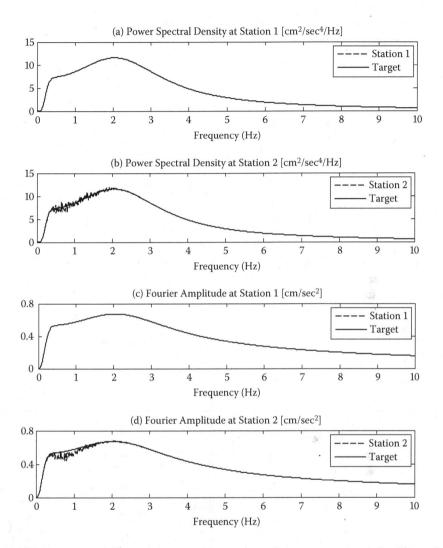

FIGURE 7.14 Comparison of the ensemble averages of the power spectral densities and Fourier amplitudes of 100 simulations generated by means of the approach introduced by Hao *et al.* [200]: Part (a) presents the comparison of the PSD at station 1 (first simulation in the sequence) with the target PSD; the results are identical. Part (b) illustrates the comparison of the PSD at station 2 (second simulation in the sequence) with the target PSD; the ensemble PSD of the simulations behaves as a random variable with mean equal to the target value and non-zero variance only over a short frequency range. Part (c) illustrates the ensemble average of the Fourier amplitudes of 100 simulations at station 1 along with the target values; again, as in subplot (a), the results are identical. Part (d) presents the ensemble average of the Fourier amplitudes of 100 simulations at station 2 along with the target values; the Fourier amplitudes behave as random variables with mean values lower than the target values and non-zero variance during a short frequency range, and, as in subplot (b), they coincide with the target values for the remaining frequencies. The characteristics of the random field are the same as those in Fig. 7.11.

motions generated by this technique at subsequent stations. In this sense, simulations based on Eq. 7.47 cannot be considered representative of a homogeneous random field, as the characteristics of the simulations at the various locations differ. It should be noted that the lack of homogeneity in these simulations is to be distinguished from the simulation of non-homogeneous random vector processes (Section 7.3.2), where the properties of the random field are deliberately selected to reflect variations in the power spectral densities of the motions at different locations.

Because of its inherent property, the Hao *et al.* [200] method also results in zero amplitude DIs in the high frequency range where coherency tends to zero. This is illustrated in Fig. 7.15(a), which presents the amplitude DIs of the 100 simulations between stations 1 and 2, and between stations 2 and 3. The amplitude DIs increase at low frequencies, but, depending on the order in which the motions are simulated in the sequence (Eq. 7.47), they start decreasing to zero values as frequency increases further. On the other hand, the phase DIs generated by this technique, which are shown in Fig. 7.15(b), follow the pattern of Fig. 7.13(b) [260]. The ergodicity properties of

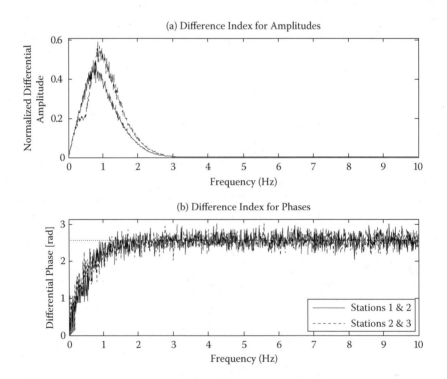

FIGURE 7.15 Comparison of the amplitude and phase difference indexes (DI in Eqs. 7.45 and 7.46) in parts (a) and (b), respectively, of 100 simulations generated by the the approach introduced by Hao *et al.* [200] for the same characteristics of the random field as in Fig. 7.11. The DIs are shown for stations 1 and 2 at a separation distance of 500 m, and stations 2 and 3, also at a separation distance of 500 m. The dotted line in subplot (b) indicates the expected value, to which the phase DI tends to in the high frequency range, where the lagged coherency tends to zero.

the time series as simulated by the Hao *et al.* [200] approach also differ: Whereas the simulation at the first location is ergodic over its period, the simulations at the subsequent locations are not. A way to remedy this difference in the characteristics of the simulations at the various stations is by utilizing a station-dependent frequency discretization, which amounts to double-indexing of the frequencies and generates time series that are ergodic over the longest period that results from the discretization [128].

The simulation schemes described in this section introduce some form of random amplitude variability in the generated motions. Some techniques, as, e.g., the CDRA approach, also generate stationary, spatially variable time series, that exhibit amplitude and phase variability consistent with the trend of recorded data (Figs. 3.19(b), 4.10 and 7.13). From a different perspective, Abrahamson [6], [7] proposed a simulation scheme that generates the Fourier phases of the spatially variable ground motions, but keeps their Fourier amplitudes constant. The approach is based on the relationship between the lagged coherency and the fraction, $\beta_{jk}(\omega)$, of random variability in the Fourier phase differences of the motions at two stations (Eqs. 2.83 and 2.84 and Fig. 2.22). The Fourier phase differences are obtained from the optimization of a penalty function, that minimizes the variance between the sample and target probability density functions of the random phase differences, and, also, penalizes the solution, if the value of the phase difference is outside the range of $[-\pi, +\pi)$ (from the product $\pi \epsilon_{jk}(\omega)$ in Eq. 2.83). The generated Fourier phase angles at each location and frequency are then combined with the predefined Fourier amplitudes, and the spatially variable time series are computed through the inverse Fourier transform. Hence, the Fourier amplitudes of the generated motions at all stations are identical and the Fourier phases conform with the prescribed coherency model.

7.3.2 NON-STATIONARY, NON-HOMOGENEOUS, RESPONSE-SPECTRUM-COMPATIBLE SIMULATIONS

Simulations in engineering applications require that the generated spatially variable seismic ground motions are non-stationary, preferably response-spectrum-compatible, and, in cases, non-homogeneous. An approach incorporating these features was presented by Deodatis [127] and further extended by Saxena *et al.* [130], [442], [443]. Their approach, which, again, is based on the spectral representation technique, is briefly presented in the following and illustrated with an example.

Consider that spatially variable seismic ground motions are to be generated at M locations on the ground surface as non-stationary random processes obeying a coherency model and power spectral densities, which differ at the various locations and are compatible with prescribed response spectra. Initially, the motions are generated as stationary based on the cross spectral density matrix, $\mathbf{S}^0(\omega)$, the elements of which are given by:

$$[\mathbf{S}^0(\omega_k)]_{lm} = \gamma(\xi_{lm}, \omega_k)\sqrt{S_{ll}(\omega_k)S_{mm}(\omega_k)} \tag{7.48}$$

The power spectral densities, $S_{ll}(\omega_k)$ and $S_{mm}(\omega_k)$, can differ at the two stations l and m, and the spatial variability term, $\gamma(\xi_{lm}, \omega_k)$, incorporates the loss of coherency and the propagation of the motions on the ground surface, i.e., the off-diagonal terms of the cross spectral density matrix in Eq. 7.48 become now complex numbers. This approach also utilizes Cholesky decomposition, i.e., $\mathbf{S}^0(\omega_k) = \mathbf{C}_k^T \mathbf{C}_k$, the elements of which are given by:

$$C_{lm}(\omega_k) = |C_{lm}(\omega_k)| \exp[i \vartheta_{lm}(\omega_k)] \tag{7.49}$$

with

$$\vartheta_{lm}(\omega_k) = \tan^{-1} \left(\frac{\Im[C_{lm}(\omega_k)]}{\Re[C_{lm}(\omega_k)]} \right) \tag{7.50}$$

and $\Im[.]$ denotes the imaginary part. The time histories are then generated according to the following expression [127]:

$$f_l(t) = 2\sqrt{\Delta\omega} \sum_{m=1}^{l} \sum_{k=1}^{N} |C_{lm}(\omega_k)| \cos[\omega_k t + \vartheta_{lm}(\omega_k) + \phi_{mk}]; \ l = 1, \dots, M \tag{7.51}$$

The deterministic phases, $\vartheta_{lm}(\omega_k)$, reflecting the wave passage effect (Eq. 7.50), can be clearly distinguished from the random phase angles, ϕ_{mk}, which follow a uniform distribution between $[0, 2\pi)$. The frequency domain discretization can be performed as, e.g., in Section 7.1.1, and the FFT approach readily incorporated in the process. Conceptually, Eq. 7.51 is an extension of the classical representation method with random phases [462], and follows the sequential simulation pattern of the approach of Hao *et al.* [200].

Once the spatially variable ground motions are generated as stationary, with assumed functional forms for the PSDs, they are multiplied with an intensity modulating function (e.g., Eqs. 7.14 or 7.15), so that non-stationarity is imposed. The intensity modulating function may be the same or differ at the various locations on the ground surface depending on the requirements of the particular problem [127]. The compatibility of the generated, non-stationary motions with prescribed response spectra is then achieved by means of an iterative process [127], [130], [443], in which the power spectral densities of the motions at each location are upgraded according to the following expression:

$$S_l^{i+1}(\omega_k) = S_l^i(\omega_k) \left[\frac{RS_l(\omega_k)}{RS_{sl}^i(\omega_k)} \right]^2 \tag{7.52}$$

In the above equation, $S_l^i(\omega_k)$ and $S_l^{i+1}(\omega_k)$ are the power spectral densities of the (stationary) ground motions at the l-th station for the i-th and $(i + 1)$-th iterations, respectively, $RS_l(\omega_k)$ is the target response spectrum at station l, and $RS_{sl}^i(\omega_k)$ is the response spectrum of the simulated motion at the l-th station for the i-th iteration. The new power spectral density, $S_l^{i+1}(\omega_k)$, is then used in the simulation of the random field, and the iterations are repeated until satisfactory agreement between the target response spectrum and the response spectrum resulting from the simulations is achieved.

To illustrate the approach, the following example is presented: Seismic ground motions are simulated at four stations along a straight line with separation distance between consecutive stations of 100 m. Stations 1 and 4 are located on stiff soil conditions (described by S_d in UBC 1997 [234]) and stations 2 and 3 on soft soil conditions (described by S_e in UBC 1997 [234]); for demonstration purposes, the target response spectra at all stations are scaled to $0.5g$. The coherency model utilized is the model of Harichandran and Vanmarcke [208] (Eq. 7.34) with parameters, as before in this chapter, $A = 0.736$, $a = 0.147$, $k = 5210$ m, $f_o = 1.09$ Hz and $b = 2.78$. The intensity modulation of the stationary time series is performed with the piecewise-continuous function described in Eq. 7.15, the parameters of which were selected as $t_2 - t_1 = 7$ sec, $I_0 = 1$ and $b = 0.4$/sec. No propagation effects are considered in this illustration. Figure 7.16 presents a set of simulated acceleration time histories at the four locations. It can be clearly seen from the figure that the frequency content of the motions at stations 1 and 4 (stiff soil conditions) is similar, as is the frequency content of the motions at stations 2 and 3 (soft soil conditions), but there are significant differences in the frequency content of the motions at different soil conditions. Differences can also be observed in the waveforms at stations located on the same soil conditions (stations 1 and 4, and stations 2 and 3), which can be attributed to the loss of coherency of the motions.

The comparison of the ensemble characteristics of 100 simulations with the target response spectra and the lagged coherency are presented in Figs. 7.17 and 7.18, respectively. Specifically, Fig. 7.17(a) presents the ensemble average of the response spectra of the 100 simulations at station 1 and station 4, located on stiff soil conditions, with the target UBC spectra at the corresponding soil conditions, and Fig. 7.17(b) the ensemble average of the response spectra of the 100 simulations at station 2 and station 3, located on soft soil conditions, with the target UBC spectra at soft soil conditions. It can be observed from the figures that, even though the average response spectra of the simulations are higher than the target spectra at short periods (high frequencies) and lower at long periods (low frequencies), the average characteristics of the simulations are in good agreement with the response spectra of the target field. It should be noted that, as indicated in Section 7.1.4, the task of converting a response spectrum to a power spectral density carries significant problems, including the fact that one frequency component of the time history can affect the value of the response spectrum at other frequencies. Convergence criteria for iterative conversion schemes, such as the approach presented herein, cannot be readily established, but, instead, the iterations are stopped when the shape of the target response spectrum is captured by the response spectrum of the simulations (Fig. 7.17). Generally, a few iterations are required to achieve this objective. Deodatis [127] and Saxena [442] stated that, if a reasonable match between the target response spectrum and the response spectrum of the simulations is not achieved within the first few iterations, the simulated time history should be discarded and a new one generated.

Figure 7.18(a) presents the ensemble average of the lagged coherency of the 100 simulations between the stations located on stiff soil conditions (stations 1 and 4) at a separation distance of 300 m along with the target coherency for the same separation distance, and Fig. 7.18(b) presents the corresponding results for stations 2 and 3, located on soft soil conditions at a separation distance of 100 m. The lagged coherency was obtained using the entire simulated time series, and the power and cross spectral densities of the simulations were smoothed with an 11-point Hamming window. The ensemble lagged coherency of the simulations follows the trend of the decay of the

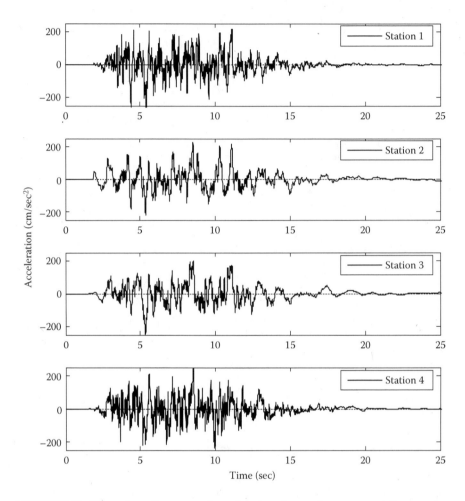

FIGURE 7.16 Illustration of one set of acceleration time series, generated from a non-stationary, non-homogeneous and response-spectrum-compatible random field at four equidistant stations along a straight line with separation distance between consecutive stations of 100 m. Stations 1 and 4 are located on UBC stiff soil conditions and stations 2 and 3 on UBC soft soil conditions. All response spectra are normalized to 0.5 g. The coherency model is that of Harichandran and Vanmarcke [208] based on the data of Event 20 at the SMART-1 array. The simulations are generated as random vector processes based on the approach presented by Deodatis [127].

coherency of the target random field, but it is lower than the target coherency at low frequencies for both soil conditions (Fig. 7.18). Clearly, the lagged coherency estimates from the stationary simulations in Fig. 7.12 are in better agreement with the target values than the lagged coherency estimates from the non-stationary and response-spectrum-compatible simulations in Fig. 7.18. This should be expected: The lagged coherency is evaluated from recorded data considering that the strong motion window used in the analysis is a segment of a stationary process (Chapter 2). Such lagged

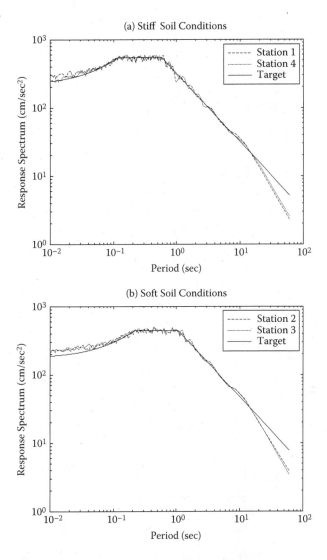

FIGURE 7.17 Comparison of the ensemble averages of the response spectra of 100 simulations, generated from the same random field characteristics as the single simulation set
presented in Fig. 7.16, with the target response spectra. Part (a) presents the comparison of the
response spectra of the simulations at stations 1 and 4, located on stiff soil conditions, along
with the target UBC response spectrum, and part (b) the corresponding results at stations 2 and
3 together with the target UBC response spectrum at soft soil conditions.

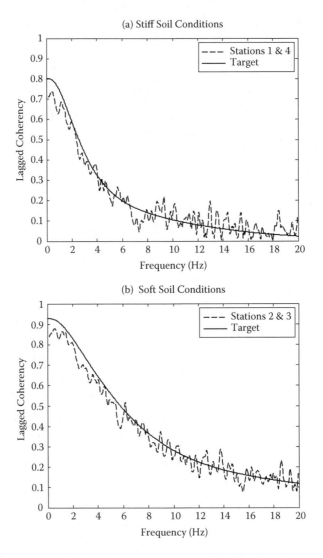

FIGURE 7.18 Comparison of the ensemble average of the lagged coherency estimates of 100 simulations, generated for the same random field characteristics as the single simulation set presented in Fig. 7.16, with the target coherency values. Part (a) presents the comparison of the lagged coherency of the simulations at the stiff soil conditions (stations 1 and 4) at a separation distance of 300 m. Part (b) presents the comparison of the lagged coherency of the simulations at the soft soil conditions (stations 2 and 3) at a separation distance of 100 m. For the evaluation of the lagged coherency of the simulations, the power and cross spectral densities were smoothed with an 11-point Hamming window (Fig. 2.4).

coherency models are used in the simulations generated initially as stationary, and, hence, the lagged coherency of the stationary simulations and the target coherency are in close agreement (Fig. 7.12). The stationary time histories are then modified to become non-stationary and iteratively altered to become compatible with the target response spectra. However, the multiplication of the stationary simulations with the intensity modulating function (Eq. 7.16 and Fig. 7.4) also affects the phases of the resulting non-stationary time series, and, hence, part of the differences between the lagged coherency of the non-stationary simulations and the target values in Fig. 7.18 may be attributed to this effect. One may argue that only the stationary part of the non-stationary simulations should be utilized in the comparisons of the lagged coherency, but, generally, the entire (non-stationary) simulated time series are used in such evaluations (e.g., [7], [443]), as was also the case in Fig. 7.18.

A problem that arises in the simulation of non-homogeneous random fields is the appropriate modeling of the amplitude and the phase variability of the motions, which are represented, respectively, by their power spectral densities (or response spectra) and their spatial variability (loss of coherency and wave passage). Physically, a non-homogeneous random field approximates the seismic motions at sites with lateral variations in their subsurface topography. As in the example application illustrated herein, it is, generally, assumed that a single coherency model, evaluated from the data at a uniform site (Section 3.4), can describe the coherency of the motions between pairs of stations located either both on stiff soil conditions, or both on soft soil conditions, or one on stiff and one on soft soil conditions. This assumption is contradicted by observations from recorded data, which suggest that: (i) The exponential decay of the lagged coherency at (uniform) "rock" sites differs from the exponential decay of the lagged coherency at (uniform) "soil" sites (Section 3.3.4 and Figs. 3.9 and 3.10); and (ii) the lagged coherency of motions between pairs of stations located one on rock and one on the softer soil deposits of a basin does not follow the trend of the lagged coherency of the motions recorded in the basin (Section 3.3.5 and Figs. 3.12–3.14). The modeling of non-homogeneous random fields further assumes that the wave passage effect at nonuniform sites can be approximated by a unidirectional wave propagating with constant velocity on the ground surface, an assumption that is valid for uniform sites, but not realistic for sites with irregular subsurface topography. Der Kiureghian [135] also suggested that an additional time delay, caused by the 1-D, vertical propagation of the waves through soil columns representing the softer sites, be incorporated in the simulations (Section 3.4.2). The same assumption, i.e., 1-D, vertical S-wave propagation through different soil columns, underlies, basically, the use of different power spectral densities for the modeling of the amplitude of the motions at these sites. Such 1-D approximations, however, cannot capture the complex 2-D and 3-D interference of the waveforms that occurs at sites with irregular subsurface topography, including basin-edge effects and generation of surface waves in the basins (Section 4.1.5). Hence, the modeling of non-homogeneous random fields as illustrated in this section and is commonly utilized in engineering applications (e.g., Section 9.4) may not suffice for the description of the amplitude and phase variation of seismic motions at sites with variable ground characteristics. These issues will be further discussed in Section 9.5.

8 Conditionally Simulated Ground Motions

As illustrated in the previous chapter, the most common approach for the generation of spatially variable ground motions is the (unconditional) simulation of random fields described by a power spectral density function and a spatial variability model. The more sophisticated schemes generate the simulated time histories from a stationary random field, then apply an intensity (envelope) modulation function to the time series to impose non-stationarity, and, finally, iteratively modify the modulated time histories to make them compatible with prescribed response spectra (e.g., Section 7.3.2). Even though such simulated spatially variable ground motions provide a most valuable tool for the evaluation of the seismic response of lifelines, their drawback is that they, generally, bear limited association with actual seismic records. On the other hand, the conditional simulation of random fields, described in this chapter, permits the use of one or more predefined time histories, and generates spatially variable ground motions compatible with these predefined time histories and the prescribed spatial variability model. The predefined time histories can be recorded data or synthetic ground motions. In this way, the generated time series exhibit spatial variability features, but, also, inherit the physical characteristics of the prescribed ground motions, such as non-stationarity in intensity and frequency content, effects of earthquake magnitude, source-site distance, local soil conditions, etc. As such, conditionally simulated spatially variable seismic ground motions can be readily incorporated into the performance-based design framework [301].

The purpose of interpolation and conditional simulation of spatially variable ground motions is similar: Given a set of seismic time histories recorded at an array of sensors, evaluate the time histories at target locations where data are not available. The simplest approach is deterministic interpolation: For example, Thráinsson et al. [521] interpolated seismic time histories by considering that the Fourier transform of the motion at a target location for each discrete frequency can be expressed as the weighted average of the Fourier transforms of the data recorded at a set of stations. The weights in their approach were the inverse squared distance of the target location to each recording station, and the phases were also modified to account for a deterministic wave passage effect. A more refined scheme was reported by Abrahamson [5], who first used a frequency-wavenumber (F-K) analysis, as described in Section 4.1, to decompose the recorded wavefield to its signal components and noise, and expressed the time history at the location of each recording station, $\vec{r}_m$, as:

$$U(\vec{r}_m, \omega) = \sum_{j=1}^{q} A_j(\omega) \exp[i\vec{\kappa}_j(\omega) \cdot \vec{r}_m + \theta_j(\omega)] + \eta(\vec{r}_m, \omega) \qquad (8.1)$$

where q is the number of plane-wave signals identified from the F-K analysis, $A_j(\omega)$, $\vec{\kappa}_j(\omega)$ and $\theta_j(\omega)$ are the amplitude, wavenumber and phase, respectively, of the

j-th signal, and $\eta(\vec{r}_m, \omega)$ is the noise at the station. Abrahamson [5] then estimated the interpolated time series at any location $\vec{r}$ by recombining the signal and noise components as:

$$U(\vec{r}, \omega) = \sum_{j=1}^{q} A_j(\omega) \exp[i\vec{\kappa}_j(\omega) \cdot \vec{r} + \theta_j(\omega)] + \mu_\eta(\vec{r}, \omega) \exp(i\phi_\eta(\vec{r}, \omega)) \quad (8.2)$$

with $\mu_\eta(\vec{r}, \omega)$ being the weighted mean amplitude of the noise, $\eta(\vec{r}_m, \omega)$, at the recording stations. The phase, $\phi_\eta(\vec{r}, \omega)$ in Eq. 8.2, was described by a normal random variable with mean value equal to the phase of the weighted sum of the noise and variance evaluated from the phases of the noise at the recording stations. The weights, in this approach, were functions of the coherency of the noise.

Conditional simulation is conceptually similar, as it denotes stochastic interpolation and extrapolation of the seismic records. The advantage, however, of conditional simulation over (deterministic) interpolation is the following: Interpolation schemes require data recorded at spatial arrays and the interpolated motions are valid for the particular site and earthquake characteristics for which they were developed. These motions can be utilized at different sites only if the soil conditions and earthquake characteristics are similar to those of the recording site. On the other hand, conditional simulations, having the advantage of incorporating information on the spatial variability of the random field, can be generated using either multiple time histories, when they are available, or a single time history. Because single seismic stations are extensively deployed around the world, single station data offer a wide selection pool for various site conditions, source-site distances and earthquake characteristics, whereas the selection pool of data from spatial arrays is more limited. This flexibility of conditional simulation over interpolation makes its applications broader and more attractive, and, for this reason, conditional simulations are described in this chapter.

Abrahamson [7] provided a six step procedure for generating conditionally simulated, response-spectrum-compatible spatially variable seismic ground motions, which he utilized to generate multiple support time histories for the seismic response evaluation/retrofitting of California bridges, e.g., among others, the San Francisco/ Oakland West Bay Bridge, the Vincent Thomas Bridge in Los Angeles and the Coronado Bridge in San Diego. The approach can be summarized as follows [7]:

1. Select the initial (or predefined or reference) seismic time history – this can be the time history in a single direction or the three component (north-south, east-west and vertical) set of time histories recorded at a single station;
2. Modify the initial time history(ies) to be compatible with the target response spectra;
3. Generate multiple support time histories at each location compatible with appropriate coherency model(s);
4. Modify the incoherent time histories to become spectrum compatible;
5. Apply a baseline correction scheme to the incoherent time histories simultaneously at all input locations; and
6. Apply the wave passage effect.

This comprehensive outline encompasses all aspects necessary for the generation of response-spectrum-compatible, conditionally simulated spatially variable seismic ground motions. In the approach, the spatially variable seismic ground motions in each direction are generated independently of the other two directions by considering that the three orthogonal components of the ground motions at each location are uncorrelated. This commonly made assumption, based on the identification of the principal directions of the seismic ground motions at each location (Section 3.3.2), will also be utilized in this chapter. The simulation scheme in Step 3, which was briefly described in Section 7.3.1, generates the Fourier phase differences of the spatially variable motions directly from the lagged coherency model by means of the minimization of a penalty function, combines the generated Fourier phase angles at each location and frequency with the predefined Fourier amplitudes, and computes the spatially variable time series through the inverse Fourier transform. Hence, the Fourier amplitudes of the generated motions at all stations are identical, and the Fourier phases conform with the prescribed coherency model [7]. The non-stationarity in the resulting time series is achieved through the utilization of multiple segments and multiple segment lengths of the reference ground motion, as will be briefly described in Section 8.1.3. In Step 2, Abrahamson [7] used Lilhand and Tseng's [304] time-domain approach to transform the initial time histories so that they become compatible to the prescribed response spectra, and then proceeded in Step 4 with the frequency-domain modification of the generated motions so that they, too, become response-spectrum-compatible. Because a direct and an iterative scheme for the generation of response-spectrum-compatible unconditional simulations were described in Sections 7.1.4 and 7.3.2, respectively, this topic will not be re-addressed in this chapter.

This chapter begins, in Section 8.1.1, with the description of a commonly used conditional simulation scheme, the conditional probability density function approach [249], [250], and continues with an example application that illustrates the characteristics of the stationary spatially variable time series resulting from this method (Section 8.1.2). Section 8.1.3 discusses approximations that are utilized for maintaining the non-stationarity of the prescribed time history in the generated motions. Of significant importance in simulating and conditionally simulating spatially variable ground motions is the processing of the resulting acceleration time series, so that they yield realistic velocity and displacement waveforms (Step 5 above). As is the case for recorded data, simulated and conditionally simulated ground motions also require processing to ensure compatibility between the corresponding acceleration, velocity and displacement time histories, zero initial values, zero residual values for accelerations and velocities and appropriate residual values for displacements. It is noted that displacement time histories are especially important for the seismic response evaluation of lifelines, when spatially variable motions are used as input excitations at their supports, because they control the pseudo-static response of these extended structures. The issue of processing simulated acceleration time histories is presented in Section 8.2. Finally, Section 8.3 presents two applications of conditional simulations: The first example (Section 8.3.1) considers as prescribed time history a synthetic one based on the physical spectrum proposed by Sabetta and Pugliese [436] from their analysis of 95 accelerograms of 17 Italian earthquakes. This application results in zero residual displacements and can be considered as representative of conditional simulation of any synthetic/recorded time history that yields zero residual displacements. The second example (Section 8.3.2) considers as prescribed time history the ground motion in the north-south direction recorded at station HEC during the 1999 Hector Mine earthquake and results in a non-zero residual displacement [75]; this application can be considered as representative of any synthetic/recorded time history that incorporates unique near-fault characteristics, such as the fling-step effect.

For clarification purposes, the definitions of the terms "synthetic" and "simulated" seismic time histories, as used herein, are explained in the following: "Simulated" ground motions indicate time series generated such that they comply with a power spectral density, e.g., the Kanai-Tajimi [253], [514] or the Clough-Penzien [107] spectrum, as described in Section 7.1.1. It is recalled that, to generate such motions, one specifies only the local site conditions (firm, medium or soft) for the power spectral density (PSD), or modifies the PSD so that it becomes compatible to a response spectrum (Section 7.1.4). It is then considered that either the amplitudes and the phases (Eq. 7.1) or only the phases (Eq. 7.3) of the simulations vary randomly. On the other hand, "synthetic" ground motions indicate time series that are based on seismological models incorporating the effects of source, path and site effects, as was described, e.g., in Eq. 3.8. Some of the approaches for generating synthetic data also consider an appropriate spectrum based on seismological observations and phase angles that are randomly distributed, and, strictly speaking, are also "simulations." Herein, the distinction is made in labeling "simulations" as those resulting from empirical power spectral densities and "synthetics" as those based on seismological observations.

8.1   CONDITIONAL SIMULATION OF RANDOM PROCESSES

Even though the conditional simulation of random processes and fields has not attracted the same wide-spread interest as the topic of unconditional simulations, several conditional simulation algorithms have been proposed. Kawakami [265] first considered conditional simulations by optimizing the harmonic amplitudes of time histories. Vanmarcke and Fenton [552] conditionally simulated both stationary and non-stationary random fields with the kriging method (linear estimation theory applied to random functions). Later, Vanmarcke et al. [553] improved the conventional kriging technique with the introduction of the multivariate linear prediction method. Heredia-Zavoni and Santa-Cruz [219] further extended the multivariate linear prediction technique to non-stationary random fields. Hoshiya [223] also improved the conventional kriging method by directly evaluating the simulation estimation error. Kameda and Morikawa [249], [250] presented an elegant approach, the Conditional Probability Density Function (CPDF) method, which provides the closed-form conditional probability function of the Fourier coefficients of the time series to be simulated. Ren et al. [422] simulated random fields conditioned by the realizations of the fields and their derivatives using the kriging method. Jin et al. [242] used filtering functions to describe the conditional statistics of a stationary random field and an updating approach for evaluating the filtering functions. Shinozuka and Zhang [469], in an insightful evaluation of the kriging and CPDF methods, concluded that they are equivalent, provided that the underlying process is Gaussian and zero-mean, but pointed out potential shortcomings of the kriging approach in terms of second moment statistics, its simulation of non-zero processes and its direct application to non-Gaussian processes. Conditional simulation of non-Gaussian random fields has been reported, among others, by Elishakoff *et al.* [147] and Hoshiya *et al.* [225].

This chapter conditionally simulates spatially variable seismic ground motions based on the CPDF technique [249], [250]. Section 8.1.1 describes the CPDF approach for the generation of Gaussian random processes, which, for all practical purposes in earthquake engineering applications, can be considered as zero-mean. Section 8.1.2 compares the behavior of time series generated as stationary by the CPDF method with the characteristics of the target random field and the trend of spatially recorded seismic data, as was conducted in Section 7.3.1 for simulated stochastic vector processes. Section 8.1.3 presents approaches utilized for the preservation of the non-stationarity of the reference time history in the conditional simulation of spatially variable seismic ground motions.

8.1.1 Description of the CPDF Technique

This section highlights the approach utilized by the CPDF method to conditionally simulate spatially variable seismic ground motions. The details and proofs of the technique can be found in Kameda and Morikawa [249], [250].

Consider L discrete spatial locations (points) of a Gaussian ground motion random field, $\mathbf{U}(\mathbf{x}, t)$, where $\mathbf{x}$ indicates location and t time. At a set of points $\{\alpha\} = \{\alpha_m, m = 1, \ldots, M\}$, $M < L$, the time histories are known, i.e., they are either recorded data or synthetic/simulated ground motions. The goal of the conditional simulation is to generate unknown ground motions at a target point set $\{\beta\} = \{\beta_n, n = M+1, \ldots, L\}$, so that they are compatible with the known time histories at the point set $\{\alpha\}$, as well as the characteristics of the random field.

The time history at any point, e.g., the i-th point, of the random field, $\mathbf{U}(\mathbf{x}, t)$, is a random process denoted by $U_i(t)$ that can be expanded as a Fourier series:

$$U_i(t) = \sqrt{\Delta\omega} \sum_{k=1}^{K} [A_{ik} \cos(\omega_k t) + B_{ik} \sin(\omega_k t)] \quad i = 1, \ldots, L \quad (8.3)$$

where ω_k represents the k-th discrete frequency, $\Delta\omega$ is the frequency step, and A_{ik} and B_{ik}, $k = 1, \ldots, K$, are the Fourier coefficients, which are uncorrelated at different frequencies. Let $\mathbf{Z} = \{\mathbf{Z}_\alpha, \ \mathbf{Z}_\beta\}^T$, with superscript T indicating transpose, denote the vector of the random Fourier coefficients at each frequency ω_k, in which the subsets $\mathbf{Z}_\alpha$ and $\mathbf{Z}_\beta$ are given by:

$$\mathbf{Z}_\alpha = \{A_{1k}, \ B_{1k}, \ A_{2k}, \ B_{2k}, \ldots, A_{Mk}, \ B_{Mk}\}^T$$

$$\mathbf{Z}_\beta = \{A_{M+1,k}, \ B_{M+1,k}, \ldots, A_{Lk}, \ B_{Lk}\}^T \quad (8.4)$$

The corresponding covariance matrix $\mathbf{C}_k = \mathrm{cov}(\mathbf{Z})$ can then be evaluated from the statistical relationships between the Fourier coefficients [249] as:

$$E[A_{ik}A_{jk}] = E[B_{ik}B_{jk}] = \begin{cases} 2S_{ii}(\omega_k) & i = j \\ 2\Re[S_{ij}(\omega_k)] & i \neq j \end{cases} \tag{8.5}$$

$$E[A_{ik}B_{jk}] = -E[B_{ik}A_{jk}] = \begin{cases} 0 & i = j \\ -2\Im[S_{ij}(\omega_k)] & i \neq j \end{cases} \tag{8.6}$$

in which E represents expectation, $\Re[.]$ and $\Im[.]$ denote the real and imaginary parts of a complex number, respectively, $S_{ii}(\omega_k)$ is the power spectral density at point i, and the cross spectral density, $S_{ij}(\omega_k)$, between points i and j is given by:

$$S_{ij}(\omega_k) = \gamma(\xi_{ij}, \omega_k)\sqrt{S_{ii}(\omega_k)S_{jj}(\omega_k)} \tag{8.7}$$

where $\gamma(\xi_{ij}, \omega_k)$ is the spatial variability function and ξ_{ij} indicates the separation distance between the stations. The covariance matrix $\mathbf{C}_k$ can be explicitly expressed in terms of submatrices as:

$$\mathbf{C}_k = \begin{bmatrix} \mathbf{C}_{\alpha\alpha} & \mathbf{C}_{\alpha\beta} \\ \mathbf{C}_{\alpha\beta}^T & \mathbf{C}_{\beta\beta} \end{bmatrix}_{2L \times 2L} \tag{8.8}$$

where the submatrices $\mathbf{C}_{\alpha\alpha} = \mathrm{cov}(\mathbf{Z}_\alpha)$, $\mathbf{C}_{\beta\beta} = \mathrm{cov}(\mathbf{Z}_\beta)$, and $\mathbf{C}_{\alpha\beta} = \mathrm{cov}(\mathbf{Z}_\alpha, \mathbf{Z}_\beta)$ describe the covariance between the Fourier coefficients of the point set $\{\alpha\}$, the point set $\{\beta\}$, and between $\{\alpha\}$ and $\{\beta\}$.

Let $U_{M+1}(t|U_i(t), i = 1, \ldots, M)$ represent a random process at location $M + 1$ to be conditioned by the time histories $U_1(t)$, $U_2(t)$, $\ldots$, $U_M(t)$ that are known at the set of points $\{\alpha\}$. The covariance matrix (Eq. 8.8) becomes then:

$$\mathbf{C}_k = \begin{bmatrix} G_{11,k} & 0 & K_{12,k} & -Q_{12,k} & \cdots & K_{1,M+1,k} & -Q_{1,M+1,k} \\ 0 & G_{11,k} & Q_{12,k} & K_{12,k} & \cdots & Q_{1,M+1,k} & K_{1,M+1,k} \\ K_{12,k} & Q_{12,k} & G_{22,k} & 0 & \cdots & K_{2,M+1,k} & -Q_{2,M+1,k} \\ -Q_{12,k} & K_{12,k} & 0 & G_{22,k} & \cdots & Q_{2,M+1,k} & K_{2,M+1,k} \\ \vdots & \vdots & \vdots & \vdots & & \vdots & \vdots \\ K_{1,M+1,k} & Q_{1,M+1,k} & K_{2,M+1,k} & Q_{2,M+1,k} & \cdots & G_{M+1,M+1,k} & 0 \\ -Q_{1,M+1,k} & K_{1,M+1,k} & -Q_{2,M+1,k} & K_{2,M+1,k} & \cdots & 0 & G_{M+1,M+1,k} \end{bmatrix} \tag{8.9}$$

where $G_{ii,k} = 2S_{ii}(\omega_k)$, $K_{ij,k} = 2\Re[S_{ij}(\omega_k)]$, $Q_{ij,k} = 2\Im[S_{ij}(\omega_k)]$ from Eqs. 8.5 and 8.6, and the dimension of the covariance matrix $\mathbf{C}_k$ is $2(M + 1) \times 2(M + 1)$. The CPDF method generates the unknown Fourier coefficients, $A_{M+1,k}$ and $B_{M+1,k}$, at the target point $M + 1$ by deriving their conditional probability density functions from the $2(M+1)$-dimensional Gaussian joint probability function of the Fourier coefficients A_{ik} and B_{ik}, $i = 1, \ldots, M+1$ [249]. The derivation determines the Fourier coefficients $A_{M+1,k}$ and $B_{M+1,k}$ as independent random variables that follow a Gaussian distribution with conditional mean values and variances evaluated as [249]:

$$\mu_{A_{M+1,k}} = E[A_{M+1,k}|\mathbf{Z}_\alpha] = -\frac{\tilde{\Phi}_k}{V_{2M+1,2M+1,k}}$$

$$\mu_{B_{M+1,k}} = E[B_{M+1,k}|\mathbf{Z}_\alpha] = -\frac{\tilde{\Gamma}_k}{V_{2M+1,2M+1,k}} \tag{8.10}$$

$$\sigma^2_{A_{M+1,k}} = \sigma^2_{B_{M+1,k}} = \frac{1}{V_{2M+1,2M+1,k}}$$

The items on the right-hand side of the equation are determined from the inverse of the covariance matrix $\mathbf{C}_k$ (Eq. 8.9), $\mathbf{V}_k = \mathbf{C}_k^{-1}$ with elements $V_{ij,k}$, as:

$$\tilde{\Phi}_k = \sum_{i=1}^{M} (V_{2i-1,2M+1,k}\, A_{ik} - V_{2i-1,2M+2,k}\, B_{ik})$$

$$\tilde{\Gamma}_k = \sum_{i=1}^{M} (V_{2i-1,2M+2,k}\, A_{ik} + V_{2i-1,2M+1,k}\, B_{ik}) \tag{8.11}$$

Based on their conditional mean and variance (Eq. 8.10), the Gaussian Fourier coefficients $A_{M+1,k}$ and $B_{M+1,k}$ are then generated at every frequency ω_k, and the conditionally simulated time series at location $M+1$ is obtained from Eq. 8.3; the FFT approach can also be readily applied in the simulation process. Once $U_{M+1}(t|U_i(t)$, $i = 1, \ldots, M)$ is conditionally simulated, the time history at the next location of the target point set $\{\beta\}$ is conditioned by the time histories recorded at the set of points $\{\alpha\}$ as well as the previously conditionally simulated ground motion and so on, with the final time history conditioned by all previous ones as $U_L(t|U_i(t)$, $i = 1, \ldots, M$; $U_{M+j}(t)$, $j = 1, \ldots, L - M - 1)$.

8.1.2 CHARACTERISTICS OF CPDF CONDITIONAL SIMULATIONS

This section illustrates the conformity of conditional simulations generated by means of the CPDF scheme with respect to the target random field and, also, the trend of spatially recorded seismic data, as was performed earlier, in Section 7.3.1, for simulated stochastic vector processes. For this purpose, the Clough and Penzien [107] spectrum and the coherency model of Luco and Wong [324] are utilized. The spectrum is given by Eq. 3.7 and repeated here for convenience:

$$S(\omega) = S_\circ \frac{1 + 4\zeta_g^2\left(\frac{\omega}{\omega_g}\right)^2}{\left[1 - \left(\frac{\omega}{\omega_g}\right)^2\right]^2 + 4\zeta_g^2\left(\frac{\omega}{\omega_g}\right)^2} \frac{\left(\frac{\omega}{\omega_f}\right)^4}{\left[1 - \left(\frac{\omega}{\omega_f}\right)^2\right]^2 + 4\zeta_f^2\left(\frac{\omega}{\omega_f}\right)^2} \tag{8.12}$$

This application considers "medium" soil conditions, the parameters of which are, as suggested by Der Kiureghian and Neuenhofer [138] and presented in Table 3.1(b): $\omega_g = 10.0$ rad/sec, $\omega_f = 1.0$ rad/sec, $\zeta_g = 0.4$ and $\zeta_f = 0.6$, and it is assumed that $S_\circ = 1$ cm^2/sec^3; the same power spectral density is used for all stations. The coherency model utilized is the model of Luco and Wong [324] (Eq. 3.27 iterated here):

$$|\gamma(\xi, \omega)| = \exp(-\alpha^2\omega^2\xi^2) \qquad (8.13)$$

with the average value for the coherency drop parameter, $\alpha = 2.5 \times 10^{-4}$ sec/m, from the ones suggested by Luco and Wong [324]. Three equidistant stations along a straight line with separation distance between adjacent stations of $\xi = 500$ m are considered. As in Section 7.3.1, the long separation distance between stations is purposely selected so that the cross spectrum of the motions clearly reflects the loss of coherency at further-away stations. The apparent propagation of the motions is not taken into consideration in this simulation illustration, because it only introduces deterministic time shifts in the time histories at different stations compatible with the apparent propagation of the waveforms on the ground surface. It is noted that, in the comparisons of this section, the motions are conditionally simulated as stationary, since the analysis of recorded data both for the evaluation of the coherency (Section 2.4) and their amplitude and phase variability (Section 4.2) considers that each window analyzed is a segment of a stationary process.

The predefined stationary and "deterministic" time history at station 1 is simulated according to the spectral representation method [463]:

$$f(t) = 2 \sum_{k=1}^{K} \sqrt{S(\omega_k)\Delta\omega} \, \cos(\omega_k t + \phi_k) \qquad (8.14)$$

where, for consistency with the subsequent conditional simulation, the frequency domain is discretized as $\omega_k = k\Delta\omega$, $S(\omega_k)$ is the power spectral density (Eq. 8.12), and the random phases, ϕ_k, $k = 1, \ldots, K$, are uniformly distributed between $[0, 2\pi)$ (cf. with Eq. 7.4). It is noted that the unsmoothed power spectral density of the simulations generated from Eq. 8.14 is identical to the prescribed PSD (Eq. 8.12) at the discrete frequencies ω_k, $k = 1, \ldots, K$, when the length of the generated time series is equal to the period of the simulations, which, for this frequency discretization, becomes $T_0 = 2\pi/\Delta\omega$. A realization of the random process (Eq. 8.14) is then considered to be the existing ("recorded") time history at station "1", i.e., the single point of the set $\{\alpha\}$. Based on each realization at station 1, conditional simulations are generated by means of the CPDF method at stations 2 and 3 (Eqs. 8.3, 8.10 and 8.11).

Figure 8.1 presents the comparison of the ensemble averages of unsmoothed power spectral densities, cross spectral densities and Fourier amplitudes of 100 conditional simulations at stations 2 and 3, for which the time history at station 1 was newly generated each time, with the characteristics of the target field. It can be seen from the figure that the average unsmoothed power and cross spectral densities of the conditional simulations (Figs. 8.1(a) and (b), respectively) behave through the entire frequency range as random functions with mean value equal to the target

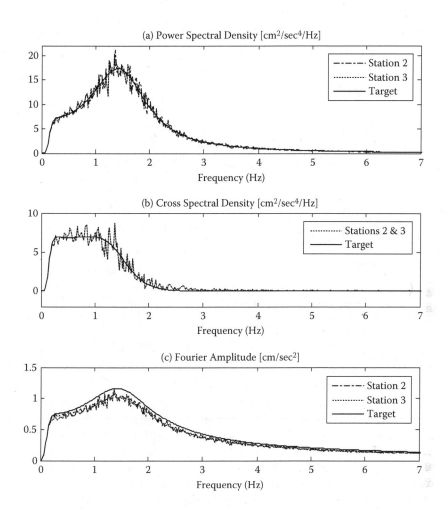

FIGURE 8.1 Comparison of the ensemble averages of the power and cross spectral densities, in parts (a) and (b), respectively, and the Fourier amplitudes, in part (c), of 100 conditional, stationary simulations generated by the CPDF method [249] with the target values. The reference time history was simulated as stationary from the Clough-Penzien spectrum for medium soil conditions (Table 3.1(b)). The distance between the stations is 500 m, and the coherency model is that of Luco and Wong with $\alpha = 2.5 \times 10^{-4}$ sec/m.

spectrum and non-zero variance. The average of the unsmoothed Fourier amplitudes (Fig. 8.1(c)) is, again, approximately 88.6% of the target values, evaluated as $\Lambda(\omega) = 2\sqrt{S(\omega)\Delta\omega}$, which follows directly from the χ^2 distribution properties of the sum of squares of normal variates [301], as already discussed in Section 7.3.1 regarding simulated seismic time histories. The ensemble averages of the lagged coherency of the 100 simulations between stations 1 and 2, and stations 2 and 3 are presented in Fig. 8.2 together with the target coherency (Eq. 8.13 for $\xi = 500$ m). For the estimation of the lagged coherency from the simulations, the power and cross spectral

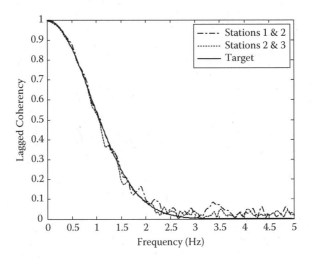

FIGURE 8.2 Comparison of the ensemble averages of the smoothed lagged coherency of 100 conditional, stationary simulations generated by the CPDF method [249] between stations 1 and 2, and stations 2 and 3, at a separation distance of $\xi = 500$ m, with the target values. The characteristics of the random field are the same as those in Fig. 8.1; for the evaluation of the lagged coherency of the simulations, the power and cross spectral densities were smoothed with an 11-point Hamming window (Fig. 2.4).

densities were smoothed with an 11-point Hamming spectral window (Eq. 2.47). The lagged coherency of the simulations is in agreement with the prescribed coherency model, suggesting that the technique preserves the characteristics of the target random field.

The behavior of the amplitudes and the phases of the spatially variable time series resulting from the CPDF method is examined in Fig. 8.3 by means of the Difference Indexes (DIs) for amplitudes and phases [259], [260], described in Eqs. 7.45 and 7.46, respectively. The figure presents the amplitude DI and phase DI (in parts (a) and (b), respectively) of 100 simulations between stations 2 and 3 generated by the CPDF technique. It is recalled from Section 7.3.1 and Fig. 7.13 that, according to observations from recorded data, the amplitude and phase DIs should increase consistently with frequency and reach constant values at higher frequencies. The DIs in Fig. 7.13 were obtained from simulated stationary motions generated by means of the CDRA technique for firm soil conditions, but for the same coherency model, the model of Luco and Wong (Eq. 8.13 with $\alpha = 2.5 \times 10^{-4}$ sec/m), and the same separation distance between stations ($\xi = 500$ m) utilized for the conditional simulations of Fig. 8.3. Figures 7.13 and 8.3 are very similar: In the low frequency range, the DIs for both amplitudes and phases are small and gradually increase with frequency in a consistent manner. When the frequency exceeds approximately 2.0 Hz, at which value the prescribed lagged coherency is below 0.1 and tends to zero afterwards, the DIs reach the expected values of 0.66 for the amplitude DI and 2.57 rad for the phase DI (dotted lines in Figs. 7.13 and 8.3), and remain fairly constant

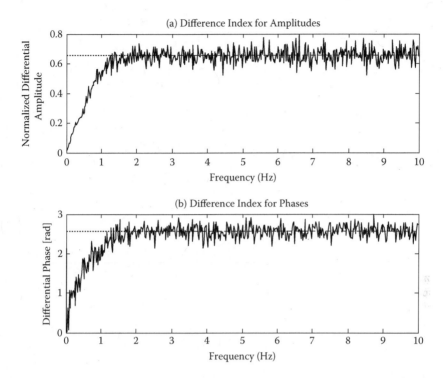

FIGURE 8.3 Comparison of the amplitude and phase difference indexes (DI in Eqs. 7.45 and 7.46) in parts (a) and (b), respectively, of 100 conditional, stationary simulations generated by the CPDF method [249] for the same characteristics of the random field as in Fig. 8.1. The DIs are shown for stations 2 and 3 at a separation distance of 500 m. The dotted lines in the subplots indicate the expected values, to which the DIs tend to in the high frequency range, where the lagged coherency tends to zero.

thereafter [301]. Even though the power spectral density of the motions differs (firm soil conditions for Fig. 7.13 and medium soil conditions for Fig. 8.3), this does not affect the trend of the DIs, which are controlled by the coherency decay. As was the case for the DIs generated by the CDRA technique (Fig. 7.13), the amplitude and phase DIs of conditional simulations based on the CPDF method will also tend to these same constant values for any power spectral density and station separation distance, as long as the coherency tends to zero [301]. Hence, the behavior of ground motions generated by the CPDF method conforms with the trend of spatially recorded data. Liao *et al.* [302] showed that conditional simulations generated by the multivariate linear prediction technique for zero-mean processes also exhibit a similar behavior.

It is noted that the CPDF technique (Section 8.1.1) generates the spatially variable time series in sequence, i.e., the ground motions at the new location are simulated to be compatible with the generated motions at all previous locations and the prescribed time series[1]. Kameda and Morikawa [250] showed that, at every time instant, the conditional mean value of the generated motion will be nearly identical to the value of the prescribed time series if the separation distance is short (i.e., the motions are highly correlated), but will tend to zero as the separation distance increases (i.e., coherency tends to zero). They further proved that the conditional mean value of the simulation at the new location depends on the number of the time series on which it is conditioned, as well as the realized values of the Fourier coefficients, whereas its conditional standard deviation depends on the number of the time series on which it is conditioned, but not on the realized values of the Fourier coefficients.

8.1.3 INTRODUCTION OF NON-STATIONARITY IN CONDITIONAL SIMULATIONS

The previous section conditionally simulated the time series as stationary. As indicated in Chapter 7, however, stationarity is an inherent feature of simulations based on the Fourier transform, but, also, an undesirable characteristic, when the time histories are used in the seismic response evaluation of structures. A similar issue regarding stationarity arises in conditional simulations. The approaches that are being used to approximately overcome this limitation consider time- and/or frequency-domain "segmentation."

In their comprehensive article on the modeling and simulation of spatially variable ground motions, Hao *et al.* [200] suggested dividing the acceleration time series into several overlapping time segments, each with similar (stationary) properties, simulating the different segments, and then piecing the segments together with a transitional weighting function. Vanmarcke and Fenton [552] also applied time-domain segmentation in conditionally simulating spatially variable ground motions and utilized linear weighting functions to recombine the different segments in the time domain. They recommended that the length of the weighting functions be selected such that it covers a few oscillations of the motions, so that it will lead to fairly smooth connections of the acceleration time histories in the transition regions [552]. The drawback of the time-domain segmentation approach, however, is that it does introduce subtle discontinuities at the segmentation locations. Abrahamson [7] utilized a hybrid time- and frequency-domain segmentation approach for the generation of non-stationary, spatially variable ground motions: To emulate the higher frequency components of the motions ($f > 5$ Hz), the reference time history was divided into multiple short segments (≤ 1 sec), each segment was conditionally simulated, and the segments were combined in the time domain to yield the simulated time series at the various locations on the ground surface. The reference time history was then divided into longer

---

[1] The difference between the CPDF conditional simulation scheme and the sequential approach of Hao *et al.* [200], described in Section 7.3.1, is that the conditional simulation scheme generates the coefficients of the Fourier series at every frequency as random variables conditioned by all pre-existing (prescribed and simulated) Fourier coefficients (Eqs. 8.3 and 8.10), whereas the technique by Hao *et al.* [200] permits the phases of the series to vary randomly, but evaluates the amplitudes in a deterministic manner to ensure that the newly generated time history satisfies the coherency relations with all previously generated simulations (Eq. 7.47). The differences in the amplitude DIs generated by the two techniques (Figs. 7.15 and 8.3) can be attributed to the same reason.

segments (e.g., approximately 5 sec for $f \approx 1$ Hz), each segment conditionally simulated, and the simulated segments recombined in the time domain. To emulate the lower frequency components ($f < 0.2$ Hz), the entire time history was conditionally simulated as stationary. The multiple spatially variable ground motions at each location, that resulted from the different time-domain segmentations, were then combined in the frequency domain by means of matched filters. To avoid time-domain segmentation, Heredia-Zavoni and Santa-Cruz [219] extended the non-stationary model proposed by Yeh and Wen [577] for simulations at a single location on the ground surface (Section 7.1.3) to conditionally simulate spatially variable time series. To estimate the parametric forms of the time-domain intensity and frequency modulating functions of the reference time history (Eqs. 7.20 and 7.21) in different frequency bands, Heredia-Zavoni and Santa-Cruz [219] used frequency-domain segmentation of the prescribed accelerogram. The drawbacks of this approach, as discussed in Section 7.1.3, are the inherent assumptions regarding the parametric forms of the modulating functions as well as the identification schemes for the evaluation of the parameters, which can affect the characteristics of the generated motions over their entire duration. It is recalled that current trends in simulating point estimates of the motions utilize approaches that incorporate both time- and frequency-domain features in their transforms (see, e.g., Fig. 7.7), and, thus, have better capabilities for maintaining the non-stationary characteristics of the reference time history. Since coherency, however, is still defined by means of (stationary) Fourier transforms, the segmentation approaches appear to be the means for addressing the non-stationarity issue.

In the examples presented herein, the time-domain segmentation is used to address the non-stationarity in the ground motions, because the consequences of the approximations in the time-domain segmentation process are localized at the transition regions. Additionally, it may be postulated that the physical rationale underlying the time-domain segmentation is the separation, in the reference time history, of time segments where wave components with different properties dominate [298]. The time-domain segmentation approach then proceeds as follows: During the conditional simulation, the reference ground motion is divided into segments with different number of time steps. The segments are determined "by observation" so that the amplitudes and zero crossings within each segment are relatively similar, i.e., the motions within each time segment are "stationary." Each segment is then used as the reference time series, conditional simulations are carried out for each segment, and the conditionally simulated segments are joined together with linear weighting functions to yield the entire spatially variable acceleration time histories.

Once the spatially variable acceleration time series are generated by simulations or conditional simulations, the corresponding velocity and displacement time histories are obtained via numerical integration. The correct evaluation of the displacement time series is essential for the seismic response analysis of lifeline systems, if spatially variable input excitations are utilized, because the pseudo-static response of the systems depends directly on displacements. Indeed, the majority of available commercial finite element codes require displacement time histories as input motions at the supports of, e.g., bridge models, when the spatial variability of the ground motions is considered. However, just as unprocessed recorded acceleration time histories can yield unrealistic velocity and displacement waveforms, direct integration of simulated accelerations can also render unrealistic velocities and displacements. The reasons, in this case, are purely numerical as, e.g., large Fourier coefficients in the very low frequency range generated randomly in the numerical simulation. Hence, simulated and conditionally simulated ground accelerations also require processing. An approach for the processing of simulated spatially variable time series is presented in the following section.

8.2 PROCESSING OF SIMULATED ACCELERATION TIME SERIES

It has long been recognized that acceleration records require proper processing to yield realistic acceleration, velocity and displacement time histories for the seismic response evaluation of structures, e.g., [239], [526], [535]. In part due to the newly identified near-fault ground motion characteristics (i.e., directivity pulse and fling-step effects) and the current trend of the earthquake engineering community to utilize more records in the seismic resistant design of structures, these processing techniques have recently attracted renewed research interest (e.g., [112]). Many schemes have been proposed in the literature for processing earthquake records (e.g., [74], [75], [101], [116], [535]), but can, basically, be categorized into two general groups: One can retain non-zero residual displacements, while the other cannot, because it uses high-pass filtering of the data. The majority of techniques currently used by various earthquake data centers, as, e.g., the California Strong Motion Instrumentation Program [91], the US Geological Survey [547], and the Pacific Earthquake Engineering Research Center [390], belong to the latter. It needs to be emphasized at this point that no single scheme is universally applicable to every case, and the selection of the appropriate processing approach depends on the specific application.

For simulated/conditionally simulated time histories, applying any available baseline adjustment scheme with high-pass filtering will lead to zero residual displacements. However, baseline adjustment algorithms inducing non-zero initial values in acceleration, velocity and displacement time histories, which can be acceptable for actual earthquake records due to, e.g., signal lost prior to the triggering threshold of accelerographs at the beginning of the record and the existence of background noise, are not suitable for simulated time series. Hence, it is desirable to enforce zero initial values for simulated accelerations, velocities and displacements. Liao and Zerva [301] suggested a processing scheme for simulations and conditional simulations, which is based on the approach proposed by Boore *et al.* [75], since their algorithm can guarantee zero initial values in the velocity and displacement time histories and negligible initial values in the acceleration time history. Additionally, it ensures compatibility among the three components, i.e., that the integration of the acceleration time history is compatible with the finally provided velocity and displacement time series. The approach, as applied to simulated spatially variable ground motions, can be summarized in the following six steps [301]:

1. Subtract the mean value from the entire acceleration time series;
2. Apply a short cosine taper function to the simulated acceleration time series to set their initial values to zero – the length of the taper function can be, usually, 50-100 steps long, depending on the specific application;

3. Integrate accelerations to velocities, and fit a quadratic function to the velocity time series in the least-squares minimization sense – zero velocity constraints are applied at the starting time;
4. Remove the derivative of the quadratic function of Step 3 from the acceleration time series;
5. Apply a high-pass filter to the acceleration time series;
6. Integrate the accelerations to get velocities and displacements.

Boore *et al.* [75] suggested that, in Step 1, if pre-event information is available, the mean value of the pre-event part of the record be subtracted from the time series, and, in the absence of pre-event information, the mean value of the entire record be used instead. The mean value of the entire series is utilized in Step 1 for simulations and conditional simulations, as, obviously, there is no pre-event information in simulated ground motions. In Step 4, subtracting the derivative of the quadratic function fitted to the velocities from the accelerations will introduce non-zero initial acceleration values, which, however, are usually negligible compared to the entire acceleration time history. The selection of the high-pass filter in Step 5 requires careful consideration and has recently received renewed interest by the engineering and seismological communities [17], [74], [75], [112], [540]. Key issues regarding the selection of the filter (filter characteristics and the use of causal and acausal filtering) are discussed in the following.

Filter Characteristics

The most commonly used filter by, e.g., the California Strong Motion Instrumentation Program [91], the US Geological Survey [547], and the Pacific Earthquake Engineering Research Center [390] is the Butterworth filter. The modulus of the frequency transfer function of the Butterworth filter is given by [614]:

$$|H_n(f)| = \left[\frac{(f/f_c)^{2n}}{1 + (f/f_c)^{2n}} \right]^{1/2} \tag{8.15}$$

where n indicates the order of the filter and f_c its corner frequency. Figure 8.4 presents the modulus of the response of a high-pass (low-cut) Butterworth filter for various orders ($n = 2, 3, 4, 6$ and 8) and a corner frequency of $f_c = 0.05$ Hz, or, equivalently, a corner period of $T_c = 1/f_c = 20$ sec, i.e., frequencies lower than 0.05 Hz (or periods longer than 20 sec) are, at least, partially removed. For the lower orders of the filter, frequencies higher than the corner frequency (or periods shorter than the corner period) are also partially removed. On the other hand, as the filter order increases, the cut-off of the lower frequencies becomes more abrupt. Following Boore *et al.* [75], a 4-th order Butterworth high-pass filter is utilized in Step 5. Boore and Bommer [74] noted, however, that data processing is not sensitive to the choice of a particular filter type, but, rather, to the value of the corner frequency of the filter. Figure 8.5, from their work, presents the effect of the selection of the corner frequency on the displacement time histories resulting from the integration of a recorded accelerogram. The displacement time history in the top subplot of Fig. 8.5 was obtained without filtering of the record and exhibits an unrealistic waveform with large residual displacements. The displacement time series shown in the second-from-top subplot was obtained

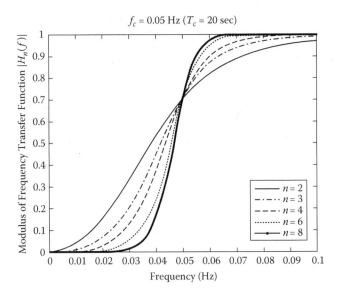

FIGURE 8.4 The modulus of the frequency transfer function, $|H_n(f)|$, of a high-pass Butterworth filter (Eq. 8.15) with corner frequency $f_c = 0.05$ Hz and orders $n = 2, 3, 4, 6$ and 8; frequencies lower than 0.05 Hz are at least partially removed. The higher the order of the filter, the more abrupt the cut-off of the lower frequencies.

by filtering the accelerations with a corner frequency of 0.05 Hz, but the waveform still looks unrealistic. As the corner frequency utilized in the acceleration processing increases (bottom three subplots), the displacement time series assume more realistic waveforms. It is noted that the displacement scale of the top two subplots differs from the scale of the bottom three. The selection of the corner frequency becomes then a critical issue in the processing of the acceleration time series.

In the processing of earthquake ground motion records, the corner frequency is determined from the signal-to-noise ratio, so that most of the background noise is removed from the signal. If such information is not available, the corner frequency f_c can be estimated, e.g., from the value of the frequency for which the Fourier amplitude spectrum of the record no longer tends to zero at low frequencies [581], since the theoretical shape of the far-field Fourier amplitude spectrum decreases as ω^2 at low frequencies, or determined iteratively, within a frequency range lower than the values of the frequencies obtained from theoretical Fourier amplitude spectra of the event, until the displacement time series exhibit realistic waveforms [17]. The situation in numerical simulations is, however, different, since there is no background noise involved in the process of simulation, but, as indicated earlier, the generated motions can be contaminated by, e.g., large low-frequency randomly simulated Fourier amplitudes. Liao and Zerva [301] proposed a purely numerical criterion to evaluate the corner frequency, f_c, of the selected filter: The corner frequency value is selected such that most of the low frequency components, whose corresponding periods are longer than the duration T of the simulated time histories, are removed. For example, if H_0 denotes the modulus of the transfer function of a continuous high-pass n-order

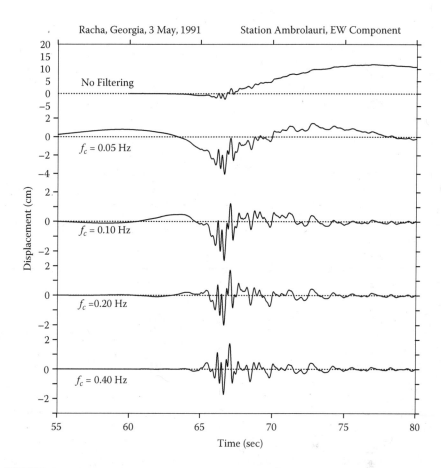

FIGURE 8.5 Illustration of the effect of the selection of the corner frequency of high-pass filters on displacement time series. The displacement time history of the top subplot was obtained without filtering; the waveform is unrealistic. The second-from-top subplot utilized a corner frequency of $f_c = 0.05$ Hz, but still looks unrealistic. More realistic displacement waveforms are obtained as the corner frequency increases further ($f_c = 0.10, 0.20$ and 0.40 Hz, in the bottom three subplots). The scale of the top two subplots differs from the scale of the bottom three. (Reprinted from *Soil Dynamics and Earthquake Engineering*, Vol. 25, D.M. Boore and J.J. Bommer, "Processing of strong-motion accelerograms: Needs, options, and consequences," pp. 93–115, Copyright ©2005, with permission from Elsevier; courtesy of D.M. Boore.)

Butterworth filter corresponding to frequency $f_0 = 1/T$ (Eq. 8.15), the corner frequency can be evaluated as:

$$f_c = \frac{1}{T[H_0^2/(1 - H_0^2)]^{1/2n}} \tag{8.16}$$

i.e., the corner frequency depends on the duration of the simulated time history and the filter amplitude threshold H_0 at $f_0 = 1/T$. Liao and Zerva [301] also suggested that the amplitude threshold be selected as $H_0 = 0.02$, which means that at least 98%

of the low frequency components with periods longer than the simulated time history duration are filtered out. Hence, if the simulated time history is 30 sec long and a 4-th order Butterworth filter is utilized, then the corner frequency is determined as 0.09 Hz. For different threshold levels and other filter types, the corner frequency can be estimated in a similar way.

Causal vs. Acausal Filtering

In addition to the aforementioned corner frequency selection considerations, the choice of causal vs. acausal filtering can also affect the waveforms: Causal filtering causes phase distortions of the time histories depending on the value of the corner frequency. On the other hand, acausal filtering is zero-phase and does not cause distortion of the waveforms. The zero-phase shift is achieved in the time domain by applying the filter one time from the beginning to the end of the record, and then another time by reversing the order and applying the filter from the end of the record to the beginning. The effect of causal vs. acausal filtering on the data recorded at the Rinaldi station during the 1994 Northridge earthquake is illustrated in Fig. 8.6 (also from the work of Boore and Bommer [74]): The top three subplots indicate accelerations, velocities and displacements processed with causal filtering and various corner periods, T_c, and the bottom three the corresponding time series when acausal filtering is used. Boore and Bommer [74] used different time scales for accelerations, velocities and displacements, so as to clearly depict the effects of the filtering process. The use of acausal filtering leaves the acceleration time histories virtually unchanged, causes small differences in the velocities and affects the displacement time histories; the latter is a consequence of the value of the corner frequency as has been discussed earlier in Fig. 8.5. On the other hand, the phase distortion of the causal filtering operation, which depends on the value of the corner frequency, is clearly identified in the acceleration time histories (top left subplot), and carried over in the velocity and displacement waveforms. As in Fig. 8.5, if the accelerations are not filtered, the resulting displacement waveforms, indicated by the lighter, dotted lines in Fig. 8.6, look unrealistic.

Considering that displacement time histories will eventually be needed in the seismic response analysis of lifelines, if spatially variable motions are utilized as input motions at their supports, the differences in the displacement time histories of Fig. 8.6 raise the question of which type of filtering process should be utilized. Obviously, the use of acausal filtering has the advantage of not causing phase distortions in the time series. However, acausal filtering requires zero-padding at both ends and can lead to large precursory transients (cf., e.g., the left top and bottom subplots of Fig. 8.6). The reason that the zero-padding is required is that, to achieve zero phase shifts, the filtering operation has to start before the beginning of the record in the forward filtering and, in the reverse process, after the end of the record. The zero-padding length depends on the value of the selected corner frequency and the order of the filter, as, e.g., suggested by Converse and Brady [116]:

$$t_{zero\_pad} = \frac{1.5 \, n_{roll}}{f_c} \qquad (8.17)$$

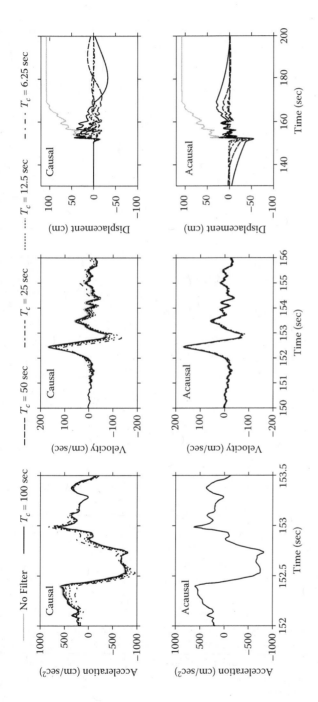

FIGURE 8.6 Illustration of the effect of the selection of causal and acausal filtering with various values of the corner period of the filter T_c ($= 1/f_c$). The ground motion is the 228° record at the Rinaldi station during the 1994 Northridge earthquake. The top subplots illustrate the acceleration, velocity and displacement time series (from left to right) obtained with causal filtering, and the bottom subplots the accelerations, velocities and displacements (also from left to right) obtained with acausal filtering. The values of T_c are 100, 50, 25, 12.5 and 6.25 sec, which correspond to corner frequencies of $f_c = 0.01$, 0.02, 0.04, 0.08 and 0.16 Hz. The effect of the phase distortion caused by causal filtering and dependent on the value of the corner frequency of the filter is clear in the top subplots. If no filtering is applied, as illustrated in both the top and bottom subplots, the displacement waveforms look unrealistic. The time scale of the acceleration, velocity and displacement subplots differs. (Reprinted from *Soil Dynamics and Earthquake Engineering*, Vol. 25, D.M. Boore and J.J. Bommer, "Processing of strong-motion accelerograms: Needs, options, and consequences," pp. 93–115, Copyright © 2005, with permission from Elsevier; courtesy of D.M. Boore.)

in which t_{zero_pad} is the zero-padding length (in [sec]) at both the beginning and the end of the time series to be processed, and n_{roll} is the roll-off parameter of the filter. Such zero-padding could significantly increase the length of the simulated time histories. On the other hand, causal filtering does not require leading zero-pads, but necessitates trailing zeros. Discarding the zero-padding steps after processing will invalidate the properties of the adjusted data [74], [116]. Because each scheme has advantages and disadvantages, both filtering operations are utilized and discussed in the subsequent applications herein.

8.3   EXAMPLE APPLICATIONS

Conditionally simulated seismic ground motions have direct applications in performance-based design. Any recorded or synthetic seismic time history selected for the performance-based design of a building structure (e.g., [508]) can be conditionally simulated with an appropriate coherency model and an apparent propagation velocity to yield spatially variable time series for the performance-based design of an extended structure, e.g., a bridge.

Recorded seismic data can be readily downloaded from a large number of sites on the World Wide Web, such as, among others, the COSMOS Virtual Data Center [111], the California Strong Motion Instrumentation Program [91], the US Geological Survey [547], the Southern California Earthquake Data Center [510], and the Pacific Earthquake Engineering Research Center [390], or are available on CD ROMs, such as the selection of European and Middle Eastern data [22], [23]. Another source of available time histories for performance-based design are synthetic ground motions for far-field, intermediate-field and near-source regions, the generation of which has also attracted significant research efforts (e.g., [485], [490]). The most commonly used techniques to generate synthetic time series are the stochastic ground motion method [54], [70], [72], [246], [474], [475], [476], the composite source method [587], [588], and the hybrid Green's function method [405], [484], [488].

Any seismic time history (recorded or synthetic) from the aforementioned pool can be used to conditionally simulate spatially variable ground motions. However, the process of conditional simulation differs if the prescribed time history results in zero or non-zero residual displacements [301], and is illustrated herein in Sections 8.3.1 and 8.3.2, respectively. Section 8.3.1 utilizes the physical spectrum developed by Sabetta and Pugliese [436] for the generation of the reference, synthetic time history. The technique belongs to the stochastic ground motion method with random phases, and generates time series with an approach similar to the one described in Section 7.1.1 and presented in Eq. 8.14. The resulting synthetic time history is then conditionally simulated; this example results in zero residual values for the displacements. Section 8.3.2 conditionally simulates the time history recorded in the north-south direction at station HEC during the 1999 Hector Mine earthquake that yields non-zero values for the residual displacements [75]; a procedure for maintaining the correct values for the residual displacements at close-by distances in the conditionally simulated ground motions is also described in this section.

8.3.1 CONDITIONAL SIMULATIONS WITH ZERO RESIDUAL DISPLACEMENTS

Sabetta and Pugliese [436], based on the analysis of 95 accelerograms from 17 Italian earthquakes with magnitudes ranging from 4.6 to 6.8, introduced their physical spectrum. The physical spectrum is a frequency-time decomposition of the time series and a natural extension of the power spectral density to the non-stationary case, incorporating non-stationarity in both intensity and frequency content. They also provided empirical predictive equations for the pseudo-velocity response spectra and the physical spectrum parameters, including ground motion duration, which depended on earthquake magnitude, source-site distance and local soil conditions [436]. For completeness, their derived expressions and parameters, used herein for the generation of the prescribed time history, are presented next; the details of the derivation can be found in Refs. [434]–[436]. Conditionally simulated spatially variable time histories (accelerations, velocities and displacements) obeying the coherency model of Luco and Wong [324] and processed with both causal and acausal filtering close the section.

Physical Spectrum

The physical spectrum of an acceleration record can be evaluated with an approach similar to the one used for the derivation of the power spectral density (e.g., Eqs. 2.44 and 2.45), only, in this case, a moving time window is utilized, that provides the physical spectrum time dependence. In other words, the physical spectrum is a series of power spectral densities that are estimated at consecutive starting times for the duration of the selected window. Figure 8.7 presents the physical spectrum of an accelerogram recorded at the Garigliano station in 1984, which was evaluated with a Gaussian moving time window of 2.5 sec duration.

The expression for the physical spectrum based on the data of the Italian earthquakes was derived as [436]:

$$PS(t, f) = \frac{P_a(t)}{\sqrt{2\pi} f \delta_S} \exp\left\{-\frac{[\ln(f) - \ln(\beta(t))]^2}{2\delta_S^2}\right\} \qquad (8.18)$$

where f indicates frequency in [Hz], $P_a(t)$ is the instantaneous power (defined subsequently), and $\beta(t)$ and δ_S are given by:

$$\ln[\beta(t)] = \ln[F_c(t)] - \delta_S^2/2 \qquad (8.19)$$

$$\delta_S = \sqrt{\ln\left[1 + F_b^2(t)/F_c^2(t)\right]} \qquad (8.20)$$

$F_c(t)$ is the central frequency and $F_b(t)$ the frequency bandwidth, which correspond to the centroid and the radius of gyration with respect to $F_c(t)$, respectively, of the physical spectrum on the frequency plane. $F_c(t)$ and $F_b(t)$ are given by the following equations [434]:

$$\ln[F_c(t)] = 3.4 - 0.35 \ln(t) - 0.218M - 0.15S_2 \qquad (8.21)$$

$$F_b(t)/F_c(t) = 0.44 + 0.07M - 0.08S_1 + 0.03S_2 \qquad (8.22)$$

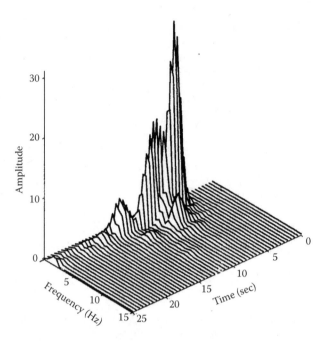

FIGURE 8.7 Illustration of the physical spectrum of an accelerogram. The ground motion was recorded at the deep alluvium site of the Garigliano station, Italy, in 1984 during an earthquake of magnitude 5.8 and epicentral distance of 52.6 km. The physical spectrum was obtained using a Gaussian moving time window of 2.5 sec duration. (Reprinted from F. Sabetta and A. Pugliese, "Estimation of response spectra and simulation of nonstationary earthquake ground motions," *Bulletin of the Seismological Society of America*, Vol. 86, pp. 337–352, Copyright ©1996 Seismological Society of America; courtesy of F. Sabetta.)

where M is magnitude, R is distance (in [km]), S_1 and S_2 assume the value of one for shallow and deep alluvium sites, respectively, and zero otherwise, $F_c(t)$ has units of [Hz], t is measured in [sec], and the ratio $F_b(t)/F_c(t)$ is dimensionless and time independent, which also renders δ_S in Eq. 8.20 to be time independent.

The instantaneous power, $P_a(t)$, in Eq. 8.18 was determined as:

$$P_a(t) = \frac{I_A}{t\sqrt{2\pi}\,s_a} \exp\left\{-\frac{[\ln(t) - m_a]^2}{2s_a^2}\right\} \qquad (8.23)$$

where,

$$m_a = \ln(T_2) + s_a^2 \qquad s_a = \ln(T_3/T_2)/2.5$$

$$T_1 = R/7 \qquad T_2 = T_1 + 0.5DV \qquad (8.24)$$

$$T_3 = T_1 + 2.5DV \qquad T_{total} = 1.3\,T_3$$

with I_A being the Arias intensity [35], and T_1 (in [sec]) the time delay between the arrival times of the S- and P-waves. T_1 was estimated as the ratio of the epicentral distance R over the factor $V_P V_S/(V_P - V_S)$, with V_P and V_S being the velocities of the P- and S-waves, respectively; in Eq. 8.24, this factor is assumed to be equal to 7 km/sec. The Arias intensity, I_A, in [cm^2/sec^3], the duration of the strong motion phase, DV, in [sec], as well as the values of the pseudo-velocity response spectrum, in [cm/sec], at each frequency are given by the following attenuation relation:

$$\log_{10}(Y) = a + bM + c \log_{10}(R^2 + h^2)^{1/2} + e_1 S_1 + e_2 S_2 \pm \sigma \qquad (8.25)$$

where Y is the parameter to be predicted, h is a fictitious depth incorporating all factors that tend to limit the motion near the source [246], σ is the standard deviation of the logarithm of Y, and the remaining parameters have been defined in Eqs. 8.21 and 8.22. The regression coefficients for the Arias intensity and the strong motion duration are provided in Table 8.1, and those for the smoothed pseudo-velocity spectrum for 5% damping as function of frequency in Table 8.2.

Equations 8.18–8.25 and Table 8.1 fully define the physical spectrum that can be used for the simulation of seismic time histories compatible with the Italian data as follows.

Generation of Spatially Variable Ground Motions

A simulation algorithm of physical-spectrum compatible time histories at a specific location can be expressed as [436]:

$$f(t) = 2 \sum_{k=1}^{K} \sqrt{PS(t, \omega_k)\Delta\omega} \, \cos(\omega_k t + \phi_k) \qquad (8.26)$$

where ϕ_k, $k = 1, \ldots, K$, are random phase angles uniformly distributed in the range $[0, 2\pi)$. It is noted that Eqs. 8.14 and 8.26 are similar, their difference being that the power spectral density, $S(\omega_k)$, in Eq. 8.14 is replaced with the physical spectrum, $PS(t, \omega_k)$, in Eq. 8.26, which is used to generate the reference time history for the subsequent conditional simulation.

The following example was presented by Liao and Zerva [301] as an illustration of conditional simulations of spatially variable ground motions with zero residual displacements. The site was assumed to be stiff, the epicentral distance $R = 25$ km and the earthquake magnitude $M = 6.8$. Acceleration time histories in the horizontal direction were generated at three equidistant stations along a straight line with separation

---

TABLE 8.1
Regression coefficients for I_A and DV of the larger horizontal component of the motions (after Sabetta and Pugliese [436], with the numbers in parentheses posted by Lestuzzi [293]); the parameters in the table are valid for $4.6 \leq M \leq 6.8$ and $R \leq 100$ km.

	a	b	c	e_1	e_2	h	σ
I_A [cm^2/sec^3]	0.729(1.679)	0.911(2.097)	−1.818	0.244	0.139	5.3	0.397
DV [sec]	−0.783(−1.802)	0.193(0.445)	0.208	−0.133	0.138	5.1	0.247

TABLE 8.2
Regression coefficients for the pseudo-velocity response spectrum (in [cm/sec]) of the larger horizontal component of the motions after Sabetta and Pugliese [436]; damping is 5%, and the parameters in the table are valid for $4.6 \leq M \leq 6.8$ and $R \leq 100$ km. (Reproduced from F. Sabetta and A. Pugliese, "Estimation of response spectra and simulation of nonstationary earthquake ground motions," *Bulletin of the Seismological Society of America*, Vol. 86, pp. 337–352, Copyright ©1996 Seismological Society of America.)

Frequency [Hz]	a	b	c	e_1	e_2	h	σ
0.25	−2.500	0.725	−1	0	0.100	2.6	0.319
0.33	−2.250	0.715	−1	0	0.108	3.0	0.319
0.50	−1.900	0.687	−1	0	0.150	3.6	0.319
0.67	−1.647	0.660	−1	0.010	0.175	4.0	0.315
1.00	−1.280	0.612	−1	0.050	0.208	4.4	0.308
1.33	−1.000	0.570	−1	0.120	0.190	4.7	0.303
2.00	−0.595	0.500	−1	0.230	0.124	5.0	0.290
2.50	−0.281	0.445	−1	0.222	0.078	5.2	0.280
3.33	0.100	0.377	−1	0.185	0.020	5.4	0.260
5.00	0.296	0.323	−1	0.161	0	5.7	0.234
6.67	0.222	0.310	−1	0.161	0	5.9	0.220
10.0	−0.019	0.304	−1	0.161	0	6.2	0.208
15.0	−0.312	0.304	−1	0.161	0	6.3	0.200
25.0	−0.817	0.330	−1	0.161	0	4.7	0.195

distance between consecutive stations of 100 m. The time history at station 1 was the synthetic ground motion obtained from Eq. 8.26, and the time series at stations 2 and 3 were conditionally simulated based on the CPDF approach (Section 8.1.1). The time step Δt was 0.02 sec, and the duration of the reference time history at station 1 was determined from Eq. 8.24 as $T_{total} = 26.32$ sec. The coherency model utilized was the model of Luco and Wong [324] with $\alpha = 2.5 \times 10^{-4}$ sec/m (Eq. 8.13). It was also considered that the motions propagate with an apparent velocity of 1000 m/sec along the line connecting the stations.

As indicated in Section 8.1.3, the non-stationarity of the reference time history cannot be readily maintained if the conditional simulation scheme is directly applied to the entire time series. This application utilized the time-domain segmentation: During the conditional simulation, the time series was divided into segments with different number of time steps. The segments were determined by observation so that the amplitudes and zero crossings within each segment were relatively similar. Each segment was then used as the reference time series, conditional simulations were carried out for each segment, and the conditionally simulated segments were joined together with linear taper functions to yield the entire acceleration time histories, into which a time shift to account for the wave passage effect was incorporated. The processing scheme with either causal or acausal filtering was then applied to the acceleration time series. The corner frequency of the filter was determined as 0.1 Hz based on the duration of the synthetic time history and Eq. 8.16. Subsequently, velocities and displacements were evaluated through consecutive integration of the accelerations.

Figures 8.8–8.10 present one sample of the acceleration, velocity and displacement time histories at the three stations; the leading zero-pads in the accelerations (Fig. 8.8) were necessary for the acausal filtering. In the figures, part (a) illustrates the results for acausal filtering and part (b) the results for causal filtering. The figures indicate that zero initial conditions and zero values at the end of the motions are satisfied for accelerations, velocities and displacements. It can also be seen from the figures that the conditionally simulated time histories at stations 2 and 3 preserve the non-stationary characteristics of the reference time history at station 1. Figure 8.8 suggests that there are no significant differences in the conditionally simulated acceleration time histories processed with either acausal or causal filtering. The differences, however, become more pronounced in the velocity and, even more, in the displacement time histories (Figs. 8.9 and 8.10, respectively). Specifically, Fig. 8.10 highlights the significance of phase distortion for causal filtering (Fig. 8.10(b)), which is absent in the acausal filtering results (Fig. 8.10(a)).

The ensemble averages of pseudo-velocity spectra of 10 simulations at the three stations, for which the reference time history was newly generated each time, are compared with the target pseudo-velocity spectrum in Fig. 8.11(a) for acausal filtering and in Fig. 8.11(b) for causal filtering; the target pseudo-velocity spectrum was evaluated from Eq. 8.25 and Table 8.2. There is satisfactory agreement between the spectra of the generated motions at the three stations and the predictive spectrum for both filtering operations. Even though it may be postulated that acausal filtering (Fig. 8.11(a)) produces a slightly better match of the spectra of the conditionally simulated motions with the target spectrum than causal filtering (Fig. 8.11(b)), the differences are very small. Figure 8.12 presents the ensemble lagged coherency of the 10 Monte Carlo simulations utilized in Fig. 8.11 together with the target coherency. The lagged coherency of the generated motions between stations 1 and 2, and stations 2 and 3, i.e., separation distance between stations of 100 m, was evaluated over their entire duration, and the spectral estimates were smoothed with an 11-point Hamming window (Eq. 2.47). The use of either acausal (Fig. 8.12(a)) or causal (Fig. 8.12(b)) filtering leads to simulated time histories that satisfactorily capture the target coherency even after the "segmenting and regrouping" operation on the time series. Figures 8.12(a) and 8.12(b) look identical, except for some very minor differences in the low frequency range (< 0.7 Hz). This should be anticipated: Coherency reflects phase variability (phase differences) between the motions at two stations. Causal filtering produces phase distortion in the low frequency range (cf. the displacement time histories in Figs. 8.10(a) and 8.10(b)). Hence, the differences in the phases between the conditionally simulated motions for each type of filtering will obey the coherency model, but the values of the estimated coherency for different types of filtering in the low frequency range can vary.

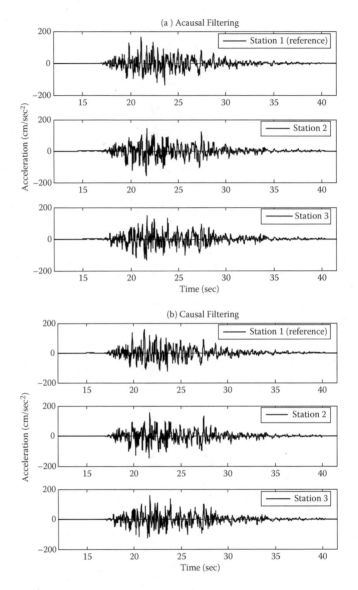

FIGURE 8.8 One sample of conditionally simulated and processed acceleration time histories. Part (a) illustrates the effect of acausal and part (b) the effect of causal filtering. The reference time history at station 1 was obtained from the physical spectrum of Sabetta and Pugliese [436] for $M = 6.8$, $R = 25$ km and stiff soil conditions. The motions were conditionally simulated at stations 2 and 3; the stations were located along a straight line with distance between consecutive stations of 100 m. The lagged coherency model of Luco and Wong [324] with $\alpha = 2.5 \times 10^{-4}$ sec/m was utilized, and the motions were considered to propagate with an apparent velocity of 1000 m/sec. (After S. Liao and A. Zerva, "Physically-compliant, conditionally simulated spatially variable seismic ground motions for performance-based design," *Earthquake Engineering and Structural Dynamics*, Vol. 35, pp. 891–919, Copyright © 2006 John Wiley & Sons Limited; reproduced with permission.)

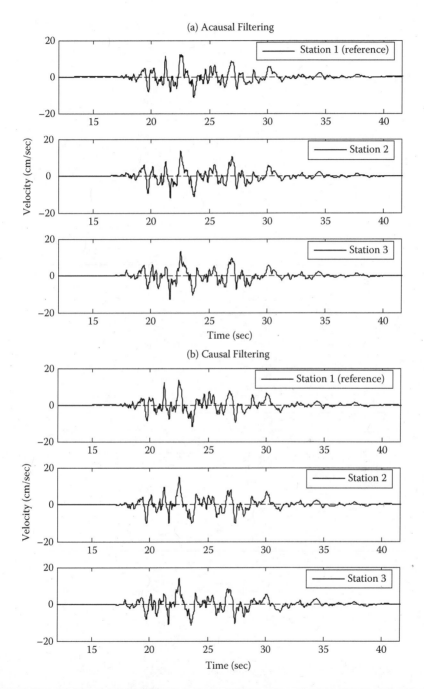

FIGURE 8.9 Velocity time histories integrated from the acceleration time series of Fig. 8.8. Part (a) illustrates the effect of acausal and part (b) the effect of causal filtering. (After S. Liao and A. Zerva, "Physically-compliant, conditionally simulated spatially variable seismic ground motions for performance-based design," *Earthquake Engineering and Structural Dynamics*, Vol. 35, pp. 891–919, Copyright © 2006 John Wiley & Sons Limited; reproduced with permission.)

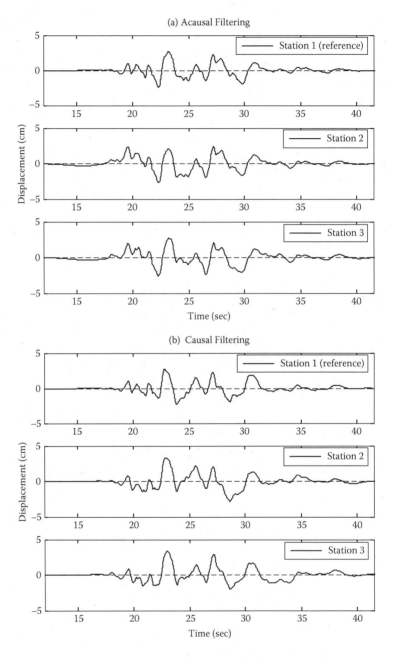

FIGURE 8.10 Displacement time histories obtained through double integration of the acceleration time series of Fig. 8.8. Part (a) illustrates the effect of acausal and part (b) the effect of causal filtering. (After S. Liao and A. Zerva, "Physically-compliant, conditionally simulated spatially variable seismic ground motions for performance-based design," *Earthquake Engineering and Structural Dynamics*, Vol. 35, pp. 891–919, Copyright © 2006 John Wiley & Sons Limited; reproduced with permission.)

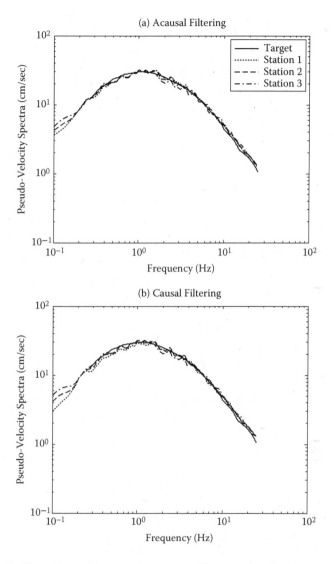

FIGURE 8.11 Comparison of the ensemble averages of the pseudo-velocity spectra of 10 simulations, generated from the same random field characteristics as the simulation set presented in Fig. 8.8, with the target pseudo-velocity spectrum. The reference time series at station 1 was newly generated each time. Part (a) illustrates the effect of acausal and part (b) the effect of causal filtering. (After S. Liao and A. Zerva, "Physically-compliant, conditionally simulated spatially variable seismic ground motions for performance-based design," *Earthquake Engineering and Structural Dynamics*, Vol. 35, pp. 891–919, Copyright © 2006 John Wiley & Sons Limited; reproduced with permission.)

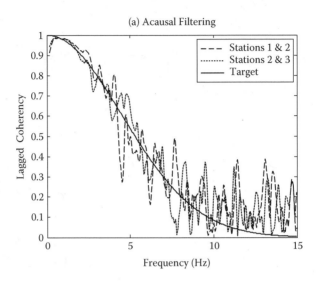

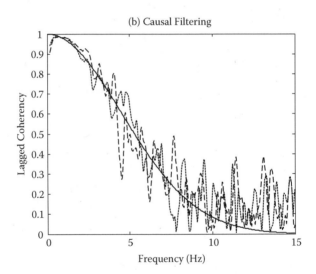

FIGURE 8.12 Comparison of the ensemble averages of the lagged coherency of 10 simulations, generated from the same random field characteristics as the simulation set presented in Fig. 8.8, with the target coherency. The separation distance between stations is 100 m; the spectral estimates for the evaluation of the lagged coherency of the time series were obtained with an 11-point Hamming window (Fig. 2.4). Part (a) illustrates the effect of acausal and part (b) the effect of causal filtering. (After S. Liao and A. Zerva, "Physically-compliant, conditionally simulated spatially variable seismic ground motions for performance-based design," *Earthquake Engineering and Structural Dynamics*, Vol. 35, pp. 891–919, Copyright © 2006 John Wiley & Sons Limited; reproduced with permission.)

8.3.2 CONDITIONAL SIMULATIONS WITH NON-ZERO RESIDUAL DISPLACEMENTS

Near-fault ground motions can be significantly affected by the directivity pulse and the fling-step [178], [394], [489]. The directivity pulse is characterized by a two-sided velocity pulse resulting in zero residual displacements, whereas the fling-step is a one-sided velocity pulse leading to non-zero residual displacements [178], [394], [489]. Graves [178] indicated that directivity effects do not, generally, require special processing algorithms for the recovery of the ground motion waveforms, but, conversely, the fling-step effect is sensitive to the processing algorithms and the true residual displacements are difficult to retrieve.

Boore et al. [75] analyzed acceleration time histories recorded at station HEC during the Hector Mine earthquake of October 16th, 1999 ($M_w = 7.1$); the closest distance of the station from the fault was 10.7 km. These near-fault motions include non-zero residual displacements, which were independently estimated from analyses of Interferometric Synthetic Aperture Radar (InSAR) data with an error estimate within 5 cm. Because of the availability of the InSAR estimation, Boore et al. [75] utilized these ground motions to compare the residual displacement retrievability of several baseline adjustment algorithms. The recorded time history in the north-south direction was corrected, and the residual displacement was reliably evaluated for both bi-linear and quadratic baseline correction schemes fitted to the velocity time series. The residual displacement in the east-west direction, however, was not so successfully recovered [75].

For corrected time histories in near-fault regions, such as the north-south component of the motions at the HEC station, or synthetic ones incorporating the fling-step effect (e.g., [394]), conditional simulations incorporating spatial variability effects can then be generated. For this purpose, Liao and Zerva [301] proposed the following approach: First, the low-frequency part of the motions leading to non-zero residual displacements is separated from the acceleration record by applying a high-pass filter with a low-valued corner frequency depending on the record. A causal filter was suggested in this stage of the process to avoid the artificial leading zero-pads required in the acausal filtering. It is then considered that these low-frequency components of the motions are highly coherent for close-by stations and, hence, regarded as deterministic; this consideration can be revisited when more information on the attenuation of low-frequency, near-fault motions becomes available. Second, it is considered that the remaining (higher-frequency) part of the acceleration time history can vary spatially on the ground surface obeying an available coherency model. This filtered acceleration time history is used as the prescribed time history to conditionally simulate spatially variable ground motions, that are, subsequently, processed to yield zero residual displacements. Finally, the low-frequency part of the motions is superimposed back to the conditionally simulated, spatially variable time series.

The approach was utilized to generate spatially variable ground motions at four equidistant stations along a straight line with separation distance between consecutive stations of 50 m [301]. The reference time history was the north-south component of the motions recorded at the HEC station with a time window of $T = 45$ sec, starting at 13.5 sec from the beginning of record; this time window includes the most significant part of the record as suggested by Boore et al. [75]. The time step Δt of the recorded time history is 0.01 sec. The unprocessed acceleration time history is available from the Southern California Earthquake Data Center [510]. For conditionally simulating motions at uniform soil conditions based on this time history, a 4-th order causal Butterworth high-pass filter with a corner frequency of 0.2 Hz was first applied to separate the low-frequency components of the recorded data. The corner frequency of 0.2 Hz was utilized, so as to ensure that the entire low-frequency part of the motions was preserved [298]. As indicated earlier, the two parts of the acceleration time history (low-frequency part and filtered part) were treated separately: The low-frequency part was considered deterministic (fully coherent) due to its low frequency content, and the remaining (higher-frequency) part of the acceleration time history was conditionally simulated based on a selected spatial coherency model. Liao and Zerva [301] indicated that, if the entire accelerogram were conditionally simulated, the resulting time histories would exhibit significantly different displacement waveforms with variable residual values even at close-by stations, which is unrealistic.

The coherency model used in this example for the filtered (higher-frequency) part of the record was the model proposed by Harichandran and Vanmarcke [208] (Eq. 3.12 repeated here):

$$|\gamma(\xi, \omega)| = A \exp\left(-\frac{2 B |\xi|}{a v(\omega)}\right) + (1 - A) \exp\left(-\frac{2 B |\xi|}{v(\omega)}\right) \qquad (8.27)$$

$$v(\omega) = k \left[1 + \left(\frac{\omega}{2\pi f_0}\right)^b\right]^{-1/2} ; \quad B = (1 - A + a A)$$

with parameters $A = 0.626$, $a = 0.022$, $k = 1.97 \times 10^4$, $f_0 = 2.02$ Hz and $b = 3.47$ evaluated by Harichandran and Wang [211] from data recorded in the radial direction during Event 20 at the SMART-1 array. The model was selected for illustration purposes, but purposely so, because it produces non-zero coherency at low frequencies and, hence, velocity and displacement time histories resulting from this model will exhibit more variability in their waveforms at shorter distances than if the model of Luco and Wong (Eq. 8.13) were utilized (see, e.g., Figs. 7.9 and 7.10). It would be interesting to examine whether the partial correlation of the motions resulting from this model at low frequencies would affect the final form of the displacements in the conditional simulation; the model of Luco and Wong [324] would yield perfect correlation at low frequencies. It is noted that when coherency models for near-fault regions are proposed, such models will be more appropriate for the generation of conditional simulations in these regions. The wave passage effect was not considered in this example application.

The filtered, higher-frequency part of the accelerogram was then divided into segments with variable number of time steps. As in the first example, the segment lengths were determined by observation, so that amplitudes and zero crossings within each segment were relatively similar. Each segment was used as the reference time history in the conditional simulation and individually simulated in order to approximately maintain the non-stationary characteristics of the filtered part of the recorded motion. The segments were then joined together with linear taper functions to construct the filtered, higher-frequency part of the acceleration time histories at the four stations.

The filtered time histories were then processed with a Butterworth filter, the corner frequency of which was evaluated as 0.06 Hz from the duration of the record and Eq. 8.16. It is noted that this additional filtering of the conditionally simulated motions, as well as the higher-frequency part of the reference ground motion, is necessary (i) to ensure that the resulting higher-frequency component of the time series will yield zero residual displacements, and (ii) for compatibility between the conditionally simulated time series with the filtered, higher-frequency part of the recorded data. Subsequently, the low-frequency part of the record, which was considered to be the same at all four stations, was superimposed to the filtered and processed higher-frequency accelerations. The corresponding velocity and displacement time histories were then obtained via integration.

One sample of the acceleration and the corresponding integrated velocity and displacement time histories at the four stations are presented in Figs. 8.13, 8.14 and 8.15, respectively, in which part (a) presents the results of the acausal filtering operation on the higher-frequency part of the series, and part (b) the results of the causal filtering operation. All velocity and displacement time histories (Figs. 8.14 and 8.15) have zero initial values. The initial values of the simulated acceleration time histories are non-zero but negligible (Fig. 8.13); these small values were caused by the removal of the derivative of the quadratic function fitted to the velocities from the accelerations. As should be the case, accelerations and velocities tend to zero values at the end of the motions (Figs. 8.13 and 8.14). For the same filtering operation, the displacement time series at the four stations (Fig. 8.15) are similar in shape and tend to the same non-zero residual value. The partial correlation of the filtered, higher-frequency part of the motions at low frequencies, caused by the model of Harichandran and Vanmarcke (Eq. 8.27), does not affect the shape of the total waveforms, which are dominated by the low-frequency part of the record that was assumed to be deterministic and identical at all stations. Figure 8.16 compares the displacement time histories at station 1, which resulted from the removal of the low-frequency component, the additional processing scheme of the filtered part, and the subsequent superposition of the low- and higher-frequency components, with the "original" time history. This original time history was processed with the approach proposed by Boore *et al.* [75] without any separation of low- and higher-frequency components. The displacement time history at station 1, resulting from the separation of the low- and higher-frequency components and the use of acausal filtering for the processing of its higher-frequency part, is almost identical with the original time history. The small differences are due to the additional (acausal) filtering required for compatibility of the filtered, higher-frequency part of the record with the conditionally simulated time series. On the other hand, the use of causal filtering in the conditional simulation process leads to a displacement waveform at station 1 that differs from the original time series, because of the phase distortion introduced by the causal filtering operation. Figure 8.17 presents the acceleration response spectra of the time histories of Fig. 8.13. The spectra of the motions at the individual stations are in good agreement irrespective of whether causal or acausal filtering was used in the processing of their higher-frequency component. Acausal filtering of the higher-frequency components of the motions (Fig. 8.17(a)) results in a better match of the response spectra of the simulations with the response spectrum of the motions at

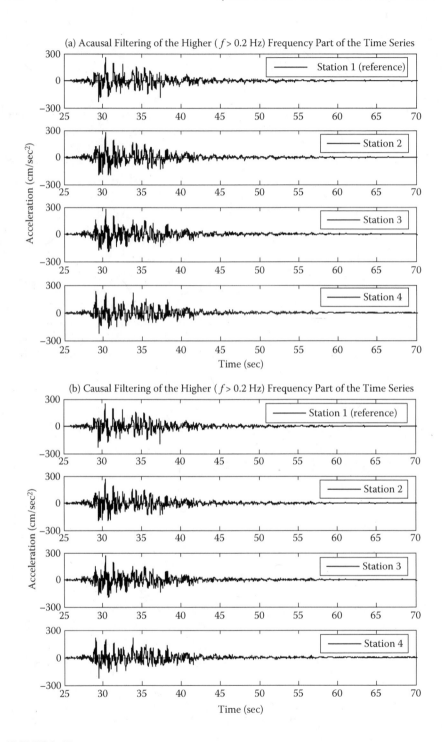

FIGURE 8.13

station 1 at lower frequencies, whereas causal filtering produces a slightly better fit at the very high frequencies (Fig. 8.17(b)); the differences, however, are not significant. Figure 8.18 compares the ensemble averages of the lagged coherency estimated from 10 simulations between stations 1 and 2, stations 2 and 3, and stations 3 and 4, at a separation distance of 50 m, with the target lagged coherency; again, the estimated lagged coherency was evaluated from the acceleration time histories and the spectral estimates were smoothed with an 11-point Hamming window (Eq. 2.47). Both filtering operations lead to consistent estimates for the lagged coherency. The only difference is in the low frequency range, where acausal filtering (Fig. 8.18(a)) results in smaller variability of the estimated coherency around the target value than causal filtering (Fig. 8.18(b)). This can also be attributed to the phase differences caused by the two filtering operations, as discussed in Fig. 8.12, but the differences are, again, not significant. Figures 8.13–8.18 then suggest that the approach of separating the low- and higher-frequency components of the reference time series and conditionally simulating its higher-frequency part provides realistic spatially variable ground motions for the seismic response of lifelines in near-fault regions. Figure 8.16, however, suggests that the use of acausal filtering in the processing of the higher-frequency part of the motions in the conditional simulation process recovers the characteristics of the original time history and, thus, is better suited for such applications.

Simulations and conditional simulations can then be utilized as input excitations for the seismic response evaluation of lifeline systems in a Monte Carlo framework. Example applications illustrating the effect of spatially variable ground motions on the response of long structures are presented in the following chapter.

FIGURE 8.13 One sample of conditionally simulated and processed acceleration time histories. The reference ground motion at station 1 is the north-south component of the motions recorded during the 1999 Hector Mine earthquake at station HEC. In the simulation process, the low-frequency ($f < 0.2$ Hz) part of the accelerogram was removed and retained, and the remaining higher-frequency part conditionally simulated at stations 2, 3 and 4. The two parts were then combined. Part (a) illustrates the effect of acausal filtering on the higher-frequency component of the motions, and part (b) the effect of causal filtering on the higher-frequency component of the motions. The stations were located along a straight line with distance between consecutive stations of 50 m; coherency was described by the model reported by Harichandran and Wang [211]. (After S. Liao and A. Zerva, "Physically-compliant, conditionally simulated spatially variable seismic ground motions for performance-based design," *Earthquake Engineering and Structural Dynamics*, Vol. 35, pp. 891–919, Copyright © 2006 John Wiley & Sons Limited; reproduced with permission.)

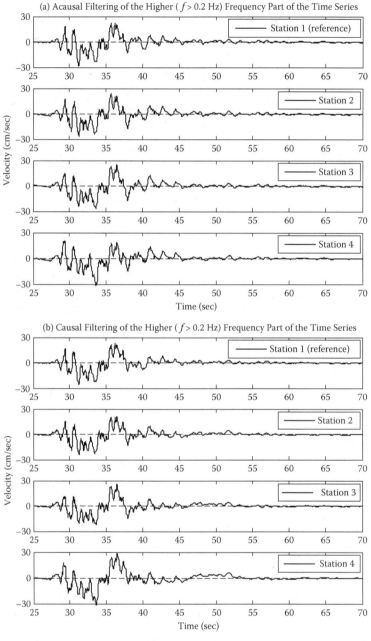

FIGURE 8.14 Velocity time histories integrated from the acceleration time series of Fig. 8.13. Part (a) illustrates the effect of acausal filtering on the higher-frequency component of the motions, and part (b) the effect of causal filtering on the higher-frequency component of the motions. (After S. Liao and A. Zerva, "Physically-compliant, conditionally simulated spatially variable seismic ground motions for performance-based design," *Earthquake Engineering and Structural Dynamics*, Vol. 35, pp. 891–919, Copyright © 2006 John Wiley & Sons Limited; reproduced with permission.)

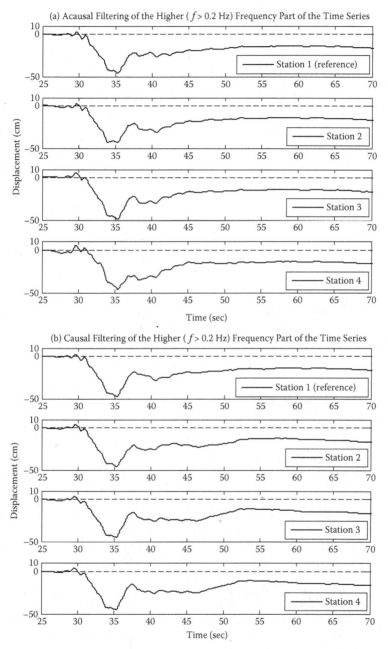

FIGURE 8.15 Displacement time histories obtained through double integration of the acceleration time series of Fig. 8.13. Part (a) illustrates the effect of acausal filtering on the higher-frequency component of the motions, and part (b) the effect of causal filtering on the higher-frequency component of the motions. (After S. Liao and A. Zerva, "Physically-compliant, conditionally simulated spatially variable seismic ground motions for performance-based design," *Earthquake Engineering and Structural Dynamics*, Vol. 35, pp. 891–919, Copyright © 2006 John Wiley & Sons Limited; reproduced with permission.)

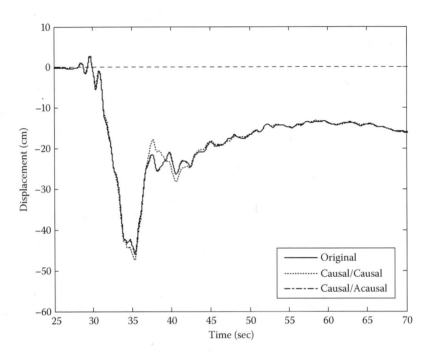

FIGURE 8.16 Illustration of the effect of the treatment of ground motions with non-zero residual displacements in conditional simulation. Labeled "original" is the time series obtained from the record with the baseline adjustment scheme suggested by Boore *et al.* [75]. Labeled "causal/causal" is the time series at station 1 (the same as that presented in the top subplot of Fig. 8.15(b)), for which the low-frequency component of the motions was removed with causal filtering and the remaining part processed, for compatibility with the conditional simulations, with causal filtering. Labeled "causal/acausal" is the time series at station 1 (the same as that presented in the top subplot of Fig. 8.15(a)), for which the low-frequency component of the motions was removed with causal filtering and the remaining part processed with acausal filtering. (After S. Liao and A. Zerva, "Physically-compliant, conditionally simulated spatially variable seismic ground motions for performance-based design," *Earthquake Engineering and Structural Dynamics*, Vol. 35, pp. 891–919, Copyright ©2006 John Wiley & Sons Limited; reproduced with permission.)

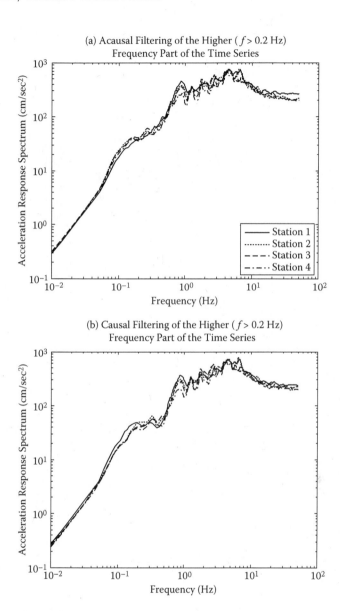

FIGURE 8.17 Comparison of acceleration response spectra of the conditionally simulated ground motions with the response spectrum of the reference time series (at station 1) based on the record of the 1999 Hector Mine earthquake at station HEC in the north-south direction; the acceleration time histories are those presented in Fig. 8.13. Part (a) illustrates the effect of acausal filtering on the higher-frequency component of the motions, and part (b) the effect of causal filtering on the higher-frequency component of the motions. (After S. Liao and A. Zerva, "Physically-compliant, conditionally simulated spatially variable seismic ground motions for performance-based design," *Earthquake Engineering and Structural Dynamics*, Vol. 35, pp. 891–919, Copyright © 2006 John Wiley & Sons Limited; reproduced with permission.)

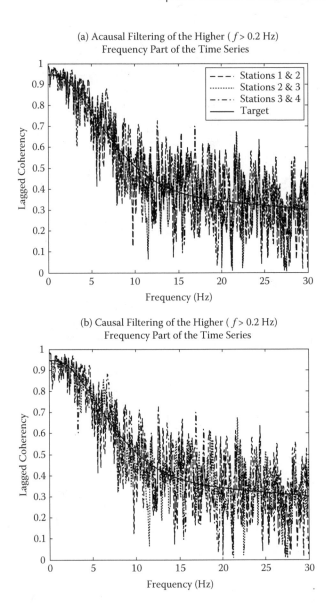

FIGURE 8.18 Comparison of the ensemble averages of the lagged coherency of 10 simulations, generated from the same random field characteristics as the simulation set presented in Fig. 8.13, with the target coherency. The separation distance between stations is 50 m; the spectral estimates for the evaluation of the lagged coherency of the time series were obtained with an 11-point Hamming window (Fig. 2.4). Part (a) illustrates the effect of acausal filtering on the higher-frequency component of the motions, and part (b) the effect of causal filtering on the higher-frequency component of the motions. (After S. Liao and A. Zerva, "Physically-compliant, conditionally simulated spatially variable seismic ground motions for performance-based design," *Earthquake Engineering and Structural Dynamics*, Vol. 35, pp. 891–919, Copyright ©2006 John Wiley & Sons Limited; reproduced with permission.)

9 Engineering Applications

Structures that extend for long distances parallel to the ground, such as pipelines, tunnels, bridges and dams, are subjected to the effect of spatially variable ground motions during an earthquake. Newmark and Rosenblueth [369] stated in 1971 regarding the seismic resistant design of bridges: "Within their diversity of dimensions, materials, and structural solutions, bridges have certain features in common from the viewpoint of their response to earthquakes. Their salient characteristic lies in the fact that their supports tend to undergo differential motions during earthquakes. This is partly due to distance between supports and partly due to the differences in geologic and topographic features at the supports or surrounding them. Therefore, even for short spans, there is a tendency for the abutments to move differentially."

Starting in the middle 1960's, pioneering studies analyzed the wave passage effect on the response of buildings, bridges and pipelines (e.g., [65], [245], [366], [367], [561], [562]). Since then, the installation of dense instrument arrays and the estimation of the coherency from seismic data have led to a plethora of studies investigating the effect of spatially variable excitations on the response of a large variety of structural systems. Studies examined the response of large, mat, rigid foundations, such as those of nuclear power plants, induced by spatially variable seismic ground motions (e.g., [203], [224], [267], [322], [324], [345], [391], [480], [554], [555]); developed response spectra incorporating spatial variability in the excitations (e.g., [534], [539], [584], [585]); analyzed the response of symmetric and asymmetric buildings (e.g., [189], [192], [198], [218]) and pounding of adjacent buildings (e.g., [49], [199]) caused by spatially variable excitations; assessed the influence of spatially variable excitations on the response of offshore structures [356], [357]; evaluated the response of buried pipelines and tunnels (e.g., [124], [217], [221], [595], [600]); considered the effect of spatial variability on the response of dams (e.g., [21], [44], [100], [272], [273], [305], [328], [378]); and investigated the response of suspension bridges (e.g., [2], [3], [306], [571]) and cable-stayed bridges (e.g., [1], [307], [362], [511]) subjected to spatially variable seismic ground motions. The structural configuration, however, that has attracted by far the most significant research interest is that of highway bridges. Early evaluations of the effect of the spatial variability of seismic ground motions on the response of highway bridges considered very simple models of the structures, namely single-span or multi-span, multiply supported beams, as, e.g., [195], [209], [210], [315], [334], [413], [590], [591], [600]. Section 6.4.5 presented an example illustration of these analyses, which, even though simple, provided insight into the complex problem of the effect of the spatial variability of the seismic ground motions on the response of the structures. The research interest on the subject increased over the years with the investigation of the linear and nonlinear response of more realistic models of highway bridges and viaducts, as, e.g., among many others, in Refs. [113], [130], [137], [247], [268], [291], [306], [319], [320], [321], [325], [347], [348], [353], [354], [355], [379], [394], [401], [442], [443], [453], [454], [455], [456], [468], [546], [575] and references therein.

This chapter presents illustrative applications of the effect of the spatial variation of the seismic ground motions on the response of long structures. It begins, in Section 9.1, with the evaluation of the response of large, mat, rigid foundations, such as those of nuclear power plants, subjected to nonuniform seismic excitations. As indicated in Section 3.2, some of the early investigations of spatial variability were geared towards the quantification of the effect of the "spatial averaging" of the incident excitations caused by the foundations of the plants, and some of the spatial arrays (e.g., the LSST array in Fig. 1.3) were deployed for this purpose. The section presents, essentially, the insightful and conclusive study of Luco and Wong [324], briefly illustrates the provisions of the ASCE 4-98 standard regarding reductions in the foundation's input motion induced by the loss of coherency in the excitations, and concludes with a recent study on the empirical estimation of these effects based on recorded data. Section 9.2 presents examples of the analysis of dams subjected to spatially variable excitations; the number of studies analyzing the effect of spatial variability on these massive structures is limited. The response of an earth dam and a concrete gravity dam are presented first; in these applications, the conventional modeling of the spatial variability, as presented herein, is utilized to describe the excitations at the base of the dam-foundation system. The significance of complex topographic effects on the 3-D response of arch dams is presented next; in this case, the conventional models for the spatial variability of the seismic ground motions are no longer applicable. The section concludes with a recent study that evaluated the response of the Pacoima Dam subjected to inferred spatially variable seismic data. Section 9.3 describes the effect of nonuniform seismic excitations on the response of suspension and cable-stayed bridges beginning with the pioneering study of Bogdanoff et al. [65] and, mainly, concentrating on the insightful work of Abdel-Ghaffar et al. [1], [2], [3], [362]. Additional studies examining the effect of nonuniform motions on the response of these structures are briefly highlighted; the majority of the studies were geared towards improvements of the random vibration approach (Chapter 6) and considered linear structural models. On the other hand, the current design/retrofit of suspension and cable-stayed bridges is based on their rigorous linear/nonlinear modeling. Furthermore, these evaluations, generally, utilize conditionally simulated spatially variable seismic ground motions as input excitations at the structures' supports, which is a realistic scenario, as spatial variability reflects the true nature of the seismic ground motions. A brief reference to a recent publication that presented in detail the modeling considerations for the design of a new suspension bridge, and a short description of the approach for the development of conditionally simulated ground motions provided to designers by the New York City Department of Transportation conclude Section 9.3. The incorporation of spatial variability effects in the design of highway bridges is, however, different. The questions posed are: Is this effect significant? Are there cases when it can be neglected? Can one establish an "equivalent" uniform excitation that captures the effect of the nonuniform motions on the structural response? An extensive number of studies conducting sensitivity analyses of the effect of the spatial variation of the motions on the response of highway bridges and addressing these issues has appeared in the literature. Some representative ones are presented and discussed to some extent in Section 9.4, as they reflect currently evolving research work for design implementations. The section also highlights the recommendations of the new European Commission code that proposed an "equivalent" approach for the consideration of spatially variable ground motions in the design of highway bridges. The approach separates the pseudo-static and the dynamic response of the structures, evaluates the pseudo-static response in terms of prescribed displacement patterns and the dynamic response based on uniform seismic excitations, and combines the two responses. It is noted that the estimation of the seismic strain field for consideration in the response evaluation of above-ground and buried structures was addressed in Chapter 5, and will not be repeated in this chapter. Finally, Section 9.5 presents some concluding remarks.

9.1   LARGE, MAT, RIGID FOUNDATIONS

It has been recognized since the early 1970's that, because of the spatially variable nature of the incident ground motions, large, mat, rigid foundations, such as those of nuclear power plants, tend to "average" the seismic excitation [86], [336], [368], [444], [518], [576]. These studies indicated that, unless the ground motion consists of vertically propagating, coherent waves, its interaction with the -massless- foundation (kinematic Soil Structure Interaction − SSI) tends to decrease the foundation's translational response as the frequency of the excitation and the foundation size increase. In other words, the foundation, subjected to spatially variable excitations, modifies the free-field motion, i.e., the ground motion in the absence of the foundation, and acts as a "filter," which smoothes the nonuniform part of the excitation, whether it is caused by inclined coherent body waves, surface waves, or motions exhibiting loss of coherency. The excitation modified by the massless, rigid foundation (Foundation Input Motion − FIM) is then obtained by means of transfer functions that filter the free-field ground motion. This section begins with the presentation of the results of the widely referenced study of Luco and Wong [324], who introduced explicitly the relative effects of loss of coherency and wave passage on the response of large, mat, rigid foundations. It is followed with a brief description of an approximate, closed-form solution to the same problem, which illustrates the effect of the foundation shape on its translational and torsional response. The incorporation of spatial variability effects in an ASCE standard for the design of power plants is presented next, and the section concludes with a fairly recent study that proposed an empirical approach for the estimation of the FIM based on recorded seismic data.

In a series of publications, Luco and Wong [324] and Luco and Mita [322], [345] examined the effect of the spatial variability in the seismic excitations on the response of rigid, rectangular and circular foundations. Their classic work [324] analyzed the response of a flat, rigid, massless foundation bonded to a viscoelastic half-space (Fig. 9.1) as a boundary value problem. Luco and Wong [324] subjected the foundation to an incident 3-D spatially variable random field and evaluated its six response components, namely, three translations, u_1, u_2, u_3, and three rotations, $u_4 = a\theta_1$, $u_5 = a\theta_2$, $u_6 = a\theta_3$, where a is half the length of the foundation along the x_1-axis (Fig. 9.1). The rotational components u_4 and u_5 indicate rocking about the x_1- and x_2-axis, respectively, whereas u_6 indicates torsion.

The spatial variability of the seismic excitation between two locations $\vec{x}$ and $\vec{x}'$ on the ground surface and between ground motion components in two orthogonal

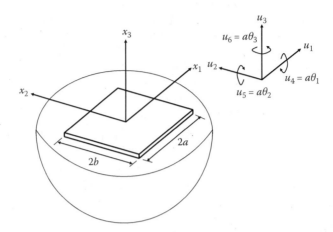

FIGURE 9.1 Set-up of a mat, massless, rigid foundation of dimensions $2a \times 2b$ bonded on a half-space that was utilized in the study by Luco and Wong [324]; the source-site direction is assumed to coincide with the x_1-axis. The insert on the top right of the figure indicates the foundation's degrees of freedom: three translations (u_1, u_2 and u_3) and three rotations (u_4, u_5 and u_6). The rotational components u_4 and u_5 indicate rocking about the x_1- and x_2-axis, respectively, whereas u_6 indicates torsion. (After J.E. Luco and H.L. Wong, "Response of a rigid foundation to a spatially random ground motion," *Earthquake Engineering and Structural Dynamics*, Vol. 14, pp. 891–908, Copyright © 1986 John Wiley & Sons Limited; reproduced with permission.)

directions m and n (m, $n = 1$, 2, 3) was given by the expression:

$$\gamma_{mn}(\vec{x}, \vec{x}', \omega) = |\gamma(\vec{x}, \vec{x}', \omega)| \exp\left[-i\omega\left(\frac{x_1}{c_m} - \frac{x_1'}{c_n}\right)\right] \tag{9.1}$$

As illustrated in Chapter 3 (Eq. 3.3), the first term on the right-hand-side of Eq. 9.1 is the lagged coherency, and the second term represents the wave passage effect. In the exponential term, x_1 and x_1' are the coordinates of the two stations in the source-site direction, and c_m and c_n indicate the apparent propagation velocity of the m and n components of the motions, respectively. The lagged coherency, $|\gamma(\vec{x}, \vec{x}', \omega)|$, was based on the analysis of wave propagation through random media by Uscinski [549], who derived the coherency function of scalar shear waves propagating a distance R through a random medium as:

$$|\gamma(\vec{x}, \vec{x}', \omega)| = \exp\left\{-\eta\left[1 - \exp\left(-|\vec{x} - \vec{x}'|^2/r_0^2\right)\right]\right\} \tag{9.2}$$

where r_0 is the scale length of random inhomogeneities along the wave path, and

$$\eta \approx \omega^2 r_0 R \mu^2 / V_s^2 \tag{9.3}$$

In the above expression, μ^2 is a measure of the relative variation of the elastic properties in the medium and V_s an estimate of the S-wave velocity (not to be identified with c_m, $m = 1$, 2, 3). For station separation distances smaller than the scale length of the random inhomogeneities, i.e., $|\vec{x} - \vec{x}'| < r_0$, Eq. 9.2 can be approximated by the following expression:

$$|\gamma(\vec{x}, \vec{x}', \omega)| = \exp\{-(\nu\omega|\vec{x} - \vec{x}'|/V_s)^2\} \qquad (9.4)$$

with $\nu = \mu(R/r_0)^{1/2}$. A simplified one-dimensional version of Eq. 9.4 was presented in Eq. 3.27, in which the coherency drop parameter, α, was defined as $\alpha = \nu/V_s$; the exponential decay of this model with separation distance and frequency for a median value of α from the ones suggested by Luco and Wong [324] was illustrated in Fig. 3.23.

Luco and Wong [324] then proceeded with the examination of the effect of loss of coherency and wave propagation on the foundation response. They presented their results in the form of transfer functions, A_{mn}^{jk}; $m, n = 1, 2, 3$; $j, k = 1, 2, 3, 4, 5, 6$, for a square foundation of dimensions $2a \times 2a$ ($a = b$ in Fig. 9.1); the subscripts m and n represent the direction of the excitation component, and the superscripts j and k the direction of the response component. The physical interpretation of the transfer functions can be illustrated as follows [324]: Consider, e.g., that the input excitation has a single component in the n-th direction, i.e., the x_1-, x_2- or x_3-direction (Fig. 9.1), which can be described by a stationary process. For $j = k$ and $n = m$, A_{nn}^{jj} represents the ratio of the power spectral density (PSD) of the j-th component of the response over the PSD of the n-th component of the excitation, and, hence, its square-root reflects the amplitude of the frequency transfer function of the foundation response (e.g., Eq. 6.18). For $j \neq k$, A_{nn}^{jk} indicates the correlation between the j-th and k-th components of the foundation response induced by the n-th component of the excitation (Fig. 9.1). When $m \neq n$, A_{mn}^{jk} represents the effect of the correlation between the components of the excitation in two different directions on the foundation response.

Figures 9.2 and 9.3 present the amplitudes of the transfer functions, $\sqrt{A_{nn}^{jj}}$, for seismic ground motions exhibiting loss of coherency only with variable ν ($\nu = 0$, i.e., fully coherent motions, and $\nu = 0.1, 0.2, 0.3, 0.4$ and 0.5) as functions of the normalized frequency $a_0 = a\omega/V_s$. In this evaluation, the seismic ground motions arrive simultaneously at all foundation locations, i.e., $c_m \rightarrow \infty$, $m = 1, 2, 3$. In particular, Fig. 9.2 presents the effect of the seismic excitation on the translational response of the foundation and Fig. 9.3 on its rotational (rocking and torsional) response. The equalities $\sqrt{A_{11}^{11}} = \sqrt{A_{22}^{22}}$ (Fig. 9.2(a)), $\sqrt{A_{22}^{11}} = \sqrt{A_{11}^{22}}$ (Fig. 9.2(b)), $\sqrt{A_{33}^{11}} = \sqrt{A_{33}^{22}}$ (Fig. 9.2(c)), $\sqrt{A_{11}^{55}} = \sqrt{A_{22}^{44}}$ (Fig. 9.3(a)), $\sqrt{A_{22}^{55}} = \sqrt{A_{11}^{44}}$ (Fig. 9.3(b)) and $\sqrt{A_{33}^{55}} = \sqrt{A_{33}^{44}}$ (Fig. 9.3(c)) are a consequence of the foundation symmetry and the directional independence of the assumed coherency function of Eq. 9.4 [324]; for the same reason, $\sqrt{A_{11}^{33}} = \sqrt{A_{22}^{33}}$ (Figs. 9.2(d) and (e)) and $\sqrt{A_{11}^{66}} = \sqrt{A_{22}^{66}}$ (Figs. 9.3(d) and (e)). It can be seen from Fig. 9.2 that the excitation in each direction affects mostly the translational response of the foundation in the same direction ($\sqrt{A_{11}^{11}}, \sqrt{A_{22}^{22}}, \sqrt{A_{33}^{33}}$ in Figs. 9.2(a) and (f)) and only minimally the translational response in the other two directions. Furthermore, the amplitudes of the transfer functions $\sqrt{A_{11}^{11}} = \sqrt{A_{22}^{22}}$ (Fig. 9.2(a)) and $\sqrt{A_{33}^{33}}$ (Fig. 9.2(f)) decrease considerably as frequency increases and the motions become less coherent (ν increases), which is caused by the kinematic constraint imposed by the rigid foundation on the free-field motion [324]. Figures 9.2(a) and (f) clearly indicate that the translational foundation response becomes lower than the incident seismic excitation, since the amplitude of

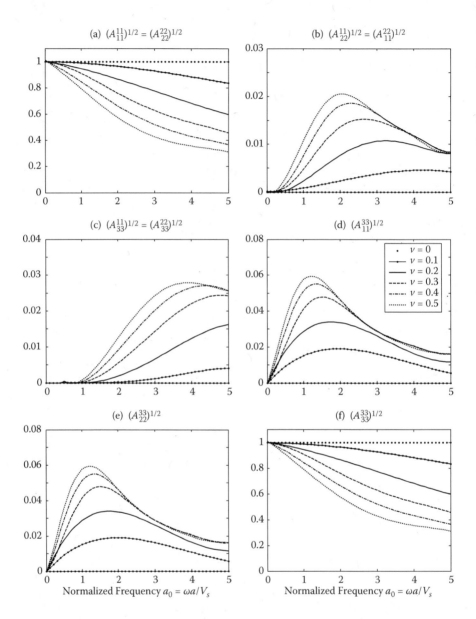

FIGURE 9.2 Translational response characteristics of the foundation illustrated in Fig. 9.1, with $a = b$, subjected to spatially variable ground motions exhibiting only loss of coherency with variable ν (Eq. 9.4) as functions of the normalized frequency $a_0 = a\omega/V_s$. The response characteristics, reported by Luco and Wong [324], are presented in terms of square roots of frequency transfer functions $\sqrt{A_{mm}^{jj}}$, where $m = 1, 2, 3$ indicates the component of the excitation and $j = 1, 2, 3$ the translational component of the response. (After J.E. Luco and H.L. Wong, "Response of a rigid foundation to a spatially random ground motion," *Earthquake Engineering and Structural Dynamics*, Vol. 14, pp. 891–908, Copyright © 1986 John Wiley & Sons Limited; reproduced with permission.)

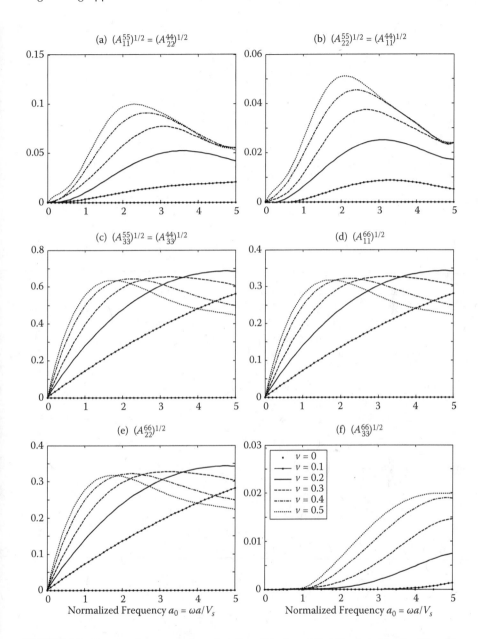

FIGURE 9.3 Rotational response characteristics of the foundation illustrated in Fig. 9.1, with $a = b$, subjected to spatially variable ground motions exhibiting only loss of coherency with variable v (Eq. 9.4) as functions of the normalized frequency $a_0 = a\omega/V_s$. The response characteristics, reported by Luco and Wong [324], are presented in terms of square roots of frequency transfer functions $\sqrt{A_{mm}^{jj}}$, where $m = 1, 2, 3$ indicates the component of the excitation and $j = 4, 5, 6$ the rotational component of the response. (After J.E. Luco and H.L. Wong, "Response of a rigid foundation to a spatially random ground motion," *Earthquake Engineering and Structural Dynamics*, Vol. 14, pp. 891–908, Copyright © 1986 John Wiley & Sons Limited; reproduced with permission.)

the transfer functions becomes less that unity. The rocking response of the foundation about the x_1- and x_2-axes (Fig. 9.1) is mainly caused by the vertical excitation, as can be seen from the comparison of the values of $\sqrt{A_{33}^{55}} = \sqrt{A_{33}^{44}}$ in Fig. 9.3(c) with those of $\sqrt{A_{11}^{55}} = \sqrt{A_{22}^{44}}$ in Fig. 9.3(a) and $\sqrt{A_{22}^{55}} = \sqrt{A_{11}^{44}}$ in Fig. 9.3(b). On the other hand, the torsional response of the system is mostly affected by the loss of coherency of the horizontal components of the motions, as the values of the transfer functions $\sqrt{A_{11}^{66}}$ and $\sqrt{A_{22}^{66}}$ in Figs. 9.3(d) and (e), respectively, are higher than those of $\sqrt{A_{33}^{66}}$ in Fig. 9.3(f). Luco and Wong [324] noted that the rotational components of the response can be significant and tend to increase at low frequencies as the loss of coherency in the motions increases.

Figure 9.4 presents the amplitudes of the non-vanishing transfer functions of the foundation response for fully coherent, but propagating motions. In this evaluation, Luco and Wong [324] considered that the motions propagate along the source-site direction, which was assumed to coincide with the x_1-axis (Fig. 9.1). It was further assumed that $c_1 = c_2 = c_3 = c$, and the ratio V_s/c was equal to 0 (motions arriving simultaneously at all foundation locations), 0.1, 0.2, 0.3, 0.4 and 0.5. Based on the comparison of Figs. 9.2 and 9.3 with Fig. 9.4, Luco and Wong [324] indicated that, qualitatively, the effect of loss of coherency of the seismic excitations on the foundation response is similar to the effect of fully coherent but propagating seismic excitations, as both tend to decrease the translational response of the system and increase its rotational response. The main differences, however, are that, when the motions experience loss of coherency only, both the x_1- and x_2-components of the motions induce torsional response ($\sqrt{A_{11}^{66}} = \sqrt{A_{22}^{66}} \neq 0$ for $v \neq 0$ in Figs. 9.3(d) and (e), respectively) and the vertical component of the motions induces rocking response about both the x_1- and x_2-axes ($\sqrt{A_{33}^{55}} = \sqrt{A_{33}^{66}} \neq 0$ for $v \neq 0$ in Fig. 9.3(c)), whereas, when the excitations propagate along the x_1-coordinate (Fig. 9.1), only the x_2-component induces torsion ($\sqrt{A_{22}^{66}} \neq 0$ for $V_s/c \neq 0$ in Fig. 9.4(h)), but $\sqrt{A_{11}^{66}} = 0$ (not shown)), and the vertical component introduces rocking only about the x_2-axis ($\sqrt{A_{33}^{55}} \neq 0$ for $V_s/c \neq 0$ in Fig. 9.4(i), but $\sqrt{A_{33}^{44}} = 0$ (not shown)). Luco and Wong [324] then concluded that, based on the results of their evaluation, the loss of coherency in the seismic ground motions produces effects that are comparable to the deterministic effects of wave passage with the former effects being perhaps slightly stronger than the latter.

$\longrightarrow$

FIGURE 9.4 Non-vanishing response characteristics of the foundation illustrated in Fig. 9.1, with $a = b$, subjected to spatially variable ground motions exhibiting only wave passage effects along the x_1-axis for variable V_s/c as functions of the normalized frequency $a_0 = a\omega/V_s$. V_s is the S-wave velocity in the medium and c the apparent propagation velocity assumed to be the same for all three components of the excitation. The response characteristics, reported by Luco and Wong [324], are presented in terms of square roots of frequency transfer functions $\sqrt{A_{mm}^{jj}}$, where $m = 1, 2, 3$ indicates the component of the excitation and $j = 1, 2, 3, 4, 5, 6$ the component of the response. (After J.E. Luco and H.L. Wong, "Response of a rigid foundation to a spatially random ground motion," *Earthquake Engineering and Structural Dynamics*, Vol. 14, pp. 891–908, Copyright © 1986 John Wiley & Sons Limited; reproduced with permission.)

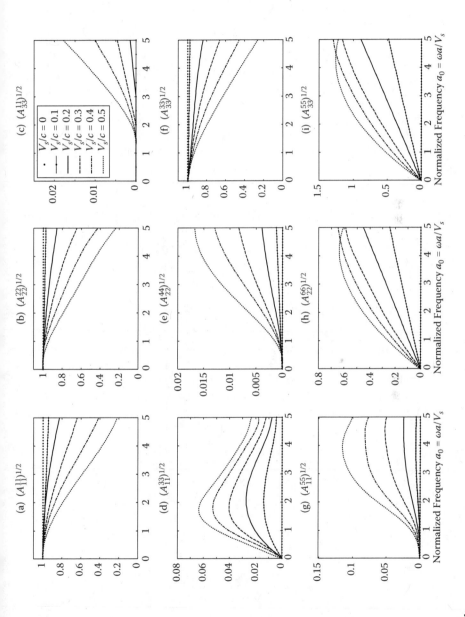

FIGURE 9.4

At approximately the same time, Harichandran [203] conducted a similar analysis for the evaluation of the translational response of square, rigid foundations of areas 2500 m^2 and 5000 m^2 subjected to spatially variable ground motions using base averaging [233], [444]. The coherency model used in this evaluation was the model developed by Harichandran and Vanmarcke [208], presented in Eq. 3.12, and the wave passage effect was considered with apparent propagation velocities of 1000 m/sec and 3600 m/sec. Harichandran [203] also concluded that spatially variable excitations resulted in the reduction of the translational response of the foundation, but that the wave passage effect was, in general, more significant than the effect of loss of coherency. The differences regarding the relative contribution of the loss of coherency and wave passage effects in the studies of Luco and Wong [324] and Harichandran [203] can be attributed to the different types of solution approaches (boundary value problem vs. base averaging), the 3-D vs. 1-D analysis, and, also, the differences in the exponential decay of the coherency models used in the evaluations (Eqs. 3.12 and 9.4).

The foundation size and the ratio of its sides (a/b in Fig. 9.1) plays also a significant role in the modification of the free-field excitation caused by the foundation. Figure 9.5 illustrates this effect for rectangular foundations using the closed-form solution provided by Veletsos *et al.* [554], [555]. Veletsos *et al.* [554], [555] used base-averaging [233], [444] and approximated the medium by an elastic half-space to evaluate the response of rigid, surface-supported circular and rectangular foundations. The coherency model utilized had the form [132]:

$$|\gamma(\vec{x}, \vec{x}', \omega)| = \exp\left\{-\left(\omega^2\left[v_1^2(x_1 - x_1')^2 + v_2^2(x_2 - x_2')^2\right]/V_s^2\right)\right\} \qquad (9.5)$$

i.e., an orthotropic version of Eq. 9.4 for the two horizontal directions, which led to closed-form expressions for the horizontal translational, u_1 and u_2, and the torsional, θ_3, foundation response (Fig. 9.1). For vertically incident shear waves and for $v_1 = v_2 = v$, for which Eqs. 9.4 and 9.5 become identical, the closed-form expressions for the power spectral densities of the translational, $u_1 = u_2 = u$, and torsional response of a rectangular foundation become:

$$S_{uu}(\omega) = f_1(\alpha_0)f_1(\beta_0)S_{gg}(\omega) \qquad (9.6)$$

$$b^2 S_{\theta_3\theta_3}(\omega) = \frac{3f_1(\alpha_0)\{f_1(\beta_0) - [1 - f_2(\beta_0)]/(2\beta_0^2)\}}{(1 + (a/b)^2)^2} S_{gg}(\omega) \qquad (9.7)$$

In the above expressions, $S_{gg}(\omega)$ is the power spectral density of the ground excitation, $\alpha_0 = v\omega a/V_s$ and $\beta_0 = v\omega b/V_s$ are normalized frequencies with a and b indicating one half of the lengths of the foundation sides (Fig. 9.1),

$$f_1(z) = \sqrt{\pi}\,\text{erf}(2z)/(2z) - f_2(z)$$
$$f_2(z) = [1 - \exp(-4z^2)]/(4z^2)$$

and erf(z) is the error function. The variation of the amplitudes of the transfer functions of the translational ($\sqrt{S_{uu}/S_{gg}}$) and torsional ($b\sqrt{S_{\theta_3\theta_3}/S_{gg}}$) components of the response as functions of the normalized frequency β_0 for various ratios of the lengths of the foundation sides ($a/b = 1/4, 1/2, 1, 2$ and 4) are presented in Figs. 9.5(a) and (b), respectively. It can be seen from the figures that, as a/b increases, both translational

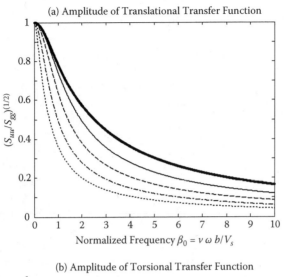

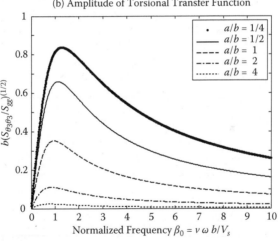

FIGURE 9.5 Response characteristics of the foundation illustrated in Fig. 9.1 subjected to spatially variable ground motions exhibiting only loss of coherency based on the closed-form expressions derived by Veletsos *et al.* [555]. Part (a) presents the variation of the amplitudes of the transfer functions of the translational ($\sqrt{S_{uu}/S_{gg}}$) and part (b) those of the torsional ($b\sqrt{S_{\theta_3\theta_3}/S_{gg}}$) components of the response as functions of the normalized frequency $\beta_0 = \nu\omega\beta/V_s$ for various aspect ratios of the foundation. The normalized frequency in this figure differs by a factor of ν from the normalized frequency utilized in Figs. 9.2–9.4. (After A.S. Veletsos, A.M. Prasad and W.H. Wu, "Transfer functions for rigid rectangular foundations," *Earthquake Engineering and Structural Dynamics*, Vol. 26, pp. 5–17, Copyright © 1997 John Wiley & Sons Limited; reproduced with permission.)

and torsional transfer functions decrease, with the decrease in the torsional response being more significant, as it is more closely related to the second moment of area of the foundation about its centroidal axis [555].

The effect of the reduction of the free-field motions caused by kinematic SSI was also addressed in the ASCE 4-98 standard [25] regarding the seismic analysis of nuclear structures. The standard notes that the assumption of vertically propagating shear and compressional waves in SSI evaluations results, usually, in conservative estimates for in-structure responses. It further suggests that, in the absence of analyses to establish the reduction of the response due to the loss of coherency in the seismic excitations, a conservative estimate would be to reduce the values of the ground response spectra by the factors presented in Table 9.1, which are applicable for vertical and horizontal spectra and all damping values. For structures with different plan dimensions, a linear interpolation scheme of the values provided in the table can be utilized. In its commentary, the standard also indicates that an accidental eccentricity of 5% of the structure's plan dimension may account for the induced torsional response of the structures due to nonvertically incident waves.

More recently, Kim and Stewart [267] utilized the amplitudes of the transfer functions for circular and rectangular foundations developed by Veletsos *et al.* [554], [555], the latter illustrated in Fig. 9.5, to propose an empirical approach for the estimation of the reduction of the input excitation caused by spatial incoherence effects on rigid foundations. Their approach was based on the analysis of 29 sites with instrumented structures and free-field accelerographs, from which the variations between the free-field ground motions and the motions at the foundation level were inferred. Kim and Stewart [267] suggested that the reduction of the free-field motion due to the base averaging of the foundation can be estimated by utilizing an effective value for the parameter ν in the coherency model of Eq. 9.4. The recommended value for ν was $7.4 \times 10^{-4} \times (V_s - 50 \text{ m/sec})$, with V_s, measured in [m/sec], being a site-depended velocity. They suggested that V_s be evaluated from the soil profile at the site over a depth equal to an effective foundation radius estimated as $r = \sqrt{A_f/\pi}$, with A_f being the foundation area. The range of values of V_s for their analyzed sites was $177 - 674$ m/sec, which leads to a coherency drop parameter α $(= \nu/V_s)$ varying

TABLE 9.1

ASCE 4-98 [25] recommendations for reduction factors of ground response spectra based on the foundation dimension. (Reproduced from *Seismic Analysis of Safety-Related Nuclear Structures and Commentary*, ASCE Standard 4-98, 2000, with permission from ASCE.)

	Reduction Factors for Plan Dimension of	
Frequency [Hz]	150 ft (45.75 m)	300 ft (91.50 m)
5	1.0	1.0
10	0.9	0.8
≥ 25	0.8	0.6

between $(5.31 - 6.85) \times 10^{-4}$ sec/m. The approximate value of v and the amplitudes of the transfer functions for circular foundations or rectangular (Eqs. 9.6 and 9.7 and Fig. 9.5) foundations provide estimates for the reduction factors of the input excitation due to kinematic SSI. Kim and Stewart [267] stated that their recommendations are valid for flat alluvial sites not near the basin edges, and provided correction factors and commentary regarding the effects of foundation embedment and stiffness. They also noted that their estimate for α differs from the average value of $(2 - 3) \times 10^{-4}$ sec/m suggested earlier by Luco and Wong [324].

Clearly, the consideration of the spatial variability in the excitations is beneficial for the response of structures on large, rigid foundations, as it reduces the foundation's translational response, even though it increases its rotational components. Figures 9.2 and 9.3, however, indicate that the controlling factor for the evaluation of the FIM is the exponential decay of the coherency model. Generally, versions of the plane-wave coherency model of Abrahamson [7], [9] (Eqs. 3.21–3.24 and Figs. 3.20(b) and 3.21(b), and Eq. 3.26) are proposed for the design of these critical structures [470], [543]. Some remarks on coherency modeling are presented in the last section of this chapter.

9.2 DAMS

The seismic response of dams has been and is currently being widely researched. A thorough and critical review of the literature related to the seismic response of arch and concrete gravity dams until 1990 (228 publications) was presented in a National Academy Press publication by the Panel on Earthquake Engineering [392]. More recently, Wieland [564], [565] presented comprehensive reviews on seismic design criteria for large concrete and embankment dams and the safety of existing dams, including references on the developments by the International Commission on Large Dams (ICOLD). He noted that ". . . it is still not possible to reliably predict the behaviour of dams during very strong ground shaking," and attributed the reasons for this to, among others, the difficulty in the modeling of the joint opening and the formation of cracks in the body of the dam, the nonlinear behavior of the foundation, and the insufficient information regarding the spatial variation of ground motions for arch dams [564]. This section highlights research efforts geared towards obtaining insight into the effect of the spatial variability in the seismic excitation on the response of these massive structures. The first two examples illustrate the effect of conventional estimates of spatial variability (i.e., loss of coherency and wave passage effects) on finite element models of an embankment dam and a concrete gravity dam. Complex topographic effects, however, give rise to additional causes for the spatial variation of the motions. Three examples briefly illustrate the effect of the angle of incidence and the type of wave on the response of arch dams using combinations of boundary elements or analytical solutions for the modeling of the infinite domain and finite element models for the dam, its foundation and the reservoir, or boundary elements for the modeling of the entire problem. A recent publication analyzing the response of the Pacoima Dam using inferred seismic data concludes the section.

Chen and Harichandran [100] conducted a sensitivity analysis of the random vibration response of the Santa Felicia earth dam, a rolled-fill earth embankment located in Southern California, which they subjected to spatially variable excitations described by various coherency models. The dam, the vertical cross section of which is shown schematically in Fig. 9.6, is 84 m (275 ft) high above its rock foundation and 137 m (450 ft) long at its base across the valley; the width of the crest is 9 m (30 ft) and its maximum length 389 m (1275 ft). A 3-D finite element model of the dam was created, for which it was assumed that the nodes on the boundary surface between the dam and the bedrock (Fig. 9.6) were completely restrained, and that the structure's response was in the linear range. The power spectral density of the ground excitation was modeled according to the Clough-Penzien spectrum [107] (Eq. 3.7) with parameters emulating the characteristics of the El Centro accelerogram. The apparent propagation velocity of the motions (Eqs. 3.3 and 3.4) was assumed to be equal to 4267 m/sec (14000 ft/sec). The coherency models utilized were the model of Harichandran and Vanmarcke [208] (Eq. 3.12) with parameters determined from the data of Event 20 at the SMART-1 array [206], the model of Abrahamson [7] (Eq. 3.21), the model of Hao *et al.* [200] of Eq. 3.14 with parameters evaluated from the data of Event 45 at the SMART-1 array, a generalized version of the model of Hindy and Novak [221], [376] (Eq. 3.9) with parameters such that, at short separation distances, it matched the trend of the exponential decay of the model of Harichandran and Vanmarcke, and the model of Luco and Wong [324] (Eq. 3.27) with a coherency drop parameter of $\alpha = 2 \times 10^{-4}$ sec/m. The qualitative evaluation of the dam's random vibration response to input excitations described by the various coherency models revealed that the distribution of the maximum shear stress, τ_{max}, depended strongly on the coherency model utilized: The models of Abrahamson [7] and Luco and Wong [324], which decay slowly with separation distance at low frequencies (e.g., Figs. 3.20 and 3.23) led to results similar to those of uniform support excitations, whereas the coherency models of Harichandran and Vanmarcke [208], Hao *et al.* [200] (e.g., Figs. 3.17 and 3.18) and Hindy and Novak [221], [376] (for the parameters selected in this study), which exhibit a strong exponential decay with separation distance in the

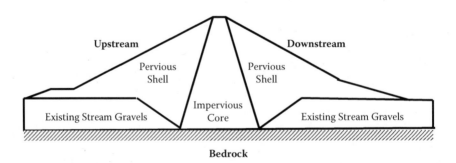

FIGURE 9.6 Schematic diagram of the vertical cross section of the Santa Felicia earth dam, located in Southern California, as utilized in the study by Chen and Harichandran [100]. The dam is 84 m (275 ft) high above its rock foundation and 137 m (450 ft) long at its base across the valley; the width of the crest is 9 m (30 ft) and its maximum length 389 m (1275 ft). (Reproduced from M.-T. Chen and R.S. Harichandran, "Response of an earth dam to spatially varying earthquake ground motion," *Journal of Engineering Mechanics, ASCE*, Vol. 127, pp. 932–939, 2001, with permission from ASCE.)

low frequency range, led to a response dominated by the pseudo-static component and to significantly higher values for the stresses[1]. The largest maximum stress occurred at the base of the dam and its values, which resulted from seismic excitations described by the five different coherency models, are presented in decreasing order in Table 9.2. The values in the table were estimated as the mean response plus three standard deviations ($\mu + 3\sigma$). It can be seen from the table that the results vary widely, and that the ratio of the highest to the lowest predicted response can be as high as 5.6. Chen and Harichandran [100] noted that, even though a nonlinear evaluation of the dam response would yield lower values for the stresses, the pseudo-static effect would not be diminished and the variability in the peak stresses caused by the different coherency models would remain. These results prompted Chen and Harichandran [100] to note that conclusions based on a single coherency model may be misleading.

In a series of papers, Bayraktar *et al.* [44], [45] analyzed the effect of wave propagation on the response of the Sariyar concrete gravity dam, which is located 120 km northwest of Ankara, Turkey. Plane strain conditions were considered in a linear finite element analysis of the dam by means of the Langrangian formulation; the 2-D cross section of the dam structure is presented in Fig. 9.7. In the finite element modeling of the dam-reservoir-foundation system, the reservoir length was three times its depth, i.e., 255 m, and the foundation depth as well as its extension on the downstream side of the dam were equal to the reservoir height (85 m). It was further considered that the nodes at the bottom and the edges of the foundation were fixed, and the nodes representing the extreme edge of the reservoir were free to move in the vertical direction. The east-west component of a recorded time history during the 1992 Erzincan earthquake was applied as input excitation to the dam. Spatial variability

---

TABLE 9.2
Peak shear stresses, τ_{max}, induced in the gravel layer of the Santa Felicia Dam (Fig. 9.6) by seismic excitations described by various coherency models, as evaluated by Chen and Harichandran [100]. (Reproduced from M.-T. Chen and R.S. Harichandran, "Response of an earth dam to spatially varying earthquake ground motion," *Journal of Engineering Mechanics, ASCE,* Vol. 127, pp. 932–939, 2001, with permission from ASCE.)

Coherency Model	τ_{max} [kPa]
Hindy and Novak [221], [376]	2650
Harichandran and Vanmarcke [208]	1670
Hao, Oliveira and Penzien [200]	1368
Abrahamson [7]	523
Luco and Wong [324]	476

---

[1] An illustration of the differences in the differential ground displacements and, consequently, the pseudo-static response of long structures resulting from the coherency models of Harichandran and Vanmarcke [208] and Luco and Wong [324] was presented earlier in Fig. 6.9(a).

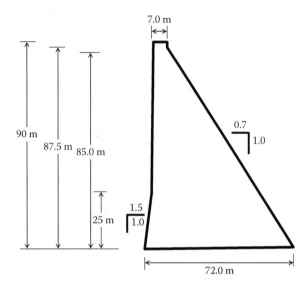

FIGURE 9.7 Schematic diagram of the 2-D cross section of the concrete structure of the Sariyar gravity dam, located 120 km northwest of Ankara, Turkey, as utilized in the studies of Bayraktar *et al.* [44], [45] for the evaluation of the effect of wave propagation on the dam response; the water elevation is at 85 m. (Reproduced from *Computers and Structures*, Vol. 58, A. Bayraktar, A.A. Dumanoğlu and Y. Calayir, "Asynchronous dynamic analysis of dam-reservoir-foundation systems by the Langrangian approach," pp. 925–935, Copyright © 1996, with permission from Elsevier.)

was introduced in the evaluation by varying the propagation velocity of the waveform from 250 m/sec to infinity, the latter value reflecting simultaneous arrival of the ground motions at the various locations of the structure's foundation. Bayraktar *et al.* [45] reported that horizontal, vertical and shear stresses in the foundation generally increased with decreasing propagation velocity, and, at a cross section close to the base of the dam, horizontal stresses also increased with decreasing velocity, but vertical and shear stresses did not exhibit a consistent pattern. Bayraktar and Dumanoğlu [44] analyzed the same dam model (Fig. 9.7), but subjected it to the north-south component of the 1992 Erzincan earthquake in order to evaluate the effect of wave propagation on the hydrodynamic pressure. They suggested that the wave passage effect influences the frequency content of the hydrodynamic pressure only slightly, but decreases considerably its amplitude.

An additional consideration in the seismic response of dams is that the complex topography of their location gives rise to significant spatial variability in the incident seismic ground motion field. Consider, e.g., the schematic diagram of an arch dam subjected to incident body (P-, SV- and SH-) and surface (Rayleigh) waves, as illustrated in Fig. 9.8 from the work of Maeso *et al.* [328]: The nonuniform reservoir geometry and the irregular local topography cause scattering of the incident waves and result in a ground motion field that is different from what would have resulted in a half-space. Notably, the variability in the seismic motions resulting from complex

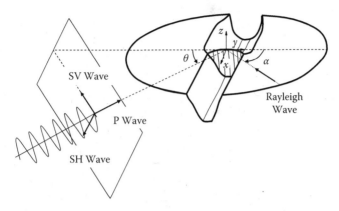

FIGURE 9.8 Schematic diagram of the complex topography and reservoir geometry of an arch dam in a canyon subjected to body (P-, SV- and SH-) waves with an incident angle of θ and Rayleigh waves with an incident angle of α. (Reprinted from O. Maeso, J.J. Aznárez and J. Domínguez, "Effects of space distribution of excitation on seismic response of arch dams," *Journal of Engineering Mechanics, ASCE*, Vol. 128, pp. 759–768, 2002, with permission from ASCE; courtesy of O. Maeso and J. Domínguez.)

topographic effects cannot be approximated by the conventional coherency models presented in Chapter 3. An approach that has been used to account for such effects evaluates first the response of the canyon, in the absence of the dam, to incident plane waves; the canyon can have an arbitrary cross section, which is, generally, assumed to extend uniformly in an infinite half-space (e.g., [513], [611], [612]). The resulting ground motions are then applied as input excitations in the evaluation of the response of arch dams (e.g., [92], [272], [273], [378]). For example, Nowak and Hall [378] analyzed the seismic response of the Pacoima Dam in two parts: In the first part, a 2-D boundary element approach was utilized to evaluate the free-field motions of the canyon walls. The canyon was considered to have a cross-section of 124 m depth and 224 m width, that extended uniformly to infinity, and was subjected to seismic waves propagating in a direction perpendicular to the axis of the canyon, as illustrated in Fig. 9.9. An incident SH-wave would then result in the "stream" component of the excitation, and incident P- and SV-waves in the in-plane (vertical and "cross-stream") components; these directions are denoted by the letters "s", "v" and "c", respectively, in Fig. 9.9. Six input motion scenarios were considered for the evaluation of the dam response: a uniform excitation in the stream direction; vertically incident SH-waves; SH-waves impinging the canyon at $\theta = 30°$; uniform excitations in the vertical and cross-stream directions; vertically incident P- and SV-waves; and P-waves impinging the canyon at an angle of $\theta = 30°$. The uniform input excitations were approximately equal to those resulting in a half-space due to the incident waves. In the second part of the analysis, the system was modeled with linear finite elements and subjected to the free-field motions developed in the previous part. The foundation effect was incorporated in the model through its stiffness, and the reservoir was either full or empty. In the former case, the water was considered to be compressible, a transmitting boundary was used at the upstream end of the water mesh to model the infinite

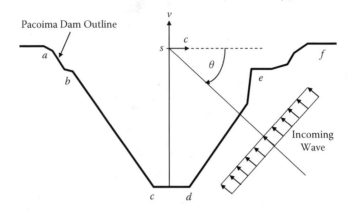

FIGURE 9.9 The cross section of the Pacoima Dam (124 m depth and 224 m width) utilized in the study by Nowak and Hall [378]. The cross section extended uniformly to infinity, and was subjected to seismic waves propagating in a direction perpendicular to the axis of the canyon. An incident SH-wave results in the "stream" component of the excitation, and incident P- and SV-waves in the in-plane (vertical and "cross-stream") components; the directions of these components are denoted by the letters "s", "v" and "c", respectively, in the figure. The dam-foundation interface is indicated by the letters b-c-d-e. (Reproduced from P.S. Nowak and J.F. Hall, "Arch dam response to nonuniform seismic input," *Journal of Engineering Mechanics, ASCE*, Vol. 116, pp. 125–139, 1990, with permission from ASCE.)

reservoir, and the water-foundation interaction was approximated by a partially absorbing boundary. The results, reproduced in Table 9.3, were presented in terms of standard deviations of total arch stresses. The input excitations were power spectral densities obtained through amplitude scaling of the average Fourier transform of horizontal ground motions recorded on rock in the vicinity of a 7.5 magnitude

TABLE 9.3

Average values of standard deviations of total arch stresses along the crest of the Pacoima Dam as percentage of the corresponding value for uniform excitations along the stream direction evaluated by Nowak and Hall [378]. (Reproduced from P.S. Nowak and J.F. Hall, "Arch dam response to nonuniform seismic input," *Journal of Engineering Mechanics, ASCE*, Vol. 116, pp. 125–139, 1990, with permission from ASCE.)

Incident Excitation	Full Reservoir	Empty Reservoir
Uniform, "s"	100	89
SH-waves at $\theta = 90°$	62	47
SH-waves at $\theta = 30°$	73	66
Uniform, "c" and "v"	78	46
P- and SV-waves at $\theta = 90°$	63	48
P-waves at $\theta = 30°$	122	62

earthquake [529]. The analysis suggested that the stresses for a full reservoir were higher than the stresses for an empty one, nonuniformity in the stream component of the excitation tended to decrease the response, but the highest stresses in the dam were obtained for the incident P-wave at an angle of $\theta = 30°$ for the full reservoir.

Shortly thereafter, Kojić and Trifunac [272], [273] presented the linear response evaluation of an idealized arch dam to incident P-, SV-, SH- and Rayleigh waves. The canyon in these studies was also assumed to have a uniform cross section extending to infinity, but the scattering of the incident waves at the canyon walls in this 2-D problem was evaluated by solving the corresponding wave equations. To simplify the problem and isolate the effects of the traveling waves on the dam response, the foundation-structure interaction was not included in the finite element modeling of the structure, and the hydrodynamic effect of the reservoir was approximately accounted for with Westergaard's added mass concept [563]. The input excitation was the scaled (by a factor of 0.34) three-component accelerogram recorded at the Pacoima Dam during the 1971 San Fernando earthquake: The S74W component was applied in the stream direction of the dam, the S16E in the cross-stream direction and the down component of the accelerogram in the vertical direction. The results revealed that, for the antiplane case (incident SH-waves resulting in excitations in the stream direction), the dam response due to nonuniform excitations decreased by approximately 25–30% as compared to the response caused by uniform motions, which is also in agreement with the results of Nowak and Hall [378] in Table 9.3. On the other hand, excitations in the vertical and cross-stream directions caused by P-, SV- and Rayleigh waves induced a different effect: Kojić and Trifunac [273] observed that vertically propagating P-waves resulted in a reduction of the total stresses obtained for uniform canyon motions, and, when their incidence angle decreased to $\theta = 30°$ (with θ measured, as in Figs. 9.8 and 9.9, from the horizontal axis), the values of the total stresses increased, but were still lower than the stresses induced by the uniform canyon motion, which is not in agreement with the observations by Nowak and Hall [378] in Table 9.3. For SV-waves impinging the canyon at an angle of $\theta = 30°$ (with θ, again, measured from the horizontal axis), the dynamic stresses increased considerably relative to stresses induced by vertically incident SV-waves as well as uniform canyon excitations. Additionally, the pseudo-static stresses for the SV-waves arriving at an angle of $\theta = 30°$ were increased by a factor of 2-10 relative to the pseudo-static stresses induced by waves impinging the site at $\theta = 90°$. For incident body waves in the nonuniform ground motion scenarios, the concentration of the pseudo-static stresses occurred near the abutment. Rayleigh waves induced lower dynamic stresses than the SV-waves arriving at $\theta = 30°$, but comparable values for the pseudo-static stresses, which, however, in this case, were located at the bottom of the dam. (The different stress distribution pattern that is caused in the body of arch dams by uniform and spatially variable ground excitations will be illustrated later herein in Fig. 9.13 for the Pacoima Dam.)

The studies of both Kojić and Trifunac [272], [273] and Nowak and Hall [378] clearly illustrate the significance of the consideration of both the angle of incidence of the waves and the wave type in the evaluation of the dam response. The differences in their results are a consequence of the ground motion characteristics and the modeling assumptions utilized (e.g., canyon cross-section geometry, boundary value problem vs. analytical solution, and differences in the finite element modeling of the system). Both approaches, however, considered that the canyon cross section extends uniformly to infinity. On the other hand, Maeso *et al.* [328], using boundary elements, analyzed the 3-D topographic effects on the seismic response of the 142 m high Morrow Point arch dam in Colorado. A schematic diagram of the boundary element model used in the approach is presented in Fig. 9.10; for simplicity, the problem was assumed to be symmetric, and, hence, only half of the discretization is shown in the figure. The boundary element model of the combined dam-foundation-reservoir problem considered the foundation rock to be a viscoelastic boundless domain, the water an inviscid compressible fluid and the concrete dam a viscoelastic solid with properties reported earlier by Hall and Chopra [191]. In a later study, the same authors [329] included in the formulation of the problem the bottom reservoir sediments, which were modeled as a two-phase poroelastic domain obeying Biot's [59] equations. The dam was subjected to incident P-, SV-, SH- and Rayleigh waves (Fig. 9.8). The analyses considered P-waves impinging the site at incident angles of $\theta = 5°$, $30°$, $60°$ and $90°$, SV-waves at angles of $\theta = 52°$ (critical angle), $60°$, $75°$ and $90°$, SH-waves at angles of $\theta = 0°$, $30°$, $60°$ and $90°$, and Rayleigh waves at incident angles of $\alpha = -90°$, $-30°$, $0°$, $30°$ and $90°$ (Fig. 9.8). Figure 9.11 illustrates their results for the extreme angles of incidence of each wave type considered in their study, when the reservoir was assumed to be full. Part (a) of the figure presents the results for the P-wave analysis, part (b) for the SV-wave, part (c) for the SH-wave and part (d) for the Rayleigh wave. The subplots of the figure present the amplitude of

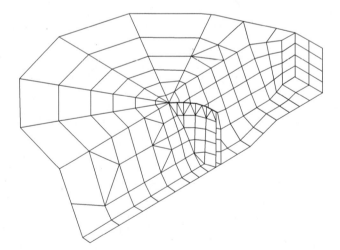

FIGURE 9.10 Schematic diagram of the boundary element model of the Morrow Point arch dam in Colorado utilized by Maeso *et al.* [328] in the investigation of 3-D effects on the response of arch dams. The incident body and Rayleigh wave components used in the analysis are illustrated in Fig. 9.8. For simplicity, the problem was assumed to be symmetric, and, hence, only half of the discretization is shown in the figure. (Reprinted from O. Maeso, J.J. Aznárez and J. Domínguez, "Effects of space distribution of excitation on seismic response of arch dams," *Journal of Engineering Mechanics, ASCE*, Vol. 128, pp. 759–768, 2002, with permission from ASCE; courtesy of O. Maeso and J. Domínguez.)

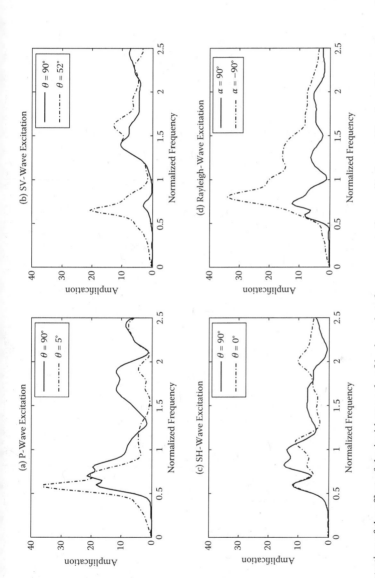

FIGURE 9.11 Illustration of the effect of the incident angle of body and surface waves on the amplitude amplification of the motions at the crest of an arch dam (Figs. 9.8 and 9.10) for a full reservoir, as reported by Maeso *et al.* [328]. Part (a) of the figure presents the results for a P-wave impinging the site at angles of $\theta = 5°$ and $90°$; part (b) for an incident SV-wave at angles $\theta = 52°$ and $90°$; part (c) for a SH-wave at angles of $\theta = 0°$ and $90°$; and part (d) for a Rayleigh wave at angles of $\alpha = -90°$ and $90°$. The amplifications in the subplots are shown as functions of normalized frequency; in parts (a), (c) and (d), frequency was normalized with respect to the fundamental natural frequency of the dam on a rigid foundation with an empty reservoir for a symmetric mode, whereas, in part (b), frequency was normalized with respect to the fundamental natural frequency of the dam on a rigid foundation with an empty reservoir for an antisymmetric mode (after Maeso *et al.* [328]).

the complex-valued frequency transfer functions for the various wave types, which amplify the reference ground motion at a dam crest location on the upstream face of the dam. The reference ground motion was caused by harmonic waves producing a unit amplitude for the motion in a uniform half-space. For P-, SH- and Rayleigh waves, the selected location was at the origin of the coordinate system (Fig. 9.8), whereas, for SV-waves, the location was at an angle of 13.25° from the $x - z$ plane due to the antisymmetry in the response of the symmetric system for vertically incident SV-waves. The amplifications in the subplots of Fig. 9.11 are shown as functions of normalized frequency. For P-, SH- and Rayleigh waves (Figs. 9.11(a), (c) and (d)), frequency was normalized with respect to the fundamental natural frequency of the dam on a rigid foundation with an empty reservoir for a symmetric mode, whereas, for SV-incident waves (Fig. 9.11(b)), the frequencies were normalized with respect to the fundamental natural frequency of the dam on a rigid foundation with an empty reservoir for an antisymmetric mode.

Figures 9.11(a) and (b) show significant increases in the motion of the dam crest at the lower frequencies as the angle of incidence of the waves becomes smaller, i.e., the incidence of the waves becomes closer to the horizontal direction (Fig. 9.8). Similar results were obtained when the reservoir was empty [328]. These differences are not observed for the SH-waves (Fig. 9.11(c)), although Maeso *et al.* [328] reported that the peak at the low frequencies for an empty reservoir showed variations of the order of 20% for various angles of incidence of the waves. For an empty reservoir, the effect of the angle of incidence of the Rayleigh waves on the crest response was significant around the fundamental frequency of the dam, but had a smaller influence at higher frequencies. On the other hand, when the reservoir was full and the waves were propagating downstream ($-90° < \alpha < 0°$), and, especially, for $\alpha = -90°$ (Fig. 9.11(d)), there was a dramatic amplification of the dam crest response over the entire frequency range shown in the figure. Maeso *et al.* [328] then concluded that the variability in the incident angle of the waves to the complex 3-D canyon topography can have a significant effect on the dam response.

From a different perspective, Alves and Hall [20], [21] analyzed the effect of spatially variable excitations on the nonlinear response of the Pacoima Dam using seismic data. After the 1971 San Fernando earthquake, an array of 17 accelerometers was installed at the Pacoima Dam. The locations of the stations are presented in Fig. 9.12: Eight channels (1-8) are located on the body of the dam and nine additional ones (9-17) at three locations on the dam-foundation interface. The array recorded only partially the motions of the Northridge earthquake, i.e., data were recorded at channels 8-11, only partially recorded at channels 1-6, 12, 13 and 15-17, and not recorded at channels 7 and 14. However, data during a magnitude 4.3 earthquake that occurred on January 13, 2001, were recorded at all stations of the array. Based on the data of this smaller event, Alves and Hall [21] evaluated frequency-dependent topographic amplification functions by considering the ratios of the response spectra of the records at the right and left abutments of the dam over the response spectrum of the records at the base in the corresponding direction (Fig. 9.12). To account for propagation effects, they also determined frequency-dependent time delay functions between two records. The time delay was evaluated from the time corresponding to the peak of the cross correlation function between the response of a single-degree-of-freedom

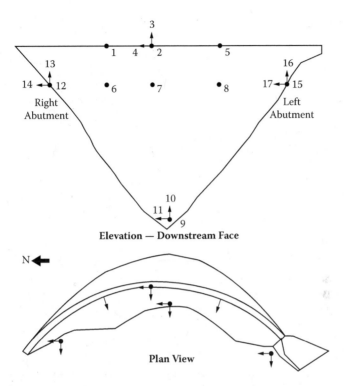

FIGURE 9.12 Elevation (top part) and plan view (bottom part) of the Pacoima Dam presenting also the locations of the 17 accelerometers of the dam's array: Channels 1–8 are located on the body of the dam and channels 9–17 at three locations on the dam-foundation interface. It should be pointed out that, when viewing the dam from downstream, as is the case for the top part of the figure, the left abutment is to the right of the dam and the right abutment to its left. (After S.W. Alves and J.F. Hall, "Generation of spatially nonuniform ground motion for nonlinear analysis of a concrete arch dam," *Earthquake Engineering and Structural Dynamics*, Vol. 35, pp. 1339–1357, Copyright ©2006 John Wiley & Sons Limited; reproduced with permission.)

oscillator with 5% damping subjected to the two analyzed time series. The delay thus determined corresponds to the natural frequency of the oscillator, and the process was repeated for all required frequencies. This information was then used to reproduce the seismic data at the array stations during the Northridge earthquake: Since the data at the base of the dam (channels 9–11 in Fig. 9.12) were recorded during this larger event, they were used, along with the topographic amplification and time delay functions of the smaller event, to reproduce the seismic motions at the abutments that were partially recorded or not recorded during the Northridge earthquake. These generated ground motions were then applied as input excitations in a finite element analysis of the dam.

The numerical model consisted of the dam, the water reservoir and the rock foundation. The evaluation utilized the finite element code SCADA [190], which is based on the smeared-crack approach for the modeling of the nonlinear behavior of the structure. The model considered an incompressible water reservoir and a massless rock foundation. The generated Northridge ground motions at the three locations along the dam-abutment boundary (Fig. 9.12) were interpolated/extrapolated and applied as input excitations at each node of the dam-foundation interface. Both spatially variable and uniform excitations were considered in the analysis. Figure 9.13 presents results of the analyses of Alves and Hall [21]. The subplots in the figure are contour plots of the distribution of the maximum compressive axial stresses on the upstream (left subplots) and the downstream (right subplots) face of the dam. Part (a) of the figure illustrates the stress distribution when the input excitation was considered to be uniform; in this case, the uniform motions were described by the time series generated at the right abutment of the structure (Fig. 9.12). Part (b) of the figure presents the corresponding stresses caused by the generated nonuniform excitations, and part (c) the pseudo-static response induced by the nonuniform motions. Alves and Hall [21] reported that, even though the response induced by uniform seismic excitations appeared to provide an upper bound for the response caused by spatially variable ground motions, the pattern of stress distribution and joint opening in the dam for the two cases of input motions differed substantially. For uniform excitations, stresses (Fig. 9.13(a)) and joint opening were largest in the central part of the dam, away from the abutments, and, also, cracks formed mostly in the central part. On the other hand, a significant part of the response due to spatially variable excitations was due to the pseudo-static component: The stresses along the abutments and in the center of the downstream face of the dam (Figs. 9.13(b) and (c)) were dominated by the pseudo-static response, whereas those near the center section of the crest by the dynamic contribution. Similarly, joint opening was caused by the pseudo-static effects along the abutments, but by the dynamic contribution at the central part of the dam. The significance of the pseudo-static effects on the dam response observed by Alves and Hall [21] was also noted earlier by Kojić and Trifunac [273], but their effect was not pronounced in the results of the study by Nowak and Hall [378]. Alves and Hall [21] concluded that, since the pseudo-static effects depend on the characteristics of the nonuniform seismic excitation, care should be taken in selecting appropriate descriptions for the input seismic ground motions in dam analyses. A similar observation was made by Chen and Harichandran [100] in their evaluation of the response of the Santa Felicia earth dam, as described earlier herein (Table 9.2).

    A qualitative conclusion that can be drawn from the diverse studies of dam structures presented in this section is that different response patterns emerge when the excitation is uniform or nonuniform. The major differences in the patterns are caused by the pseudo-static response, that is induced only by nonuniform excitations, and the significance of which depends, for each structure, on the characteristics of the spatially variable ground motions. Additionally, the angle of incidence of body and surface waves to the complex local topography of arch dams was shown to play a significant role on the response of the structures. These results, however, are still qualitative, and further investigations of the response of dams subjected to uniform and nonuniform seismic ground motions are warranted.

(a)

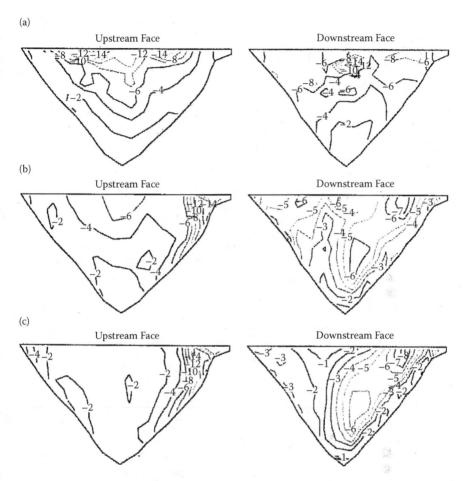

FIGURE 9.13 Stress distribution induced in the Pacoima Dam (Fig. 9.12) by uniform and spatially variable seismic ground motions as reported by Alves and Hall [21]; the input excitations were data inferred for the Northridge earthquake. The subplots in the figure are contour plots of the distribution of the maximum compressive axial stresses on the upstream (left subplots) and the downstream (right subplots) face of the dam. Part (a) presents the stress distribution when the input excitation was considered to be uniform and described by the time series at the right abutment of the structure. Part (b) of the figure illustrates the corresponding stresses caused by nonuniform excitations, and part (c) the pseudo-static response induced by the nonuniform motions. (After S.W. Alves and J.F. Hall, "Generation of spatially nonuniform ground motion for nonlinear analysis of a concrete arch dam," *Earthquake Engineering and Structural Dynamics*, Vol. 35, pp. 1339–1357, Copyright © 2006 John Wiley & Sons Limited; reproduced with permission.)

9.3 SUSPENSION AND CABLE-STAYED BRIDGES

The effect of the spatial variation of seismic ground motions on long-span bridges has been recognized since the middle 1960's, when pioneering research work was conducted by Bogdanoff *et al.* [65]. The bridge model used in their analysis (Fig. 9.14)

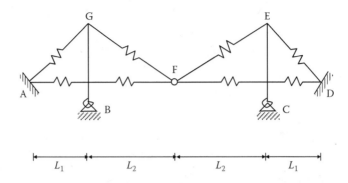

FIGURE 9.14 Schematic diagram of the long-span bridge utilized in the pioneering work of Bogdanoff *et al.* [65]. The model consisted of two rigid towers (GB and EC), an extensible cable (AGFED) and a deck (AFD). The towers were pinned at their supports (B and C), but elastically restrained in rotation by linear springs. (Reproduced from J.L. Bogdanoff, J.E. Goldberg and A.J. Schiff, "The effect of ground transmission time on the response of long structures," *Bulletin of the Seismological Society of America*, Vol. 55, pp. 627–640, Copyright © 1965 Seismological Society of America.)

consisted of two rigid towers (GB and EC), an extensible cable (AGFED) and a deck (AFD). The towers were pinned at their supports (B and C), but elastically restrained in rotation by linear springs. The seismic excitation was approximated by "packets" of damped oscillatory waves with random amplitudes, frequencies, arrival times and velocities of propagation [64], [169]. The linear response of the model was evaluated in terms of extreme value statistics. The results for a 1951 m (6400 ft) long bridge with end spans of $L_1 = 335$ m (1100 ft), a center span of $2L_2 = 1280$ m (4200 ft) and tower heights of 214 m (700 ft) suggested that the consideration of the delayed arrival of the excitation at the bridge supports led to a significantly higher response than the response induced by seismic motions arriving simultaneously. Bogdanoff *et al.* [65], cautioning that their evaluation is valid for their assumed model, concluded that the "transmission time of a seismic disturbance" should not be ignored when evaluating the response of a long structure.

Pioneering work on the response of suspension and cable-stayed bridges subjected to recorded spatially variable ground motions was conducted by Abdel-Ghaffar *et al.* [1], [2], [3], [362]. In 1982, Abdel-Ghaffar and Rubin [2] introduced a random vibration approach for the evaluation of the vertical response of suspension bridges. Their example application considered the Vincent Thomas Bridge in Los Angeles, California, and, as seismic excitations, they utilized seismic data recorded during the 1971 San Fernando earthquake in the Los Angeles area at distances comparable to the length of the center span of the bridge (457 m or 1500 ft). The results of their evaluation suggested coupling between the vertical and longitudinal directions of the bridge, and, also, that the incorporation of higher modes was necessary for the reliable estimation of the response. The study concluded that the response induced by uniform motions at the supports was significantly different from the response caused by the partially correlated multi-support excitations. In the following year, Abdel-Ghaffar and Rubin [3] extended their approach to the evaluation of the lateral seismic response of suspension bridges, and applied it in the analysis of the Golden Gate Bridge in San Francisco, California, a bridge also analyzed in further random vibration studies (e.g., [207], [359]). The schematic diagram of the bridge model used by Abdel-Ghaffar and Rubin [3] is presented in Fig. 9.15. As input excitations at the structure's supports, these authors used the seismic data recorded during the 1979 Imperial Valley earthquake at stations with separation distances comparable to the distances between the bridge supports. In their evaluations, they first used the motions at stations, referred to as "Arrays," 4, 5, 6 and 7 as input excitations at supports A, B, C and D, respectively, of the bridge (Fig. 9.15), they then utilized the data at Arrays 5, 6, 7 and 8, and, in their last analysis, they used the data at Array 5 as uniform input excitations at all supports.

Table 9.4 presents a summary of their results: Part (a) of the table provides the root-mean-square (rms) displacements of the suspended structure, part (b) the rms displacements of the cable, and part (c) the rms flexural seismic stresses at various locations along the structure (Fig. 9.15) resulting from the three input motion scenarios. Abdel-Ghaffar and Rubin [3] observed that the uniform support motions did not excite the antisymmetric or the higher modes of the structure, which, however, were strongly excited by the spatially variable excitations. The comparison of the first two rows in Tables 9.4(a) and (c) (response caused by spatially variable excitations) with the last row (response induced by uniform motions) suggests that significantly higher values can result from the spatially variable excitations along the end spans of the structure; the differences are not so pronounced for the displacements induced in the cable (Table 9.4(b)). Additionally, the results in the table indicate that the pattern of displacement and stress distribution along the bridge deck depends strongly on the characteristics of the seismic excitation. Displacements and stresses at the right span of the structure are significantly increased when the motions at Arrays 4, 5, 6 and 7 are utilized as input excitations, whereas displacements and stresses at the left span of the structure are significantly increased when the motions at Arrays 5, 6, 7 and 8 are utilized as input excitations at the supports. Abdel-Ghaffar and Rubin [3] concluded

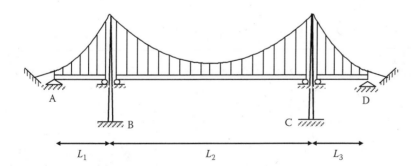

FIGURE 9.15 Schematic diagram of the Golden Gate Bridge in San Francisco, California, utilized in the pioneering work by Abdel-Ghaffar and Rubin [3] for the evaluation of spatial variability effects on the response of suspension bridges. (Reproduced from A.M. Abdel-Ghaffar and L.I. Rubin, "Lateral earthquake response of suspension bridges," *Journal of Structural Engineering, ASCE*, Vol. 109, pp. 665–675, 1983, with permission from ASCE.)

TABLE 9.4
Root-mean-square response induced in the Golden Gate Bridge (Fig. 9.15) by the ground motions of the 1979 Imperial Valley earthquake applied in the lateral direction according to the study by Abdel-Ghaffar and Rubin [3]. (Reproduced from A.M. Abdel-Ghaffar and L.I. Rubin, "Lateral earthquake response of suspension bridges," *Journal of Structural Engineering, ASCE,* Vol. 109, pp. 665–675, 1983, with permission from ASCE.)

(a) Displacements (in [cm]) of the Suspended Structure

Motions Recorded at Arrays:	Left Span			Center Span			Right Span		
	$L_1/4$	$L_1/2$	$3L_1/4$	$L_2/4$	$L_2/2$	$3L_2/4$	$L_3/4$	$L_3/2$	$3L_3/4$
4,5,6,7	5.51	7.27	6.01	6.32	4.37	7.24	21.62	29.98	21.27
5,6,7,8	18.98	26.76	19.11	8.28	8.14	7.49	11.65	15.76	11.64
5,5,5,5	15.13	20.88	15.20	8.32	9.72	8.32	15.20	20.88	15.13

(b) Displacements (in [cm]) of the Cable

Motions Recorded at Arrays:	Left Span			Center Span			Right Span		
	$L_1/4$	$L_1/2$	$3L_1/4$	$L_2/4$	$L_2/2$	$3L_2/4$	$L_3/4$	$L_3/2$	$3L_3/4$
4,5,6,7	3.16	4.05	4.24	9.32	4.68	9.18	5.98	6.91	5.46
5,6,7,8	5.89	7.98	6.10	10.46	8.08	8.73	4.03	3.27	4.03
5,5,5,5	5.08	6.57	5.10	8.32	9.65	8.32	5.10	6.57	5.08

(c) Flexural Stresses ($\times 10^3$ kN/m^2)

Motions Recorded at Arrays:	Left Span			Center Span			Right Span		
	$L_1/4$	$L_1/2$	$3L_1/4$	$L_2/4$	$L_2/2$	$3L_2/4$	$L_3/4$	$L_3/2$	$3L_3/4$
4,5,6,7	10.82	15.51	11.03	8.48	11.03	9.58	48.61	68.46	47.71
5,6,7,8	43.50	62.40	44.33	11.10	13.79	10.14	24.06	33.85	23.58
5,5,5,5	33.03	47.30	33.65	11.51	16.00	11.51	33.65	47.30	33.03

that the assumption of uniform excitations at the supports is unconservative, as it does not represent the worst loading scenario over the entire length of the Golden Gate Bridge. It should be noted at this point that a possible reason for the very significant differences in the displacements of the suspended structure and its flexural stresses (Table 9.4(a) and (c)) is that Arrays 4, 5 and 6 are to the east of the Imperial fault, whereas Arrays 7 and 8 are to the west of the causative fault. This implies that, in the first nonuniform ground motion scenario presented in Table 9.4, the anchorage support at D (Fig. 9.15) is located on one side of the fault but the rest of the supports of the structure are located on the other side, whereas, for the second nonuniform ground motion scenario in the table, the fault is located between the two towers.

In the early 1990's, Abdel-Ghaffar and Nazmy [1] and Nazmy and Abdel-Ghaffar [362] evaluated the effect of spatial variability on the seismic response of cable-stayed bridges. A 2-D schematic diagram of their 3-D bridge models is presented in Fig. 9.16. Two different bridge lengths were considered [362]: The first bridge had a central span of 335 m (1100 ft) and side spans that were 146 m (480 ft) long. The second bridge had span lengths twice the size of the first bridge model, i.e., 671 m (2200 ft) for the central span and 293 m (960 ft) for the side spans. For each bridge, two alternate designs, a concrete and a steel design, were analyzed; the concrete models considered a rigid connection between the bridge deck and the tower shafts, whereas the steel models had a free or floating connection between the deck and the towers. Nazmy and Abdel-Ghaffar [362] first performed a nonlinear static analysis of the 3-D models of the bridges to evaluate their tangent stiffness at the deformed state due to the dead load, and, subsequently, conducted a linear dynamic analysis with the estimated tangent stiffness. The models were subjected to three input excitation scenarios: In the first case, as in the earlier suspension bridge study [3], the spatially variable data recorded at Arrays 4, 5, 6 and 7 during the 1979 Imperial Valley earthquake were utilized as input excitations at supports A, B, C and D, respectively, of the structure (Fig. 9.16). In the second case, the data at Array 6 were applied as uniform excitations at the supports of the structure. In the third case, the ground motions recorded at Array 6 were assumed to propagate in the longitudinal direction of the structure with 6 different propagation velocities ranging from 61 m/sec (200 ft/sec) to 1951 m/sec (6400 ft/sec). All three components of the motions were simultaneously applied as input excitations at the structure's supports.

Figures 9.17–9.20 illustrate representative results of their evaluation for the first two cases of support excitations. The figures present the response of the steel design alternative for the shorter structure. Figure 9.17 presents the absolute maximum displacement (part (a)), axial force (part (b)), vertical shear force (part (c)), bending moment (part (d)) and torsional moment (part (e)) when uniform excitations are applied at the structure's supports, and Fig. 9.18 the corresponding results when nonuniform seismic ground motions are utilized as excitations. Figure 9.19 presents the absolute maximum response of a representative tower in terms of displacements (part (a)), axial forces (part (b)) and torsional moments (part (c)) for the uniform input motion scenario, and Fig. 9.20 the corresponding results for the nonuniform input excitations.

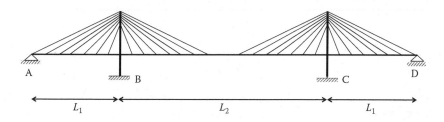

FIGURE 9.16 Schematic diagram of a cable-stayed bridge utilized in the pioneering work by Abdel-Ghaffar and Nazmy [1], [362] for the evaluation of spatial variability effects on the response of cable-stayed bridges; two alternate designs, a concrete and a steel design, were analyzed. (After A.S. Nazmy and A.M. Abdel-Ghaffar, "Effect of ground motion spatial variability on the response of cable-stayed bridges," *Earthquake Engineering and Structural Dynamics*, Vol. 21, pp. 1–20, Copyright © 1992 John Wiley & Sons Limited; reproduced with permission.)

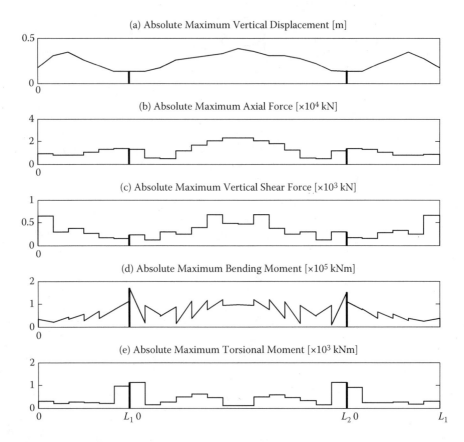

FIGURE 9.17 The response of the deck of the steel design alternative of the shorter cable-stayed bridge (Fig. 9.16) analyzed by Nazmy and Abdel-Ghaffar [362] when subjected to uniform seismic ground motions; the central span of the bridge was 335 m long and the end spans had a length of 146 m. Part (a) of the figure presents the maximum displacement, part (b) the maximum axial force, part (c) the maximum vertical shear force, part (d) the maximum bending moment, and part (e) the maximum torsion. All three components of the motions were simultaneously applied as input excitations at the structure's supports. The vertical, thicker lines in the subplots indicate the location of the towers. (After A.S. Nazmy and A.M. Abdel-Ghaffar, "Effect of ground motion spatial variability on the response of cable-stayed bridges," *Earthquake Engineering and Structural Dynamics*, Vol. 21, pp. 1–20, Copyright © 1992 John Wiley & Sons Limited; reproduced with permission.)

Figures 9.17–9.20 present, perhaps, the clearest illustration of the differences in the pattern of the response induced by uniform and spatially variable excitations. The values of the displacements of the bridge deck (Figs. 9.17(a) and 9.18(a)) for the two input motion scenarios are very different; the displacements caused by the uniform excitations contain the pseudo-static (rigid-body) motions of the supports, whereas the significantly higher displacements induced by the spatially variable excitations are controlled by the dynamic part of the response. The rigid-body, pseudo-static excitation of the uniform input motions does not contribute to the internal forces in

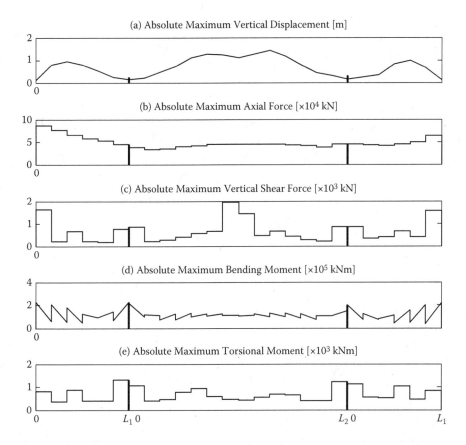

FIGURE 9.18 The response of the deck of the steel design alternative of the shorter cable-stayed bridge (Fig. 9.16) analyzed by Nazmy and Abdel-Ghaffar [362] when subjected to spatially variable seismic ground motions; the central span of the bridge was 335 m long and the end spans had a length of 146 m. Part (a) of the figure presents the maximum displacement, part (b) the maximum axial force, part (c) the maximum vertical shear force, part (d) the maximum bending moment, and part (e) the maximum torsion. All three components of the motions were simultaneously applied as input excitations at the structure's supports. The vertical, thicker lines in the subplots indicate the location of the towers. (After A.S. Nazmy and A.M. Abdel-Ghaffar, "Effect of ground motion spatial variability on the response of cable-stayed bridges," *Earthquake Engineering and Structural Dynamics*, Vol. 21, pp. 1–20, Copyright © 1992 John Wiley & Sons Limited; reproduced with permission.)

the structure, which are composed solely of the dynamic response (Figs. 9.17(b)–(e)). However, the very significant increase in the axial forces, the shear forces in the vicinity of the towers and the bending moments at the end supports induced by the spatially variable ground motions (Figs. 9.18(b), (c) and (d)), as compared to the corresponding response for the uniform excitations (Figs. 9.17(b), (c) and (d)), is mainly caused by the pseudo-static contribution. On the other hand, the significantly higher shear forces and torsional moments at the end supports induced by the spatially variable

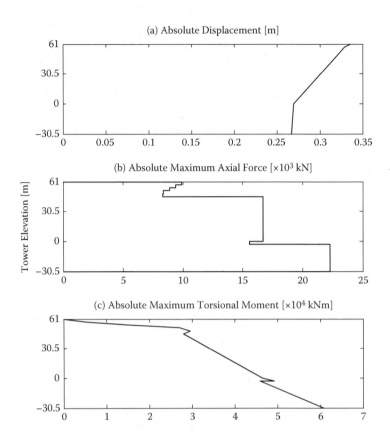

FIGURE 9.19 The response of a tower of the steel design alternative of the shorter cable-stayed bridge (Fig. 9.16) analyzed by Nazmy and Abdel-Ghaffar [362] when subjected to uniform seismic ground motions; the central span of the bridge was 335 m long and the end spans had a length of 146 m. Part (a) of the figure presents the maximum displacement, part (b) the maximum axial force, and part (c) the maximum torsion. All three components of the motions were simultaneously applied as input excitations at the structure's supports. (After A.S. Nazmy and A.M. Abdel-Ghaffar, "Effect of ground motion spatial variability on the response of cable-stayed bridges," *Earthquake Engineering and Structural Dynamics*, Vol. 21, pp. 1–20, Copyright © 1992 John Wiley & Sons Limited; reproduced with permission.)

excitations (Figs. 9.18(c) and (e)), as compared to the response caused by uniform motions (Figs. 9.17(c) and (e)), are a consequence of the contribution of the anti-symmetric and higher modes of the structure. The displaced configuration of the tower for the uniform input motion scenario (Fig. 9.19(a)) is dominated by the pseudo-static, rigid-body motion, whereas the above-ground displacements of the tower for the spatially variable excitations (Fig. 9.20(a)) are dominated by the dynamic response and the below-ground displacements by the pseudo-static effect. Axial forces in the tower are also higher for the nonuniform input ground motions (Fig. 9.20(b)) than the uniform excitations (Fig. 9.19(b)); in this case, the pseudo-static and dynamic axial forces caused by the multi-support excitations (Fig. 9.20(b)) interfere destructively,

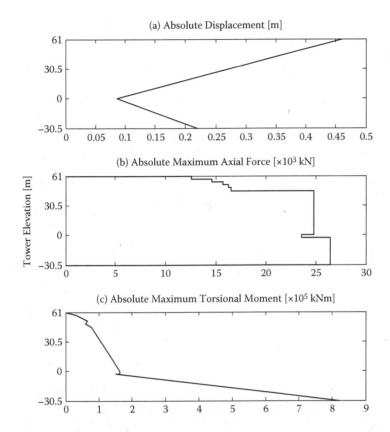

FIGURE 9.20 The response of a tower of the steel design alternative of the shorter cable-stayed bridge (Fig. 9.16) analyzed by Nazmy and Abdel-Ghaffar [362] when subjected to spatially variable seismic ground motions; the central span of the bridge was 335 m long and the end spans had a length of 146 m. Part (a) of the figure presents the maximum displacement, part (b) the maximum axial force, and part (c) the maximum torsion. All three components of the motions were simultaneously applied as input excitations at the structure's supports. (After A.S. Nazmy and A.M. Abdel-Ghaffar, "Effect of ground motion spatial variability on the response of cable-stayed bridges," *Earthquake Engineering and Structural Dynamics*, Vol. 21, pp. 1–20, Copyright © 1992 John Wiley & Sons Limited; reproduced with permission.)

especially below ground level, but the total axial force is still higher than the dynamic axial force induced by the uniform input motions. The torsional moment in the above-ground part of the tower is also higher for the spatially variable (Fig. 9.20(c)) than the uniform (Fig. 9.19(c)) excitations. However, below ground level, an excessive pseudo-static response renders a significantly higher total torsional moment for the multi-support excitations (Fig. 9.20(c)) than the uniform motions (Fig. 9.19(c)). Less dramatic differences between the responses induced by the two input motion scenarios were observed for the longer bridge. Nazmy and Abdel-Ghaffar [362] noted, as Abdel-Ghaffar and Rubin [3] did earlier in their evaluation of the Golden Gate Suspension Bridge, that spatially variable excitations induced higher modes in the structure than uniform motions at the supports; the same observation has been made for short-span bridges, as will be illustrated in Section 9.4. It is noted, again, that the variability in the response is also conditional on the characteristics of the ground motions used in its evaluation. As indicated earlier, in the presentation of the suspension bridge analyzed by Abdel-Ghaffar and Rubin [3], the ground motions used in the studies were on the two sides of the causative fault. For this cable-stayed bridge, the support at D (Fig. 9.16) was located on the west part of the Imperial fault, whereas the other supports on its east side.

In this extensive, quantitative study, Nazmy and Abdel-Ghaffar [362] concluded that spatially variable excitations can have a significant effect on the displacement and member forces of cable-stayed bridges, especially for the more rigid structures and for structures that have poor or variable soil conditions underneath their supports. They emphasized, however, that the degree of the increase in the response induced by spatially variable excitations depended on the structural characteristics, such as span length, rigidity and structural redundancy. Regarding the wave passage effect, Nazmy and Abdel-Ghaffar [362] reported that time delays/phase differences at the supports of the structures can be especially important for low apparent propagation velocities and when the structural redundancy is high, because they induce significant pseudo-static member forces. Concluding their insightful study, the authors recommended that, for the rational seismic design of a cable-stayed bridge, at least three different types of seismic motions, consistent with the characteristics of the location of the bridge site, should be applied as input excitations at the bridge supports in the evaluation of its response.

After the pioneering studies of Abdel-Ghaffar et al. [1], [2], [3], [362], the effect of spatially variable ground motions on the seismic response of suspension and, more recently, cable-stayed bridges attracted research interest: In presenting a response-spectrum random vibration approach for the multi-support excitation of lifelines, Yamamura and Tanaka [571] investigated the response of a suspension bridge model. To account for the spatial variability in the excitations, they assumed that the motions within a group of adjacent supports were uniform, but that the motions between different groups were uncorrelated. The analysis indicated that the spatially variable multi-support excitations can significantly affect the response of the structure, as they excite both its symmetric and antisymmetric modes. Hyun et al. [232] introduced non-stationarity in the vertical seismic excitations by means of intensity modulating functions (Section 7.1.2) to evaluate the response of suspension bridges, utilized the approach in an example application of the vertical response of the Golden Gate Bridge for apparent propagation velocities ranging between 60 m/sec and infinity, and concluded that the propagation of the motions can have a significant effect on the structural response. Allam and Datta [19] also presented a non-stationary random vibration response evaluation of a cable-stayed bridge model subjected to spatially variable excitations. They concluded that the shape of the intensity modulating function (Section 7.1.2) can affect the structural response considerably, and that displacements induced by fully correlated motions were lower than displacements caused by spatially variable excitations. In presenting and validating the pseudo-excitation method (Section 6.1.2), Lin et al. [307] evaluated the response of the cable-stayed Canton Jin-Ma bridge subjected to spatially variable excitations incorporating loss of coherency according to the model of Qu *et al.*[2] [417] and wave passage effects. They concluded that the spatial variability in the excitations, and, especially, the apparent propagation of the motions, cannot be neglected in the seismic response evaluation of the structure. Extending the approach further, Lin and Zhang [306] evaluated the response of the Song-Hua-Jiang suspension bridge in China considering the effects of wave passage and loss of coherency, the latter described by the models of Loh and Yeh [318] (the second expression in the set of Eqs. 3.10) and Qu *et al.* [417]. They also concluded that the wave passage effect was more important than the effect of the loss of coherency, but that the latter effect depended on the coherency model used in the evaluation. The dependence of the structural response on the selected coherency model, noted earlier by Chen and Harichandran [100] in their evaluation of the response of the Santa Felicia earth dam (Table 9.2), was also observed by Soyluk and Dumanoğlu [511]. Extending their earlier study on the stochastic response of the Jindo Bridge subjected to spatially variable seismic excitations [144], Soyluk and Dumanoğlu [511] utilized the coherency models of Harichandran and Vanmarcke [208] and Luco and Wong [324] to investigate the effect of the coherency decay of the models on the structural response. They concluded that the coherency model of Harichandran and Vanmarcke [208] resulted in a higher response than the model of Luco and Wong [324], and attributed this observation to the fact that the two models yield different degrees of correlation in the low frequency range (Figs. 3.16 and 3.23) that controlled the response of their model.

The aforementioned qualitative studies reach, basically, the same conclusion, namely that the spatial variability in the seismic ground motions cannot be neglected in the seismic response evaluation of suspension and cable-stayed bridges, as spatially variable excitations induce the pseudo-static response as well as additional modes in the structures than uniform support motions. The majority of these investigations are based on the stationary or non-stationary random vibration approach, and were geared towards improvement or validation of random-vibration-approach related methodologies. The structural models varied from the consideration of the first few modes of the structures for the evaluation of its response to more elaborate linear, finite element models. As indicated in Chapter 6, the major advantage of the random vibration approach is that a single evaluation suffices for the stochastic characterization of the stationary structural response, but its disadvantage is that it is still limited to the linear response of the structures.

On the other hand, the design and retrofit of these majestic structures involves their detailed linear and nonlinear modeling (e.g., [37], [122], [152]), and, generally, considers spatially variable excitations at their supports (e.g., [7], [37], [284]). An informative description of the modeling approach for the performance-based design of the Tacoma Narrows Parallel Crossing in the State of Washington was recently

---

[2] The coherency model of Qu *et al.* [417] was developed by fitting a functional form to the average of parametric coherency models evaluated at the site of the SMART-1 and LSST arrays by various investigators. As indicated in Section 3.4.1, the exponential decay of the fitted parametric forms to the lagged coherency of recorded data depends, to a large degree, on the data processing approaches as well as the mathematical expressions for the models. Hence, a parametric model fitted to parametric estimates represents more an "averaging" of the processing approaches and mathematical expressions used by the individual investigators, but does not necessarily reflect an "average" trend of the coherency of the data.

presented by Arzoumanidis *et al.* [37]. The new suspension bridge has a 1647 m (5400 ft) superstructure with a main span of 854 m (2800 ft), side spans of 366 m (1200 ft) on the Tacoma side and 427 m (1400 ft) on the Gig Harbor side, and two 154 m (505 ft) tall reinforced concrete towers. Geometric and material nonlinearities were considered in the modeling of the structure, and stress, deformation and ductility criteria were utilized to verify the post-earthquake level of service of the bridge; in the absence of well-established criteria, as those relating the response of the tower shafts to specific levels of performance, capacity analyses were conducted to verify that the design satisfies the performance objectives. The details of the modeling of the structure can be found in the extensive description of Arzoumanidis *et al.* [37].

Spatially variable seismic ground motions that are used in these analyses are, generally, conditionally simulated (Chapter 8). For example, Appendix D of the seismic design criteria of the New York City Department of Transportation (NYCDOT) [364] provides the designers with three sets of three-component spatially variable ground motions for earthquake return periods of 500 and 2500 years at rock outcrop conditions. The ground motions were generated at 17 equidistant locations with separation distance between consecutive locations of 100 m. These locations reflect the locations of the structure's supports, and it is assumed that the bridge foundations have the same local soil conditions and are at the same elevation. The spatially variable ground motions were conditionally simulated with the six-step approach of Abrahamson [7], which was briefly presented in the introductory remarks of Chapter 8. The coherency model utilized is a later (unpublished) version of the plane-wave coherency model of Abrahamson [7], which was presented in Eqs. 3.21–3.24 and Figs. 3.20(b) and 3.21(b); these coherency models were developed from spatial array data recorded at multiple sites. It was also assumed that the apparent propagation of the motions was 2.5 km/sec, and the direction of propagation was parallel to the axis of the bridge. In the conditional simulation approach, a set of recorded data was first selected, and then modified so that the time series conformed with the rock uniform hazard spectra for the specific return period and direction of motion. These response-spectrum-compatible motions were conditionally simulated, and the simulated motions were also modified to become compatible with the target response spectra. The systematic shift to account for the wave passage effect was then introduced in the generated data. NYCDOT provides the time series (accelerations, velocities and displacements) in digital form for use as input excitations in a time-domain structural response evaluation. It is also recommended that, for bridges at variable site conditions, the amplification effect and arrival time delays should be considered by utilizing 1-D wave propagation through the soil column of the softer site. The appendix also indicates that, since each of the three sets of ground motions conforms with the target response spectra, a single time history analysis including all three components of the excitation may be performed. It suggests, however, that all three sets be utilized in three time history evaluations, as Nazmy and Abdel-Ghaffar [362] commented earlier, and that the average of the absolute value of the response from the three evaluations be used instead. The code recommends against interpolation of the time series for the evaluation of ground motions at intermediate locations, as interpolation will modify their characteristics. Instead, it suggests that the designer select a subset of available time series with spacing such that it provides the closest match to the footprint of the bridge.

It should be noted that this is, perhaps, the optimal approach for the incorporation of spatial variability in the seismic excitation for the response evalution of lifeline systems, namely that the ground motions are conditionally simulated so as to be compatible with prescribed time series, design response spectra and appropriate coherency models. Some comments regarding coherency modeling at sites with uniform and variable ground characteristics, as utilized in the aforementioned conditional simulation approach, are presented in the last section of this chapter.

9.4 HIGHWAY BRIDGES

As illustrated in the previous section, spatially variable ground motions are, generally, applied as input excitations at the supports of suspension and cable-stayed bridges in their seismic design and retrofit. On the other hand, the approach for the incorporation of spatial variability effects in the design of highway bridges differs (e.g., [30], [31], [32], [109]). The objectives here are, first, to establish when and to what degree is the effect of the spatial variation of the seismic ground motions significant for the seismic response of the structures, and, second, to provide, if possible, an alternative approach for its incorporation in design utilizing an "equivalent" uniform excitation that captures the effects of spatially variable ground motions. Selective sensitivity analyses of the effect of the spatial variation of the seismic ground motions on the response of highway bridges that were geared towards design recommendations are illustrated in this section. The section concludes with a brief presentation of the new European Commission code [109], that, recently, recommended such an equivalent approach; the code recommends that the response of bridge structures be obtained through the combination of a pseudo-static response induced by prescribed displacement patterns at its supports and a dynamic response induced by uniform excitations.

One of the first sensitivity evaluations of the nonlinear response of Reinforced Concrete (RC) highway bridges subjected to a range of spatially variable excitations was reported by Monti *et al.* [347]. The question posed in the study was: If bridges are designed without consideration for spatial variability, will they be able to withstand spatially variable excitations? To address this question, Monti *et al.* [347] conducted a sensitivity analysis of the bridge model illustrated in Fig. 9.21. The bridge consisted of a six-span continuous deck over five piers of the same height H and diameter equal to 2.5 m. Each span was 50 m long, leading to a total length for the bridge of 300 m. The deck was transversely hinged to the piers and the abutments, and the piers had fixed supports and acted as cantilever beams. The bridge models were designed for uniform support excitations with a PGA of $0.42g/q$, with q indicating the behavior factor [108], and subjected to seismic excitations in the transverse direction. The ground motions were generated with an approach similar to the simulation scheme presented in Section 7.3.2. The sensitivity analyses considered variability in the soil type, the exponential decay of the coherency model, the value of the apparent propagation velocity, the structural stiffness, represented by the height of the piers, and the behavior factor q. The soil conditions underneath the bridge were assumed to be uniform, either firm or medium, and described by the Clough-Penzien spectrum [107] of Eq. 3.7 with parameters as presented in Table 3.1(b). The coherency model used in the evaluation was the model of Luco and Wong [324] with three values for the coherency

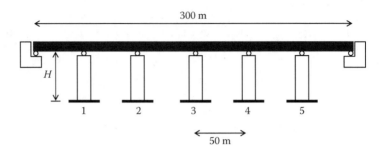

FIGURE 9.21 Schematic diagram of the bridge model utilized by Monti *et al.* [347]. The bridge consists of a six-span continuous deck over five piers of the same height H. The height of the piers considered in the analysis assumed the values of 7.5, 10 and 15 m. (Reproduced from G. Monti, C. Nuti and P.E. Pinto, "Nonlinear response of bridges under multi-support excitation," *Journal of Structural Engineering, ASCE*, Vol. 122, pp. 1147–1159, 1996, with permission from ASCE.)

drop parameter (Eq. 3.27): $\alpha = 0$, i.e., fully coherent motions at all supports, and $\alpha = 1.67 \times 10^{-3}$ and 3.3×10^{-3} sec/m. The apparent propagation velocity of the motions (Eq. 3.4) assumed four values: $c = 300, 600$ and 1200 m/sec and $c \to \infty$, the latter value representing simultaneous arrival of the motions at all supports. The height of the piers was 7.5, 10 and 15 m, and the behavior factor was assumed to be $q = 2, 4$ and 6. The analysis was conducted using the code ASPIDEA [168].

Monti *et al.* [347] first performed a linear evaluation of the structures and concluded that the spatial variation of the motions (whether loss of coherency or wave passage or their combined effect) led, consistently, to a beneficial response of the structures in terms of shear forces at the piers. They further noted, that the shear force distribution along the piers followed, basically, the pattern of the first mode of the structure when uniform excitations were applied at the supports, whereas, for partially correlated motions, the variation of the response was flatter, suggesting that higher modes, in addition to the pseudo-static component, contributed to the response. The nonlinear behavior of the bridges was modeled next with Takeda-type plastic hinges [515] at the piers. Figures 9.22(a) and (b) illustrate representative results of the study. The figures present the displacement ductility demand for bridges with pier heights of 10 m located on medium soil conditions. In part (a) of the figure the coherency drop parameter of the model of Luco and Wong (Eq. 3.27) was $\alpha = 3.3 \times 10^{-3}$ sec/m, and, in part (b), the motions were fully coherent ($\alpha = 0$). Both subplots of the figure present results for three values of the apparent propagation velocity ($c = 300$ and 600 m/sec and $c \to \infty$) and the three behavior factors. For bridges subjected to the highly incoherent motions (i.e., $\alpha = 3.3 \times 10^{-3}$ sec/m in Fig. 9.22(a)), the loss of coherency effect overshadows the effect of wave passage. In this case, the displacement ductility demand becomes lower for the central piers than the lateral piers. Since the bridges were designed for uniform support motions, the lateral piers were subjected to lower forces and, hence, were weaker than the central ones. On the other hand, the consideration of spatial variability, which excites the pseudo-static response as well as additional modes, imposed almost equal distortions

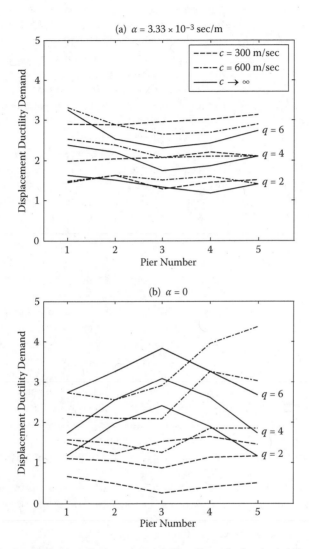

FIGURE 9.22 Representative displacement ductility demands for the bridges analyzed by Monti *et al.* [347], illustrated in Fig. 9.21; the height of the piers was 10 m and the soil conditions were assumed to be medium. Part (a) presents the displacement ductility demand when the coherency drop parameter of the model of Luco and Wong (Eq. 3.27) used in the simulation of the spatially variable excitations was $\alpha = 3.3 \times 10^{-3}$ sec/m, and part (b) when the motions were fully coherent ($\alpha = 0$). Both subplots presents results for three values of the apparent propagation velocity ($c = 300$ and 600 m/sec and $c \to \infty$) and three behavior factors ($q = 2, 4$ and 6). (Reproduced from G. Monti, C. Nuti and P.E. Pinto, "Nonlinear response of bridges under multi-support excitation," *Journal of Structural Engineering, ASCE*, Vol. 122, pp. 1147–1159, 1996, with permission from ASCE.)

to all piers, thus leading to a higher ductility demand from the lateral piers [347]. The wave passage effect became pronounced for the fully coherent and propagating motions, especially for the value of $c = 600$ m/sec (Fig. 9.22(b)), and the direction of propagation of the motions led to a significantly higher ductility demand from the last encountered piers. Monti *et al.* [347] then concluded that the overall effect of spatial variability was to reduce the ductility demand at the central piers and increase it at the lateral ones, but, in all cases, the ductility demand was close to the value of the behavior factor. They further suggested that, for the bridge configurations and seismic ground motion models used in their study, the consideration of uniform excitations at the structure's supports resulted in a conservative design.

Different conclusions, however, were reached by Tzanetos *et al.* [546] for a shorter bridge model. Tzanetos *et al.* [546] analyzed two different bridge configurations based on the model shown schematically in Fig. 9.23. The five-span structure had a total length of 184 m, with the end-spans being 32 m and the spans between piers 40 m long. The superstructure was a twin-box prestressed concrete girder supported on single column piers. The two middle columns were 10.5 m high and the end columns were 6.5 m high; all columns were considered to be fixed at the ground level. The bearings at the top of the piers restrained fully the motions in the longitudinal and transverse directions, but allowed rotations about all directions. The left abutment (A1 in Fig. 9.23) was designed to restrict the relative displacement of the deck end in both horizontal directions, whereas the right abutment (A2) provided no restraint in the longitudinal direction. At both abutments, rotations about the longitudinal axis were constrained, whereas rotations about the transverse axis were free. Tzanetos *et al.* [546] investigated two different models with respect to the rotational component about the vertical axis at abutment A1: The first model permitted rotation in this direction, whereas the second had a rotational restraint. The study used the finite element code ADAPTIC [238]. The piers were modeled by cubic elasto-plastic elements that permitted the spread of plasticity both along the length of the element and across its cross section, and the deck was assumed to respond in the elastic range. The analysis evaluated and compared pier drifts, maximum shear forces at the base of the piers and maximum deck moments.

Two different sets of uniform and spatially variable motions were employed as excitations at the structure's supports. The first case utilized four recorded time

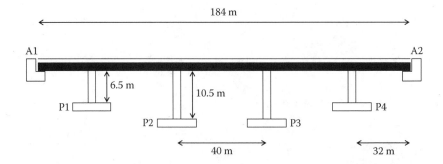

FIGURE 9.23 Schematic diagram of the bridge model used by Tzanetos *et al.* [546]. The abutment A1 was considered to be either "pinned" or "fixed" (after Tzanetos *et al.* [546]).

histories, which were considered to propagate with velocities of $c = 160, 400, 800$ and 1600 m/sec on the ground surface. The second set of input excitations consisted of simulated motions based on the coherency model of Luco and Wong [324] for five values of the coherency drop parameter ($\alpha = 0, 3.33 \times 10^{-4}, 5 \times 10^{-4}, 1 \times 10^{-3}$, and 3.3×10^{-3} sec/m in Eq. 3.27) in combination with three values for the apparent propagation velocity of the motions (Eq. 3.4), $c = 600$ and 1200 m/sec and $c \to \infty$. Two sets of time histories at uniform soil conditions were generated for each case with the approach utilized by Monti $et\ al.$ [347]. Each simulated time series at abutment A1 was scaled to $0.175g$, and the other records in the realization were scaled accordingly.

Table 9.5 presents illustrative results of their evaluation. The quantities in the table represent the transverse response of the two structural configurations (i.e., pinned and fixed abutment at A1), subjected to two scenarios of spatially variable excitations ($\alpha = 3.3 \times 10^{-3}$ sec/m and $c = 600$ m/sec, i.e., strongly incoherent

TABLE 9.5
Comparison of the transverse response of the two configurations of the bridge model of Fig. 9.23 subjected to highly ($\alpha = 3.3 \times 10^{-3}$ sec/m and $c = 600$ m/sec) and moderately ($\alpha = 5 \times 10^{-4}$ sec/m and $c = 1200$ m/sec) varying seismic excitations as reported by Tzanetos $et\ al.$ [546]; "pinned" and "fixed" indicate a pinned and a fixed abutment at A1. The results are presented as percentage differences of the response due to spatially varying excitations relative to the response due to uniform motions at the supports. (After N. Tzanetos, A.S. Elnashai, F.H. Hamdan and S. Antoniou, "Inelastic dynamic response of RC bridges subjected to spatially non-synchronous earthquake motion," $Advances\ in\ Structural\ Engineering$, Vol. 3, pp. 191–214, Copyright © 2000 Multi-Science Publishing Co Ltd.; reproduced with permission.)

| | $\alpha = 3.3 \times 10^{-3}$ sec/m $c = 600$ sec/m | | | | $\alpha = 5 \times 10^{-4}$ sec/m $c = 1200$ sec/m | | | |
| | Record Set 1 | | Record Set 2 | | Record Set 1 | | Record Set 2 | |
	Pinned	Fixed	Pinned	Fixed	Pinned	Fixed	Pinned	Fixed
Drift								
Pier 1	−33%	+50%	0%	+178%	+33%	+14%	+20%	+22%
Pier 2	−38%	−19%	−16%	−38%	+21%	−7%	+12%	+10%
Pier 3	−30%	−16%	−15%	−36%	+11%	−6%	+7%	+13%
Pier 4	−30%	−2%	+4%	−39%	+16%	−11%	+8%	+22%
Shear demand								
Pier 1	−15%	+20%	0%	+90%	0%	+8%	+8%	+25%
Pier 2	−26%	−14%	−14%	−4%	+4%	−1%	+6%	+13%
Pier 3	−23%	−15%	−6%	−23%	+3%	−2%	+4%	+16%
Pier 4	−5%	−5%	0%	−28%	+5%	−5%	+2%	+11%
Moment								
Deck	−19%	−9%	+20%	+21%	+19%	−8%	+6%	+41%

motions propagating slowly on the ground surface, and $\alpha = 5 \times 10^{-4}$ sec/m and $c = 1200$ m/sec, i.e., more coherent ground motions propagating faster on the ground surface) in the transverse horizontal direction, and two different time history sets generated for each ground motion scenario. The results presented in the table are the differences between the response induced by spatially variable excitations and the response induced by uniform ground motions as a percentage of the latter. The table suggests that spatially variable ground motions lead to a different seismic demand pattern in the structures than uniform support excitations, and, can, in cases, cause a significant increase in the structural response [546]. Similar results were reported for the response of the structures in the longitudinal direction. It was also noted that, in general, once the loss of coherency in the excitations was taken into consideration, it overshadowed the wave passage effects. Tzanetos *et al.* [546] observed that the findings of Monti *et al.* [347] cannot be generalized due to the symmetric bridge configuration considered in their (Monti *et al.* [347]) analysis, and concluded that, depending on the characteristics of the spatially variable excitations, the bridge configuration and its boundary conditions, spatially variable ground motions can induce a higher response in the structure than the response resulting from uniform motions at its supports. They further suggested that the spatially variable excitations may cause one of the following two effects on bridges: (i) For bridges with stiff decks and flexible piers, the piers will be subjected to very high flexural, shear force and deformation demands; or (ii) For bridges with flexible decks and stiff piers, the deck will be subjected to excessive in-plane bending (twisting). Concluding their insightful study, Tzanetos *et al.* [546] recommended, as a first step in the implementation of spatially variable excitations in design codes, the modification of the participation factor of the "second mode" of the structure to account for the increase in the contribution of higher modes to the structural response when spatially variable motions excite the structure.

Two additional observations can be made from the results of Tzanetos *et al.* [546] in Table 9.5. The first is that the less coherent motions do not necessarily induce the worst loading conditions for the bridge. This can be seen from the comparison of the response of the bridge with the pinned abutment at A1: The less coherent motions ($\alpha = 3.3 \times 10^{-3}$ sec/m and $c = 600$ m/sec on the left side of the table) produce a response for the piers that is consistently less or equal to the response induced by the uniform excitations. On the other hand, the more coherent motions ($\alpha = 5 \times 10^{-4}$ sec/m and $c = 1200$ m/sec on the right side of the table) yield a consistently higher response for the piers than the uniform input motions. The second observation is that the variability in the structural response for the two sets of each spatially variable ground motion scenario, especially for the less coherent excitations on the left part of the table, can be very significant. This clearly suggests that one simulation (or a small number of simulations) can only be an indication of the trend of the response, but cannot suffice for its reliable characterization. This latter observation was addressed in the following study by Shinozuka *et al.* [468].

Shinozuka *et al.* [468] conducted nonlinear analyses of seven bridges in an effort to correlate the bridge length with the significance of the spatial variability of the excitation on the structures' response. The number of spans for the bridges was in the range of 3-12, and the total length of the structures ranged between 34 m (111 ft) and 242 m (795 ft) for six of the bridges, and was 500 m (1640.5 ft) for the longest bridge examined, the Santa Clara Bridge in California (Fig. 9.24). The structure is a twelve-span RC bridge with interior spans 44 m (143 ft) long and exterior spans 32 m (105.25 ft) long. The superstructure is a four-cell concrete box girder supported on single-column piers, all of height equal to 12 m (39.35 ft), and, at the two ends, on abutments. The response of the bridges, subjected to excitations in the longitudinal direction, was evaluated using the finite element code SAP2000 [110]. The structures were modeled in two dimensions using frame elements. The analysis considered that the piers were fixed at the ground level and the deck, and nonlinearity was introduced at each end of the columns through the formation of plastic hinges, the characteristics of which were determined using the Column Ductility Program COLx [90] of the California Department of Transportation (Caltrans). Time histories were generated based on the approach presented by Deodatis [127], and illustrated herein in Section 7.3.2. The bridge supports were considered to be located either on uniform or variable site conditions described by the UBC acceleration response spectra [234] for soils of Type I (rock and stiff soils), II (deep cohesionless or stiff clay soils) and III (soft to medium clays and sands). The motions were considered to be either fully coherent or obeying the coherency model of Harichandran and Vanmarcke [208], provided in Eq. 3.12, with parameters $A = 0.626, a = 0.022, k = 1.97 \times 10^4, f_0 = 2.02$ Hz and $b = 3.47$ [211]. The apparent propagation velocity was assumed to be $c = 300$ and 1000 m/sec and $c \to \infty$. The statistics (mean value and standard deviation) of the peak ductility demand were obtained from the ensemble average of the response induced by 20 sets of time histories generated for each case.

Shinozuka *et al.* [468] concluded that the peak ductility demand at the piers increased substantially when spatially variable excitations were considered, especially when the supports of the bridge were located on variable site conditions; for bridges located on uniform sites, the increase in the ductility demand was smaller. They also observed that, in general, the effect of loss of coherency was more important than the effect of wave passage, but noted that the contribution of the wave passage effect was still substantial, and, hence, should not be neglected. Based on their evaluation they recommended that, for bridges of total length less than approximately $300 - 450$ m $(1000 - 1500$ ft) located on uniform site conditions, the small increase of the response

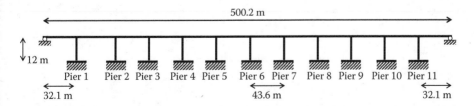

FIGURE 9.24 Schematic diagram of the Santa Clara Bridge utilized by Shinozuka *et al.* [468] and Saxena *et al.* [442], [443] for the evaluation of the effect of spatially variable ground motions on the response statistics of the structure. (After V. Saxena, G. Deodatis, M. Shinozuka and M. Feng, "Development of fragility curves for multi-span reinforced concrete bridges," in *Monte Carlo Simulation: Proceedings of the International Conference on Monte Carlo Simulation, MCS-2000*, G.I. Schuëller and P.D. Spanos editors, Monte Carlo, Monaco, 2000, Copyright © Balkema; reproduced with permission.)

due to spatial variability can be accommodated by the various safety margins built in design codes, and that these bridges can be designed according to current practice, a result also in agreement with the observations by Monti *et al.* [347]. On the other hand, for bridges longer than $300 - 450$ m with their supports located on uniform site conditions, as well as for all bridges located on variable site conditions, the code provisions may compromise the overall safety of the structures. In this case, they suggested that reliable estimates of the peak structural response should be evaluated by means of nonlinear dynamic time history analyses of the bridge based on, at least, 20 sets of spatially variable time series.

The results of Shinozuka *et al.* [468] were also reflected in code provisions. For example, the joint venture of the Multidisciplinary Center for Earthquake Engineering Research (MCEER) and the Applied Technology Council (ATC), publication ATC-49 [32], on guidelines for the seismic design of highway bridges suggested that, for long bridges, all causes of the spatial variability in the seismic ground motions may be important. On the other hand, they suggested that, for short bridges, the effects of wave passage and loss of coherency are, in general, relatively unimportant in comparison to the effects of variable site conditions underneath the structure's supports. In the latter case, they recommended a uniform seismic ground motion evaluation of the structures, but with design response spectra estimated from the envelope of the individual response spectra of the abutments and intermediate piers that have different site classifications. They further suggested that, if it is assessed that the bridge response is dominated by the abutment ground motions, then only the abutment response spectra need to be taken into consideration.

Extending the study of Shinozuka *et al.* [468], Saxena *et al.* [442], [443] analyzed the significance of the spatial variability of the seismic ground motions on the Santa Clara Bridge (Fig. 9.24) by means of fragility curves. Five ground motion scenarios were considered: (1) spatially variable ground motions at different local site conditions with piers 1-4 and 8-11 (Fig. 9.24) located on UBC Soil Type II ("medium"), and piers 5-7 on UBC Soil Type III ("soft"). This scenario was termed "Case 1" (variable site-wave passage-coherency); (2) spatial variability in the excitation but all bridge supports on the same (medium) local soil conditions, termed "Case 2" (medium site-wave passage-coherency); (3) uniform support ground motions with all bridge supports on the same (medium) local soil conditions, termed "Case 3" (medium site-uniform motions); (4) spatial variability in the excitation but with all bridge supports on the same (soft) local soil conditions, termed "Case 4" (soft site-wave passage-coherency); and (5) uniform support ground motions with all bridge supports on the same (soft) local soil conditions, termed "Case 5" (soft site-uniform motions). For the cases incorporating spatial variability, Saxena *et al.* [443] utilized the coherency model of Harichandran and Vanmarcke [208] with the same parameters as in their previous study [468], and the apparent propagation velocity of the motions assumed the value of 1000 m/sec.

Fragility curves for the bridge based on the five scenario input motions were then established. A fragility curve describes the exceedance probability corresponding to a specific damage state for a range of seismic intensities, which were represented in the study by the peak ground acceleration. Five different damage states were utilized in terms of the ductility demand on the piers: Damage State 1 (no damage); Damage

State 2 (slight damage); Damage State 3 (moderate damage); Damage State 4 (extensive damage); and Damage State 5 (complete damage). A total of 1500 sets of ground motion time histories for each one of the five scenarios of support excitations were generated for the evaluation of the fragility curves. Figures 9.25(a) and (b) present the fragility curves for Damage State 3 (moderate) and Damage State 5 (collapse), respectively, caused by each scenario support excitation. It is noted that the dashed lines in the figures present the results due to uniform motions, whereas the continuous lines the results induced by spatially variable excitations. The results in Fig. 9.25 show very significant differences in bridge fragility due to wave passage and loss of coherency: Case 2 (medium site-wave passage-coherency) and Case 4 (soft site-wave passage-coherency) result in significantly higher values for the exceedance probability of both damage states than the values induced by the uniform input motion scenarios at the corresponding sites, i.e., Case 3 (medium site-uniform motions) and Case 5 (soft site-uniform motions), respectively. Case 1 (variable site-wave passage-coherency) also yields higher fragilities for both damage states than either scenario of uniform support excitations (Cases 3 and 5). Interestingly, however, the worst scenario for the bridge was not the case considering seismic excitations at variable site conditions (Case 1), but the case of uniform, soft soil conditions incorporating wave passage and loss of coherency effects (Case 4). Saxena et $al.$ [443] concluded that a design based on uniform input motions is clearly unconservative for the Santa Clara Bridge.

Lupoi et $al.$ [325] also conducted an extensive analysis of bridges in the nonlinear range using fragility curves. A schematic diagram of the bridge configuration used in this study is presented in Fig. 9.26. The bridges analyzed had a continuous deck over three cantilever piers; the span lengths in all cases were equal to 50 m, yielding a total bridge length of 200 m. The deck was considered to be a prestressed box girder and the concrete piers had a hollow cross section. Twenty seven different bridge models were created from the basic configuration of Fig. 9.26 by varying the cross sectional characteristics of the deck and the piers, the height of the piers, and the behavior factor q [108]. The bridges were designed to withstand uniform seismic excitations at their supports described by a uniform elastic response spectrum of intensity $S_e = 2.5 \times \text{PGA}$, where $\text{PGA} = 0.35g$. As in the work of Monti et $al.$ [347], a Takeda-type [515] plastic hinge model was assumed at the base of the columns. The bridges were considered to fail in bending and their damage state was described by the curvature ductilities at the base of the piers. For all piers, the curvature ductility capacity was described by a log-normal random variable with median value equal to 8 and coefficient of variation equal to 0.4. The seismic excitation was applied in the transverse direction.

In the evaluation of the fragility curves for the bridges, four site condition scenarios were assumed at the bridge supports: "FFFFF", "FMMMF", "FFMFF" and "FMFMF", where "F" and "M" indicate firm and medium soil conditions, respectively. The first and last letter in these notations indicate the soil type underneath the left and right abutment, respectively, and the middle three letters the soil type underneath the piers (Fig. 9.26). The study used the coherency model of Luco and Wong [324] with coherency drop parameters of $\alpha = 0$, 1.1×10^{-3} and 3.3×10^{-3} sec/m (Eq. 3.27), and three values for the apparent propagation velocity of the motions

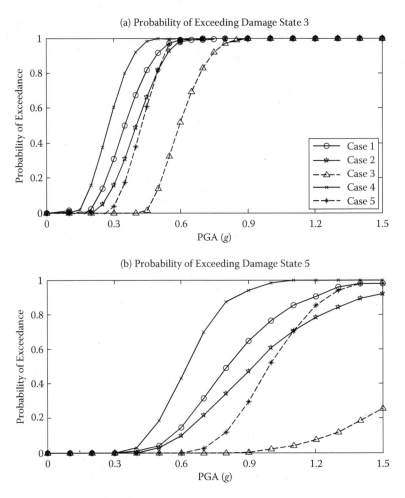

FIGURE 9.25 Fragility curves for Damage State 3 (moderate damage), in part (a), and for Damage State 5 (complete damage), in part (b), evaluated by Saxena *et al.* [442], [443] for the Santa Clara Bridge (Fig. 9.24). In the subplots, "Case 1" corresponds to variable site-wave passage-coherency; "Case 2" to medium site-wave passage-coherency; "Case 3" to medium site-uniform motions; "Case 4" to soft site-wave passage-coherency; and "Case 5" to soft site-uniform motions. (After V. Saxena, G. Deodatis, M. Shinozuka and M. Feng, "Development of fragility curves for multi-span reinforced concrete bridges," in *Monte Carlo Simulation: Proceedings of the International Conference on Monte Carlo Simulation, MCS-2000*, G.I. Schuëller and P.D. Spanos editors, Monte Carlo, Monaco, 2000, Copyright © Balkema; reproduced with permission.)

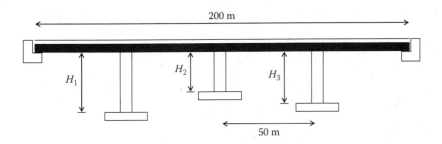

FIGURE 9.26 Schematic diagram of the basic bridge model used for the 27 bridge configurations analyzed by Lupoi *et al.* [325]. (After A. Lupoi, P. Franchin, P.E. Pinto and G. Monti, "Seismic design of bridges accounting for spatial variability of ground motion," *Earthquake Engineering and Structural Dynamics*, Vol. 34, pp. 327–348, Copyright © 2005 John Wiley & Sons Limited; reproduced with permission.)

(Eq. 3.4), $c = 300$ and 900 m/sec and $c \to \infty$. Each bridge configuration was subjected to 36 scenarios of combinations of site conditions, coherency decay and apparent propagation velocities, and, for each scenario, 20 realizations of the input excitations were generated.

Figure 9.27 illustrates the fragility curves presented by Lupoi *et al.* [325] for the ground motion scenarios FFFFF (in part(a)) and FMMMF (in part (b)) underneath the supports of the bridge configuration of Fig. 9.26, pier heights of 10 m and behavior factor equal to 4. Lupoi *et al.* [325] also indicated that similar results were obtained for the other two scenarios of site conditions underneath this bridge, as well as all other examined cases. It can be clearly seen from the figure that uniform excitations at the structure's supports (case FFFFF, $\alpha = 0$ and $c \to \infty$ presented by a thicker line in Fig. 9.27(a)) do not represent the worst loading condition for the bridge. Instead, the worst loading condition on the bridge was induced by the less coherent ground motions ($\alpha = 3.3 \times 10^{-3}$ sec/m). Furthermore, Figs. 9.27(a) and (b) indicate that, for such incoherent excitations, the wave passage effects assume a secondary role, as can be seen from the essential overlap of the fragility curves for $\alpha = 3.3 \times 10^{-3}$ sec/m and all values for the apparent propagation velocity at both sets of site conditions. The most favorable ground motion scenario for the bridge was the case of fully coherent motions propagating with the lowest apparent propagation velocity ($c = 300$ m/sec).

Figure 9.28 summarizes the results obtained by Lupoi *et al.* [325] for all sets of bridge configurations analyzed: For the 36 ground excitation scenarios, the value of the probability of failure, P_f, at the design PGA for each bridge configuration was divided by the corresponding probability of failure of the bridge subjected to uniform support excitations, and the statistics of the ratio (mean value and coefficient of variation) were evaluated. The ordinates in Fig. 9.28 were truncated at the value of 10; the actual values of the ratio for the two bridge configurations, for which the values were truncated, were 47.6 and 25.8, respectively. The figure indicates that, in the majority of cases, the mean value of the ratio is greater than one. For the cases where the ratio is equal to one, the probability of failure of the bridges due to uniform excitations was already close to unity, and, hence, the probability of failure due to

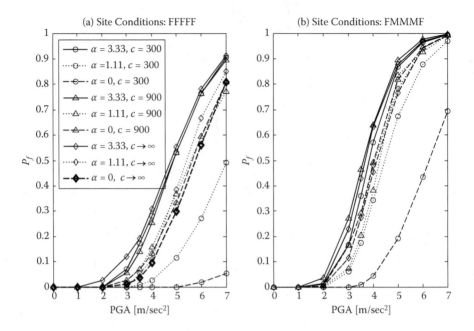

FIGURE 9.27 Fragility curves presented by Lupoi *et al.* [325] for various ground motion scenarios exciting their reference bridge, termed "A1"; the pier heights for this bridge (Fig. 9.26) were 10 m and it was designed for $q = 4$. Part (a) of the figure presents the case of uniform (firm) soil conditions underneath the bridge supports ("FFFFF"), and part (b) the case of firm soil conditions underneath the abutments and medium soil conditions underneath the piers ("FM-MMF"). In both subplots, α [$\times 10^{-3}$ sec/m] indicates the value of the coherency drop parameter of the model of Luco and Wong [324] and c [m/sec] the value of the apparent propagation velocity of the motions. (After A. Lupoi, P. Franchin, P.E. Pinto and G. Monti, "Seismic design of bridges accounting for spatial variability of ground motion," *Earthquake Engineering and Structural Dynamics*, Vol. 34, pp. 327–348, Copyright © 2005 John Wiley & Sons Limited; reproduced with permission.)

spatially variable excitations could not increase further. On the other hand, for the two cases where the ratio is an order of magnitude higher than the rest, the probability of failure for the bridge subjected to uniform support excitations was particularly low. Lupoi *et al.* [325] then concluded that, for the bridge configurations and input excitations considered in their study, the spatial variability of the seismic excitations decreases the level of safety provided by standard design procedures that assume uniform motions at the supports. They further noted that the variability in the local site conditions has a more pronounced effect on the value of the failure probability than the effects of loss of coherency and wave passage only, even though the two soil types considered in their study were not dramatically different (i.e., firm and medium).

All aforementioned studies considered that the column supports were fixed at the ground surface. This condition was relaxed in one of the bridge response evaluations reported by Sextos *et al.* [455], [456]. Sextos *et al.* [455], [456] evaluated the response of 20 bridge models subjected to various spatially variable ground motion scenarios.

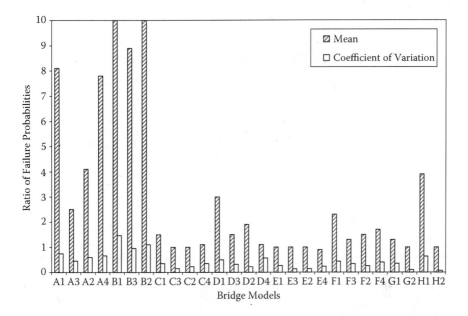

FIGURE 9.28 Mean and coefficient of variation of the ratio of the failure probability, P_f, at the design PGA for each of the 27 bridge configurations (Fig. 9.26) analyzed by Lupoi *et al.* [325] when the bridges were subjected to spatially variable ground motions, over the failure probability of the structures caused by uniform support excitations. The numbers 1, 2, 3 and 4 combined with letters A-F represent pier heights of 10-10-10 m, 8-12-8 m, 8-10-12 m and 12-8-12 m for H_1-H_2-H_3 (Fig. 9.26). Bridges A were designed for $q = 4$, bridges B for $q = 2.5$, and bridges C-F for more flexible decks. The numbers 1 and 2 for cases G and H represents pier heights of 14-21-14 m and 14-7-21 m for H_1-H_2-H_3 (Fig. 9.26). Bridges G were designed for an alternate deck and pier cross sectional configuration than bridges A-F, and bridges H were stiffer than bridges G. The ordinates in the figure are truncated at the value of 10; the actual values of the ratios for bridges B1 and B2 are 47.6 and 25.8, respectively. (After A. Lupoi, P. Franchin, P.E. Pinto and G. Monti, "Seismic design of bridges accounting for spatial variability of ground motion," *Earthquake Engineering and Structural Dynamics*, Vol. 34, pp. 327–348, Copyright © 2005 John Wiley & Sons Limited; reproduced with permission.)

Their reference model, presented in Fig. 9.29, consisted of four spans of equal length of 50 m, leading to a total length for the bridge of 200 m. The deck had a box girder section, which was monolithically connected to rectangular hollow piers of unequal heights (Fig. 9.29). The abutment bearings were assumed to be pinned in the transverse direction and free to slide in the longitudinal direction. Each pier was supported on a 3×4 pile group foundation. The ground motion was the recording at Kallithea during the 1999 Athens, Greece, earthquake ($M_w = 5.9$), scaled to the design peak acceleration of $0.24g$. The study utilized the coherency model of Luco and Wong [324] and wave passage effects.

The bridge models used in the study were created from the reference configuration of Fig. 9.29 by varying the effective rigidity of the piers, their heights, the

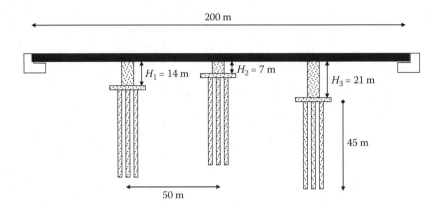

FIGURE 9.29 Schematic diagram of the reference bridge model (Model A) utilized by Sextos *et al.* [456] in their evaluation of the effect of spatially variable excitations on 20 bridge configurations. (After A.G. Sextos, A.J. Kappos and K.D. Pitilakis, "Inelastic dynamic analysis of RC bridges accounting for spatial variability of ground motion, site effects and soil-structure interaction phenomena. Part 2: Parametric study," *Earthquake Engineering and Structural Dynamics*, Vol. 32, pp. 629–652, Copyright © 2003 John Wiley & Sons Limited; reproduced with permission.)

boundary conditions at the abutments, the number of spans and the span length. The excitation was applied in the transverse direction, except for one model, which was subjected to longitudinal ground motions. In one case, the transverse excitation was described by the Gilroy bedrock motion recorded during the 1989 Loma Prieta earthquake. The response evaluation of the bridges was conducted with the numerical code SAP2000 [110]. The bridge configurations were first assumed to respond in the linear range with their supports fixed on the ground surface. They were subjected to uniform excitations, propagating motions, and propagating motions exhibiting loss of coherency. Site effects were approximately incorporated in the evaluation by simulating the motions as spatially variable in the bedrock and then permitting 1-D wave propagation through highly variable soil columns underneath each support. The fixed support conditions were then relaxed and kinematic and inertial SSI effects were introduced. The last scenario considered that the structures respond in the nonlinear range. One set of simulated ground motions was generated from the target acceleration response spectrum of the recorded time history for each ground motion scenario, i.e., motions incorporating wave passage effects only, propagating motions exhibiting loss of coherency, and motions evaluated at variable soil conditions with the 1-D approach indicated earlier. The time series were simulated at the locations of the 13 supports of the longest bridge model (total length of 600 m) with the shortest span length (50 m). The same simulated ground motions were then used for all other bridge models with proper consideration for their respective support separation distances.

The linear analysis of the bridge models subjected to spatially variable excitations (wave passage effects, and wave passage and loss of coherency effects) suggested that these effects become significant for the longer (total length $\geq$ 400 m) bridge models. Sextos *et al.* [456] reported that the consideration of wave passage and coherency in the excitations could increase the maximum bending moments, and absolute and relative pier displacements for the longer structures by as much as 60%, 40% and 350%, respectively, compared to the values obtained for the uniform input motion scenario at the bridge supports. It should be noted, however, that, since the value of the coherency drop parameter of the model of Luco and Wong [324] and the value of the apparent propagation velocity were not disclosed in this study, it cannot be assessed whether the increase in the reported response quantities is a result of moderate, intermediate or extreme variations in the input excitations. Variable site conditions at the bridge supports were then incorporated in the simulations, and their effect on the bridge response was compared to that induced by the scenario of propagating motions exhibiting loss of coherency at uniform site conditions. The bridge models in this comparison were also allowed to respond in the linear range with their supports fixed on the ground. The results suggested that the incorporation of site effects increased all response quantities by an average of approximately 50%, with cases, however, where transverse displacements at the top of the piers and bending moments increased by 100%. In their last evaluation, Sextos *et al.* [456] permitted the bridges to respond in the nonlinear range by scaling the Kallithea record to PGA = 0.72g. Nonlinearity was introduced in their models through an elastoplastic rotational spring at the bottom of the piers, that represented the plastic rotation of the RC section and the soil flexibility at the bottom of the piers. Figure 9.30 presents the range of the ratio of the rotation ductility demand at the base of the piers obtained from the nonlinear analysis of each bridge configuration subjected to spatially variable excitations (including variable site conditions, wave passage and loss of coherency) and incorporating SSI effects over the response obtained from the nonlinear response evaluation of the bridge with fixed supports subjected to uniform ground motions. It can be seen from the figure that, for the majority of the bridges, the ductility demand ratio increases and, in an extreme case, it reaches a factor of 3. It is noted again, however, that results based on a single simulation are indicative, but not a reliable measure of the stochastic characteristics of the response.

Each aforementioned study considered different ground motion scenarios, different structural models, different modeling assumptions and different numerical codes to conduct the analysis. To examine the effect of the latter in the numerical investigation of the effect of spatially variable ground motions on the response of highway bridges, Lou and Zerva [319], [321] conducted a numerical experiment. They investigated the response of two specific, code-prescribed bridge configurations subjected to various scenarios of spatially variable and uniform excitations at their supports utilizing two different modeling assumptions/numerical tools. The two bridges were selected from the seven seismic design examples developed for the Federal Highway Administration (FHWA) to illustrate applications of the seismic analysis and design requirements of the American Association of State Highway and Transportation Officials (AASHTO). The first model (Example 1 [156]) was a two-span bridge with a total length of 74 m, and the second (Example 4 [157]) a three-span, skewed bridge with a total length of 98 m. For illustration purposes, the results for the three-span structure are highlighted in the following. Details of the extensive study can be found in Lou [319].

The plan and elevation of Example 4 of the FHWA design manual [157] is shown in Fig. 9.31. Its superstructure was a cast-in-place reinforced concrete box girder with

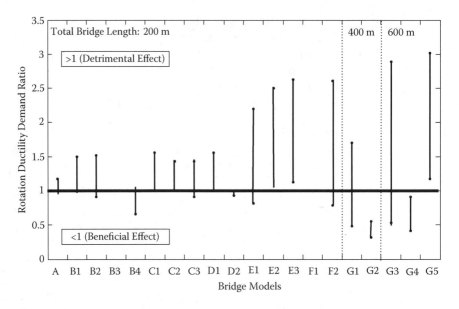

FIGURE 9.30 Variation of the ratio of the rotation ductility demand at the bottom of the piers obtained from the nonlinear analysis of each of the 20 bridge configurations analyzed by Sextos *et al.* [456] when subjected to spatially variable excitations (incorporating variable site conditions at the supports and the effects of wave passage and loss of coherency) and considering SSI effects, over the rotation ductility demand obtained from the nonlinear response evaluation of the bridge configurations on fixed supports subjected to uniform ground motions. The reference bridge model (Model A) is shown in Fig. 9.29. Bridges denoted as B had variable effective vs. gross rigidity ratio, bridges C had variable pier heights leading to symmetric structural configurations, bridges D had variable pier heights leading to irregular structural configurations, bridges E had various abutment-deck connections, bridges F considered different excitations, and bridges G had variable span lengths and total bridge lengths. (After A.G. Sextos, A.J. Kappos and K.D. Pitilakis, "Inelastic dynamic analysis of RC bridges accounting for spatial variability of ground motion, site effects and soil-structure interaction phenomena. Part 2: Parametric study," *Earthquake Engineering and Structural Dynamics*, Vol. 32, pp. 629–652, Copyright © 2003 John Wiley & Sons Limited; reproduced with permission.)

two interior webs; the length of the end spans was 30.5 m and that of the middle span 36.6 m. Seat-type abutments were used to allow free longitudinal movement of the superstructure. All substructure elements were oriented at a skew of 30°. The bent columns were pinned at the top of square spread footing foundations. The stiffness of each bent foundation was modeled by six soil springs at the lower end of the footing elements, which were determined using an elastic half-space approach [157]. The support at each abutment was represented by two elastic, fully plastic spring elements in parallel with initial gaps; their force-deformation relationship was determined according to Caltrans' Seismic Design Criteria (SDC) [89]. Parametric studies were conducted for different gap sizes (0, 5, 10 and 15 cm) between the ends of the bridge deck and the abutments. The nonlinear behavior of the piers was modeled with two approaches: The first approach permitted the formation of plastic hinges,

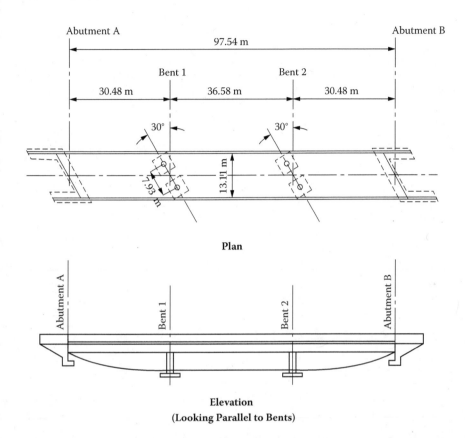

FIGURE 9.31 The plan (top part) and elevation (bottom part) of Example 4 of the FHWA design manual [157] utilized by Lou [319] in the numerical experiment of the effect of the use of different numerical tools in the evaluation of the response of the skewed bridge induced by spatially variable and uniform ground motions (after FHWA [157]).

as suggested earlier by Shinozuka *et al.* [468], and utilized the finite element code DRAIN-3DX [410], [412]. The second approach modeled the nonlinear behavior of the piers with fiber elements and used the object-oriented software framework OpenSEES [337], [338].

Various scenarios of spatially variable and uniform seismic ground motions were applied as input excitations at the structure's supports in its longitudinal direction (Fig. 9.31). The acceleration time histories were generated by means of the simulation scheme of Deodatis [127], as illustrated herein in Section 7.3.2, and integrated to yield displacements with the approach of Liao and Zerva [301], presented in Section 8.2. For the spatially variable excitations, the coherency model of Harichandran and Vanmarcke [208] was utilized with parameters $A = 0.736$, $a = 0.147, k = 5210$ m, $f_o = 1.09$ Hz and $b = 2.78$ (Eq. 3.12). An illustrative, slow apparent propagation velocity of $c = 750$ m/sec was assumed due to the bridge's short length (< 100 m). Table 9.6 presents the results of the analysis for uniform and

TABLE 9.6

Comparison of the peak response quantities of the three-span bridge of Fig. 9.31 modeled with the numerical code DRAIN-3DX and the numerical framework OpenSEES; uniform and spatially variable ground motions at soft soil conditions were used as input excitations at the bridge supports (after Lou [319]).

	Response Quantity	Uniform Ground Motions	Spatially Variable Ground Motions
DRAIN-3DX	Deck axial force [N]	2.10×10^6	3.91×10^6
	Column shear force [N]	1.26×10^6	1.30×10^6
	Bearing shear force [N]	1.41×10^5	1.56×10^5
	Pounding force [N]	1.01×10^6	2.34×10^6
	Bearing shear deformation [m]	0.070	0.088
	Rotational ductility demand	1.83	3.05
OpenSEES	Deck axial force [N]	1.40×10^6	2.54×10^6
	Column shear force [N]	1.34×10^6	1.36×10^6
	Bearing shear force [N]	1.47×10^5	1.66×10^5
	Pounding force [N]	0.78×10^6	1.09×10^6
	Bearing shear deformation [m]	0.078	0.104
	Column drift ratio (%)	1.44	1.50

spatially variable ground motions when all bridge supports are located at soft soil conditions described by the corresponding UBC spectrum [234]; to ensure that the structure responds in the nonlinear range, the target PGA was assumed to be 0.5 g. The table presents the comparison of the results obtained from the nonlinear evaluation of the model using DRAIN-3DX and OpenSEES when the gap size between the ends of the bridge deck and the abutments was 5 cm (2 in). The response quantities presented in the table include the peak deck axial force, peak column shear force, peak bearing shear force, peak pounding force and peak bearing shear deformation. Additionally, the table presents the maximum rotational ductility demand obtained with DRAIN-3DX and the column drift ratio obtained with OpenSEES.

Even though significant effort was placed in consistently modeling the bridges with the two numerical tools and the analyses were conducted by the same investigator, Table 9.6 indicates that the analyses with DRAIN-3DX and OpenSEES resulted in a similar trend for the seismic response of the bridge models, but some of the response quantities varied significantly [319]. For the example illustration in the table, both DRAIN-3DX and OpenSEES predict similar values for the column and bearing shear forces. However, whereas both models lead to significantly higher deck axial forces and pounding forces when spatially variable input excitations are utilized, the DRAIN-3DX results are significantly higher than the results of OpenSEES. The model in DRAIN-3DX also indicates that, even for this short bridge, the rotational ductility demand is higher for spatially variable than uniform motions at the structure's supports. The differences in the results of Table 9.6 may be attributed to the different mechanical models used to emulate the nonlinear inelastic response of the pier columns (i.e., plastic hinge model vs. fiber element model), and to the differences in the material models employed in the two numerical tools [319]: The properties of the rotational spring of the plastic hinge for the DRAIN-3DX model were determined from Mander's concrete model [331], [332] and the USC-RC steel model [153], whereas the properties of the OpenSEES fiber element model were obtained from the high-strain-rate Mander's concrete model [331], [332] and a bi-linear steel model [89]; as a consequence, different damping properties were also generated for the finite element models from the two numerical tools. Lou [319] also reported that the results were sensitive to the selected value of the gap size. Generally, as the gap size increased and the structure became more flexible, the differences between the response caused by uniform and spatially variable excitations decreased, and, in cases, the uniform excitations led to a higher rotational ductility demand. This numerical experiment then indicates that the modeling assumptions can influence considerably the estimated values of the resulting response quantities.

The trend of the results of the nonlinear response evaluation of this short bridge (Table 9.6) also indicates that spatially variable input excitations can significantly affect the deck axial forces and pounding forces between the bridge superstructure and the seat-type abutments. This observation is also in agreement with extensive studies on the pounding of bridge structures by Hao and his collaborators [103], [104], [580], who concluded that the consideration of spatial variability is necessary for the realistic assessment of pounding and the induced bearing forces, as well as the potential for the unseating of the girders. Indeed, this is the most widely accepted consequence of the spatial variability in the seismic ground motions to damage observed in bridges during earthquakes: The relative displacements induced by spatially variable excitations can lead to pounding between girders and possible unseating of the deck due to insufficient seating length. Design codes provide specific recommendations regarding minimum seat length requirements as, e.g., the provisions of AASHTO [24], ATC [32], Caltrans [89] and the Japan Road Association (JRA) [241]; an insightful evalution of these provisions was presented by Park et al. [395] and Yen et al. [578].

The seismic response of curved bridges was also given attention, e.g., [83], [453], [454]. Recently, Burdette et al. [82], [83] analyzed the effect of the spatial variation of seismic ground motions on the response of a curved bridge with a total length of 344 m rotating 98°, and compared the results with the response of an equivalent straight bridge of equal length. The schematic diagrams of the plan of the curved bridge and the elevation of the bridges are presented in Figs. 9.32(a) and (b), respectively. The single-column bents supported a prestressed concrete twin box girder superstructure with sliding bearings, which allowed rotation and longitudinal motion, but were restrained transversely. The deck was similarly attached to abutment 10, but fixed at abutment 1. The bridges were analyzed using the inelastic analysis program ZEUS-NL [150]. The piers and abutments were considered to be fixed to the ground. Cubic elasto-plastic 3-D beam-column elements were used for the deck, piers and pier cap elements, the deck and pier cap masses were considered using lumped mass elements, and the bearings were modeled as springs with sliding friction. Pier displacements were evaluated for seismic ground motions applied simultaneously in the two horizontal directions. The study utilized the coherency model of Luco and Wong [324] for three values of the coherency drop parameter

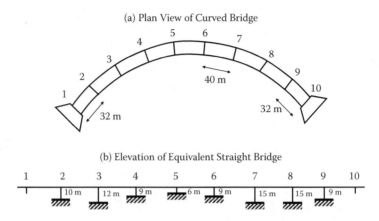

FIGURE 9.32 Schematic diagram of the bridges used in the study by Burdette *et al.* [82], [83]. Part (a) of the figure illustrates the plan of the curved bridge; the typical length of the intermediate spans is 40 m. Part (b) presents the elevation of the straight bridge with the values of the pier heights; the height of each pier (2-9) was the same for the curved and the straight bridge. (After N.J. Burdette, "The effect of asynchronous earthquake motion on complex bridges," M.Sc. Thesis, University of Illinois at Urbana-Champaign, Urbana, IL, Copyright © 2006 MAE Center; reproduced with permission.)

($\alpha = 0$, 1.1×10^{-3} and 3.3×10^{-3} sec/m), and the wave passage effect was modeled with three values for the apparent propagation velocity ($c = 300$ and 900 m/sec and $c \rightarrow \infty$). Ten sets of simulated ground motions were generated for each combination of coherency drop parameter and apparent propagation velocity; nine sets were obtained with the approach used by Lupoi *et al.* [325] and one set with the approach used by Sextos *et al.* [456]. The results of this evaluation also pointed to the significance of the pseudo-static response induced by spatially variable excitations, and the differences in the dynamic response of the structures caused by uniform and nonuniform ground motions. An interesting observation made in this study was that the pier displacements of the equivalent straight bridge model caused by spatially variable excitations were higher than the corresponding displacements of the curved bridge model in the longitudinal direction, whereas the opposite was observed in the transverse direction. These differences were attributed to the interaction of the deck with the piers [82], [83]: In the longitudinal direction, the curved bridge resists the forces by a combination of the flexural and axial stiffness of the deck, whereas the straight bridge deck resists these forces in pure axial tension or compression. Hence, the deck of the straight bridge is stiffer in the longitudinal direction than the deck of the curved bridge, and, in order to accommodate the larger relative pseudo-static motion between the deck and the ground, the piers of the straight bridge have to move significantly. On the other hand, in the transverse direction, the curved deck, acting as an arch, provides greater stiffness, whereas the straight deck resists transverse forces in flexure. Hence, part of the pseudo-static displacements are absorbed by the deck of the straight bridge, leading to lower transverse displacements for the straight bridge than the curved bridge. Curvature becomes then an additional parameter that needs to be investigated in the analyses of the effect of the spatial variability of the seismic ground motions on the response of curved bridges.

From a different perspective, very few, large-scale experiments examining the effect of the spatial variation of seismic ground motions on the response of high-way bridges have been conducted, and these were mostly in support of the new EUROCODE 8, e.g., [399], [404], [452]. One of the first experiments, at a 1:8 scale, was conducted at the three shake tables of the Instituto Sperimentale Modelli e Strutture (ISMES) in Bergamo, Italy. The scope of the experiment was to test the models first under uniform excitations in the transverse direction, and then under spatially variable ground motions incorporating wave passage effects. Severn [452], however, reported that the control systems of the shake tables differed in quality, and, eventually, the model was only subjected to spatially variable excitations that differed in amplitude and phase. Pegon and Pinto [399] reported large scale (1:2.5) experimental investigations at the European Laboratory for Structural Assessment (ELSA) at Ispra, Italy, using pseudo-dynamic testing with substructuring. The geometric configuration of the bridge model used in this study was similar to the model of Lupoi et al. [325], illustrated in Fig. 9.26, with pier heights of $H_1 = 14$ m, $H_2 = 7$ m and $H_3 = 21$ m. The findings of the experimental study were similar to the analytical results, namely that nonuniform seismic excitations induce a different response pattern in the bridge, which was also attributed to the different modes excited by uniform and nonuniform excitations. More recently, Pinto et al. [404] reported the pseudo-dynamic testing at ELSA of a large-scale (1:2.5) model of an existing RC highway bridge, located in Austria, using substructuring. In this case, the structure was subjected to synthetic, spatially variable excitations appropriate for the bridge's site.

A clear conclusion that can be drawn from all studies on highway bridges in this section is that spatially variable excitations cause a different response pattern in these long structures than uniform ground motions. They induce pseudo-static internal forces in the structures, which the uniform ground motions do not, as the latter excitations result in a rigid-body, pseudo-static deformation. Furthermore, spatially variable excitations induce a different dynamic response than uniform ground motions at the structure's supports. An obvious example of such differences is the response of symmetric structures, for which uniform ground motions excite only the symmetric modes, whereas spatially variable ground motions excite both the symmetric and the antisymmetric modes. In addition, spatially variable seismic ground motions excite the higher modes of the structures more significantly than uniform motions. These conclusions point directly to the fact that the effect of spatially variable excitations on the response of these structures cannot be isolated from their properties and structural configuration. The studies described herein used different modeling assumptions, including the modeling of the abutment-deck and deck-pier connections, the relative rigidity of the deck and the piers, the consideration or not of SSI effects, 2-D or 3-D modeling, and the modeling of material and geometric nonlinearities. Examples of the significance of the modeling assumptions were illustrated in the results of Tzanetos et al. [546] in Table 9.5, where the modeling of the abutment resulted in different distributions of the demand, and in the results of Lou [319] in Table 9.6, where the modeling of the same structure with different numerical tools led to significantly varying results. Whether different modeling assumptions would tilt the conclusions of the aforementioned studies towards a more or a less beneficial/detrimental effect of spatially variable ground motions on the response of the bridge structures cannot be readily assessed. Hence, the following remarks are qualitative.

All aforementioned studies agree that, as the loss of coherency in the motions increases, it overshadows the wave passage effect. The differences in the waveforms of spatially variable excitations caused by loss of coherency and wave passage effects were illustrated in Figs. 7.10(c) and 7.10(d), respectively. The figures present the (stationary) displacement time series at five locations on the ground surface with distance between consecutive locations of 100 m. Figure 7.10(c) illustrates motions that were simulated with the coherency model of Luco and Wong [324] for a coherency drop parameter of $\alpha = 1 \times 10^{-3}$ sec/m; the exponential decay of this model with frequency and separation distance is shown in Fig. 7.9(c). The majority of the studies presented herein utilized a similar value for the coherency drop parameter ($\alpha = 1.1 \times 10^{-3}$ or 1.67×10^{-3} sec/m), as well as a higher value ($\alpha = 3.3 \times 10^{-3}$ sec/m). Figure 7.10(d) presents strongly coherent (stationary) displacement waveforms propagating on the ground surface with a velocity of $c = 500$ m/sec, which is an average value from the ones utilized in the studies. The variability in the seismic ground motions caused by loss of coherency and wave passage is different: The wave passage effect introduces a systematic time shift in the time series, whereas the loss of coherency leads to sudden large differential motions. Clearly, the differential motions resulting from either loss of coherency or wave passage will significantly excite the higher modes of the structures, as observed in the majority of the studies for highway, as well as suspension and cable-stayed bridges. For the values of the coherency drop parameter and the apparent propagation velocity utilized in the spatially variable excitations of Figs. 7.10(c) and 7.10(d), both the loss of coherency and the apparent propagation of the motions will contribute to the response of the structures, as both introduce significant differential motions, but they may interfere constructively or destructively when their effects are combined. Constructive interference was observed by Burdette et al. [83], who noted that the combined effects of loss of coherency and wave passage appeared like the "superposition" of the contribution of the individual causes of differential motions, whereas destructive interference may be observed in the fragility curves of Lupoi et al. [325] in Fig. 9.27, where the slower apparent propagation velocity of $c = 300$ m/sec appears to lead to a decrease in the response when combined with, essentially, any of the considered coherency drop parameters. It should be emphasized at this point that the value of $\alpha = 1 \times 10^{-3}$ sec/m indicates that the seismic ground motions are, already, strongly incoherent. This can be clearly recognized from the comparison of the exponential decay of the model for $\alpha = 1 \times 10^{-3}$ sec/m in Fig. 7.9(c) with the empirical plane-wave coherency model of Abrahamson [7] illustrated in Fig. 3.20(b). Even though the latter coherency model was based on data recorded at various site conditions, the approach used for its derivation was the most robust of the approaches utilized in the literature (Section 3.4.1). The coherency model of Luco and Wong [324] for $\alpha = 1 \times 10^{-3}$ sec/m in Fig. 7.9(c) tends to zero at a frequency of approximately 3 Hz when the separation distance is 100 m, whereas the coherency of the data tends to zero at approximately 18 Hz for the same separation distance. Similar observations can be made for the longer separation distances. If, instead, the coherency drop parameter assumes the value of

$\alpha = 3.3 \times 10^{-3}$ sec/m, the coherency model of Luco and Wong [324] tends to zero from a frequency of 1 Hz at the separation distance of 100 m. The motions in this case become so incoherent, that the wave passage effect can affect the results only minimally. Sensitivity analyses utilizing values for the coherency drop parameter of $\alpha = 0$ and $\alpha > 1 \times 10^{-3}$ sec/m examine, basically, the response of the structures subjected to fully coherent and strongly/fully incoherent excitations, but not to any "intermediate" spatially variable ground motions. It should be noted, however, that the worst loading scenario for the structures is not necessarily the case of the strongly incoherent/slowly propagating motions, as already discussed in the presentation of the results of Tzanetos et al. [546] in Table 9.5. Additionally, for the bridges analyzed by Monti et al. [347], the highest ductility demand was observed for fully coherent motions propagating with the intermediate value of the velocity ($c = 600$ m/sec in Fig. 9.22(b)), whereas the most favorable ground motion scenario for the bridges analyzed by Lupoi et al. [325] were time series propagating with the slowest considered velocity ($c = 300$ m/sec in Fig. 9.27). The selection of coherency models from the ones available in the literature, the degree of their exponential decay with separation distance and frequency, and the values for the apparent propagation velocity of the motions are key factors for a more reliable estimation of the structural response characteristics of long structures; these aspects are discussed in the following section.

All aforementioned studies also agree that, generally, the effects of loss of coherency and apparent propagation of the motions are overshadowed by the effect of variable site conditions underneath the bridge supports. This may also be anticipated: Variable site conditions underneath the structure's supports introduce significant amplitude variability in the motions, in addition to the phase variability introduced by the coherency. An illustration of the differences of (acceleration) time series at such nonuniform sites was presented in Fig. 7.16: Stations 1 and 4 were located at stiff soil conditions, and stations 2 and 3 at soft soil conditions. Interestingly, however, essentially all bridges presented herein had total lengths shorter than $300 - 450$ m, which, at least according to Shinozuka et al. [468], is the limiting length below which spatial variability effects at uniform site conditions can be neglected; this limiting length becomes a function of ground classification in the new EUROCODE 8, as will be presented in Table 9.7. In addition, the majority of the bridges were located on "firm" or "medium" soil conditions. The only longer highway bridge analyzed in the nonlinear range and considered to be located at either uniform or variable site conditions was the 500 m long Santa Clara Bridge (Fig. 9.24) reported by Saxena et al. [443]. (The longer bridges analyzed by Sextos et al. [456] (Fig. 9.30) were permitted to respond nonlinearly only when the site conditions underneath their supports were variable.) Saxena et al. [443] indicated that the worst loading scenario for the bridge was not the case of variable site conditions underneath the structure's supports, but the case of uniform ("soft") soil conditions incorporating loss of coherency and wave passage effects (Fig. 9.25). Whether this result was caused by the 2-D modeling of the structure, the consideration of its longitudinal response, the low-frequency content of the soft soil conditions (Fig. 3.1) and the use of the coherency model of Harichandran and Vanmarcke [208], which is only partially correlated at low frequencies (e.g., Fig. 3.16), cannot be readily assessed. Further investigations of the effects of all causes of the spatial variation of the seismic ground motions on the response of longer highway bridges will shed more light into their relative contribution to the response. In addition, the modeling of spatially variable excitations at sites with irregular subsurface topography needs to be readdressed, as it is based on simplifying assumptions that can underestimate the seismic loads at such sites. This issue was discussed in Sections 5.6 and 7.3.2 and will be further elaborated upon in the next section.

It should also be noted that, except for the last study by Burdette et al. [83], which analyzed the combined effect of both horizontal components of the seismic excitation on the structural response, all other studies presented in this section applied the excitation in a single direction. Considering that, for the evaluation of the nonlinear response of structures, the three seismic ground motion components should be applied simultaneously at the structure's supports, the full effect of spatially variable ground motions on the response of highway bridges is yet to be quantified[3]. The modeling of the 3-D random field and issues regarding the derivation of displacement time series from simulated accelerations will be briefly addressed in the concluding remarks of this chapter.

The results of the studies presented herein indicate that the unique interaction of the structural models with the spatially variable excitations makes the establishment of design criteria difficult, especially if the goal is to specify an equivalent uniform excitation that can be used in lieu of the spatially variable ground motions at the bridge supports. In spite of this difficulty, the new EUROCODE 8 (EC8) [109] took the bold step forward and proposed such an approach. For this reason, the relevant provisions of the code are presented next and briefly discussed.

"EUROCODE 8 - Part 2: Bridges" [109] specifies that spatial variability should be considered in the design of bridge sections with a continuous deck when one or both of the following two conditions apply:

(i) The soil properties vary along the bridge and more than one ground types correspond to the supports of the bridge deck;

(ii) The soil properties are approximately uniform, but the length of the continuous deck exceeds a limiting length, $L_{\lim}$; a recommended value for the limiting length is $L_{\lim} = L_g / 1.5$, where L_g is the distance beyond which the motions can be considered to be uncorrelated and its value depends on the ground type. L_g and $L_{\lim}$ for various ground types are presented in Table 9.7.

The EUROCODE 8 - Part 2 provisions state that the spatial variability of the motions should incorporate, even if only in a simplified way, the effects of loss of coherency, apparent propagation of the motions and variable site conditions underneath the structures' supports. For this purpose, it recommends either a rigorous or a simplified approach, which are highlighted in the following.

---

[3] It is noted that, perhaps, the most complete 3-D modeling of a highway bridge subjected to all three ground motion components was reported by Mylonakis et al. [355] in their investigation of the performance of the southbound separation and overhead bridge at the SR14/I-5 interchange during the Northridge earthquake. The bridge was located right above the ruptured zone (approximately 10 km from the epicenter); two piers of the bridge failed and three spans supported by those piers collapsed.

TABLE 9.7

Soil characterization for ground types A-E after EC8 - Part 1 [108]. S is the soil factor, T_C the upper limit of the period of the constant spectral acceleration branch, and T_D the value defining the beginning of the constant displacement response range of the spectrum. L_g is the length beyond which the motions are considered to be uncorrelated according to EC8 - Part 2 [109], and L_{lim} is the limiting bridge length below which the code recommends that spatial variability effects at uniform site conditions can be neglected. (The European Standards EN 1998-1:2004 and EN 1998-2:2005 have the status of British Standards. Permission to reproduce extracts from British Standards is granted by BSI[4].)

Ground Type	Description	S^*	T_C^* [sec]	T_D^* [sec]	L_g [m]	L_{lim} [m]
A	Rock or other rock-like geological formation, including at most 5 m of weaker material at the surface	1.0	0.4	2.0	600	400
B	Deposits of very dense sand, gravel, or very stiff clay, at least several tens of meters in thickness, characterized by a gradual increase of mechanical properties with depth	1.2	0.5	2.0	500	333
C	Deep deposits of dense or medium-dense sand, gravel or stiff clay with thickness from several tens to many hundreds of meters	1.15	0.6	2.0	400	267
D	Deposits of loose-to-medium cohesionless soil (with or without some soft cohesive layers), or of predominantly soft-to-firm cohesive soil	1.35	0.8	2.0	300	200
E	A soil profile consisting of a surface alluvium layer with V_s values of type C or D and thickness varying between about 5 m and 20 m, underlain by stiffer material with $V_s > 800$ m/sec	1.4	0.5	2.0	500	333

* Soil parameters are provided in the table for "Type 1" elastic response spectra. Type 1 and Type 2 elastic response spectra are specified by the code with the recommendation that [108]: "If the earthquakes that contribute most to the seismic hazard defined for the site for the purpose of probabilistic hazard assessment have a surface-wave magnitude, M_S, not greater than 5.5, it is recommended that the Type 2 spectrum is adopted."

[4] British Standards can be obtained in PDF format from the BSI online shop: http://www.bsi-global.com/en/Shop/ or by contacting BSI Customer Services for hardcopies: Tel: +44 (0)20 8996 9001, Email: mailto: cservices@bsi-global.com.

The rigorous approach, presented in an informative annex, recommends time history or random vibration analyses of the bridge subjected to spatially variable excitations, described by Eqs. 3.1 and 3.3. As an example coherency model, EC8 - Part 2 suggests the model of Luco and Wong [324]. The wave passage effect should be incorporated in the analysis with the exponential term of Eqs. 3.3 and 3.4. Variable site conditions underneath the structures' supports should be modeled by means of power spectral densities compatible with the different response spectra provided by the code for each ground type. Initially, the provisions indicate that an additional time delay be incorporated in the spatial variability expression (Eq. 3.3) to account for the vertical propagation of the waves though soil columns with different characteristics underneath the bridge supports, as proposed by Der Kiureghian [135] and discussed in Section 3.4.2. However, at the end of the annex, it is suggested that, because this term has a small influence on the results, it can be neglected. The analysis of bridges can then be conducted in the time or frequency domain. For the time-domain evaluations, EC8 - Part 2 suggests that the spatially variable ground motions be generated with a simulation scheme similar to the one presented by Deodatis [127] and illustrated herein in Section 7.3.2. Linear or nonlinear analyses can be conducted with a sufficient number of sample input motions, such that the mean values of the maximum response quantities of interest are stable. If a linear analysis is performed, the design values should be determined by dividing the elastic action (load) effects by the appropriate behavior factor q, and ensuring the conformity of the ductile response of the structure with the code's provisions. Alternatively, a random vibration analysis of the bridge can be performed with an approach similar to the methodology suggested by Der Kiureghian and Neuenhofer [138], presented in Section 6.5.1. In this latter case, nonlinearity should be accounted for in a manner similar to that suggested for linear time history evaluations.

If, on the other hand, a simplified evaluation of the effect of the spatially variable ground motions on the structural response is to be performed, the code recommends that the designer conduct first a dynamic analysis of the structure caused by uniform seismic excitations based on prescribed response spectra, then a pseudo-static analysis based on two patterns of prescribed displacements at the bridge supports, and then combine the dynamic response with the worst scenario pseudo-static response by means of the Square Root of Sum of Squares (SRSS) rule. These steps are further highlighted in the following.

The code recommends that the dynamic ("inertia") response of the structures be evaluated by means of code-specific approaches considering a "single seismic action for the entire structure," i.e., a single response spectrum or corresponding accelerogram sets. For variable site conditions, the analysis should be conducted for the most severe ground type underneath the supports.

The spatial variability of the seismic motions should then be taken into consideration through the evaluation of the pseudo-static response induced in the structure by two scenarios of horizontal support displacements, termed "Set A" and "Set B" (Fig. 9.33). These sets of displacements should be applied separately in each horizontal direction on the relevant support foundations or on the soil end of the relevant springs representing the soil stiffness. The two displacement scenarios should not be combined, but the most severe one should be used in the calculation of the total response. The two displacement sets are as follows:

(a) Displacement Configuration for Set A

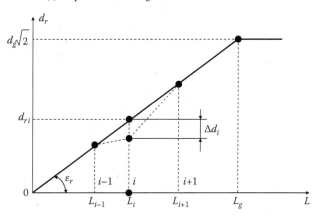

(b) Displacement Configuration for Set B

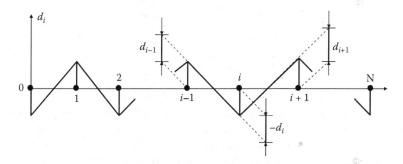

FIGURE 9.33 The two sets of bridge support displacement patterns suggested by EU-ROCODE 8 - Part 2: Bridges [109] for the evaluation of the pseudo-static response of the structures. Part (a) of the figure presents the support displacement pattern of Set A, which corresponds to a "linear" increase of displacements from the utmost left support (indicated by "0" in the subplots) up to the limiting value $d_g\sqrt{2}$. Part (b) of the figure presents the support displacement pattern of Set B, which corresponds to ground displacements occurring in opposite directions at all supports. (The European Standard EN 1998-2:2005 has the status of a British Standard. Permission to reproduce extracts from British Standards is granted by BSI[5].)

Set A considers relative displacements of the form:

$$d_{ri} = \varepsilon_r L_i \le d_g\sqrt{2} \tag{9.8}$$

$$\varepsilon_r = d_g\sqrt{2}/L_g \tag{9.9}$$

[5] British Standards can be obtained in PDF format from the BSI online shop: http://www.bsi-global.com/en/Shop/ or by contacting BSI Customer Services for hardcopies: Tel: +44 (0)20 8996 9001, Email: mailto:cservices@bsi-global.com.

to be applied simultaneously with the same sign $(+$ or $-)$ to all supports of the bridge in the horizontal direction considered, as illustrated in Fig. 9.33(a). L_i (Eq. 9.8) is the distance (projection on the horizontal plane) of support i from a reference support $i = 0$, which, for convenience, can be one of the end supports (Fig. 9.33(a)). In the above expressions, d_g is the design ground displacement corresponding to the soil conditions underneath support i and provided in Part 1 of the EC8 code as [108]:

$$d_g = 0.025 \, a_g \, S \, T_C T_D \qquad (9.10)$$

where a_g is the design ground acceleration on type A ground, S is the soil factor, T_C is the upper limit of the period of the constant spectral acceleration branch, and T_D is the value defining the beginning of the constant displacement response range of the spectrum. The ground classifications A-E according to EC8 - Part 1 [108], the values of the parameters in Eq. 9.10, and the suggested values for L_g and $L_{\lim}$ were presented in Table 9.7.

Set B considers the effect of ground displacements occurring in opposite directions at adjacent supports as illustrated in Fig. 9.33(b). The differential displacements, Δd_i, are first evaluated at each intermediate support i considering that its adjacent supports $i - 1$ and $i + 1$ are undisplaced (Fig. 9.33(a)) as:

$$\Delta d_i = \pm \beta_r \varepsilon_r L_{\mathrm{av},i} \qquad (9.11)$$

where $L_{\mathrm{av},i}$ is the average of the distances $L_{i-1,i}$ and $L_{i,i+1}$ between the intermediate support i and its adjacent supports $i - 1$ and $i + 1$, respectively. For the end supports, 0 and N, the average distances should be taken as $L_{\mathrm{av},0} = L_{01}$ and $L_{\mathrm{av},N} = L_{N-1,N}$ (Fig. 9.33(b)). β_r is a factor to account for the amplitude of the ground displacement occurring in opposite directions at the adjacent supports; the recommended values are $\beta_r = 0.5$ when all three supports are located on the same site conditions, and $\beta_r = 1.0$ if one of the supports is located on a ground type different than the ground type of the other two supports. ε_r is as defined in Eq. 9.9 for the Set A displacements. For the Set B displacements, it is further specified that, if there is a difference in the ground type underneath the two supports, then the maximum value of ε_r should be used. The Set B displacement configuration is obtained by imposing the following absolute displacements with opposite signs at adjacent supports i and $i + 1$ (Fig. 9.33(b)):

$$d_i = \pm \Delta d_i / 2$$
$$d_{i+1} = \pm \Delta d_{i+1} / 2 \qquad (9.12)$$

In each horizontal direction, the most severe pseudo-static response resulting from Set A or Set B displacement configurations should then be combined, by means of the SRSS rule, with the corresponding purely dynamic response caused by uniform support excitations compatible with the response spectra of the worst ground conditions underneath the supports. The results of this evaluation constitute the response of the structure in the direction considered. The total response of the structure is then obtained either by using the SRSS rule for the two horizontal and the vertical components, or through appropriate combination rules [108].

The code's recommendations are, undoubtedly, innovative, and will attract significant renewed research interest for the examination and validation of its provisions. Some comments to this effect are as follows: (i) The trend for the limiting length, $L_{\lim}$ in Table 9.7, appears to be realistic, as seismic ground motions at "rock" sites are more highly correlated than motions at softer soil sites. In addition, their values are more conservative than the limiting length proposed earlier by Shinozuka *et al.* [468], who suggested that spatial variability effects can be neglected for bridges with total length of 300 − 450 m located at uniform site conditions. The rationale, however, for the selection of the particular values for $L_{\lim}$, irrespective of the structural configuration, is unclear; (ii) The two sets of support displacements (Fig. 9.33), that account for the pseudo-static response, appear to be based on the wavelength of the seismic excitation: Set A (Fig. 9.33(a)) assumes that the dominant displacement waveform has a very long wavelength (low frequency), approximately four times the total length of the bridge, so that the supports move almost "linearly" (in-phase) with large absolute displacements. Set B (Fig. 9.33(b)) considers that the dominant displacement waveform has a short wavelength (high frequency), which, if the spans had equal length, would correspond to twice the span length, such that the supports move completely out-of-phase, and leads to large relative displacements. Most probably, the two extreme displacement patterns were conceived so as to induce extreme values for the pseudo-static response of the structures, in an effort to counteract the lack of consideration of the effects of the spatially variable excitations on their dynamic response. It is recalled, however, that the dynamic response induced by spatially variable excitations is very different from the dynamic response caused by uniform excitations. Whether the demand pattern along the structure induced by this approximation reflects or envelops the demand pattern of the ensemble average of nonlinear time history analyses cannot be readily assessed; (iii) Before the introduction of spatial variability aspects, the code's provisions suggest that, for nonlinear time history analyses, the ground motion time series be selected from recorded data appropriate for the site and event, and, only if such data are not available, modified recorded time series or simulations be used instead. On the other hand, if spatial variability is considered in nonlinear time history evaluations, simulated time series are recommended as the only alternative, and, as illustrated in this section, all studies evaluating the response characteristics of highway bridges utilized simulated spatially variable excitations. It is noted that simulations (Chapter 7) are, indeed, a most versatile tool for the generation of spatially variable ground motions, but, they are, to a certain degree, "artificial" in comparison to recorded data. Conditional simulations of spatially variable excitations (Chapter 8) can be used instead for a more realistic representation of the input ground motions; (iv) The approach for the selection of the coherency drop parameter for the model of Luco and Wong [324] and the value of the apparent propagation velocity is not clearly stated in the code, even though the designer will be faced with the difficulty of this decision when utilizing the suggested rigorous approach in the response evaluation of the structures; and (v) The modeling of the seismic ground motions at variable site conditions requires further consideration, as the use of vertical (1-D) S-wave propagation at sites where the soil conditions vary considerably over distances of a few meters is not a valid approximation. Some additional remarks regarding the aforementioned issues are presented next.

9.5   SOME CONCLUDING REMARKS

If the results of the diverse studies analyzing the effect of spatially variable seismic ground motions on the response of long structures can be summarized, the following general conclusions can be drawn: The consideration of kinematic SSI for spatially variable excitations has a beneficial effect on the response of nuclear power plants, as the foundation of these structures filters the spatially variable ground motions, thus reducing the translational response at higher frequencies, but introducing a rotational response. On the other extreme, for buried pipelines and tunnels, the spatial variability in the excitation constitutes, essentially, the seismic load for the structures. For pipelines and tunnels, that are buried in uniform site conditions and subjected to uniform excitations, the response is a stress-free rigid body motion. In-between these two extreme cases, lie the response patterns of dams and bridges. The pattern that emerges in the comparison of the response of these structures to uniform and spatially variable excitations is complex. Spatially variable excitations induce pseudo-static internal forces in the structures, which the uniform excitations do not, because they result in a pseudo-static, rigid-body deformation. For the dynamic response of lifeline systems, all studies conclude that spatially variable seismic ground motions excite the dynamic modes of the structures differently than uniform excitations. The most obvious difference, as illustrated in the simple beam example of Section 6.4.5, is for symmetric structures, for which uniform ground motions excite only the symmetric modes, whereas spatially variable excitations induce the anti-symmetric modes of the structure as well. In addition, as indicated in the previous sections, spatially variable seismic ground motions excite the higher modes of the structures more significantly than uniform motions. In all cases, the distribution pattern of the demand along the structure differs when the excitation is assumed to be uniform or nonuniform. It is not uncommon that studies investigating the effects of spatial variability even on simple structural models conclude that, depending on the response quantity evaluated, its position along the structure and the natural frequencies of the system, spatially variable excitations may or may not induce a higher response than uniform motions (e.g., [193], [194], [590], [591]).

The approaches for the incorporation of spatial variability in the design of these diverse structures also differ. Spatial variability effects are considered in the design of nuclear power plants through reduction factors for the foundation's input motion (e.g., [25]). The traveling-wave assumption still defines the seismic loads for buried pipelines and tunnels (e.g., [235]). For suspension and cable-stayed bridges, spatially variable excitations provided in codes are applied, in most cases, as input excitations at their supports (e.g., [364]). Highway bridges are designed, when applicable, for uniform excitations that attempt to "mimic" the effect of spatially variable ground motions (e.g., [32], [109]). On the other hand, the effect of spatial variability on dams, and, especially, the effect of the complex 3-D spatial variation of the incident excitation caused by the canyon topography on the response of arch dams, still requires further investigation (e.g., [564]).

The examination of the effect of spatially variable excitations on the response of lifeline systems is a topic of active research investigation. The vast majority of the studies are analytical/numerical. Obviously, the results of the investigations will depend on the structural configuration (and its modeling) and the input excitation (and its modeling). From the example applications presented herein, the one that best highlights the differences caused by the structural and ground motion modeling on the response of the same structure was described in Section 9.2 for the Pacoima Dam: Whereas the linear modeling of the dam, the assumption that the canyon is infinite, and the approximation of the incident excitation by plane waves did not indicate any significant pseudo-static response in the structure [378], the investigation of the same structure permitted to respond in the nonlinear range and subjected to inferred spatially variable seismic data indicated a controlling pseudo-static response at certain locations on the dam [21], as illustrated in Fig. 9.13. In concluding this chapter, aspects of analyses approaches, and structural and ground motion modeling assumptions are given some additional attention.

9.5.1 ANALYSIS APPROACH

Random vibrations or Monte Carlo simulations can be used to analyze the dynamic response of any structural system, including structures subjected to spatially variable excitations. If the linear, stationary structural response is of interest, the random vibration approach surpasses Monte Carlo simulations, as a single evaluation suffices for the estimation of the stochastic characteristics of the response. In contrast, Monte Carlo simulations require the generation of a significant number of spatially variable time series and the computational evaluation ("run") of the response of the structure for each generated ground motion set. If, on the other hand, the linear, non-stationary structural response is of interest, essentially all studies indicate that the results of the non-stationary random vibration approach will be affected by the selection of the intensity modulating function (Section 7.1.2), and, hence, its selection should be given proper consideration. It is also iterated, that since the nonlinear random vibration approach has not yet fully evolved (Section 6.5.2), the nonlinear structural response evaluation of the systems requires Monte Carlo simulations.

9.5.2 SIMULATION AND INTEGRATION

The most commonly used approach for the simulation of spatially variable ground motions is the spectral representation method (Chapter 7). The majority of the studies that consider simulated spatially variable time series utilize approaches similar to the one presented by Deodatis [127], illustrated herein in Section 7.3.2, and describe the "point" estimates of the motions by either a power spectral density or a target response spectrum. As indicated in Chapter 7, simulated time series based on any spatially variable random field are sensitive to a number of parameters that control their convergence to the target characteristics, and possess different properties that depend on the approach used for their generation. Additionally, simulated time series, just as the non-stationary random vibration approach, are affected by the shape of the intensity modulating function, and, consequently, the structural response will be conditional on this selection. Another consideration in the generation of spatially variable time series is the selection of the duration of the strong motion part of the excitation, which can affect the nonlinear response of the structures. EC8 - Part 2 [109] recommends that the duration of the stationary part of the excitation be compatible with the duration of the seismic event used to establish the design ground acceleration.

Simulated ground motions are generated from stationary random fields, and, hence, do not possess the same properties as recorded data (Section 7.1.3), as, e.g., the variability with time of the frequency content of recorded accelerograms. On the other hand, conditionally simulated ground motions (Chapter 8) do not suffer from this "artificiality," as they are compatible with a predefined recorded (or synthetic) time history, and, hence, maintain the non-stationary characteristics of the predefined time series. The major advantage of simulations is that they can be generated quickly, which is, essentially, the argument for their use, especially in sensitivity analyses that require a large number of "runs." The development of conditional simulations is more elaborate, as it requires the simulation of segments of the time series and their subsequent combination in the time or frequency domains (Section 8.1.3), and, also, involves the selection of proper reference records, on which the conditional simulation will be based. Still, the nonlinear response of the structures may be affected by the varying frequency content of actual records, and, if the sensitivity analyses are geared towards recommendations for design implementation, the evaluation of the structural response caused by conditionally simulated spatially variable ground motions is a more realistic alternative. Spatially variable seismic ground motions generated for use in design, as, e.g., those provided by the NYCDOT provisions [364] (Section 9.3), are conditional simulations that comply with recorded accelerograms and the target response spectra at the site.

It is emphasized again that the response induced by a single simulated time history (or a single set of spatially variable ground motions) may indicate a trend, but is not conclusive, as the results of a single "run" do not suffice for the stochastic characterization of the response. A clear illustration of this effect is presented in the results of Tzanetos et al. [546] in Table 9.5: For one simulation of the highly incoherent motions, the percentage difference of the drift at Pier 1 due to spatially variable excitations relative to the response due to uniform motions is 50%, and for the second simulation it is 178%; similarly, the corresponding percentages for the shear demand are 20% and 90%, respectively. EC8 - Part 2 [109] suggests that the number of Monte Carlo simulations performed should be "sufficient," so that the mean of the response quantity of interest is stable, and, earlier, Shinozuka et al. [468] recommended that at least 20 simulations be utilized for the estimation of the peak structural response. The EC8 - Part 2 recommendations [109] seem appropriate with the consideration of the 20 simulations suggested by Shinozuka et al. [468] being the lower bound for the number of "runs" conducted. It is noted that Nazmy and Abdel-Ghaffar [362], in their evaluation of the response of cable-stayed bridges, suggested that at least three different ground motion scenarios be selected, and the results of the response evaluations averaged. The appendix of the NYCDOT seismic design criteria [364] also provides three sets of conditionally simulated spatially variable ground motions. Similarly, EC8 - Part 2 [109], before the introduction of spatial variability issues, recommends that, at least, three pairs of horizontal time history components be selected from recorded data for the site and the appropriate earthquake characteristics, and utilized in the nonlinear response evaluations of the structures. The smaller number of simulations suggested by these recommendations is due to the fact that the ground motions, in this case, are either recorded or conditionally simulated, and, hence, reflect better the actual characteristics of the seismic events at the site. As indicated in Chapter 8, any recorded or synthetic seismic time history selected for the performance-based design of a building structure (e.g., [508]) can be conditionally simulated with an appropriate coherency model and an apparent propagation velocity to yield spatially variable time series for the performance-based design of a long structure.

Simulation and conditional simulation algorithms, usually, generate acceleration time series. An important consideration in the analysis of the seismic response of lifeline systems is the correct evaluation of the corresponding displacement time histories. Numerical codes permitting the use of spatially variable excitations as input motions at the supports of lifeline models, generally, require displacement rather than acceleration time histories. Indeed, simulated spatially variable acceleration time histories, that are not properly processed, will lead to unrealistic displacements, and, subsequently, to an incorrect evaluation of the system's pseudo-static component. As indicated in Chapter 8, simulated and conditionally simulated acceleration time series also require processing, but for reasons different than seismic records. The seismological community has placed a lot of emphasis on the correct processing of recorded accelerograms. Recent evaluations, e.g., [73], also indicated that the selection of the high-pass corner frequency in the processing scheme can influence the response of linear and, even more significantly, the response of nonlinear single-degree-of-freedom oscillators. Considering that, in addition, the pseudo-static response of lifeline systems is controlled by the displacements (lower frequency components of the motions), the processing of simulated and conditionally simulated time series becomes an important issue. As noted by Alves and Hall [21] in their evaluation of the response of the Pacoima Dam subjected to spatially variable excitations, "in particular, the displacements should be carefully integrated from the accelerations." The conditional simulations used in the response evaluation and retrofit of, e.g., the great California bridges [7] are processed with seismological standards. On the other hand, the approach used for the processing of simulated time series in the sensitivity analyses of highway bridges (Section 9.4) is, in most cases, not disclosed. A methodology for the processing of conditionally simulated acceleration time histories, which can be readily extended to simulated spatially variable ground motions, was presented in Sections 8.2 and 8.3.

9.5.3 STRUCTURAL MODELING

In the studies presented herein, the structural configurations and modeling approximations within each general structural category (i.e., dams, suspension, cable-stayed and highway bridges) varied considerably. However, the structural response, especially in the nonlinear range, will be conditional on the modeling assumptions[6]. An example of this effect was presented in Table 9.6, that illustrated the numerical experiment of Lou [319]. The use of two different numerical tools for the same structural system by

---

[6] Clearly, this is not a novel observation nor does it apply only to the effect of spatially variable seismic ground motions on the response of lifeline systems. Past and current efforts on "blind predictions" of the response of any structure are a clear manifestation of this issue.

the same investigator, and with considerable care so that the two finite element models were as similar as possible, led to significant differences in the structural response. The results can also be sensitive to the modeling "details," as, e.g., the selected gap size between the ends of the bridge deck and the abutments [319] or whether the support at an abutment was considered pinned or fixed [546]. For the design of, e.g., suspension and cable-stayed bridges, as reported in the study of the new Tacoma Narrows Bridge by Arzoumanidis *et al*. [37], presented in Section 9.3, great emphasis is placed on the modeling details. Indeed, Arzoumanidis *et al*. [37] further reported that the analysis and design of the structure was reviewed at every step, and that an independent team also performed an independent analysis and validation. On the other hand, a variety of modeling assumptions are utilized in the evaluation of the effects of spatially variable ground motions on the response of lifeline systems. The limited analyses examining these effects on the diverse structural configurations of dams (Section 9.2) used different approximations for the modeling of the dam-foundation-reservoir problem, as, e.g., the modeling of the infinite domain, the contact between the structure and the foundation, and the interaction between the dam and the reservoir, and between the foundation and the reservoir. In addition, as Wieland [564] indicated, it is still difficult to model the joint opening and the formation of cracks in the body of the dam, as well as the nonlinear behavior of the foundation. A variety of modeling assumptions were also used in the sensitivity analyses of the highway bridges presented in Section 9.4, as, e.g., 2-D or 3-D analyses, consideration or not of SSI effects, modeling of supports and connections, and modeling of geometric and material nonlinearities. It is noted that sensitivity analyses require a significant number of "runs," and, hence, the more elaborate the structural modeling, the more time-consuming the effort, and a trade-off between accuracy and efficiency needs to be made. Considering, however, the dependence of the outcome of the evaluations on the structural configuration and the modeling assumptions, it may be appropriate that sensitivity analyses examining the effect of the spatial variation of the seismic ground motions on the response of these structures, especially when they are geared towards design implementations, also elaborate on the impact of their modeling assumptions. Additionally, as indicated in Section 9.3, the most "complete" comparison of the effects of spatially variable and uniform excitations on a lifeline's response was conducted by Nazmy and Abdel-Ghaffar [362] in their analyses of cable-stayed bridges. Figures 9.17–9.20 presented comparisons of displacements, axial and shear forces, and bending and torsional moments induced in the structure by the two types of excitations. Generally, however, the sensitivity analyses of the systems presented herein provided comparisons of selected response quantities only. It would be interesting, if such analyses also conduct a more detailed response evaluation, so that the full effect of spatially variable ground motions on the structure's response can be quantified.

An issue that requires further consideration in the sensitivity analyses of highway bridges is the fully 3-D nonlinear response analysis of the structures, in which the excitations in the two horizontal and the vertical directions are applied simultaneously. The majority of the results reported in the literature for these structures (e.g., Section 9.4) considered input excitations in a single (longitudinal or transverse) direction. Since the principle of superposition does not apply in the nonlinear range, the bridge response in the three orthogonal directions cannot be examined independently.

For such 3-D evaluations, it can be considered that the motions along their principal directions (i.e., epicentral, normal to epicentral and vertical) are uncorrelated, as discussed in Section 3.3.2.

It should be emphasized at this point that the accuracy in the numerical modeling of the structural systems will always be conditional on the accuracy in the specification of the ground excitation. This is especially true for spatially variable seismic excitations, since, in addition to the description of the "point" estimates of the motions, i.e., a recorded time history and/or a response spectrum, they also require the description of the variation of the time series in space. There are three distinct, but highly interrelated, causes underlying the spatial variation of the seismic ground motions: amplitude variability, wave passage and loss of coherency. The modeling of these causes and their use in the analysis and design of lifeline systems differs significantly if the soil conditions at the site are uniform or variable, as illustrated next.

9.5.4 MODELING OF SPATIALLY VARIABLE GROUND MOTIONS AT UNIFORM SITES

The spatial variation of seismic ground motions has been investigated essentially exclusively from spatially recorded data at uniform sites, and the resulting models reflect the characteristics of data at uniform sites. The amplitude variability, and the modeling of the apparent propagation velocity and the loss of coherency of the motions at such sites are addressed in the following.

Amplitude Variability

It is recalled that coherency is mostly affected by the phase variability of spatially recorded data rather than their amplitude variability (Fig. 2.21). Whereas the phase variability of spatially recorded data has been extensively investigated through the coherency, the amplitude variability of the seismic ground motions has not received significant attention. The few analyses of the amplitude variability of spatially recorded data (Figs. 3.19(b), 4.8 and 4.10) indicate that amplitudes, just as phases, can vary over distance and frequency. Presently, however, there are no explicit models for the amplitude variability of the seismic motions. The formulation of certain simulation algorithms permits the amplitude of the generated motions at different locations to vary, and, in cases, results in amplitude variations consistent with the trend of the amplitude variation of recorded data (Figs. 7.13 and 8.3). Other algorithms are designed so that the amplitude of the generated motions is the same at all locations and the phase differences conform with the coherency model. The latter approach is also used in the conditional simulation of spatially variable ground motions for design applications, e.g., [7], [364]; the generated spatially variable time series are then modified in the frequency domain to become compatible with the target response spectra.

Apparent Propagation Velocity

The first recognized cause for the spatial variation of the seismic ground motions was their apparent propagation on the ground surface. The geotechnical community placed a significant effort in the estimation of appropriate values for the apparent propagation of the waveforms, depending on the type of wave that controls the motions during the window analyzed, as the peak ground strain used in the design of buried pipelines and tunnels is inversely proportional to its value (e.g., Eq. 5.38). Approaches geared towards the estimation of the apparent propagation velocity of the motions were presented in Chapters 4 and 5. On the other hand, sensitivity analyses conducted for the evaluation of the effect of loss of coherency and wave passage on the response of, e.g., bridges assume a range of apparent propagation velocities from very low (a few hundred m/sec) to infinity, thus reflecting propagation velocity characteristics of different types of waves (surface or body waves). It is recalled that the vast majority of coherency models were developed from the strong motion, horizontal S-wave window of seismic data. For consistency, if the wave passage effect is considered along with the loss of coherency in the motions, the apparent propagation velocity should reflect the value of the apparent propagation velocity of the incident S-waves at the site, which is very different from the shear wave velocity of the top layer (e.g., Fig. 5.1). For example, the conditionally simulated time series provided by the NYCDOT seismic design criteria [364] utilize a value for the apparent propagation velocity of 2.5 km/sec.

Loss of Coherency

The selection of a coherency model for the evaluation of the seismic response of lifeline systems is an important task. Section 3.4 presented the most commonly used coherency models with brief descriptions of the assumptions utilized in their derivations. The variability in the exponential decay with frequency and separation distance of the various coherency models affects the response of the systems, as illustrated in the evaluations of the (linear) pseudo-static and dynamic response of a generic lifeline model in Fig. 6.9, the response of the Santa Felicia dam (Table 9.2) by Chen and Harichandran [100], the response of the Song-Hua-Jiang suspension bridge by Lin and Zhang [306], and the response of the cable-stayed Jindo Bridge by Soyluk and Dumanoğlu [511]. In particular, the behavior of the coherency model at low frequencies affects significantly the pseudo-static response of the structures. The most frequently utilized models in sensitivity analyses are the models of Harichandran and Vanmarcke [208], which is partially correlated at low frequencies (Fig. 3.16), and of Luco and Wong [324], which is fully correlated in the low frequency range (Fig. 3.23). Physically, the coherency model should tend to unity as frequency and separation distance tend to zero, as, e.g., the semi-empirical coherency model of Luco and Wong [324] in Fig. 3.23. Empirical coherency models evaluated from recorded data, on the other hand, cannot yield values equal to unity at low frequencies, unless the model is "forced" to tend to 1 as $\omega \rightarrow 0$, as is, e.g., the coherency model of Abrahamson [7] presented in Figs. 3.20 and 3.21. Furthermore, displacement time series integrated from recorded accelerations often show close agreement at distances pertinent for engineering applications [75], suggesting that the low frequency components of the data are highly correlated. Hence, the use of coherency models that tend to unity as frequency and separation distance decrease are recommended in spatial variability studies. Indeed, the majority of the sensitivity analyses of highway bridges in Section 9.4 utilized the coherency model of Luco and Wong [324], and the conditionally simulated spatially variable ground motions generated for use in design utilize versions of the coherency model developed by Abrahamson [7].

Another important consideration in the selection of the coherency model is the site classification. The lagged coherency at "rock" sites has a flatter exponential decay than the lagged coherency at the softer "soil" sites (Figs. 3.9 and 3.10), and Section 3.3.4 noted that different coherency models should describe the coherency at "soil" and "rock" sites. For example, in sensitivity analyses, the versatile semi-empirical coherency model of Luco and Wong [324] with a range of slower exponential decays, compatible with the trend of recorded data (as, e.g., in Fig. 3.10), can be used as representative for the coherency at "rock" sites, and the same model with a range of faster exponential decays, compatible, again, with the trend of recorded data (as, e.g., in Fig. 3.9), as representative for the coherency at "soil" sites. From the empirical models, the coherency model of Abrahamson [7], presented in Eqs. 3.21 and 3.22 and Fig. 3.20, may be used as a proxy for the coherency at "soil" sites. As indicated in Section 3.4.1, even though the model was developed from data recorded at multiple sites, the majority of the data were from arrays located at soil conditions. Hence, it may be postulated that the exponential decay of the few data at the rock sites has not overly affected the overall trend of the exponential decay of the model, and the model may be considered as representative of the loss of coherency at "soil" sites. For the exponential decay at rock sites, an empirical model that can be used is the newly developed model by Abrahamson [9], presented in Eq. 3.26, from data recorded at the rock site of the Pinyon Flat array, for which the slow lagged coherency decay was illustrated in Fig. 3.10. The last two suggestions on the use of empirical coherency models, however, are made with caution, as the description of the coherency may require a site classification more detailed than simply "rock" and "soil." This observation is also corroborated by the recent results of Kim and Stewart [267], presented in Section 9.1, who observed dependency of the exponential decay of the coherency on the S-wave velocity of the top layers of alluvial sites.

It is also noted that, in simulations and conditional simulations, it is considered that coherency, evaluated from the strong motion, S-wave window of the recorded data, is applicable to the entire duration of the simulated motions, and, furthermore, that the entire time histories propagate with a constant velocity on the ground surface. The rationale behind this assumption is that, since it is implicitly considered that the destructive energy is carried by the S-wave, it suffices to utilize the loss of coherency and apparent propagation velocity of the S-wave for the entire ground motion duration, instead of complicating the problem further[7].

As indicated in Section 3.4.1, the coherency of the vertical motions has not been extensively analyzed. The coherency of the vertical motions during the P-wave window is higher than the coherency of the horizontal motions during the strong-motion, S-wave window [4], [480], [608], as was also illustrated in the results of the early studies on the spatial variation of the motions in Figs. 3.2 and 3.5. For the description of the coherency of the vertical motions in sensitivity analyses, the semi-empirical model of Luco and Wong [324] may be utilized, again, but with a range of exponential

---

[7] The correct approach would be to consider the various wave components contributing to the motions, evaluate spatially variable time series for each component with different (and, in cases, not yet available) coherency models, propagate each wave component on the ground surface with its appropriate propagation velocity, and then combine the time series of all components.

decays slower than the ones selected for the horizontal motions. Alternatively, for consistency with the estimation of the coherency of the horizontal components of the motions during the S-wave window, the coherency of the vertical motions may also be evaluated during the same window. This consideration was observed by Abrahamson [7] in the development of his coherency models for the vertical component of the motions (Fig. 3.21). With the reservations indicated earlier, the coherency models of Abrahamson [7], Eqs. 3.23 and 3.24 and Fig. 3.21, can be used as a proxy for the coherency of the vertical motions at "soil" sites, and the newly developed coherency model by Abrahamson [9] from data recorded at the rock site of the Pinyon Flat array (Eq. 3.26) can be utilized for the modeling of the coherency of the vertical motions at "rock" sites.

9.5.5 MODELING OF SPATIALLY VARIABLE GROUND MOTIONS AT NONUNIFORM SITES

If the modeling of spatially variable ground motions at uniform sites is not a straight-forward task, the issue becomes even more complex when the local site conditions underneath the structures' supports vary. The highest seismic response of, essentially, all above-ground and buried structures that may encounter variations in the subsurface soil conditions along their length, was observed when the ground characteristics at their supports varied. Considerations regarding the modeling of the amplitude variability, the apparent propagation velocity and the loss of coherency of the motions at such sites is presented in the following.

Amplitude Variability

The amplitude variability in this case is caused by the differences in the amplification of the waveforms at the variable soil conditions of the site. Essentially all approaches for the estimation of this amplification are based on the assumption of 1-D, S-wave propagation through soil columns representing the softer soil deposits, e.g., [109], [364]. This approximation may be valid for ground-surface locations that are far apart from each other and in the absence of surface waves. However, such 1-D wave propagation analyses cannot capture the complex constructive interference of the waveforms occurring at sites with irregular surface and subsurface topography, nor the generation of additional, e.g., surface waves. Section 4.1.5 illustrated, through F-K analyses, the complex propagation pattern that can result in an alluvium valley, i.e., body waves and surface waves propagating with different velocities and directions of propagation, and dominating the records during different windows. An excellent review on 2-D and 3-D site effects, including field observations and numerical methods with an extensive list of references, was presented by Bard and Riepl-Thomas [42]. Indeed, ATC-32 [30], [31] states that "The records obtained in basin-type deposits are quite different from those obtained in a half-space topography. Theoretical studies and actual recordings show that both long-period energy and duration of shaking are significantly enhanced in basins, due to trapped energy ... Proper estimation of such enhancements in ground motion is crucial in estimating the performance of structures. Since such effects result from laterally varying geologic structure, they cannot be estimated from site-to-site using conventional one dimensional response-analysis methods." The recent ISO 23469:2005 standard [235] recommends two types of evaluations for the estimation of seismic ground motions at sites with significant lateral variations in their subsurface conditions: The "site-specific simplified dynamic analysis" recommended by the standard is also based on the 1-D wave propagation applied at multiple locations of these sites. On the other hand, the "site-specific de-tailed dynamic analysis" recommends that spatially variable seismic ground motions be evaluated based on the effects of the lateral geotechnical heterogeneities incorpo-rating the deep basin effects below the local soil deposits. The provision also suggests that a complete analysis of the seismic wave propagation from the source to the site be performed by considering a fault rupture model as well as 2-D or 3-D models of the soil profile. An effort in this direction was reported by Yang *et al.* [575], who analyzed the response of the Humboldt Bay Middle Channel Bridge, near Eureka, California, subjected to spatially variable excitations. Two-dimensional plane-strain conditions were used in a nonlinear finite element model of the bridge site, which included the bridge structure, pile groups, and supporting foundation soil extending over an area of 1045 m (along the bridge length) $\times$ 220 m (in depth). The excitation was approximated by a Ricker wavelet with a maximum acceleration of 2 m/sec^2 and dominant frequency of 2 Hz, which was assumed to propagate either vertically or at 60° from the horizontal axis. The authors concluded that the moment-curvature response at the bottom of the bridge piers was similar for the vertical incidence of the wavelet, but showed noticeable differences in the case of the inclined incidence due to the greater spatial variation of the excitations resulting at the bridge supports. It is noted, however, that this type of approach requires extensive computational re-sources, may carry modeling simplifications due to the complexity and size of the problem, and requires detailed geotechnical measurements to properly account for the soil variability at the site.

Apparent Propagation Velocity

The wave propagation pattern at these sites is also very different from the pattern of the propagation of the waveforms at uniform sites, for which, as indicated earlier, the velocity can be approximated by the constant apparent propagation velocity of the incident S-wave. A constant S-wave apparent propagation velocity, however, cannot reflect the complex reflection/refraction of the waveforms at the basin edges nor the propagation velocity of surface waves that can form in the valley. Indeed, as illustrated in Section 5.6, very significant strain amplifications occur at the interface between the two media, which can have a devastating effect on long buried structures. On the other hand, when simulations of spatially variable seismic ground motions are generated at sites with irregular subsurface topography for the seismic response of bridges (e.g., [109], [364]), it is, generally, assumed that the wave passage effect can be approximated with a constant (finite) propagation velocity, as was the case for uniform site conditions, which is not realistic. It is further recommended (e.g., [109], [364]) that the time delay in the arrival of the waveforms at locations on the softer soil deposits be accounted for by means of vertical S-wave propagation through the soil layers, even though EC8 - Part 2 [109] suggests that this time delay is not significant and may be neglected. It should be noted, however, that if vertical incidence through different soil columns is assumed in the evaluation of the characteristics of ground motions at variable site conditions, then the use of an apparent propagation velocity of the waveforms on the ground surface other than infinity is inconsistent.

Loss of Coherency

Essentially all empirical coherency models have been determined from spatial array data recorded at "uniform" site conditions, such as, e.g., the site of the SMART-1 array, and, hence, are valid only for the description of homogeneous random fields. However, such coherency estimates are also used in the simulation of non-homogeneous random fields, and considered, as in the example application of Figs. 7.16–7.18 and the estimation of the response of above-ground lifelines at variable site conditions, as representative of the coherency of the motions between pairs of stations located either both on uniform (firm, medium or soft) soil conditions, or each at different soil conditions. As indicated in Section 3.3.4 and Figs. 3.9 and 3.10, the exponential decay of the lagged coherency depends on the type of (uniform) site conditions. It follows then, that different coherency models should be used to represent the correlation of the motions at the softer and the stiffer ground types. Additionally, Section 3.3.5 and Figs. 3.12–3.14 suggested that the trend of the lagged coherency of the motions between the rock and the soil stations of the Parkway Valley array (Fig. 3.11) does not follow the trend of the lagged coherency of the motions recorded at the stations in the sediments. The loss of coherency of the seismic motions was also different during different time windows, depending on the type of wave (body or surface wave) that dominated the motions [298]. Hence, lagged coherency models evaluated at uniform sites are not appropriate for the description of the coherency between stations located at variable site conditions.

An additional important consideration in this case is that local site effects are difficult to generalize [415]. The recommendation in the ISO 23469:2005 standard [235] regarding the "site-specific detailed dynamic analysis" is, presently, the most rigorous approach for the estimation of ground motions at such sites, given, however, its associated constraints indicated earlier (i.e., computational resources, modeling approximations and geotechnical information). At this point, if simulations are to be used in the evaluation of the lifeline response, a possibly more viable, but still crude alternative for the modeling of the spatially variable random field at such sites would be the use of different response spectra underneath the structure's supports that are located at different ground types, different coherency models and apparent propagation velocities to describe the spatial variability of the motions at the locations of the supports on the "stiffer" site and the spatial variability of the motions at the locations of the supports on the "softer" site, and, as a first approximation, zero coherency between the motions if the supports are located at different subsurface conditions. If, on the other hand, recorded data for the appropriate earthquake characteristics are available at the site or at sites with similar lateral variations in their subsurface conditions, then a more realistic approach would be the conditional simulation of the motions based on one record at the "stiffer" site (e.g., rock outcrop) for the simulation of the motions at the supports located on the "stiffer" soil conditions, and another record on the surface of the "softer" site (e..g., sediments) for the simulation of the motions at the supports located on the "softer" soil conditions, and with the spatial variability of the motions modeled as previously described for the case of simulations. Undoubtedly, more refined approaches will emerge as additional seismic data, extensive computational resources and more advanced numerical tools become available.

References

1. A.M. Abdel-Ghaffar and A.S. Nazmy, "3-D nonlinear seismic behavior of cable-stayed bridges," *Journal of Structural Engineering, ASCE*, Vol. 117, pp. 3456–3476, 1991.
2. A.M. Abdel-Ghaffar and L.I. Rubin, "Suspension bridge response to multiple support excitations," *Journal of the Engineering Mechanics Division, ASCE*, Vol. 108, pp. 419–435, 1982.
3. A.M. Abdel-Ghaffar and L.I. Rubin, "Lateral earthquake response of suspension bridges," *Journal of Structural Engineering, ASCE*, Vol. 109, pp. 665–675, 1983.
4. N.A. Abrahamson, "Estimation of seismic wave coherency and rupture velocity using the SMART-1 strong motion recordings," Earthquake Engineering Research Center Report No. UCB/EERC-85-02, University of California, Berkeley, CA, 1985.
5. N.A. Abrahamson, "Spatial interpolation of array ground motions for engineering analysis," *Proceedings of the Ninth World Conference on Earthquake Engineering*, Tokyo, Japan, 1988.
6. N.A. Abrahamson, "Generation of spatially incoherent strong motion time histories," *Proceedings of the Tenth World Conference on Earthquake Engineering*, Madrid, Spain, 1992.
7. N.A. Abrahamson, "Spatial variation of multiple support inputs," *Proceedings of the 1st U.S. Seminar on Seismic Evaluation and Retrofit of Steel Bridges. A Caltrans and University of California at Berkeley Seminar*, San Francisco, CA, 1993.
8. N.A. Abrahamson, "Program on technology innovation: Spatial coherency models for soil-structure interaction," Electric Power Research Institute Report No. EPRI-1012968, EPRI, Palo Alto, CA, and U.S. Department of Energy, Washington, D.C., 2006.
9. N.A. Abrahamson, "Hard-rock coherency functions based on the Pinyon Flat array data," Draft report to EPRI, Electric Power Research Institute, Palo Alto, CA, 2007.
10. N.A. Abrahamson and B.A. Bolt, "Array analysis and synthesis mapping of strong seismic motion," in *Seismic Strong Motion Synthetics*, B.A. Bolt editor, Academic Press Inc., New York, NY, 1987.
11. N.A. Abrahamson, B.A. Bolt, R.B. Darragh, J. Penzien and Y.B. Tsai, "The SMART-1 accelerograph array (1980–1987): A review," *Earthquake Spectra*, Vol. 3, pp. 263–287, 1987.
12. N.A. Abrahamson, J.F. Schneider and J.C. Stepp, "Spatial variation of strong ground motion for use in soil-structure interaction analyses," *Proceedings of the Fourth U.S. National Conference on Earthquake Engineering*, Palm Springs, CA, 1990.
13. N.A. Abrahamson, J.F. Schneider and J.C. Stepp, "Spatial coherency of shear waves from the Lotung, Taiwan large-scale seismic test," *Structural Safety*, Vol. 10, pp. 145–162, 1991.
14. N.A. Abrahamson, J.F. Schneider and J.C. Stepp, "Empirical spatial coherency functions for applications to soil-structure interaction analyses," *Earthquake Spectra*, Vol. 7, pp. 1–27, 1991.
15. K. Aki (editor), "Technical glossary of earthquake and engineering seismology," in *International Handbook of Earthquake & Engineering Seismology*, W. Lee, H. Kanamori, P. Jennings and C. Kisslinger editors, Elsevier Science and Technology, London, United Kingdom, 2003.
16. K. Aki and P.G. Richards, *Quantitative Seismology*, Second Edition, University Science Books, Sausalito, CA, 2002.

17. S. Akkar and J.J. Bommer, "Influence of long-period filter cut-off on elastic spectral displacements," *Earthquake Engineering and Structural Dynamics*, Vol. 35, pp. 1145–1165, 2006.

18. S.M. Allam and T.K. Datta, "Analysis of cable-stayed bridges under multi-component random ground motion by response spectrum method," *Engineering Structures*, Vol. 22, pp. 1367–1377, 2000.

19. S.M. Allam and T.K. Datta, "Seismic response of a cable-stayed bridge deck under multi-component non-stationary random ground motion," *Earthquake Engineering and Structural Dynamics*, Vol. 33, pp. 375–393, 2004.

20. S.W. Alves and J.F. Hall, "System identification of a concrete arch dam and calibration of its finite element model," *Earthquake Engineering and Structural Dynamics*, Vol. 35, pp. 1321–1337, 2006.

21. S.W. Alves and J.F. Hall, "Generation of spatially nonuniform ground motion for nonlinear analysis of a concrete arch dam," *Earthquake Engineering and Structural Dynamics*, Vol. 35, pp. 1339–1357, 2006.

22. N. Ambraseys, P. Smit, R. Beradi, D. Rinaldis, F. Cotton and C. Berge, "European strong-motion database," CD ROM, European Council, Environmental and Climate Programme, Brussels, Belgium, 2000.

23. N. Ambraseys, J. Douglas, R. Sigbjörnsson, C. Berge-Thierry, P. Suhadolc, G. Costa and P. Smit, "European strong-motion database. Volume 2," CD ROM, Engineering and Physical Sciences Research Council, United Kingdom, 2004.

24. American Association of State Highway and Transportation Officials (AASHTO), *AASHTO LRFD Bridge Design Specifications, 2006 Interim Revisions*, Washington, D.C., 2006.

25. American Society of Civil Engineers (ASCE), *Seismic analysis of safety-related nuclear structures and commentary*, ASCE Standard, ASCE 4-98, Reston, VA, 2000.

26. M. Amin and A. H-S. Ang, "A nonstationary stochastic model for strong-motion earthquakes," Civil Engineering Studies, Structural Research Series No. SRS-86, University of Illinois at Urbana-Champaign, Urbana, IL, 1966.

27. M. Amin and A. H-S. Ang, "Non-stationary stochastic model of earthquake motion," *Journal of the Engineering Mechanics Division, ASCE*, Vol. 94, pp. 559–583, 1968.

28. A. H-S. Ang and W.H. Tang, *Probabilistic Concepts in Engineering Planning and Design. Volume I - Basic Principles*, John Wiley & Sons, Inc., New York, NY, 1975.

29. ANSYS Inc., http://www.ansys.com/, accessed 2008.

30. Applied Technology Council (ATC), *Improved Seismic Design Criteria for California Bridges: Provisional Recommendations*, ATC-32, Redwood City, CA, 1996.

31. Applied Technology Council (ATC), *Improved Seismic Design Criteria for California Bridges: Resource Document*, ATC-32.1, Redwood City, CA, 1996.

32. Applied Technology Council (ATC), *Recommended LRFD Guidelines for the Seismic Design of Highway Bridges, Part I: Specifications* and *Part II: Commentary and Appendices*, ATC/MCEER Joint Venture, ATC-49, Redwood City, CA, 2003.

33. R.J. Archuleta, S.H. Seale, P.V. Sangas, L.M. Baker and S.T. Swain, "Garner Valley downhole array of accelerometers: Instrumentation and preliminary data analysis," *Bulletin of the Seismological Society of America*, Vol. 82, pp. 1592–1621, 1992.

34. R.J. Archuleta, J.H. Steidl and L.F. Bonilla, "Engineering insights from data recorded on vertical arrays," *Proceedings of the 12th World Conference on Earthquake Engineering*, Auckland, New Zealand, 2000.

35. A. Arias, "A measure of earthquake intensity," in *Seismic Design of Nuclear Power Plants*, R. Hansen editor, M.I.T. Press, Cambridge, MA, 1969.

36. T. Ariman and G.E. Muleski, "A review of the response of buried pipelines under seismic excitation," *Earthquake Engineering and Structural Dynamics*, Vol. 9, pp. 133–151, 1981.

37. S. Arzoumanidis, A. Shama and F. Ostadan, "Performance-based seismic analysis and design of suspension bridges," *Earthquake Engineering and Structural Dynamics*, Vol. 34, pp. 349–367, 2005.

38. T.S. Atalik and S. Utku, "Stochastic linearization of multi-degree-of-freedom nonlinear systems," *Earthquake Engineering and Structural Dynamics*, Vol. 4, pp. 411–420, 1976.

39. T.T. Baber and Y.K. Wen, "Random vibration of hysteretic, degrading systems," *Journal of the Engineering Mechanics Division, ASCE*, Vol. 107, pp. 1069–1087, 1981.

40. L.G. Baise and S.D. Glaser, "Fundamental aspects of site response determined from inversion of vertical array data," *Proceedings of the 12th World Conference on Earthquake Engineering*, Auckland, New Zealand, 2000.

41. P.-Y. Bard, "Seismic input motion for large structures," *18ieme Seminaire Regional Europeen de Génie Parasismique*, Ecole Centrale de Lyon, France, 1995.

42. P.-Y. Bard and J. Riepl-Thomas, "Wave propagation in complex geological structures and their effects on strong ground motion," in *Wave Motion in Earthquake Engineering*, E. Kausel and G.D. Manolis editors, Volume in the Series Advances in Earthquake Engineering, WIT Press, Southampton, United Kingdom, 2000.

43. M.S. Bartlett, *An Introduction to Stochastic Processes with Special Reference to Methods and Applications*, Cambridge University Press, Cambridge, MA, 1953.

44. A. Bayraktar and A.A. Dumanoğlu, "The effect of the asynchronous ground motion on hydrodynamic pressures," *Computers and Structures*, Vol. 68, pp. 271–282, 1998.

45. A. Bayraktar, A.A. Dumanoğlu and Y. Calayir, "Asynchronous dynamic analysis of dam-reservoir-foundation systems by the Langrangian approach," *Computers and Structures*, Vol. 58, pp. 925–935, 1996.

46. L.K. Bear, G.L. Pavlis and G.H.R. Bokelmann, "Multi-wavelet analysis of three-component seismic arrays: Application to measure effective anisotropy at Piñon Flats, California," *Bulletin of the Seismological Society of America*, Vol. 89, pp. 693–705, 1999.

47. J.L. Beck, "Determining models of structures from earthquake records," Report No. EERL 78-01, Earthquake Engineering Research Laboratory, California Institute of Technology, Pasadena, CA, 1978.

48. J.L. Beck and L.S. Katafygiotis, "Updating models and their uncertainties. I: Bayesian statistical framework," *Journal of Engineering Mechanics, ASCE*, Vol. 124, pp. 455–461, 1998.

49. F. Behnamfar and Y. Sugimura, "Dynamic response of adjacent structures under spatially variable seismic waves," *Probabilistic Engineering Mechanics*, Vol. 14, pp. 33–44, 1999.

50. H. Benaroya, *Mechanical Vibration: Analysis, Uncertainties and Control*, Prentice-Hall, Inc., Upper Saddle River, NJ, 1998.

51. H. Benaroya and S.M. Han, *Probability Models in Engineering and Science*, CRC Press, Taylor and Francis Group, Boca Raton, FL, 2005.

52. J.S. Bendat, *Principles and Applications of Random Noise Theory*, Robert E. Krieger Publishing Co., Inc., Huntington, NY, 1958.

53. J.S. Bendat and A.G. Piersol, *Random Data: Analysis and Measurement Procedures*, John Wiley & Sons, Inc., New York, NY, 1971.

54. I.A. Beresnev and G.M. Atkinson, "Generic finite-fault model for ground-motion prediction in Eastern North America," *Bulletin of the Seismological Society of America*, Vol. 83, pp. 608–625, 1999.

55. J. Berger, D.C. Agnew, R.L. Parker and W.E. Farrell, "Seismic system calibration: 2. Cross-spectral calibration using random binary signals," *Bulletin of the Seismological Society of America*, Vol. 69, pp. 271–288, 1979.

56. M.K. Berrah and E. Kausel, "Response spectrum analysis of structures subjected to spatially varying motions," *Earthquake Engineering and Structural Dynamics*, Vol. 21, pp. 461–470, 1992.

57. M.K. Berrah and E. Kausel, "A modal combination rule for spatially varying seismic motions," *Earthquake Engineering and Structural Dynamics*, Vol. 22, pp. 791–800, 1993.

58. B.K. Bhartia and E.H. VanMarcke, "Associate linear system approach to nonlinear random vibration," *Journal of Engineering Mechanics, ASCE*, Vol. 117, pp. 2407–2428, 1991.

59. M.A. Biot, "The theory of propagation of elastic waves in a fluid-saturated porous solid. I. Low-frequency range," *Journal of the Acoustical Society of America*, Vol. 28, pp. 168–178, 1956.

60. R.B. Blackman and J.W. Tukey, *The Measurement of Power Spectra from the Point of View of Communications Engineering*, Dover Publications Inc., New York, NY, 1958.

61. P. Bloomfield, *Fourier Analysis of Time Series: An Introduction*, John Wiley & Sons, New York, NY, 2000.

62. J. Boatwright, J.B. Fletcher and T.E. Fumal, "A general inversion scheme for source, site and propagation characteristics using multiply recorded sets of moderate sized earthquakes," *Bulletin of the Seismological Society of America*, Vol. 81, pp. 1754–1782, 1991.

63. P. Bodin, J. Gomberg, S.K. Singh and M. Santoyo, "Dynamic deformations of shallow sediments in the valley of Mexico. Part I: Three-dimensional strains and rotations recorded on a seismic array," *Bulletin of the Seismological Society of America*, Vol. 87, pp. 528–539, 1997.

64. J.L. Bogdanoff, J.E. Goldberg and M.C. Bernard, "Response of a simple structure to a random earthquake-type disturbance," *Bulletin of the Seismological Society of America*, Vol. 51, pp. 293–310, 1961.

65. J.L. Bogdanoff, J.E. Goldberg and A.J. Schiff, "The effect of ground transmission time on the response of long structures," *Bulletin of the Seismological Society of America*, Vol. 55, pp. 627–640, 1965.

66. H.-P. Boissières and E.H. Vanmarcke, "Estimation of lags for a seismograph array: Wave propagation and composite correlation," *Soil Dynamics and Earthquake Engineering*, Vol. 14, pp. 5–22, 1995.

67. G.H.R. Bokelmann and S. Baisch, "Nature of narrow-band signals at 2.083 Hz," *Bulletin of the Seismological Society of America*, Vol. 89, pp. 156–164, 1999.

68. B.A. Bolt, C.H. Loh, J. Penzien, Y.B. Tsai and Y.T. Yeh, "Preliminary report on the SMART-1 strong motion array in Taiwan," Earthquake Engineering Research Center Report No. UCB/EERC-82/13, University of California, Berkeley, CA, 1982.

69. G. Bongiovanni, P. Marsan and R.W. Romeo, "Combined geological and geophysical investigations for site effect analysis in seismic zoning perspective," *Proceedings of the Fifth International Conference on Seismic Zonation*, Nice, France, 1995.

70. D.M. Boore, "Stochastic simulation of high-frequency ground motions based on seismological models of the radiated spectra," *Bulletin of the Seismological Society of America*, Vol. 73, pp. 1865–1894, 1983.

71. D.M. Boore, "Phase derivatives and simulation of strong ground motions," *Bulletin of the Seismological Society of America*, Vol. 93, pp. 1132–1143, 2003.

72. D.M. Boore, "Simulation of ground motion using the stochastic method," *Pure and Applied Geophysics*, Vol. 160, pp. 635–676, 2003.

73. D.M. Boore and S. Akkar, "Effect of causal and acausal filters on elastic and inelastic response spectra," *Earthquake Engineering and Structural Dynamics*, Vol. 32, pp. 1729–1748, 2003.

74. D.M. Boore and J.J. Bommer, "Processing of strong-motion accelerograms: Needs, options, and consequences," *Soil Dynamics and Earthquake Engineering*, Vol. 25, pp. 93–115, 2005.

75. D.M. Boore, C.D. Stephens and W.B. Joyner, "Comments on baseline correction of digital strong-motion data: Examples from the 1999 Hector Mine, California, earthquake," *Bulletin of the Seismological Society of America*, Vol. 92, pp. 1543–1560, 2002.

76. M. Bouchon, "The motion of the ground during an earthquake. Part 1: The case of a strike-slip fault," *Journal of Geophysical Research*, Vol. 85, pp. 356–366, 1980.

77. M. Bouchon, "The motion of the ground during an earthquake. Part 2: The case of a dip-slip fault," *Journal of Geophysical Research*, Vol. 85, pp. 367–375, 1980.

78. M. Bouchon and K. Aki, "Strain, tilt, and rotation associated with strong ground motion in the vicinity of earthquake faults," *Bulletin of the Seismological Society of America*, Vol. 72, pp. 1717–1738, 1982.

79. M. Bouchon, K. Aki and P.-Y. Bard, "Theoretical evolution of differential ground motions produced by earthquakes," *Proceedings of the Third International Conference on Earthquake Microzonation*, Seattle, WA, 1982.

80. R.N. Bracewell, *The Fourier Transform and Its Applications*, McGraw-Hill, Inc., New York, NY, 2000.

81. D.R. Brillinger, *Time Series. Data Analysis and Theory*, McGraw-Hill, Inc., New York, NY, 1981.

82. N.J. Burdette, "The effect of asynchronous earthquake motion on complex bridges," M.Sc. Thesis, University of Illinois at Urbana-Champaign, Urbana, IL, 2006.

83. N.J. Burdette, A.S. Elnashai, A. Lupoi and A.G. Sextos, "The effect of asynchronous earthquake motion on complex bridges," MAE Report No. 06-03, Mid-America Earthquake Center (MAE), University of Illinois at Urbana-Champaign, Urbana, IL, 2006.

84. J.P. Burg, "Three dimensional filtering with an array of seismometers," *Geophysics*, Vol. 29, pp. 693–713, 1964.

85. G.N. Bycroft, "White-noise representation of earthquakes," *Journal of the Engineering Mechanics Division, ASCE*, Vol. 86, pp. 1–16, 1960.

86. G.N. Bycroft, "Soil-foundation interaction and differential ground motions," *Earthquake Engineering and Structural Dynamics*, Vol. 8, pp. 397–404, 1980.

87. G.N. Bycroft, "El Centro, California differential ground motion array," USGS Open-File Report 80-919, U.S. Geological Survey, Denver, CO, 1980.

88. California Department of Transportation (Caltrans), *Bridge Design Specifications*, http://www.dot.ca.gov/hq/esc/techpubs/manual/bridgemanuals/bridge-design-specifications/bds.html, accessed 2008.

89. California Department of Transportation (Caltrans), *Seismic Design Criteria, SDC*, Version 1.4, http://www.dot.ca.gov/hq/esc/techpubs/manual/othermanual/other-engin-manual/seismic-design-criteria/sdc.html, accessed 2008.

90. California Department of Transportation (Caltrans), *User's Manual for COLx, Column Ductility Program*, Sacramento, CA, 1993.

91. California Strong Motion Instrumentation Program (CSMIP), http://www.consrv. ca.gov/cgs/smip/, accessed 2008.

92. R.J. Câmara, "A method for coupled arch dam-foundation-reservoir seismic behaviour analysis," *Earthquake Engineering and Structural Dynamics*, Vol. 29, pp. 441–460, 2000.

93. J. Capon, "High-resolution frequency-wavenumber spectrum analysis," *Proceedings of the IEEE*, Vol. 57, pp. 1408–1418, 1969.

94. J. Capon, "Signal processing and frequency-wavenumber spectrum analysis for a large aperture seismic array," *Methods in Computational Physics*, Vol. 13, pp. 1–59, 1973.

95. F. Casciati, M. Grigoriu, A. Der Kiureghian, Y.K. Wen and T. Vrouwenvelder, "Report of the working group on dynamics," *Structural Safety*, Vol. 19, pp. 271–282, 1997.

96. A. Castellani and G. Boffi, "Rotational components of the surface ground motion during an earthquake," *Earthquake Engineering and Structural Dynamics*, Vol. 14, pp. 751–767, 1986.

97. A. Castellani and G. Boffi, "On rotational components of seismic motion," *Earthquake Engineering and Structural Dynamics*, Vol. 18, pp. 785–797, 1989.

98. T.K. Caughey, "Equivalent linearization technique," *Journal of the Acoustical Society of America*, Vol. 35, pp. 1706–1711, 1963.

99. F.J. Chávez-García, J. Castillo and W.R. Stephenson, "3D site effects: A thorough analysis of a high-quality dataset," *Bulletin of the Seismological Society of America*, Vol. 92, pp. 1941–1951, 2002.

100. M.-T. Chen and R.S. Harichandran, "Response of an earth dam to spatially varying earthquake ground motion," *Journal of Engineering Mechanics, ASCE*, Vol. 127, pp. 932–939, 2001.

101. H.C. Chiu, "Stable baseline correction of digital strong-motion data," *Bulletin of the Seismological Society of America*, Vol. 87, pp. 932–944, 1997.

102. H. C. Chiu, "Data files from the SMART-2 strong-motion array for the Chi-Chi earthquake," *Bulletin of the Seismological Society of America*, Vol. 91, pp. 1391–1392, 2001.

103. N. Chouw and H. Hao, "Study of SSI and non-uniform ground motion effect on pounding between bridge girders," *Soil Dynamics and Earthquake Engineering*, Vol. 25, pp. 717–728, 2005.

104. N. Chouw, H. Hao and S. Su, "Multi-sided pounding response of bridge structures with non-linear bearings to spatially varying ground excitation," *Advances in Structural Engineering*, Vol. 9, pp. 55–66, 2006.

105. J.T. Christian, "Generating seismic design power spectral density functions," *Earthquake Spectra*, Vol. 5, pp. 351–368, 1989.

106. J.P. Claassen, "The estimation of seismic spatial coherence and its application to a regional event," Report SAND85-2093, Sandia National Laboratory, Albuquerque, NM, 1985.

107. R.W. Clough and J. Penzien, *Dynamics of Structures*, McGraw-Hill, Inc., New York, NY, 1975.

108. Comité Européen de Normalisation (CEN), *EUROCODE 8: Design of Structures for Earthquake Resistance – Part 1: General Rules, Seismic Actions and Rules for Buildings*, EN 1998-1:2004, Brussels, Belgium, 2005.

109. Comité Européen de Normalisation (CEN), *EUROCODE 8: Design of Structures for Earthquake Resistance – Part 2: Bridges*, EN 1998-2:2005, Brussels, Belgium, 2005.

110. Computers and Structures, Inc., *SAP2000 v11 User Reference Manuals*, http://www.comp-engineering.com/SAP2000V11EN.html, accessed 2008.

111. Consortium of Organizations for Strong Motion Observation Systems (COSMOS), http://www.cosmos-eq.org, accessed 2008.

112. Consortium of Organizations for Strong Motion Observation Systems (COSMOS), *Proceedings of the Workshop on Strong-Motion Record Processing*, http://www.cosmos-eq.org/recordProcessingPapers.html, accessed 2008.

113. J.P. Conte, A. Elgamal, Z. Yang, Y. Zhang, G. Acero and F. Seible, "Nonlinear seismic analysis of a bridge ground system," *Proceedings of the 15th ASCE Engineering Mechanics Conference*, New York, NY, 2002.

114. J.P. Conte and B.F. Peng, "Fully nonstationary analytical earthquake ground-motion model," *Journal of Engineering Mechanics, ASCE*, Vol. 123, pp. 15–24, 1997.

115. J.P. Conte, K.S. Pister and S.A. Mahin, "Nonstationary ARMA modeling of seismic ground motions," *Soil Dynamics and Earthquake Engineering*, Vol. 11, pp. 411–426, 1992.

116. A.M. Converse and A.G. Brady, "BAP-basic strong-motion acceleration processing software; Version 1.0," USGS Open-File Report 92-296A, U.S. Geological Survey, Denver, CO, 1992.

117. C.A. Cornell, "Stochastic process models in structural engineering," Technical Report No. 84, Department of Civil Engineering, Stanford University, Stanford, CA, 1964.

118. C. Cornou, P.-Y. Bard and M. Dietrich, "Contribution of dense array analysis to the identification and quantification of basin-edge induced waves. Part I: Methodology," *Bulletin of the Seismological Society of America*, Vol. 93, pp. 2604–2623, 2003.

119. C. Cornou, P.-Y. Bard and M. Dietrich, "Contribution of dense array analysis to the identification and quantification of basin-edge induced waves. Part II: Application to Grenoble Basin (French Alps)," *Bulletin of the Seismological Society of America*, Vol. 93, pp. 2624–2648, 2003.

120. S.H. Crandall and W.D. Mark, *Random Vibration in Mechanical Systems*, Academic Press, New York, NY, 1963.

121. E. Cranswick, "The information content of high-frequency seismograms and the near-surface geologic structure of "hard rock" recording sites," *Pure and Applied Geophysics*, Vol. 128, pp. 333–363, 1988.

122. R.A. Dameron, S.G. Arzoumanidis, S.W. Bennett and A. Malik, "Seismic analysis and displacement-based evaluation of the Brooklyn-Queens Expressway, New York," *Transportation Research Record 1845*, Paper No. 03-4528, pp. 213–225, 2003.

123. R.B. Darragh, "Analysis of near-source waves: Separation of wave types using strong motion array recordings," Ph.D. Dissertation, Department of Earth and Planetary Science, University of California, Berkeley, CA, 1987.

124. T.K. Datta and E.A. Mashaly, "Transverse response of offshore pipelines to random ground motion," *Earthquake Engineering and Structural Dynamics*, Vol. 19, pp. 217–228, 1990.

125. A.G. Davenport, "Note on the distribution of the largest value of a random function with application to gust loading," *Proceedings of the Institution of Civil Engineers*, Vol. 28, pp. 187–196, 1964.

126. W.B. Davenport, *Probability and Random Processes. An Introduction for Applied Scientists and Engineers*, McGraw Hill, Inc., New York, NY, 1970.

127. G. Deodatis, "Non-stationary stochastic vector processes: Seismic ground motion applications," *Probabilistic Engineering Mechanics*, Vol. 11, pp. 149–168, 1996.

128. G. Deodatis, "Simulation of ergodic multivariate stochastic processes," *Journal of Engineering Mechanics, ASCE*, Vol. 122, pp. 778–787, 1996.

129. G. Deodatis and R.C. Micaletti, "Simulation of highly skewed non-Gaussian stochastic processes," *Journal of Engineering Mechanics, ASCE*, Vol. 127, pp. 1284–1295, 2001.

130. G. Deodatis, V. Saxena and M. Shinozuka, "Effect of spatial variability of ground motion on bridge fragility curves," *Proceedings of the Eighth ASCE Specialty Conference on Probabilistic Mechanics and Structural Reliability*, University of Notre Dame, Notre Dame, IN, 2000.

131. G. Deodatis and M. Shinozuka, "Auto-regressive model for non-stationary stochastic processes," *Journal of Engineering Mechanics, ASCE*, Vol. 114, pp. 1995–2012, 1988.

132. Z.A. Der, R.H. Shumway and A.C. Lees, "Frequency domain coherent processing of regional seismic signals at small arrays," *Bulletin of the Seismological Society of America*, Vol. 78, pp. 326–338, 1988.

133. A. Der Kiureghian, "Structural response to stationary excitation," *Journal of the Engineering Mechanics Division, ASCE*, Vol. 106, pp. 1195–1213, 1980.

134. A. Der Kiureghian, "CQC modal combination rule for high-frequency modes," *Transactions of the 11th International Conference in Reactor Technology*, Tokyo, Japan, 1991.

135. A. Der Kiureghian, "A coherency model for spatially varying ground motions," *Earthquake Engineering and Structural Dynamics*, Vol. 25, pp. 99–111, 1996.

136. A. Der Kiureghian and J. Crempien, "An evolutionary model for earthquake ground motion," *Structural Safety*, Vol. 6, pp. 235–246, 1989; "Discussion" by J.D. Riera and G.I. Schuëller, *Structural Safety*, Vol. 9, pp. 75–77, 1990.

137. A. Der Kiureghian, P. Keshishian and A. Hakobian, "Multiple support response spectrum analysis of bridges including the site response effect and the MSRS code," Earthquake Engineering Research Center Report No. UCB/EERC-97/02, University of California, Berkeley, CA, 1997.

138. A. Der Kiureghian and A. Neuenhofer, "Response spectrum method for multisupport seismic excitation," *Earthquake Engineering and Structural Dynamics*, Vol. 21, pp. 713–740, 1992.

139. C. Desceliers, C. Soize and S. Cambier, "Non-parametric - parametric model for random uncertainties in non-linear structural dynamics: Application to earthquake engineering," *Earthquake Engineering and Structural Dynamics*, Vol. 33, pp. 315–327, 2003.

140. H-P. Ding, Q-F. Liu, X. Jin and Y-F. Yuan, "A coherency function model of ground motion at base rock corresponding to strike-slip fault," *Acta Seismologica Sinica*, Vol. 17, pp. 64–69, 2004.

141. M. Di Paola and M. Zingales, "Digital simulation of multivariate earthquake ground motions," *Earthquake Engineering and Structural Dynamics*, Vol. 29, pp. 1011–1027, 2000.

142. K.K. Dong and M. Wieland, "Application of response spectrum method to a bridge subjected to multiple support excitation," *Proceedings of the Ninth World Conference on Earthquake Engineering*, Tokyo, Japan, 1988.

143. D.E. Dudgeon and R.M. Mersereau, *Multidimensional Digital Signal Processing*, Prentice-Hall, Inc., Englewood Cliffs, NJ, 1984.

144. A.A. Dumanoğlu and K. Soyluk, "A stochastic analysis of long span structures subjected to spatially varying ground motions including the site-response effect," *Engineering Structures*, Vol. 25, pp. 1301–1310, 2003.

145. A.-W. Elgamal, Z. Yang and J.C. Stepp, "Seismic downhole arrays and applications in practice," *Proceedings of the International Workshop for Site Selection, Installation and Operation of Geotechnical Strong-Motion Arrays*, Consortium of Organizations for Strong-Motion Observation Systems (COSMOS), Los Angeles, CA, 2004.

146. I. Elishakoff, *Probabilistic Methods in the Theory of Structures*, John Wiley & Sons, New York, NY, 1983.

147. I. Elishakoff, Y.J. Ren and M. Shinozuka, "Conditional simulation of non-Gaussian random fields," *Engineering Structures*, Vol. 16, pp. 558–563, 1994.

148. B.R. Ellingwood and M.E. Batts, "Characterization of earthquake forces for probability-based design of nuclear structures," Technical Report BNL-NUREG-51587, NUREG/CR-2945, Department of Nuclear Energy, Brookhaven National Laboratoty, Upton, NY, 1982.

149. G.W. Ellis and A.S. Cakmak, "Time series modeling of strong ground motion from multiple event earthquakes," *Soil Dynamics and Earthquake Engineering*, Vol. 10, pp. 42–54, 1991.

150. A.S. Elnashai, V.K. Papanikolaou and D-H. Lee, "ZEUS-NL, A system for inelastic analysis of structures, Version 1.7.4," Mid-America Earthquake Center (MAE), University of Illinois at Urbana-Champaign, Urbana, IL, http://mae.cee.uiuc.edu/software_and_tools/zeus_nl.html, accessed 2008.

151. H.H. Emam, H.J. Pradlwarter and G.I. Schuëller, "A computational procedure for the implementation of equivalent linearization in finite element analysis," *Earthquake Engineering and Structural Dynamics*, Vol. 29, pp. 1–17, 2000.

152. M. Erdik and N. Apaydin, "Earthquake response of suspension bridges," in *Vibration Problems ICOVP 2005*, E. Iman and A. Kiris editors, Springer Verlag, New York, NY, 2007.

153. A. Esmaeily, "USC-RC," http://www.usc.edu/dept/civil_eng/structural_lab/software.html, accessed 2008.

154. J.E. Evans, J.R. Johnson and D.F. Sun, "High resolution angular spectrum estimation techniques for terrain scattering analysis and angle of arrival estimation," *Proceedings of the First ASSP Workshop on Spectral Estimation*, Hamilton, Ontario, Canada, 1981.

155. F.G. Fan and G. Ahmadi, "Nonstationary Kanai-Tajimi models for El Centro 1940 and Mexico City 1985 earthquakes," *Probabilistic Engineering Mechanics*, Vol. 5, pp. 171–181, 1990.

156. Federal Highway Administration (FHWA), "Seismic design of bridges. Design example No. 1: Two-span continuous CIP concrete box bridge," Report No. FHWA-SA-97-006, National Technical Information Service, Springfield, VA, 1996.

157. Federal Highway Administration (FHWA), "Seismic design of bridges. Design example No. 4: Three-span continuous CIP concrete bridge," Report No. FHWA-SA-97-009, National Technical Information Service, Springfield, VA, 1996.

158. G.A. Fenton, "Error evaluation of three random-field generators," *Journal of Engineering Mechanics, ASCE*, Vol. 120, pp. 2478–2497, 1994.

159. G.A. Fenton, "Estimation for stochastic soil models," *Journal of Geotechnical and Geoenvironmental Engineering, ASCE*, Vol. 125, pp. 470–485, 1999.

160. G.A. Fenton and E.H. Vanmarcke, "Simulations of random fields via local average subdivision," *Journal of Engineering Mechanics, ASCE*, Vol. 116, pp. 1733–1749, 1990.

161. E.H. Field, "Spectral amplification in a sediment-filled valley exhibiting clear basin-edge induced effects," *Bulletin of the Seismological Society of America*, Vol. 86, pp. 991–1005, 1996.

162. J.B. Fletcher, L.M. Baker, P. Spudich, P. Goldstein, J.D. Sims and M. Hellweg, "The USGS Parkfield, California, dense seismograph array: UPSAR," *Bulletin of the Seismological Society of America*, Vol. 82, pp. 1041–1070, 1992.

163. A. Frankel, D. Carver, E. Cranswick, T. Bice, R. Sell and S. Hanson, "Observations of basin ground motions from a dense seismic array in San Jose, California," *Bulletin of the Seismological Society of America*, Vol. 91, pp. 1–12, 2001.

164. A. Frankel, S. Hough, P. Friberg and R. Busby, "Observations of Loma Prieta aftershocks from a dense array in Sunnyvale, California," *Bulletin of the Seismological Society of America*, Vol. 81, pp. 1900–1922, 1991.

165. S. Gaffet, C. Larroque, A. Deschamps and F. Tressols, "A dense array experiment for the observation of waveform perturbations," *Soil Dynamics and Earthquake Engineering*, Vol. 17, pp. 475–484, 1998.

166. D. Gasparini and E.H. Vanmarcke, "Simulated earthquake ground motions compatible with prescribed response spectra," Technical Report No. R76-4, Department of Civil Engineering, Massachusetts Institute of Technology, Cambridge, MA, 1976.

167. L. Geli, P.-Y. Bard and B. Jullien, "The effect of topography on earthquake ground motion: A review and new results," *Bulletin of the Seismological Society of America*, Vol. 78, pp. 42–63, 1988.

168. R. Giannini, G. Monti, C. Nuti and T. Pagnoni, "ASPIDEA. A program for nonlinear analysis of isolated bridges under nonsynchronous seismic excitations," Report No. 5/92, Dipartimento delle Strutture delle Acque e del Terreno, Università dell'Aquila, L'Aquila, Italy, 1992.

169. J.E. Goldberg, J.L. Bogdanoff and D.R. Sharpe, "The response of simple nonlinear systems to a random disturbance of the earthquake type," *Bulletin of the Seismological Society of America*, Vol. 54, pp. 263–276, 1964.

170. P. Goldstein, "Array measurements of earthquake rupture," Ph.D. Dissertation, Department of Physics, University of California, Santa Barbara, CA, 1988.

171. P. Goldstein and R.J. Archuleta, "Deterministic frequency-wavenumber methods and direct measurements of rupture propagation during earthquakes using a dense array: Theory and methods," *Journal of Geophysical Research*, Vol. 96, pp. 6173–6185, 1991.

172. P. Goldstein and R.J. Archuleta, "Deterministic frequency-wavenumber methods and direct measurements of rupture propagation during earthquakes using a dense array: Data Analysis," *Journal of Geophysical Research*, Vol. 96, pp. 6187–6198, 1991.

173. J. Gomberg, "Dynamic deformation and the M6.7, Northridge, California, earthquake," *Soil Dynamics and Earthquake Engineering*, Vol. 16, pp. 471–494, 1997.

174. J. Gomberg and D. Agnew, "The accuracy of seismic estimates of dynamic strains: An evaluation using strainmeter and seismometer data from Piñon Flat Observatory, California," *Bulletin of the Seismological Society of America*, Vol. 86, pp. 212–220, 1996.

175. J. Gomberg, P. Bodin and P.A. Reasenberg, "Observing earthquakes triggered in the near field by dynamic deformations," *Bulletin of the Seismological Society of America*, Vol. 93, pp. 118–138, 2003.

176. J. Gomberg, G. Pavlis and P. Bodin, "The strain in the array is mainly in the plane (waves below $\sim$ 1Hz)," *Bulletin of the Seismological Society of America*, Vol. 89, pp. 1428–1438, 1999.

177. H. Goto and K. Toki, "Structural response to nonstationary random excitation," *Proceedings of the Fourth World Conference on Earthquake Engineering*, Santiago, Chile, 1969.

178. R.W. Graves, "Processing issues for near source strong motion recordings," *Proceedings of the Workshop on Strong-Motion Record Processing*, Consortium of Organizations for Strong-Motion Observation Systems (COSMOS), Richmond, CA, 2004.

179. M. Grigoriu, "On the spectral representation method in simulation," *Probabilistic Engineering Mechanics*, Vol. 8, pp. 75–90, 1993.

180. M. Grigoriu, *Applied Non-Gaussian Processes: Examples, Theory, Simulation, Linear Random Vibration, and MATLAB Solutions*, Prentice-Hall, Inc., Englewood Cliffs, NJ, 1995.

181. M. Grigoriu, "A spectral representation based model for Monte Carlo simulation," *Probabilistic Engineering Mechanics*, Vol. 15, pp. 365–370, 2000.

182. M. Grigoriu, "Spectral representation for a class of non-Gaussian processes," *Journal of Engineering Mechanics, ASCE*, Vol. 130, pp. 541–546, 2004.

183. M. Grigoriu, "Evaluation of Karhunen-Loéve spectral and sampling representations for stochastic processes," *Journal of Engineering Mechanics, ASCE*, Vol. 132, pp. 179–189, 2006.

184. M. Grigoriu, S.E. Ruiz and E. Rosenblueth, "Nonstationary models of seismic ground acceleration," Technical Report NCEER-88-0043, National Center for Earthquake Engineering Research (NCEER), State University of New York at Buffalo, Buffalo, NY, 1988.

185. I.D. Gupta and M.D. Trifunac, "Investigation of nonstationarity in stochastic seismic response of structures," Report No. CE 96-01, Department of Civil Engineering, University of Southern California, Los Angeles, CA, 1996.

186. I.D. Gupta and M.D. Trifunac, "Defining equivalent stationary PSDF to account for nonstationarity of earthquake ground motion," *Soil Dynamics and Earthquake Engineering*, Vol. 17, pp. 89–99, 1997.

187. K. Gurley and A. Kareem, "On the analysis and simulation of random processes utilizing higher order spectra and wavelet transforms," *Proceedings of the Second International Conference on Computational Stochastic Mechanics*, Athens, Greece, 1994.

188. A.H. Hadjian, "On the correlation of the components of strong motion," *Proceedings of the Second International Conference on Microzonation*, San Francisco, CA, 1978.

189. G.D. Hahn and X. Liu, "Torsional response of unsymmetric buildings to incoherent ground motions," *Journal of Structural Engineering, ASCE*, Vol. 120, pp. 1158–1181, 1994; and "Erratum," Vol. 120, pp. 3101–3104, 1994.

190. J.F. Hall, "Efficient nonlinear seismic analysis of arch dams, users manual for SCADA (Smeared Crack Arch Dam Analysis)," Report No. EERL 96-01, Earthquake Engineering Research Laboratory, California Institute of Technology, Pasadena, CA, 1996 (Modified July 1997).

191. J.F. Hall and A.K. Chopra, "Dynamic analysis of arch dams including hydrodynamic effects," *Journal of Engineering Mechanics, ASCE*, Vol. 109, pp. 149–167, 1983.

192. H. Hao, "Response of multiply supported rigid plate to spatially correlated seismic excitations," *Earthquake Engineering and Structural Dynamics*, Vol. 20, pp. 821–838, 1991.

193. H. Hao, "Arch responses to correlated multiple excitations," *Earthquake Engineering and Structural Dynamics*, Vol. 22, pp. 389–404, 1993.

194. H. Hao, "Ground-motion spatial variation effects on circular arch responses," *Journal of Engineering Mechanics, ASCE*, Vol. 120, pp. 2326–2341, 1994.

195. H. Hao, "Stability of simple beam subjected to multiple seismic excitations," *Journal of Engineering Mechanics, ASCE*, Vol. 123, pp. 739–742, 1997.

196. H. Hao, "Characteristics of torsional ground motions," *Earthquake Engineering and Structural Dynamics*, Vol. 25, pp. 599–610, 1996.

197. H. Hao, "A parametric study of the required seating length for bridge decks during earthquake," *Earthquake Engineering and Structural Dynamics*, Vol. 27, pp. 91–103, 1998.

198. H. Hao and X.N. Duan, "Seismic response of asymmetric structures to multiple ground motions," *Journal of Structural Engineering, ASCE*, Vol. 121, pp. 1557–1564, 1995.

199. H. Hao and L. Gong, "Analysis of coupled lateral-torsional-pounding responses of one-storey asymmetric adjacent structures subjected to bi-directional ground

motions - Part II: Spatially varying ground motion input," *Advances in Structural Engineering*, Vol. 8, pp. 481–496, 2005.

200. H. Hao, C.S. Oliveira and J. Penzien, "Multiple-station ground motion processing and simulation based on SMART-1 array data," *Nuclear Engineering and Design*, Vol. 111, pp. 293–310, 1989.

201. T. Harada, "Probabilistic modeling of spatial variation of strong earthquake ground displacement," *Proceedings of the Eighth World Conference on Earthquake Engineering*, San Francisco, CA, 1984.

202. R.S. Harichandran, "Space-time variation of earthquake ground motion," Ph.D. Dissertation, Department of Civil Engineering, Massachusetts Institute of Technology, Cambridge, MA, 1985.

203. R.S. Harichandran, "Stochastic analysis of rigid foundation filtering," *Earthquake Engineering and Structural Dynamics*, Vol. 15, pp. 889–899, 1987.

204. R.S. Harichandran, "Correlation analysis in space-time modeling of strong ground motion," *Journal of Engineering Mechanics, ASCE*, Vol. 113, pp. 629–634, 1987.

205. R.S. Harichandran, "Local spatial variation of earthquake ground motion," in *Earthquake Engineering and Soil Dynamics II-Recent Advances in Ground Motion Evaluation*, J.L. Von Thun editor, Geotechnical Special Publication No. 20, ASCE, New York, NY, 1988.

206. R.S. Harichandran, "Estimating the spatial variation of earthquake ground motion from dense array recordings," *Structural Safety*, Vol. 10, pp. 219–233, 1991.

207. R.S. Harichandran, A. Hawwari and B.N. Sweidan, "Response of long-span bridges to spatially varying ground motion," *Journal of Structural Engineering, ASCE*, Vol. 122, pp. 476–484, 1996.

208. R.S. Harichandran and E.H. Vanmarcke, "Stochastic variation of earthquake ground motion in space and time," *Journal of Engineering Mechanics, ASCE*, Vol. 112, pp. 154–174, 1986; "Discussion" by M. Novak (also in Ref. [375]) and "Closure" by Authors, *Journal of Engineering Mechanics, ASCE*, Vol. 113, pp. 1267–1270, 1987.

209. R.S. Harichandran and W. Wang, "Response of simple beam to spatially varying earthquake excitation," *Journal of Engineering Mechanics, ASCE*, Vol. 114, pp. 1526–1541, 1988.

210. R.S. Harichandran and W. Wang, "Response of indeterminate two-span beam to spatially varying earthquake excitation," *Earthquake Engineering and Structural Dynamics*, Vol. 19, pp. 173–187, 1990.

211. R.S. Harichandran and W. Wang, "Effect of spatially varying seismic excitation on surface lifelines," *Proceedings of the Fourth U.S. National Conference on Earthquake Engineering*, Palm Springs, CA, 1990.

212. N. Harnpornchai, H.J. Pradlwarter and G.I. Schuëller, "Stochastic analysis of dynamical systems by phase-space-controlled Monte Carlo simulation," *Computer Methods in Applied Mechanics and Engineering*, Vol. 168, pp. 273–283, 1999.

213. G.C. Hart, M. DiJulio, and M. Lew, "Torsional response of high-rise buildings," *Journal of the Structural Division, ASCE*, Vol. 101, pp. 397–414, 1975.

214. S.H. Hartzell, "Site response estimation from earthquake data," *Bulletin of the Seismological Society of America*, Vol. 82, pp. 2308–2327, 1992.

215. S. Hartzell, D. Carver, R.A. Williams, S. Harmsen and A. Zerva, "Site response, shallow shear-wave velocity, and wave propagation at the San Jose, California, dense seismic array," *Bulletin of the Seismological Society of America*, Vol. 93, pp. 443–464, 2003.

216. S. Hartzell and D.V. Helmberger, "Strong-motion modeling of the Imperial Valley earthquake of 1979," *Bulletin of the Seismological Society of America*, Vol. 72, pp. 571–596, 1982.

217. Y.M.A. Hashash, J.J. Hook, B. Schmidt and J.I-C. Yao, "Seismic design and analysis of underground structures," *Tunneling and Underground Space Technology*, Vol. 16, pp. 247–293, 2001.

218. E. Heredia-Zavoni and F. Barranco, "Torsion in symmetric buildings due to ground-motion spatial variation," *Journal of Engineering Mechanics, ASCE*, Vol. 122, pp. 834–843, 1996.

219. E. Heredia-Zavoni and S. Santa-Cruz, "Conditional simulation of a class of nonstationary space-time random fields," *Journal of Engineering Mechanics, ASCE*, Vol. 126, pp. 398–404, 2000.

220. E. Heredia-Zavoni and E.H. Vanmarcke, "Seismic random-vibration analysis of multisupport-structural systems," *Journal of Engineering Mechanics, ASCE*, Vol. 120, pp. 1107–1128, 1994; "Discussion" by A. Der Kiureghian and A. Neuenhofer and "Closure" by Authors, *Journal of Engineering Mechanics, ASCE*, Vol. 121, pp. 1037–1038, 1995.

221. A. Hindy and M. Novak, "Pipeline response to random ground motion," *Journal of the Engineering Mechanics Division, ASCE*, Vol. 106, pp. 339–360, 1980.

222. M. Horike and Y. Takeuchi, "Possibility of spatial variation of high-frequency seismic motions due to random-velocity fluctuations of sediments," *Bulletin of the Seismological Society of America*, Vol. 90, pp. 48–65, 2000.

223. M. Hoshiya, "Kriging and conditional simulation of Gaussian field," *Journal of Engineering Mechanics, ASCE*, Vol. 121, pp. 181–186, 1995.

224. M. Hoshiya and K. Ishii, "Evaluation of kinematic interaction of soil-foundation systems by a stochastic model," *Soil Dynamics and Earthquake Engineering*, Vol. 2, pp. 128–134, 1983.

225. M. Hoshiya, S. Noda and H. Inada, "Estimation of conditional non-Gaussian translation stochastic fields," *Journal of Engineering Mechanics, ASCE*, Vol. 124, pp. 435–445, 1998.

226. S.E. Hough, "Triggered earthquakes and the 1811-1812 New Madrid, Central United States, earthquake sequence," *Bulletin of the Seismological Society of America*, Vol. 91, pp. 1574–1581, 2001.

227. G.W. Housner, "Properties of strong ground motion earthquakes," *Bulletin of the Seismological Society of America*, Vol. 45, pp. 187–218, 1955.

228. G.W. Housner and P. C. Jennings, "Generation of artificial earthquakes," *Journal of the Engineering Mechanics Division, ASCE*, Vol. 90, pp. 113–150, 1964.

229. B.S. Huang, "Ground rotational motions of the 1999 Chi-Chi, Taiwan earthquake as inferred from dense array observations," *Geophysical Research Letters*, Vol. 30, Art. No. 1307, 2003.

230. D.E. Hudson, "Response spectrum techniques in engineering seismology," *Proceedings of the World Conference on Earthquake Engineering*, Berkeley, CA, 1956.

231. L. Hutchings and S. Jarpe, "Ground motion variability at the Highway 14 and I-5 interchange in the northern San Fernando Valley," *Bulletin of the Seismological Society of America*, Vol. 86, pp. S289–S299, 1996.

232. C.-H. Hyun, C.-B. Yun and D.-G. Lee, "Nonstationary response analysis of suspension bridges for multiple support excitations," *Probabilistic Engineering Mechanics*, Vol. 7, pp. 27–35, 1992.

233. M. Iguchi, "Earthquake response of embedded cylindrical foundation to SH and SV waves," *Proceedings of the Eighth World Conference on Earthquake Engineering*, San Francisco, CA, 1984.

234. International Code Council (ICC), *Uniform Building Code (UBC)*, Washington, D.C., 1997.

235. International Organization for Standardization (ISO), *Basis for Design of Structures - Seismic Actions for Designing Geotechnical Works*, ISO 23469:2005, Geneva, Switzerland, 2005.

236. International Working Group on Rotational Seismology (IWGoRS), http://www.rotational-seismology.org/, accessed 2008.

237. R.N. Iyengar and K.T.S. Iyengar, "A non-stationary random process model for earthquake acceleration," *Bulletin of the Seismological Society of America*, Vol. 59, pp. 1163–1188, 1969.

238. B.A. Izzuddin and A.S. Elnashai, "ADAPTIC program for static and dynamic large displacement nonlinear analysis of space frames by adaptive mesh refinement. User manual," Imperial College of Science, Technology and Medicine, London, United Kingdom, 1991.

239. W.D. Iwan, M.A. Moser and C.-Y. Peng, "Some observations on strong motion earthquake measurement using a digital accelerograph," *Bulletin of the Seismological Society of America*, Vol. 75, pp. 1225–1246, 1985.

240. R.S. Jangid, "Response of SDOF system to non-stationary earthquake excitation," *Earthquake Engineering and Structural Dynamics*, Vol. 33, pp. 1417–1428, 2004.

241. Japan Road Association, *Specifications for Highway Bridges. Part V. Seismic Design*, Tokyo, Japan, 2002.

242. S. Jin, L.D. Lutes and S. Sarkani, "Efficient simulation of multidimensional random fields," *Journal of Engineering Mechanics, ASCE*, Vol. 123, pp. 1082–1089, 1997.

243. G.W. Jenkins and D.G. Watts, *Spectral Analysis and its Applications*, Holden-Day, San Francisco, CA, 1969.

244. P.C. Jennings, G.W. Housner and N.C. Tsai, "Simulated earthquake motions," Technical Report, Earthquake Engineering Research Laboratory, California Institute of Technology, Pasadena, CA, 1968.

245. N.E. Johnson and R.D. Galletly, "The comparison of the response of a highway bridge to uniform ground shock and moving ground excitation," *The Shock and Vibration Bulletin*, Vol. 42, pp. 75–85, 1972.

246. W.B. Joyner and D.M. Boore, "Measurement, characterization and prediction of strong ground motion," *Earthquake Engineering and Soil Dynamics II-Recent Advances in Ground Motion Evaluation*, J.L. Von Thun editor, Geotechnical Special Publication No. 20, ASCE, New York, NY, 1988.

247. M. Kahan, R.-J. Gibert and P.-Y. Bard, "Influence of seismic waves spatial variability on bridges: A sensitivity analysis," *Earthquake Engineering and Structural Dynamics*, Vol. 25, pp. 795–814, 1996.

248. H. Kameda, "Evolutionary spectra of seismogram by multifilter," *Journal of the Engineering Mechanics Division, ASCE*, Vol. 101, pp. 787–801, 1975.

249. H. Kameda and H. Morikawa, "An interpolating stochastic process for simulation of conditional random fields," *Probabilistic Engineering Mechanics*, Vol. 7, pp. 243–254, 1992.

250. H. Kameda and H. Morikawa, "Conditioned stochastic processes for conditional random fields," *Journal of Engineering Mechanics, ASCE*, Vol. 120, pp. 855–875, 1994.

251. M. Kamiyama, M.J. O'Rourke and R. Flores-Berrones, "A semi-empirical analysis of strong-motion peaks in terms of seismic source, propagation path and local site conditions," Technical Report NCEER-92-0023, National Center for Earthquake Engineering Research (NCEER), State University of New York at Buffalo, Buffalo, NY, 1992.

252. M. Kamiyama and T. Satoh, "Seismic response of laterally inhomogeneous ground with emphasis on strains," *Soil Dynamics and Earthquake Engineering*, Vol. 22, pp. 877–884, 2002.

253. K. Kanai, "Semi-empirical formula for the seismic characteristics of the ground," *Bulletin of the Earthquake Research Institute*, University of Tokyo, Japan, Vol. 35, pp. 309–325, 1957.

254. K. Kanda, "Seismic responses of structures subjected to incident incoherent waves considering a layered media with irregular interfaces," *Proceedings of the 12th World Conference on Earthquake Engineering*, Auckland, New Zealand, 2000.

255. A. Kareem, J. Zhao and M.A. Tognarelli, "Surge response statistics of tension leg platforms under wind and wave loads: A statistical quadratization approach," *Probabilistic Engineering Mechanics*, Vol. 10, pp. 225–240, 1995.

256. L.S. Katafygiotis and J.L. Beck, "Updating models and their uncertainties. II: Model identifiability," *Journal of Engineering Mechanics, ASCE*, Vol. 124, pp. 463–467, 1998.

257. L.S. Katafygiotis and K.-V. Yuen, "Bayesian spectral density approach for modal updating using ambient data," *Earthquake Engineering and Structural Dynamics*, Vol. 30, pp. 1103–1123, 2001.

258. L.S. Katafygiotis, A. Zerva and A.A. Malyarenko, "Simulations of homogeneous and partially isotropic random fields," *Journal of Engineering Mechanics, ASCE*, Vol. 125, pp. 1180–1189, 1999.

259. L.S. Katafygiotis, A. Zerva and D. Pachakis, "An efficient approach for the simulation of spatially variable motions for the seismic response of lifelines," *Proceedings of the 13th Engineering Mechanics Conference, ASCE*, Baltimore, MD, 1999.

260. L.S. Katafygiotis, A. Zerva and D. Pachakis, "Simulation of homogeneous and partially isotropic random fields," *Proceedings of the International Conference on Applications of Statistics and Probability, ICASP·8*, Sydney, Australia, 1999.

261. T. Katayama, "Use of dense array data in the determination of engineering properties of strong motions," *Structural Safety*, Vol. 10, pp. 27–51, 1991.

262. H. Katsukura, T. Watabe and M. Izumi, "A study on the Fourier analysis of nonstationary seismic wave," *Proceedings of the Eighth World Conference on Earthquake Engineering*, San Francisco, CA, 1984.

263. M.K. Kaul, "Stochastic characterization of earthquakes through their response spectrum," *Earthquake Engineering and Structural Dynamics*, Vol. 6, pp. 497–509, 1978.

264. E. Kausel and A. Pais, "Stochastic deconvolution of earthquake motions," *Journal of Engineering Mechanics, ASCE*, Vol. 113, pp. 266–277, 1987.

265. H. Kawakami, "Simulation of space-time variation of earthquake ground motion including a recorded time history," *Structural Engineering and Earthquake Engineering, Proceedings of Japan Society of Civil Engineers*, No. 410/I-12, pp. 435–443, 1989 (in Japanese).

266. H. Kawase, "Site effects of strong ground motions," in *International Handbook of Earthquake & Engineering Seismology*, W. Lee, H. Kanamori, P. Jennings and C. Kisslinger editors, Elsevier Science and Technology, London, United Kingdom, 2003.

267. S. Kim and J.P. Stewart, "Kinematic soil-structure interaction from strong motion recordings," *Journal of Geotechnical and Geoenvironmental Engineering, ASCE*, Vol. 129, pp. 323–335, 2003.

268. S.H. Kim and M.Q. Feng, "Fragility analysis of bridges under ground motion with spatial variation," *International Journal of Non-Linear Mechanics*, Vol. 38, pp. 705–721, 2003.

269. J.L. King, "Observations on the seismic response of sediment-filled valleys," Ph.D. Dissertation, Department of Earth Science, University of California, San Diego, CA, 1981.

270. J.L. King and B.E. Tucker, "Analysis of differential array data from El Centro, USA and Garm, USSR," *Proceedings of the International Conference on Microzonation for Safer Construction – Research and Application*, Seattle, WA, 1982.

271. G. Kjell, "Predicting response spectra for earthquake signals generated as filtered noise," *Probabilistic Engineering Mechanics*, Vol. 17, pp. 241–252, 2002.

272. S.B. Kojić and M.D. Trifunac, "Earthquake stresses in arch dams. I: Theory and antiplane excitation," *Journal of Engineering Mechanics, ASCE*, Vol. 117, pp. 532–552, 1991.

273. S.B. Kojić and M.D. Trifunac, "Earthquake stresses in arch dams. II: Excitation by SV-, P-, and Rayleigh waves," *Journal of Engineering Mechanics, ASCE*, Vol. 117, pp. 553–574, 1991.

274. P.S. Koutsourelakis and G. Deodatis, "Simulation of binary random fields with applications to two-phase random media," *Journal of Engineering Mechanics, ASCE*, Vol. 131, pp. 397–412, 2005.

275. F. Kozin, "Auto-regressive moving-average models of earthquake records," *Probabilistic Engineering Mechanics*, Vol. 3, pp. 58–63, 1988.

276. S.L. Kramer, *Geotechnical Earthquake Engineering*, Prentice-Hall, Inc., Upper Saddle River, NJ, 1996.

277. T. Kubo and J. Penzien, "Simulation of three-dimensional strong ground motions along principal axes, San Fernando earthquake," *Earthquake Engineering and Structural Dynamics*, Vol. 7, pp. 279–294, 1979.

278. K. Kudo, E. Shima and M. Sakaue, "Digital strong motion accelerographs array in Ashigara valley - seismological and engineering prospects of strong motion observation," *Proceedings of the Ninth World Conference on Earthquake Engineering*, Tokyo, Japan, 1988.

279. R.T. Lacoss, E.J. Kelly and M.N. Toksöz, "Estimation of seismic noise structure using arrays," *Geophysics*, Vol. 34, pp. 21–38, 1969.

280. N.D. Lagaros, G. Stefanou and M. Papadrakakis, "An enhanced hybrid method for the simulation of highly skewed non-Gaussian stochastic fields," *Computer Methods in Applied Mechanics and Engineering*, Vol. 194, pp. 4824–4844, 2005.

281. N. Laouami and P. Labbe, "Analytical approach for evaluation of the seismic ground motion coherency function," *Soil Dynamics and Earthquake Engineering*, Vol. 21, pp. 727–733, 2001.

282. N. Laouami and P. Labbe, "Experimental analysis of seismic torsional ground motion recorded by the LSST-Lotung array," *Earthquake Engineering and Structural Dynamics*, Vol. 31, pp. 2141–2148, 2002.

283. N. Laouami, P. Labbe and R. Bahar, "Stochastic model of seismic torsional ground motion: Applications to Lotung soft site," *Journal of Seismology*, Vol. 9, pp. 463–472, 2005.

284. H. Law, I.P. Lam, B. Maroney and S. Mohan, "Development of seismic ground motion criteria for new San Francisco-Oakland Bay Bridge," in *Geotechnical Engineering for Transportation Projects (GSP No. 126), GeoTrans 2004*, M.K. Yegian and E. Kavazanjian editors, ASCE, New York, NY, 2004.

285. M.-C. Lee and J. Penzien, "Stochastic analysis of structures and piping systems subjected to stationary multiple support excitations," *Earthquake Engineering and Structural Dynamics*, Vol. 11, pp. 91–110, 1983.

286. V.W. Lee, "Empirical scaling of strong earthquake ground motion - Part I: Attenuation and scaling of response spectra," *ISET Journal of Earthquake Technology*, Vol. 39, pp. 219–254, 2002.

287. V.W. Lee, "Empirical scaling of strong earthquake ground motion - Part II: Duration of strong motion," *ISET Journal of Earthquake Technology*, Vol. 39, pp. 255–271, 2002.

288. V.W. Lee, "Empirical scaling of strong earthquake ground motion - Part III: Synthetic strong motion," *ISET Journal of Earthquake Technology*, Vol. 39, pp. 273–310, 2002.

289. V.W. Lee and M.D. Trifunac, "Torsional accelerograms," *Soil Dynamics and Earthquake Engineering*, Vol. 4, pp. 132–139, 1985.

290. V.W. Lee and M.D. Trifunac, "Rocking strong earthquake accelerations," *Soil Dynamics and Earthquake Engineering*, Vol. 6, pp. 75–89, 1987.

291. P. Léger, I.M. Idé and P. Paultre, "Multiple-support seismic analysis of large structures," *Computers and Structures*, Vol. 36, pp. 1153–1158, 1990.

292. S. Levy and J.P.D. Wilkinson, "Generation of artificial time histories, rich in all frequencies, from given response spectra," *Nuclear Engineering and Design*, Vol. 32, pp. 148–155, 1976.

293. P. Lestuzzi, http://imacwww.epfl.ch/Team/Lestuzzi/lestuzzi.jsp, accessed 2008.

294. Y. Li and A. Kareem, "Simulation of multivariate nonstationary random processes by FFT," *Journal of Engineering Mechanics, ASCE*, Vol. 117, pp. 1037–1058, 1991.

295. Y. Li and A. Kareem, "Simulation of multivariate nonstationary random processes: Hybrid DFT and digital filtering approach," *Journal of Engineering Mechanics, ASCE*, Vol. 123, pp. 1302–1310, 1997.

296. J-H. Li and J. Li, "A response spectrum method for seismic response analysis of structures under multi-support excitations," *Structural Engineering and Mechanics*, Vol. 21, pp. 255–273, 2005.

297. J. Liang and S. Sun, "Site effects on seismic behavior of pipelines: A review," *Journal of Pressure Vessel Technology, ASME*, Vol. 122, pp. 469–475, 2000.

298. S. Liao, "Physical characterization of ground motion spatial variation and conditional simulation for performance-based design," Ph.D. Dissertation, Department of Civil, Architectural and Environmental Engineering, Drexel University, Philadelphia, PA, 2006.

299. S. Liao and J. Li, "A stochastic approach to site-response component in seismic ground motion coherency model," *Soil Dynamics and Earthquake Engineering*, Vol. 22, pp. 813–820, 2002.

300. S. Liao and A. Zerva, "Amplitude and phase variability from analyses of spatially recorded data," *Proceedings of Geo-Congress 2006, ASCE*, Atlanta, Georgia, 2006.

301. S. Liao and A. Zerva, "Physically-compliant, conditionally simulated spatially variable seismic ground motions for performance-based design," *Earthquake Engineering and Structural Dynamics*, Vol. 35, pp. 891–919, 2006.

302. S. Liao, A. Zerva and H. Morikawa "Conditional simulation of nonstationary random fields based on physical spectra," *Proceedings of the Sixth European Conference on Structural Dynamics, EURODYN 2005*, Paris, France, 2005.

303. S. Liao, A. Zerva and W.R. Stephenson, "Seismic spatial coherency at a site with irregular subsurface topography," *Proceedings of the Geo-Denver 2007 Congress, ASCE*, Denver, CO, 2007.

304. K. Lilhand and W.S. Tseng, "Development and application of realistic earthquake time histories compatible with multiple damping response spectra," *Proceedings of the Ninth World Conference on Earthquake Engineering*, Tokyo, Japan, 1988.

305. G. Lin, J. Zhou and J. Wang, "Seismic response of arch dams to wave scattering and spatial variation of ground motions," *Proceedings of the 11th World Conference on Earthquake Engineering*, Oxford, England, 1996.

306. J.H. Lin and Y.H. Zhang, "Seismic random vibration of long-span structures," in *Vibration and Shock Handbook*, C.W. de Silva editor, CRC Press, Boca Raton, FL, 2005.

307. J.H. Lin, Y.H. Zhang, Q.S. Li and F.W. Williams, "Seismic spatial effects for long-span bridges using the pseudo excitation method," *Engineering Structures*, Vol. 26, pp. 1207–1216, 2004.

308. Y.K. Lin, *Probabilistic Theory of Structural Dynamics*, Robert E. Krieger Publishing Company, Huntington, NY, 1976.

309. Y.K. Lin and G.Q. Cai, *Probabilistic Structural Dynamics: Advanced Theory and Applications*, McGraw-Hill, Inc., New York, NY, 1995.

310. Y.K. Lin and Y. Yong, "Evolutionary Kanai-Tajimi earthquake models," *Journal of Engineering Mechanics, ASCE*, Vol. 113, pp. 1119–1137, 1987.

311. S.C. Liu, "Evolutionary power spectral density of strong-motion earthquakes," *Bulletin of the Seismological Society of America*, Vol. 60, pp. 891–900, 1970.

312. C.H. Loh, "Analysis of the spatial variation of seismic waves and ground movements from SMART-1 data," *Earthquake Engineering and Structural Dynamics*, Vol. 13, pp. 561–581, 1985.

313. C.H. Loh, "Study of spatial variation of SMART-1 array data, and the effects of wave propagation on buried pipelines," Civil Engineering Studies, National Central University, Department of Civil Engineering, Chung-Li, Taiwan, 1984.

314. C.H. Loh, A. H-S. Ang and Y.K. Wen, "Spatial correlation study of strong motion array data with application to lifeline earthquake engineering," Civil Engineering Studies, Structural Research Series No. SRS-503, University of Illinois at Urbana-Champaign, Urbana, IL, 1983.

315. C.H. Loh and B.D. Ku, "An efficient analysis of structural response for multiple-support seismic excitations," *Engineering Structures*, Vol. 17, pp. 15–26, 1995.

316. C.H. Loh and S.G. Lin, "Directionality and simulation in spatial variation of seismic waves," *Engineering Structures*, Vol. 12, pp. 134–143, 1990.

317. C.H. Loh, J. Penzien and Y.B. Tsai, "Engineering analysis of SMART-1 array accelerograms," *Earthquake Engineering and Structural Dynamics*, Vol. 10, pp. 575–591, 1982.

318. C.H. Loh and Y.T. Yeh, "Spatial variation and stochastic modeling of seismic differential ground movement," *Earthquake Engineering and Structural Dynamics*, Vol. 16, pp. 583–596, 1988.

319. L. Lou, "Effect of the spatial variability of ground motions on the seismic response of reinforced concrete highway bridges," Ph.D. Dissertation, Department of Civil, Architectural and Environmental Engineering, Drexel University, Philadelphia, PA, 2006.

320. L. Lou and A. Zerva, "Effects of spatially variable ground motions on the seismic response of a skewed, multi-span bridge," *Soil Dynamics and Earthquake Engineering*, Vol. 25, pp. 729–740, 2005.

321. L. Lou and A. Zerva, "Influence of spatial variation of ground motions on the nonlinear response of a multi-span bridge," *Proceedings of the Fourth European Workshop on the Seismic Behaviour of Irregular and Complex Structures*, Thessaloniki, Greece, 2005.

322. J.E. Luco and A. Mita, "Response of circular foundation to spatially random ground motion," *Journal of Engineering Mechanics, ASCE*, Vol. 113, pp. 1–15, 1987.

323. J.E. Luco and D.A. Sotiropoulos, "Local characterization of free-field ground motion and effects of wave passage," *Bulletin of the Seismological Society of America*, Vol. 70, pp. 2229–2244, 1980.

324. J.E. Luco and H.L. Wong, "Response of a rigid foundation to a spatially random ground motion," *Earthquake Engineering and Structural Dynamics*, Vol. 14, pp. 891–908, 1986.

325. A. Lupoi, P. Franchin, P.E. Pinto and G. Monti, "Seismic design of bridges accounting for spatial variability of ground motion," *Earthquake Engineering and Structural Dynamics*, Vol. 34, pp. 327–348, 2005.

326. L.D. Lutes and S. Sarkani, *Stochastic Analysis of Structural and Mechanical Vibrations*, Prentice-Hall, Inc., Englewood Cliffs, NJ, 1997.

327. L.D. Lutes and J. Wang, "Simulation of improved Gaussian time history," *Journal of Engineering Mechanics, ASCE*, Vol. 117, pp. 218–224, 1991; "Discussion" by O. Ditlevsen and "Closure" by Authors, *Journal of Engineering Mechanics, ASCE*, Vol. 118, pp. 1276–1277, 1992.

328. O. Maeso, J.J. Aznárez and J. Domínguez, "Effects of space distribution of excitation on seismic response of arch dams," *Journal of Engineering Mechanics, ASCE*, Vol. 128, pp. 759–768, 2002.

329. O. Maeso, J.J. Aznárez and J. Domínguez, "Three-dimensional models of reservoir sediment and effects on the seismic response of arch dams," *Earthquake Engineering and Structural Dynamics*, Vol. 33, pp. 1103–1123, 2004.

330. K. Makra, F.J. Chavez-García, D. Raptakis and K. Pitilakis, "Parametric analysis of the seismic response of a 2D sedimentary valley: Implications for code implementations of complex site effects," *Soil Dynamics and Earthquake Engineering*, Vol. 25, pp. 303–315, 2005.

331. J.B. Mander, M.J.N. Priestley and R. Park, "Theoretical stress-strain model for confined concrete," *Journal of Structural Engineering, ASCE*, Vol. 114, pp. 1804–1826, 1988.

332. J.B. Mander, M.J.N. Priestley and R. Park, "Observed stress-strain model for confined concrete," *Journal of Structural Engineering, ASCE*, Vol. 114, pp. 1827–1849, 1988.

333. A. Mantoglou and J.L. Wilson, "Simulation of random fields with the turning bands method," Report No. 264, Department of Civil Engineering, Massachusetts Institute of Technology, Cambridge, MA, 1981.

334. S.F. Masri, "Response of beams to propagating boundary excitation," *Earthquake Engineering and Structural Dynamics*, Vol. 4, pp. 497–507, 1976.

335. F. Masters and K.R. Gurley, "Non-Gaussian simulation: Cumulative distribution function map-based spectral correction," *Journal of Engineering Mechanics, ASCE*, Vol. 129, pp. 1418–1428, 2003.

336. Y. Matsushima, "Stochastic response of structure due to spatially variant earthquake excitations," *Proceedings of the Sixth World Conference on Earthquake Engineering*, New Delhi, India, 1977.

337. S. Mazzoni, F. McKenna, M.H. Scott, G.L. Fenves *et al.*, "Open system for earthquake engineering simulation. User command-language manual. OpenSees version 1.7.3," Pacific Earthquake Engineering Research Center (PEER), University of California, Berkeley, CA, http://opensees.berkeley.edu/OpenSees/manuals/usermanual/index.html, accessed 2008.

338. F. McKenna and G.L. Fenves, "An object-oriented software design for parallel structural analysis," *Proceedings of Structures Congress 2000 - Advanced Technology in Structural Engineering, ASCE*, Philadelphia, PA, 2000.

339. K.L. McLaughlin, "Spatial coherency of seismic waveforms," Ph.D. Dissertation, Department of Earth and Planetary Science, University of California, Berkeley, CA, 1983.

340. K.L. McLaughlin, L.R. Johnson and T.V. McEvilly, "Two-dimensional array measurements of near-source ground accelerations," *Bulletin of the Seismological Society of America*, Vol. 73, pp. 349–375, 1983.

341. W. Menke, A.L. Lerner-Lam, B. Dubendorff and J. Pacheco, "Polarization and coherence of 5 to 30 Hz seismic wave fields at a hard-rock site and their relevance to velocity heterogeneities in the crust," *Bulletin of the Seismological Society of America*, Vol. 80, pp. 430–449, 1990.

342. M. Meremonte, A. Frankel, E. Cranswick, D. Carver and D. Worly, "Urban seismology-Northridge aftershocks recorded by multi-scale arrays of portable digital

seismographs," *Bulletin of the Seismological Society of America*, Vol. 86, pp. 1350–1363, 1996.

343. M.P. Mignolet and M.V. Harish, "Comparison of some simulation algorithms on basis of distribution," *Journal of Engineering Mechanics, ASCE*, Vol. 122, pp. 172–176, 1996.

344. M.P. Mignolet and P.D. Spanos, "Simulation of homogeneous two-dimensional random fields: Part I - AR and ARMA models," *Journal of Applied Mechanics, ASME*, Vol. 59, pp. 260–269, 1992.

345. A. Mita and J.E. Luco, "Response of structures to a spatially random ground motion," *Proceedings of the Third U.S. National Conference on Earthquake Engineering*, Charleston, SC, 1986.

346. V. Montaldo, A.S. Kiremidjian, H. Thráinsson and G. Zonno, "Simulation of the Fourier phase spectrum for the generation of synthetic accelerograms," *Journal of Earthquake Engineering*, Vol. 7, pp. 427–445, 2003.

347. G. Monti, C. Nuti and P.E. Pinto, "Nonlinear response of bridges under multi-support excitation," *Journal of Structural Engineering, ASCE*, Vol. 122, pp. 1147–1159, 1996.

348. G. Monti and P.E. Pinto, "Effects of multi-support excitation on isolated bridges," Technical Report MCEER 98/0015, Multidisciplinary Center for Earthquake Engineering Research (MCEER), University at Buffalo, State University of New York, Buffalo, NY, 1998.

349. J. Mori, J. Filson, E. Cranswick, R. Borcherdt, R. Amirbekian, V. Aharonian and L. Hachverdian, "Measurements of P and S wave fronts from the dense three-dimensional array at Garni, Armenia," *Bulletin of the Seismological Society of America*, Vol. 84, pp. 1089–1096, 1994.

350. H. Morikawa, S. Sawada, K. Toki, K. Kawasaki and Y. Kaneko, "Phase characteristics of source time function modeled by stochastic impulse train," *Proceedings of the 12th World Conference on Earthquake Engineering*, Auckland, New Zealand, 2000.

351. H. Morikawa, S. Sawada and Y. Ono, "Detailed analysis for earthquake source spectrum represented by stochastic impulse train and its applications," *Journal of Seismology*, Vol. 9, pp. 151–170, 2005.

352. A. Moya, J. Aguirre and K. Irikura, "Inversion of source parameters and site effects from strong ground motion records using genetic algorithms," *Bulletin of the Seismological Society of America*, Vol. 90, pp. 977–992, 2000.

353. G. Mylonakis, S. Nikolaou, D. Papastamatiou and J. Psycharis, "Simple model for response of bridges to non-uniform seismic excitation," *Proceedings of the Eighth Canadian Conference on Earthquake Engineering*, Vancouver, Canada, 1999.

354. G. Mylonakis, D. Papastamatiou, J. Psycharis and K. Mahmoud, "Simplified modeling of bridge response on soft soil to nonuniform seismic excitation," *Journal of Bridge Engineering*, Vol. 6, pp. 587–597, 2001.

355. G. Mylonakis, V. Simeonov, A.M. Reinhorn and I.G. Buckle, "Implications of spatial variation of ground motion on the collapse of SR14/I5 southbound separation and overhead bridge in the Northridge earthquake," *ACI International* - Special Publication SP-187, pp. 299–327, 1999.

356. F. Nadim, E.H. Vanmarcke, O.T. Gudmestad and S. Hetland, "Influence of spatial variation of earthquake motion on response of gravity base platforms," *Structural Safety*, Vol. 10, pp. 113–128, 1991.

357. F. Nadim and O.T. Gudmestad, "Reliability of an engineering system under a strong earthquake with application to offshore platforms," *Structural Safety*, Vol. 14, pp. 203–217, 1994.

358. A. Naess, H.C. Karlsen and P.S. Teigen, "Numerical methods for calculating the crossing rate of high and extreme response levels of compliant offshore structures subjected to random waves," *Applied Ocean Research*, Vol. 28, pp. 1–8, 2006.

359. Y. Nakamura, A. Der Kiureghian and D. Liu, "Multiple-support response spectrum analysis of the Golden Gate Bridge," Earthquake Engineering Research Center Report No. UCB/EERC-93-05, University of California, Berkeley, CA, 1993.

360. J. Náprstek and C. Fischer, "Analysis of non-stationary response of structures due to seismic random processes of evolutionary type," *Proceedings of the 12th World Conference on Earthquake Engineering*, Auckland, New Zealand, 2000.

361. N.D. Nathan and J.R. MacKenzie, "Rotational components of earthquake motion," *Canadian Journal of Civil Engineering*, Vol. 2, pp. 430–436, 1975.

362. A.S. Nazmy and A.M. Abdel-Ghaffar, "Effect of ground motion spatial variability on the response of cable-stayed bridges," *Earthquake Engineering and Structural Dynamics*, Vol. 21, pp. 1–20, 1992.

363. I. Nelson and P. Weidlinger, "Development of interference response spectra for lifelines seismic analysis," Grant Report No. 2, Weidlinger Associates, New York, NY, 1977.

364. New York City Department of Transportation (NYCDOT), *Seismic Criteria Guidelines*, New York, NY, 1998.

365. D.E. Newland, *An Introduction to Random Vibrations and Spectral Analysis*, Longman Inc., New York, NY, 1984.

366. N.M. Newmark, "Problems in wave propagation in soil and rocks," *Proceedings of the International Symposium on Wave Propagation and Dynamic Properties of Earth Materials*, University of New Mexico Press, Albuquerque, NM, 1967.

367. N.M. Newmark, "Torsion in symmetrical buildings," *Proceedings of the Fourth World Conference on Earthquake Engineering*, Santiago, Chile, 1969.

368. N.M. Newmark, W.J. Hall and J.K. Morgan, "Building response and free field motion in earthquakes," *Proceedings of the Sixth World Conference on Earthquake Engineering*, New Delhi, India, 1977.

369. N.M. Newmark and E. Rosenblueth, *Fundamentals of Earthquake Engineering*, Prentice-Hall, Inc., Englewood Cliffs, NJ, 1971.

370. M. Niazi, "Spatial coherence of the ground motion produced by the 1979 Imperial Valley earthquake across El Centro differential array," *Physics of the Earth and Planetary Interiors*, Vol. 38, pp. 162–173, 1985.

371. M. Niazi, "Inferred displacements, velocities and rotations of a long rigid foundation located at El Centro differential array site during the 1979 Imperial Valley, California, earthquake," *Earthquake Engineering and Structural Dynamics*, Vol. 14, pp. 531–542, 1986.

372. N.C. Nigam, "Phase properties of a class of random processes," *Earthquake Engineering and Structural Dynamics*, Vol. 10, pp. 711–717, 1982.

373. N.C. Nigam and S. Narayanan, *Applications of Random Vibrations*, Springer Verlag, New York, NY, 1994.

374. R.L. Nigbor, "Six-degree-of-freedom ground-motion measurement," *Bulletin of the Seismological Society of America*, Vol. 84, pp. 1665–1669, 1994.

375. M. Novak, Discussion on "Stochastic variation of earthquake ground motion in space and time" by R.S. Harichandran and E.H. Vanmarcke, *Journal of Engineering Mechanics, ASCE*, Vol. 113, pp. 1267–1270, 1987.

376. M. Novak and A. Hindy, "Seismic response of buried pipelines," *Proceedings of the Third Canadian Conference on Earthquake Engineering*, Montreal, Canada, 1979.

377. M. Novak, T. Nogami and F. Aboul-Ella, "Dynamic soil reactions for plane strain case," *Journal of the Engineering Mechanics Division, ASCE*, Vol. 104, pp. 953–959, 1978.

378. P.S. Nowak and J.F. Hall, "Arch dam response to nonuniform seismic input," *Journal of Engineering Mechanics, ASCE*, Vol. 116, pp. 125–139, 1990.

379. C. Nuti and I. Vanzi, "Influence of earthquake spatial variability on differential soil displacements and SDF system response," *Earthquake Engineering and Structural Dynamics*, Vol. 34, pp. 1353–1374, 2005.

380. Y. Ohsaki, "On the significance of the phase content in earthquake ground motions," *Earthquake Engineering and Structural Dynamics*, Vol. 7, pp. 427–439, 1979.

381. C.S. Oliveira and B.A. Bolt, "Rotational components of surface strong ground motion," *Earthquake Engineering and Structural Dynamics*, Vol. 18, pp. 517–526, 1989.

382. C.S. Oliveira, H. Hao and J. Penzien, "Ground motion modeling for multiple-input structural analysis," *Structural Safety*, Vol. 10, pp. 79–93, 1991.

383. A.V. Oppenheim and R.W. Schafer, *Discrete-Time Signal Processing*, Prentice-Hall, Inc., Upper Saddle River, NJ, 1999.

384. M.J. O'Rourke, M.C. Bloom and R. Dobry, "Apparent propagation velocity of body waves," *Earthquake Engineering and Structural Dynamics*, Vol. 10, pp. 283–294, 1982.

385. M.J. O'Rourke and G. Castro, "Design of buried pipelines for wave propagation," in *Lifeline Earthquake Engineering*, D.J. Smith editor, ASCE, New York, NY, 1981.

386. M.J. O'Rourke, G. Castro and N. Centola, "Effects of seismic wave propagation upon buried pipelines," *Earthquake Engineering and Structural Dynamics*, Vol. 8, pp. 455–467, 1980.

387. M.J. O'Rourke, G. Castro and I. Hossain, "Horizontal soil strain due to seismic waves," *Journal of Geotechnical Engineering, ASCE*, Vol. 110, pp. 1173–1187, 1984.

388. M.J. O'Rourke and K. El Hmadi, "Analysis of continuous buried pipelines for seismic wave effects," *Earthquake Engineering and Structural Dynamics*, Vol. 16, pp. 917–929, 1988.

389. T.D. O'Rourke, H.E. Stewart and S.-S. Jeon, "Geotechnical aspects of lifeline engineering," *Proceedings of the Institution of Civil Engineers: Geotechnical Engineering*, Vol. 149, pp. 13–26, 2001.

390. Pacific Earthquake Engineering Research Center (PEER), http://peer.berkeley.edu/nga/index.html, accessed 2008.

391. A.L. Pais and E. Kausel, "Stochastic response of rigid foundations," *Earthquake Engineering and Structural Dynamics*, Vol. 19, pp. 611–622, 1990.

392. Panel on Earthquake Engineering of Concrete Dams, Committee on Earthquake Engineering, Division on Natural Hazard Mitigation, Commission on Engineering and Technical Systems (CETS), and National Research Council, *Earthquake Engineering for Concrete Dams: Design, Performance and Research Needs*, National Academies Press, Washington, D.C., 1990.

393. A. Papoulis, *Probability, Random Variables and Stochastic Processes*, McGraw Hill, Inc., New York, NY, 1984.

394. S.W. Park, H. Ghasemi, J. Shen, P.G. Somerville, W.P. Yen and M. Yashinsky, "Simulation of the seismic performance of the Bolu Viaduct subjected to near-fault ground motions," *Earthquake Engineering and Structural Dynamics*, Vol. 33, pp. 1249–1270, 2004.

395. S.W. Park, W.P. Yen, J.D. Cooper, S. Unjoh, T. Terayama and H. Otsuka, "A comparative study of U.S. Japan seismic design of highway bridges: II. Shake-table model tests," *Earthquake Spectra*, Vol. 19, pp. 933–958, 2003.

396. Y.J. Park, "New conversion method from response spectrum to PSD functions," *Journal of Engineering Mechanics, ASCE*, Vol. 121, pp. 1391–1392, 1995.

397. Y.J. Park, Y.K. Wen and A. H-S. Ang, "Random vibration of hysteretic systems under bi-directional ground motions," *Earthquake Engineering and Structural Dynamics*, Vol. 14, pp. 543–557, 1986.

398. E. Parzen, *Stochastic Processes*, Holden-Day, Inc., San Francisco, CA, 1962.

399. P. Pegon and A.V. Pinto, "Pseudo-dynamic testing with substructuring at the ELSA Laboratory," *Earthquake Engineering and Structural Dynamics*, Vol. 29, pp. 905–925, 2000.

400. J. Penzien and M. Watabe, "Characteristics of 3-dimensional earthquake ground motions," *Earthquake Engineering and Structural Dynamics*, Vol. 3, pp. 365–373, 1975.

401. F. Perotti, "Structural response to non-stationary multiple-support random excitation," *Earthquake Engineering and Structural Dynamics*, Vol. 19, pp. 513–527, 1990.

402. D.D. Pfaffinger, "Calculation of power spectra from response spectra," *Journal of Engineering Mechanics, ASCE*, Vol. 109, pp. 357–372, 1983.

403. K.K. Phoon, H.W. Huang and S.T. Quek, "Simulation of strongly non-Gaussian processes using Karhunen-Loeve expansion," *Probabilistic Engineering Mechanics*, Vol. 20, pp. 188–198, 2005.

404. A.V. Pinto, P. Pegon, G. Magonette and G. Tsionis, "Pseudo-dynamic testing of bridges using non-linear substructuring," *Earthquake Engineering and Structural Dynamics*, Vol. 33, pp. 1125–1146, 2004.

405. A. Pitarka, P. Somerville, Y. Fukushima, T. Uetake and K. Irikura, "Simulation of near-fault strong-ground motion using hybrid Green's functions," *Bulletin of the Seismological Society of America*, Vol. 90, pp. 566–586, 2000.

406. N.W. Polhemus and A.S. Cakmak, "Simulation of earthquake ground motion using autoregressive moving average (ARMA) models," *Earthquake Engineering and Structural Dynamics*, Vol. 9, pp. 343–354, 1981.

407. R. Popescu, G. Deodatis and J.H. Prevost, "Simulation of homogeneous non-Gaussian stochastic vector fields," *Probabilistic Engineering Mechanics*, Vol. 13, pp. 1–13, 1998.

408. B. Porat, *Digital Signal Processing of Random Signals. Theory and Methods*, Prentice-Hall, Inc., Englewood Cliffs, NJ, 1994.

409. B. Porat, *A Course in Digital Signal Processing*, John Wiley & Sons, New York, NY, 1997.

410. G.H. Powell and S. Campbell, "DRAIN-3DX: Element description and user guide for element type01, type04, type05, type08, type09, type15, type17," Report No. UCB/SEMM-94/08, University of California, Berkeley, CA, 1994.

411. H.J. Pradlwarter and G.I Schuëller, "On advanced Monte Carlo simulation procedures in stochastic structural dynamics," *International Journal of Non-Linear Mechanics*, Vol. 32, pp. 735–744, 1997.

412. V. Prakash, G.H. Powell and S. Campbell, "DRAIN-3DX: Base program description and user guide," Report No. UCB/SEMM-94/07, University of California, Berkeley, CA, 1994.

413. T.E. Price and M.O. Eberhard, "Effects of varying ground motions on short bridges," *Journal of Structural Engineering, ASCE*, Vol. 124, pp. 948–955, 1998.

414. M.B. Priestley, *Non-Linear and Non-Stationary Time Series Analysis*, Academic Press, New York, NY, 1988.

415. M.J.N. Priestley, F. Seible and G.M. Calvi, *Seismic Design and Retrofit of Bridges*, John Wiley & Sons, New York, NY, 1996.

416. C. Proppe, H.J. Pradlwarter and G.I. Schuëller, "Equivalent linearization and Monte Carlo simulation in stochastic dynamics," *Probabilistic Engineering Mechanics*, Vol. 18, pp. 1–15, 2003.

417. T.J. Qu, J.J. Wang and Q.X. Wang, "A practical model for the power spectrum of spatially variant ground motion," *Acta Seismologica Sinica*, Vol. 9, pp. 69–79, 1996.

418. O. Ramadan and M. Novak, "Coherency functions for spatially correlated seismic ground motions," Geotechnical Research Centre Report No. GEOT-9-93, The University of Western Ontario, London, Canada, 1993.

419. O. Ramadan and M. Novak, "Simulation of spatially incoherent random ground motions," *Journal of Engineering Mechanics, ASCE*, Vol. 119, pp. 997–1016, 1993.

420. O. Ramadan and M. Novak, "Simulation of multidimensional anisotropic ground motions," *Journal of Engineering Mechanics, ASCE*, Vol. 120, pp. 1773–1785, 1994.

421. M. Rassem, A. Ghobarah and A.C. Heidebrecht, "Site effects on the seismic response of a suspension bridge," *Engineering Structures*, Vol. 18, pp. 363–370, 1996.

422. Y.J. Ren, I. Elishakoff and M. Shinozuka, "Simulation of multivariate Gaussian fields conditioned by realizations of the fields and their derivatives," *Journal of Applied Mechanics, ASME*, Vol. 63, pp. 758–765, 1996.

423. S.O. Rice, "Mathematical analysis of random noise," *Bell Systems Technology Journal*, Vol. 23, pp. 282–332, 1944, and Vol. 24, pp. 46–156, 1945.

424. S.O. Rice, "Mathematical analysis of random noise," in *Selected Papers on Noise and Stochastic Processes*, N. Wax editor, Dover, NY, 1954.

425. J. Riepl, C.S. Oliveira and P.-Y. Bard, "Spatial coherence of seismic wave fields across an alluvial valley (weak motion)," *Journal of Seismology*, Vol. 1, pp. 253–268, 1997.

426. J.B. Roberts and P.D. Spanos, *Random Vibration and Statistical Linearization*, John Wiley & Sons, New York, NY, 1990.

427. M. Rosenblatt, *Stationary Sequences and Random Fields*, Birkhäuser Boston, Inc., Boston, MA, 1985.

428. E. Rosenblueth, "Some applications of probability theory on aseismic design," *Proceedings of the World Conference on Earthquake Engineering*, Berkeley, CA, 1956.

429. E. Rosenblueth and J. Elorduy, "Response of linear systems to certain linear disturbances," *Proceedings of the Fourth World Conference on Earthquake Engineering*, Santiago, Chile, 1969.

430. D.H. Rothman, "Nonlinear inversion, statistical mechanics, and residual statics estimation," *Geophysics*, Vol. 50, pp. 2784–2796, 1985.

431. D.H. Rothman, "Automatic estimation of large residual statics corrections," *Geophysics*, Vol. 51, pp. 332–346, 1986.

432. P. Ruiz and J. Penzien, "Probabilistic study of the behavior of structures during earthquakes," Engineering Research Center, University of California, Berkeley, CA, 1969.

433. A. Rutenberg and A.C. Heidebrecht, "Response spectra for rocking, torsion and rigid foundations," *Earthquake Engineering and Structural Dynamics*, Vol. 13, pp. 543–557, 1985.

434. F. Sabetta, R. Masiani and A. Giuffrè, "Frequency nonstationarity in Italian strong motion accelerograms," *Proceedings of the Eighth European Conference on Earthquake Engineering*, Lisbon, Portugal, 1986.

435. F. Sabetta and A. Pugliese, "Attenuation of peak horizontal acceleration and velocity from Italian strong-motion records," *Bulletin of the Seismological Society of America*, Vol. 77, pp. 1491–1513, 1987.

436. F. Sabetta and A. Pugliese, "Estimation of response spectra and simulation of nonstationary earthquake ground motions," *Bulletin of the Seismological Society of America*, Vol. 86, pp. 337–352, 1996.

437. A. Sakurai and T. Takahashi, "Dynamic stresses of underground pipelines during earthquakes," *Proceedings of the Fourth World Conference on Earthquake Engineering*, Santiago, Chile, 1969.

438. H. Sandi, "Conventional seismic forces corresponding to non-synchronous ground motion," *Proceedings of the Third European Symposium on Earthquake Engineering*, Sofia, Bulgaria, 1970.

439. G.R. Saragoni and G.C. Hart, "Simulation of artificial earthquakes," *Earthquake Engineering and Structural Dynamics*, Vol. 2, pp. 249–267, 1974.

440. T. Sato and Y. Murono, "Simulation of earthquake motion from phase information," *Journal of Natural Disaster Science*, Vol. 2, pp. 93–101, 2003.

441. S. Sawada, H. Morikawa, K. Toki and K. Yokoyama, "Identification of path and local site effects on phase spectrum of seismic ground motion," *Proceedings of the 12th World Conference on Earthquake Engineering*, Auckland, New Zealand, 2000.

442. V. Saxena, "Spatial variation of earthquake ground motion and development of bridge fragility curves," Ph.D. Dissertation, Department of Civil Engineering and Operations Research, Princeton University, Princeton, NJ, 2000.

443. V. Saxena, G. Deodatis, M. Shinozuka and M. Feng, "Development of fragility curves for multi-span reinforced concrete bridges," in *Monte Carlo Simulation: Proceedings of the International Conference on Monte Carlo Simulation, MCS-2000*, G.I. Schuëller and P.D. Spanos editors, Monte Carlo, Monaco, 2000.

444. R.H. Scanlan, "Seismic wave effects on soil-structure interaction," *Earthquake Engineering and Structural Dynamics*, Vol. 4, pp. 379–388, 1976.

445. R.H. Scanlan and K. Sachs, "Earthquake time-histories and response spectra," *Journal of the Engineering Mechanics Division*, ASCE, Vol. 100, pp. 635–655, 1974.

446. R.J. Scherer, J.D. Riera and G.I. Schuëller, "Estimation of time-dependent frequency content of earthquake accelerations," *Nuclear Engineering and Design*, Vol. 71, pp. 301–310, 1982.

447. R.O. Schmidt, "A signal subspace approach to multiple emitter location and spectral estimation," Ph.D. Dissertation, Department of Computer Science, Stanford University, Stanford, CA, 1981.

448. R.O. Schmidt, "Multiple emitter location and signal parameter estimation," *IEEE Transactions on Antennas and Propagation*, Vol. AP-34, pp. 276–280, 1986.

449. J.F. Schneider, J.C. Stepp and N.A. Abrahamson, "The spatial variation of earthquake ground motion and effects of local site conditions," *Proceedings of the Tenth World Conference on Earthquake Engineering*, Madrid, Spain, 1992.

450. G.I. Schuëller, M.D. Pandey and H.J. Pradlwarter, "Equivalent linearization (EQL) in engineering practice for aseismic design," *Probabilistic Engineering Mechanics*, Vol. 9, pp. 95–102, 1994.

451. G.I. Schuëller and P.D. Spanos (editors), *Proceedings of Monte Carlo Simulation 2000*, Monte Carlo, Monaco, 2000.

452. R.T. Severn, "European experimental research in earthquake engineering for Eurocode 8," *Proceedings of the Institute of Civil Engineers, Structures and Buildings*, Vol. 134, pp. 205–217, 1999.

453. A.G. Sextos and A.J. Kappos, "Evaluation of the new Eurocode 8 - Part 2 provisions regarding asynchronous excitation of irregular bridges," *Proceedings of the Fourth European Workshop on the Seismic Behaviour of Irregular and Complex Structures*, Thessaloniki, Greece, 2005.

454. A.G. Sextos, A.J. Kappos and P. Mergos, "Effect of soil-structure interaction and spatial variability of ground motion on irregular bridges: The case of the Krystallopigi Bridge," *Proceedings of the 13th World Conference on Earthquake Engineering*, Vancouver, Canada, 2004.

455. A.G. Sextos, K.D. Pitilakis and A.J. Kappos, "Inelastic dynamic analysis of RC bridges accounting for spatial variability of ground motion, site effects and soil-structure interaction phenomena. Part 1: Methodology and analytical tools," *Earthquake Engineering and Structural Dynamics*, Vol. 32, pp. 607–627, 2003.

456. A.G. Sextos, A.J. Kappos and K.D. Pitilakis, "Inelastic dynamic analysis of RC bridges accounting for spatial variability of ground motion, site effects and soil-structure interaction phenomena. Part 2: Parametric study," *Earthquake Engineering and Structural Dynamics*, Vol. 32, pp. 629–652, 2003.

457. M. Seyyedian-Choobi and A.R. Robinson, "Motion on the surface of a layered elastic half-space produced by a buried dislocation pulse," Civil Engineering Studies, Structural Research Series No. SRS-421, University of Illinois at Urbana-Champaign, Urbana, IL, 1975.

458. T.-J. Shan, M. Wax and T. Kailath, "On spatial smoothing for direction-of-arrival estimation of coherent signals," *IEEE Transactions on Acoustics, Speech and Signal Processing*, Vol. ASSP-33, pp. 806–811, 1985.

459. M. Shinozuka, "Simulation of multivariate and multidimensional random processes," *Journal of the Acoustical Society of America*, Vol. 49, pp. 357–367, 1971.

460. M. Shinozuka, "Monte Carlo solution of structural dynamics," *Computers and Structures*, Vol. 2, pp. 855–874, 1972.

461. M. Shinozuka, "Digital simulation of random processes in engineering mechanics with the aid of FFT technique," in *Stochastic Problems in Mechanics*, S.T. Ariaratnam and H.H.E. Leipholz editors, University of Waterloo Press, Ontario, Canada, 1974.

462. M. Shinozuka, "Stochastic fields and their digital simulation," in *Stochastic Methods in Structural Dynamics*, G.I. Schuëller and M. Shinozuka editors, Martinus Nijhoff, Dordrecht, The Netherlands, 1987.

463. M. Shinozuka and G. Deodatis, "Simulation of stochastic processes by spectral representation," *Applied Mechanics Reviews, ASME*, Vol. 44, pp. 191–203, 1991.

464. M. Shinozuka and G. Deodatis, "Stochastic wave models for stationary and homogeneous seismic ground motion," *Structural Safety*, Vol. 10, pp. 235–246, 1991.

465. M. Shinozuka and G. Deodatis, "Simulation of multidimensional Gaussian stochastic fields by spectral representation," *Applied Mechanics Reviews, ASME*, Vol. 49, pp. 29–53, 1996.

466. M. Shinozuka and C.-M. Jan, "Digital simulation of random processes and its applications," *Journal of Sound and Vibration*, Vol. 25, pp. 111–128, 1972.

467. M. Shinozuka and Y. Sato, "Simulation of non-stationary random process," *Journal of the Engineering Mechanics Division, ASCE*, Vol. 93, pp. 11–40, 1967.

468. M. Shinozuka, V. Saxena and G. Deodatis, "Effect of spatial variation of ground motion on highway structures," Technical Report MCEER-00-0013, Multidisciplinary Center for Earthquake Engineering Research (MCEER), University at Buffalo, State University of New York, Buffalo, NY, 2000.

469. M. Shinozuka and R. Zhang, "Equivalence between kriging and CPDF methods for conditional simulation," *Journal of Engineering Mechanics, ASCE*, Vol. 122, pp. 530–538, 1996.

470. S. Short, G. Hardy, K. Merz and J. Johnson, "Effect of seismic wave incoherence on foundation and building response," Electric Power Research Institute Report No. EPRI-1013504, EPRI, Palo Alto, CA, and U.S. Department of Energy, Washington, D.C., 2006.

471. M. Shrikhande and V.K. Gupta, "Synthesizing ensembles of spatially correlated accelerograms," *Journal of Engineering Mechanics, ASCE*, Vol. 124, pp. 1185–1192, 1998.

472. M. Shrikhande and V.K. Gupta, "On the characterisation of the phase spectrum for strong motion synthesis," *Journal of Earthquake Engineering*, Vol. 5, pp. 465–482, 2001.

473. M. Shrikhande and V.K. Gupta, "Dynamic soil-structure interaction effects on the seismic response of suspension bridges," *Earthquake Engineering and Structural Dynamics*, Vol. 28, pp. 1383–1403, 1999.

474. W.J. Silva, N.A. Abrahamson, G.M. Toro and C.J. Costantino, "Description and validation of the stochastic ground motion model," Final Report, Brookhaven National Laboratory, Associated Universities, Inc., Upton, NY, 1997.

475. W.J. Silva, R.B. Darragh, C. Stark, I. Wong, J.C. Stepp, J.F. Schneider and S-J. Chiou, "A methodology to estimate design response spectra in the near-source region of large earthquakes using the band-limited-white-noise ground motion model," *Proceedings of the Fourth U.S. Conference on Earthquake Engineering*, Palm Springs, CA, 1990.

476. W.J. Silva, N.J. Gregor and R.B. Darragh, "Evaluation of numerical procedures for simulating near-fault long-period ground motions using Silva method," PEER Utilities Program, Report No. 2000/02, Pacific Earthquake Engineering Research Center (PEER), University of California, Berkeley, CA, 2000.

477. S.K. Singh, J. Lermo, T. Dominguez, M. Ordaz, J.M. Espinosa, E. Mena and R. Quaas, "A study of amplification of seismic waves in the Valley of Mexico with respect to the hill zone site," *Earthquake Spectra*, Vol. 4, pp. 652–673, 1988.

478. S.K. Singh, E. Mena and R. Castro, "Some aspects of the source characteristics and ground amplifications in and near Mexico City from acceleration data of the September, 1985, Michoacan, Mexico earthquakes," *Bulletin of the Seismological Society of America*, Vol. 78, pp. 451–477, 1988.

479. S.K. Singh, M. Santoyo, P. Bodin and J. Gomberg, "Dynamic deformations of shallow sediments in the Valley of Mexico. Part II: Single-station estimates," *Bulletin of the Seismological Society of America*, Vol. 87, pp. 540–550, 1997.

480. S.W. Smith, J.E. Ehrenberg and E.N. Hernandez, "Analysis of the El Centro differential array for the 1979 Imperial Valley earthquake," *Bulletin of the Seismological Society of America*, Vol. 72, pp. 237–258, 1982.

481. K. Sobczyk, *Stochastic Wave Propagation*, Elsevier Science Ltd., Boston, MA, 1985.

482. L. Socha and T.T. Soong, "Linearization in analysis of nonlinear stochastic systems," *Applied Mechanics Reviews, ASME*, Vol. 44, pp. 399–422, 1991.

483. J. Solnes, *Stochastic Processes and Random Vibrations: Theory and Practice*, John Wiley & Sons, New York, NY, 1997.

484. P.G. Somerville, "Engineering applications of strong ground motion simulation," *Tectonophysics*, Vol. 218, pp. 195–219, 1993.

485. P. Somerville, D. Anderson, J. Sun, S. Punyamurthula and N. Smith, "Generation of ground motion time histories for performance-based design," *Proceedings of the Sixth U.S. National Conference on Earthquake Engineering*, Seattle, WA, 1998.

486. P.G. Somerville, J.P. McLaren, C.K. Saikia and D.V. Helmberger, "Site-specific estimation of spatial incoherence of strong ground motion," in *Earthquake Engineering and Soil Dynamics II-Recent Advances in Ground Motion Evaluation*, J.L. Von Thun editor, Geotechnical Special Publication No. 20, ASCE, New York, NY, 1988.

487. P.G. Somerville, J.P. McLaren, M.K. Sen and D.V. Helmberger, "The influence of site conditions on the spatial incoherence of ground motions," *Structural Safety*, Vol. 10, pp. 1–13, 1991.

488. P.G. Somerville, C. Saikia, D. Wald and R. Graves, "Implications of the Northridge earthquake for strong ground motions from thrust faults," *Bulletin of the Seismological Society of America*, Vol. 86, pp. S115–S125, 1996.

489. P.G. Somerville, N.F. Smith, R.W. Graves and N.A. Abrahamson, "Modification of empirical strong ground motion attenuation relations to include the amplitude and duration effects of rupture directivity," *Seismological Research Letters*, Vol. 68, pp. 180–203, 1997.

490. P. Somerville, N. Smith, S. Punyamurthula and J. Sun, "Development of ground motion time histories for Phase 2 of the FEMA/SAC steel project," CUREE Publication No. SAC/BD-97/04, Consortium of Universities for Research in Earthquake Engineering, Richmond, CA, 1997.

491. J. Song and A. Der Kiureghian, "Generalized Bouc-Wen model for highly asymmetric hysteresis," *Journal of Engineering Mechanics, ASCE*, Vol. 132, pp. 610–618, 2006.

492. T.T. Soong and M. Grigoriu, *Random Vibration of Mechanical and Structural Systems*, Prentice-Hall, Inc., Englewood Cliffs, NJ, 1993.

493. P.D. Spanos and G. Failla, "Evolutionary spectra estimation using wavelets," *Journal of Applied Mechanics, ASME*, Vol. 130, pp. 952–960, 2004.

494. P.D. Spanos and M.P. Mignolet, "Simulation of homogeneous two-dimensional random fields: Part II-MA and ARMA models," *Journal of Applied Mechanics, ASME*, Vol. 59, pp. 270–277, 1992.

495. P.T.D. Spanos and L.M. Vargas Loli, "A statistical approach to generation of design spectrum compatible earthquake time histories," *Soil Dynamics and Earthquake Engineering*, Vol. 4, pp. 2–8, 1985.

496. P.D. Spanos and B.A. Zeldin, "Efficient iterative ARMA approximation of multivariate random processes for structural dynamics applications," *Earthquake Engineering and Structural Dynamics*, Vol. 25, pp. 497–507, 1996.

497. P.D. Spanos and B.A. Zeldin, "Monte Carlo treatment of random fields: A broad perspective," *Applied Mechanics Reviews, ASME*, Vol. 51, pp. 219–237, 1998.

498. P. Spudich, "Recent seismological insights into the spatial variation of earthquake ground motions," in *New Developments in Earthquake Ground Motion Estimation and Implications for Engineering Design Practice*, ATC 35-1, Redwood City, CA, 1994.

499. P. Spudich and E. Cranswick, "Direct observation of rupture propagation during the 1979 Imperial Valley earthquake using a short baseline accelerometer array," *Bulletin of the Seismological Society of America*, Vol. 74, pp. 2083–2114, 1984.

500. P. Spudich, M. Hellweg and W.H.K. Lee, "Directional topographic site response at Tarzana observed in aftershocks of the 1994 Northridge, California, earthquake: Implications for main shock motions," *Bulletin of the Seismological Society of America*, Vol. 86, pp. 193–208, 1996.

501. P. Spudich and D. Oppenheimer, "Dense seismograph array observations of earthquake rupture dynamics," in *Earthquake Source Mechanics*, Geophysical Monograph 37, S. Das, J. Boatwright and C.H. Scholz editors, American Geophysical Union, Washington, D.C., 1986.

502. P. Spudich, L.K. Steck, M. Hellweg, J.B. Fletcher and L.M. Baker, "Transient stresses at Parkfield, California, produced by the M 7.4 Landers earthquake of June 28, 1992: Observations from the UPSAR dense seismograph array," *Journal of Geophysical Research*, Vol. 100, pp. 675–690, 1995.

503. C.M. St. John and T.F. Zahrah, "Aseismic design of underground structures," *Tunneling and Underground Space Technology*, Vol. 2, pp. 165–197, 1987.

504. J.H. Steidl and R.L. Nigbor, "Executive Summary," *Proceedings of the SCEC/ROSRINE Workshop on Borehole Array Data Utilization*, Palm Springs, CA, 2000.

505. J.H. Steidl, A.G. Tumarkin and R.J. Archuleta, "What is a reference site?," *Bulletin of the Seismological Society of America*, Vol. 86, pp. 1733–1748, 1996.

506. W.R. Stephenson, "The dominant resonance response of Parkway basin," *Proceedings of 12th World Conference on Earthquake Engineering*, Auckland, New Zealand, 2000.

507. W.R. Stephenson, "Visualisation of resonant basin response at the Parkway array, New Zealand," *Soil Dynamics and Earthquake Engineering*, Vol. 27, pp. 487–496, 2007.

508. J.P. Stewart, S.-J. Chiou, J.D. Bray, R.W. Graves, P.G. Somerville and N.A. Abrahamson, "Ground motion evaluation procedures for performance-based design," Report No. PEER 2001/09, Pacific Earthquake Engineering Research Center (PEER), University of California, Berkeley, CA, 2001.

509. J.P. Stewart, R.B. Seed and J.D. Bray, "Incidents of ground failure from the 1994 Northridge earthquake," *Bulletin of the Seismological Society of America*, Vol. 86, pp. S300–S318, 1996.

510. Southern California Earthquake Data Center (SCEDC), http://www.data.scec.org/, accessed 2008.

511. K. Soyluk and A.A. Dumanoğlu, "Spatial variability effects of ground motions on cable-stayed bridges," *Soil Dynamics and Earthquake Engineering*, Vol. 24, pp. 241–250, 2004.

512. L. Su, S.L. Dong and S. Kato, "A new average response spectrum method for linear response analysis of structures to spatial earthquake ground motions," *Engineering Structures*, Vol. 28, pp. 1835–1842, 2006.

513. T. Szczesiak, B. Weber and H. Bachmann, "Nonuniform earthquake input for arch-dam foundation interaction," *Soil Dynamics and Earthquake Engineering*, Vol. 18, pp. 487–493, 1999.

514. H. Tajimi, "A statistical method of determining the maximum response of a building structure during an earthquake," *Proceedings of the Second World Conference on Earthquake Engineering*, Tokyo and Kyoto, Japan, 1960.

515. T. Takeda, M.A. Sozen and N. Nielsen, "Reinforced concrete response to simulated earthquakes," *Journal of the Structural Division, ASCE*, Vol. 96, pp. 2557–2573, 1970.

516. M. Takeo, "Ground rotational motions recorded in near-source region of earthquakes," *Geophysical Research Letters*, Vol. 25, pp. 789–792, 1998.

517. M. Takeo and H.M. Ito, "What can be learned from rotational motions excited by earthquakes?," *Geophysical Journal International*, Vol. 129, pp. 319–329, 1997.

518. C. Tamura, T. Noguchi and K. Kato, "Earthquake observation along measuring lines on the surface of alluvial soft ground," *Proceedings of the Sixth World Conference on Earthquake Engineering*, New Delhi, India, 1977.

519. T.L. Teng and J. Qu, "Long-period ground motions and dynamic strain field of Los Angeles basin during large earthquakes," *Bulletin of the Seismological Society of America*, Vol. 86, pp. 1417–1433, 1996.

520. H. Thráinsson and A.S. Kiremidjian, "Simulation of digital earthquake accelerograms using the inverse Fourier transform," *Earthquake Engineering and Structural Dynamics*, Vol. 31, pp. 2023–2048, 2002.

521. H. Thráinsson, A.S. Kiremidjian and S.R. Winterstein, "Modeling of earthquake ground motion in the frequency domain," Report No. 134, The John A. Blume Earthquake Engineering Center, Department of Civil and Environmental Engineering, Stanford University, Stanford, CA, 2000.

522. C.W.S. To, *Nonlinear Random Vibration: Analytical Techniques and Applications*, Swets and Zeitlinger, Lisse, The Netherlands, 2000.

523. M.I. Todorovska and M.D. Trifunac, "Hazard mapping of normalized peak strains in soils during earthquakes: Microzonation of a metropolitan area," *Soil Dynamics and Earthquake Engineering*, Vol. 15, pp. 321–329, 1996.

524. M.N. Toksöz, A.M. Dainty and E.E. Charrette III, "Spatial variation of ground motion due to lateral heterogeneity," *Structural Safety*, Vol. 10, pp. 53–77, 1991.

525. M.D. Trifunac, "Response envelope spectrum and interpretation of strong earthquake ground motion," *Bulletin of the Seismological Society of America*, Vol. 61, pp. 343–356, 1971.

526. M.D. Trifunac, "Zero baseline correction of strong-motion accelerograms," *Bulletin of the Seismological Society of America*, Vol. 61, pp. 1201–1211, 1971.

527. M.D. Trifunac, "A method for synthesizing realistic strong ground motion," *Bulletin of the Seismological Society of America*, Vol. 61, pp. 1739–1753, 1971.

528. M.D. Trifunac, "Preliminary analysis of peaks of strong earthquake ground motion - dependence of peaks on earthquake magnitude, epicentral distance, and recording site conditions," *Bulletin of the Seismological Society of America*, Vol. 66, pp. 189–219, 1976.

529. M.D. Trifunac, "Preliminary empirical model for scaling Fourier amplitude spectra of strong ground acceleration in terms of earthquake magnitude, source-to-station distance, and recording site conditions," *Bulletin of the Seismological Society of America*, Vol. 66, pp. 1343–1373, 1976.

530. M.D. Trifunac, "A note on surface strains associated with incident body waves," *Bulletin of the European Association of Earthquake Engineering*, Vol. 5, pp. 85–95, 1979.

531. M.D. Trifunac, "A note on rotational components of earthquake motions for incident body waves," *Soil Dynamics and Earthquake Engineering*, Vol. 1, pp. 11–19, 1982.

532. M.D. Trifunac, "Curvograms of strong ground motion," *Journal of Engineering Mechanics, ASCE*, Vol. 116, pp. 1426–1432, 1990.

533. M.D. Trifunac and A.G. Brady, "A study on the duration of strong earthquake ground motion," *Bulletin of the Seismological Society of America*, Vol. 65, pp. 581–626, 1975.

534. M.D. Trifunac and V. Gicev, "Response spectra for differential motion of columns paper II: Out-of-plane response," *Soil Dynamics and Earthquake Engineering*, Vol. 26, pp. 1149–1160, 2006.

535. M.D. Trifunac and V.W. Lee, "Routine processing of strong-motion accelerograms," in Report No. EERL 70-07, Earthquake Engineering Research Laboratory, California Institute of Technology, Pasadena, CA, 1973.

536. M.D. Trifunac and V.W. Lee, "Peak surface strains during strong earthquake motion," *Soil Dynamics and Earthquake Engineering*, Vol. 15, pp. 311–319, 1996.

537. M.D. Trifunac and E.I. Novikova, "Duration of earthquake fault motion in California," *Earthquake Engineering and Structural Dynamics*, Vol. 24, pp. 781–799, 1995.

538. M.D. Trifunac and M.I. Todorovska, "Northridge, California, earthquake of 1994: Density of pipe breaks and surface strains," *Soil Dynamics and Earthquake Engineering*, Vol. 16, pp. 193–207, 1997.

539. M.D. Trifunac and M.I. Todorovska, "Response spectra for differential motion of columns," *Earthquake Engineering and Structural Dynamics*, Vol. 26, pp. 251–268, 1997.

540. M.D. Trifunac and M.I. Todorovska, "A note on the usable dynamic range of accelerographs recording translation," *Soil Dynamics and Earthquake Engineering*, Vol. 21, pp. 275–286, 2001.

541. M.D. Trifunac, M.I. Todorovska and S.S. Ivanovic, "Peak velocities and peak surface strains during Northridge, California, earthquake of 17 January 1994," *Soil Dynamics and Earthquake Engineering*, Vol. 15, pp. 301–310, 1996.

542. I. Tromans, "Behaviour of buried water supply pipelines in earthquake zones," Ph.D. Dissertation, Department of Civil and Environmental Engineering, Imperial College of Science, Technology and Medicine, London, United Kingdom, 2004.

543. W.S. Tseng and K. Lilhanand, "Soil-structure interaction analysis incorporating spatial incoherence of ground motions," Pacific Gas & Electric Company, San Francisco, CA,

and Electric Power Research Institute, EPRI, Palo Alto, CA, Report No. TR-102631, 1997.

544. W.K. Tso and T.-I. Hsu, "Torsional spectrum for earthquake motions," *Earthquake Engineering and Structural Dynamics*, Vol. 6, pp. 375–382, 1978.

545. K. Tsuda, R.J. Archuleta, and K. Koketsu, "Quantifying the spatial distribution of site response by use of the Yokohama high-density strong-motion network," *Bulletin of the Seismological Society of America*, Vol. 96, pp. 926–942, 2006.

546. N. Tzanetos, A.S. Elnashai, F.H. Hamdan and S. Antoniou, "Inelastic dynamic response of RC bridges subjected to spatially non-synchronous earthquake motion," *Advances in Structural Engineering*, Vol. 3, pp. 191–214, 2000.

547. United States Geological Survey (USGS), http://www.usgs.gov/, accessed 2008.

548. J.F. Unruh and D.D. Kana, "An iterative procedure for the generation of consistent power/response spectrum," *Nuclear Engineering and Design*, Vol. 66, pp. 427–435, 1981.

549. B.J. Uscinski, *The Elements of Wave Propagation in Random Media*, McGraw-Hill, Inc., New York, NY, 1977.

550. E.H. Vanmarcke, "On the distribution of first-passage time for normal stationary processes," *Journal of Applied Mechanics, ASME*, Vol. 42, pp. 215–220, 1975.

551. E.H. Vanmarcke, *Random Fields. Analysis and Synthesis*, MIT Press, Cambridge, MA, 1983.

552. E.H. Vanmarcke and G.A. Fenton, "Conditioned simulation of local fields of earthquake ground motion," *Structural Safety*, Vol. 10, pp. 247–264, 1991.

553. E.H. Vanmarcke, E. Heredia-Zavoni and G.A. Fenton, "Conditional simulation of spatially correlated earthquake ground motion," *Journal of Engineering Mechanics, ASCE*, Vol. 119, pp. 2333–2352, 1993.

554. A.S. Veletsos and A.M. Prasad, "Seismic interaction of structures and soils: Stochastic approach," *Journal of Structural Engineering, ASCE*, Vol. 115, pp. 935–956, 1989.

555. A.S. Veletsos, A.M. Prasad and W.H. Wu, "Transfer functions for rigid rectangular foundations," *Earthquake Engineering and Structural Dynamics*, Vol. 26, pp. 5–17, 1997.

556. F. Vernon, J. Fletcher, L. Carroll, A. Chave and E. Sembera, "Coherence of seismic body waves from local events as measured by a small aperture array," *Journal of Geophysical Research*, Vol. 96, pp. 11981–11996, 1991.

557. D.J. Wald, T.H. Heaton and K.W. Hudnut, "The slip history of the 1994 Northridge, California, earthquake determined from strong motion, teleseismic, GPS, and leveling data," *Bulletin of the Seismological Society of America*, Vol. 86, pp. S49–S70, 1996.

558. J. Wang, S. Hu and X. Wei, "Effects of engineering geological condition on response of suspension bridges," *Soil Dynamics and Earthquake Engineering*, Vol. 18, pp. 297–304, 1999.

559. Y.K. Wen, "Equivalent linearization for hysteretic systems under random excitation," *Journal of Applied Mechanics, ASME*, Vol. 47, pp. 150–154, 1980.

560. Y.K. Wen and P. Gu, "Description and simulation of nonstationary processes based on Hilbert spectra," *Journal of Engineering Mechanics, ASCE*, Vol. 130, pp. 942–951, 2004.

561. S.D. Werner and L.C. Lee, "The three-dimensional response of structures subjected to traveling wave excitation," *Proceedings of the Second U.S. National Conference on Earthquake Engineering*, Stanford, CA, 1979.

562. S.D. Werner, L.C. Lee, H.L. Wong and M.D. Trifunac, "Structural response to traveling seismic waves," *Journal of the Structural Division, ASCE*, Vol. 105, pp. 2547–2564, 1979.

563. H.M. Westergaard, "Water pressure on dams during earthquakes," *Transactions, ASCE*, Vol. 98, pp. 418–432, 1933.

564. M. Wieland, "Review of design criteria of large concrete and embankment dams," *Proceedings of the 73rd Annual Meeting of ICOLD*, Tehran, Iran, 2005.

565. M. Wieland, "Earthquake safety of existing dams," *Proceedings of the First European Conference on Earthquake Engineering and Seismology*, Geneva, Switzerland, 2006.

566. P.H. Wirsching, T.L. Paez and K. Ortiz, *Random Vibrations: Theory and Practice*, John Wiley & Sons, New York, NY, 1995.

567. H.L. Wong and M.D. Trifunac, "Generation of artificial strong motion accelerograms," *Earthquake Engineering and Structural Dynamics*, Vol. 7, pp. 509–527, 1979.

568. Z. Xu, S.Y. Schwartz and T. Lay, "Seismic wave-field observations at a dense, small-aperture array located on a landslide in the Santa Cruz Mountains, California," *Bulletin of the Seismological Society of America*, Vol. 86, pp. 655–669, 1996.

569. A.M. Yaglom, *An Introduction to the Theory of Stationary Random Functions*, Prentice-Hall, Inc., Englewood Cliffs, NJ, 1962.

570. A.M. Yaglom, *Correlation Theory of Stationary and Related Random Functions*, Springer-Verlag, New York, NY, 1987.

571. N. Yamamura and H. Tanaka, "Response analysis of flexible MDF systems for multiple-support seismic excitations," *Earthquake Engineering and Structural Dynamics*, Vol. 19, pp. 345–357, 1990.

572. F. Yamazaki and T. Türker, "Spatial variation study on earthquake ground motion observed by the Chiba array," *Proceedings of the Tenth World Conference on Earthquake Engineering*, Madrid, Spain, 1992.

573. J.N. Yang, "Simulations of random envelope processes," *Journal of Sound and Vibration*, Vol. 25, pp. 73–85, 1972.

574. Q.S. Yang and Y.J. Chen, "A practical coherency model for spatially varying ground motions," *Structural Engineering and Mechanics*, Vol. 9, pp. 141–152, 2000.

575. Z. Yang, L. He, J. Bielak, Y. Zhang, A. Elgamal and J. Conte, "Nonlinear seismic response of a bridge site subject to spatially varying ground motion," *Proceedings of the 16th Engineering Mechanics Conference, ASCE*, Seattle, WA, 2003.

576. H. Yamahara, "Ground motion during earthquakes and the input loss of earthquake power to excitation of buildings," *Journal of Soils and Foundations*, Vol. 10, pp. 145–161, 1970.

577. C.-H. Yeh and Y.K. Wen, "Modeling of nonstationary ground motion and analysis of inelastic structural response," *Structural Safety*, Vol. 8, pp. 281–298, 1990.

578. W.P. Yen, J.D. Cooper, S.W. Park, S. Unjoh, T. Terayama and H. Otsuka, "A comparative study of U.S.-Japan seismic design of highway bridges: I. Design methods," *Earthquake Spectra*, Vol. 19, pp. 913–932, 2003.

579. K.-V. Yuen, J.L. Beck and L.S. Katafygiotis, "Spectral density estimation of stochastic vector processes," *Probabilistic Engineering Mechanics*, Vol. 17, pp. 265–272, 2002.

580. G. Zanardo, H. Hao and C. Modena, "Seismic response of multi-span simply supported bridges to a spatially varying earthquake ground motion," *Earthquake Engineering and Structural Dynamics*, Vol. 31, pp. 1325–1345, 2002.

581. M. Zaré and P.-Y. Bard, "Strong motion dataset of Turkey: data processing and site classification," *Soil Dynamics and Earthquake Engineering*, Vol. 22, pp. 703–718, 2002.

582. B.A. Zeldin and P.D. Spanos, "Random field representation and synthesis using wavelet bases," *Journal of Applied Mechanics, ASME*, Vol. 63, pp. 946–952, 1996.

583. Z. Zembaty, A. Castellani and G. Boffi, "Spectral analysis of the rotational component of seismic ground motion," *Probabilistic Engineering Mechanics*, Vol. 8, pp. 5–14, 1993.

584. Z. Zembaty and S. Krenk, "Spatial seismic excitations and response spectra," *Journal of Engineering Mechanics, ASCE*, Vol. 119, pp. 2449–2460, 1993.

585. Z. Zembaty and A. Rutenberg, "Spatial response spectra and site amplification effects," *Engineering Structures*, Vol. 24, pp. 1484–1496, 2002.

586. D. Zendagui, M.K. Berrah and E. Kausel, "Stochastic deamplification of spatially varying seismic motions," *Soil Dynamics and Earthquake Engineering*, Vol. 18, pp. 409–421, 1999.

587. Y. Zeng, "A realistic synthetic Green's function calculation using a combined deterministic and stochastic modeling approach," *EOS, Transactions of American Geophysical Union*, Vol. 76, F357, 1995.

588. Y. Zeng, J.G. Anderson and G. Yu, "A composite source model for computing realistic synthetic strong ground motions," *Geophysical Research Letters*, Vol. 21, pp. 725–728, 1994.

589. A. Zerva, "Stochastic differential ground motion and structural response," Ph.D. Thesis, Department of Civil Engineering, University of Illinois at Urbana-Champaign, Urbana, IL, 1986.

590. A. Zerva, "Response of multi-span beams to spatially incoherent seismic ground motions," *Earthquake Engineering and Structural Dynamics*, Vol. 19, pp. 819–832, 1990.

591. A. Zerva, "Effect of spatial variability and propagation of seismic ground motions on the response of multiply supported structures," *Probabilistic Engineering Mechanics*, Vol. 6, pp. 212–221, 1991.

592. A. Zerva, "Seismic ground motion simulations from a class of spatial variability models," *Earthquake Engineering and Structural Dynamics*, Vol. 21, pp. 351–361, 1992.

593. A. Zerva, "Spatial incoherence effects on seismic ground strains," *Probabilistic Engineering Mechanics*, Vol. 7, pp. 217–226, 1992.

594. A. Zerva, "Seismic loads predicted by spatial variability models," *Structural Safety*, Vol. 11, pp. 227–243, 1992.

595. A. Zerva, "Pipeline response to directionally and spatially correlated seismic ground motions," *Journal of Pressure Vessel Technology, ASME*, Vol. 115, pp. 53–58, 1993.

596. A. Zerva, "On the spatial variation of seismic ground motions and its effects on lifelines," *Engineering Structures*, Vol. 16, pp. 534–546, 1994.

597. A. Zerva, "Spatial variability of seismic motions recorded over extended ground surface areas," in *Wave Motion in Earthquake Engineering*, E. Kausel and G.D. Manolis editors, Volume in the Series Advances in Earthquake Engineering, WIT Press, Southampton, United Kingdom, 2000.

598. A. Zerva, A. H-S. Ang and Y.K. Wen, "Study of seismic ground motion for lifeline response analysis," Civil Engineering Studies, Structural Research Series No. SRS-521, University of Illinois at Urbana-Champaign, Urbana, IL, 1985.

599. A. Zerva, A. H-S. Ang and Y.K. Wen, "Development of differential response spectra for lifeline seismic analysis," *Probabilistic Engineering Mechanics*, Vol. 1, pp. 208–218, 1987.

600. A. Zerva, A. H-S. Ang and Y.K. Wen, "Lifeline response to spatially variable ground motions," *Earthquake Engineering and Structural Dynamics*, Vol. 16, pp. 361–379, 1988.

601. A. Zerva and J.L. Beck, "Updating stochastic models for spatially variable seismic ground motions," *Proceedings of the International Conference on Applications of Statistics and Probability, ICASP-8*, Sydney, Australia, 1999.

602. A. Zerva and J.L. Beck, "Identification of parametric ground motion random fields from spatially recorded data," *Earthquake Engineering and Structural Dynamics*, Vol. 32, pp. 771–791, 2003.

603. A. Zerva and T. Harada, "Effect of surface layer stochasticity on seismic ground motion coherence and strain estimates," *Soil Dynamics and Earthquake Engineering*, Vol. 16, pp. 445–457, 1997.

604. A. Zerva and S. Liao, "Physically-based methodology to characterize the spatial variation of seismic ground motions and its validation," Research Series Report No. 2002/01, Department of Civil and Architectural Engineering, Drexel University, Philadelphia, PA, 2002.

605. A. Zerva and M. Shinozuka, "Stochastic differential ground motion," *Structural Safety*, Vol. 10, pp. 129–143, 1991.

606. A. Zerva and V. Zervas, "Spatial variation of seismic ground motions: An overview," *Applied Mechanics Reviews, ASME*, Vol. 55, pp. 271–297, 2002.

607. A. Zerva and O. Zhang, "Estimation of signal characteristics in seismic ground motions," *Probabilistic Engineering Mechanics*, Vol. 11, pp. 229–242, 1996.

608. A. Zerva and O. Zhang, "Correlation patterns in characteristics of spatially variable seismic ground motions," *Earthquake Engineering and Structural Dynamics*, Vol. 26, pp. 19–39, 1997.

609. B. Zhang and A.S. Papageorgiou, "Estimation of the differential ground motions induced in the near field by the 1994 Northridge, California, earthquake," *Seismological Research Letters*, Vol. 67, p. 63, 1996.

610. J. Zhang and B. Ellingwood, "Orthogonal series expansions of random fields in reliability analysis," *Journal of Engineering Mechanics, ASCE*, Vol. 120, pp. 2660–2677, 1994.

611. L. Zhang and A.K. Chopra, "Three-dimensional analysis of spatially varying ground motions around a uniform canyon in a homogeneous half-space," *Earthquake Engineering and Structural Dynamics*, Vol. 20, pp. 911–926, 1991.

612. L. Zhang and A.K. Chopra, "Impedance functions for three-dimensional foundation supported on an infinitely long canyon of uniform cross-section in a homogeneous half-space," *Earthquake Engineering and Structural Dynamics*, Vol. 20, pp. 1011–1027, 1991.

613. O. Zhang, "Estimation of signal characteristics and coherent patterns in spatially variable seismic ground motions," Ph.D. Dissertation, Department of Civil and Architectural Engineering, Drexel University, Philadelphia, PA, 1997.

614. A.I. Zverev, *Handbook of Filter Synthesis*, John Wiley & Sons, New York, NY, 1967.

Index